Introduction to Geobiology

Introduction to Geobiology

Xingliang Zhang
Department of Geology, Northwest University, Xi'an, China

Zisheng Guo
College of Life Sciences, Northwest University, Xi'an, China

Wei Liu
Northwest University Museum, Northwest University, Xi'an, China

Ruliang He
Department of Geology, Northwest University, Xi'an, China

Luoyang Li
College of Marine Geosciences, Ocean University of China, Qingdao, China

Weiduo Hao
Department of Geology, Northwest University, Xi'an, China

WILEY

This edition first published 2026

Registered Offices
John Wiley & Sons, Inc., 111 River Street, Hoboken, NJ 07030, USA
John Wiley & Sons Ltd, New Era House, 8 Oldlands Way, Bognor Regis, West Sussex, PO22 9NQ, UK

For details of our global editorial offices, customer services, and more information about Wiley products visit us at www.wiley.com.

The manufacturer's authorized representative according to the EU General Product Safety Regulation is Wiley-VCH GmbH, Boschstr. 12, 69469 Weinheim, Germany, e-mail: Product_Safety@wiley.com.

Wiley also publishes its books in a variety of electronic formats and by print-on-demand. Some content that appears in standard print versions of this book may not be available in other formats.

Library of Congress Cataloging-in-Publication Data Applied for:

Hardback ISBN: 9781394346196

Cover Design: Wiley
Cover Images: © moonoi172/Shutterstock, Courtesy of Xingliang Zhang, Courtesy of Xingliang Zhang

Set in 9.5/12.5pt STIXTwoText by Straive, Pondicherry, India

Printed and bound by CPI Group (UK) Ltd, Croydon, CR0 4YY

C9781394346196_030326

Contents

Preface

Geobiology, sharing structural similarities with geochemistry and geophysics in terminology, is an intersecting scientific field between earth science and life science. The term *geobiology* (Dutch *Geobiologie*) has traditionally been attributed to the Dutch microbiologist Lourens Baas Becking, who introduced it in 1934—approximately a century after the terms *geophysics* (German *Geophysik* by Julius Fröbel in 1834) and *geochemistry* (German *Geochemie* by Carl Friedrich Schönbein in 1838). He defined geobiology as an attempt to describe the relationship between organisms and the Earth, emphasizing that organisms are part of the Earth and their fates are intertwined. Baas Becking's thought was influenced by earlier scientists such as Martinus Beijerinck, Vladimir Vernadsky, and Sergei Vinogradsky. Vernadsky, in particular, argued in his 1926 book *The Biosphere* that life significantly alters the Earth's surface environment. However, the *Oxford English Dictionary* notes that the earliest known use of the term *geobiology* in English can be found in *Studies in Indiana Geography* by Charles Redway Dryer (1897, page 11), who used the term *geobiology* as part of a conceptual framework for geography and outlined as the third of four pyramids of geographic sciences, "The third course is geobiology, which deals with the vegetable and animal elements of the earth organism."

Although geobiology was formalized in science in 1934, it took several decades for geobiology to become a firmly rooted scientific discipline. Peter C. Sylvester-Bradley identified geobiology as one of the three hybrid Earth sciences—alongside geophysics and geochemistry in "Geobiology and the Future of Palaeontology" (*Journal of the Geological Society*, 1972, 128: 109–117). The field of geobiology gained more prominence in the late 20th century with advancements in geochemistry and genetics, which allowed scientists to better understand the interactions between life and the Earth's physical environment. In 2000, a symposium held by the American Academy of Microbiology officially described geobiology as "an interdiscipline that attempts to understand the interactions between the biosphere and the geosphere." In 2001, the Geological Society of America established a geobiology section, followed by foundation of the International Geobiology Society in 2016 and the Geobiology Sub-society of China Paleontological Society in 2018. It is evident that Geobiology is becoming increasingly popular worldwide and is now formalized as a subdiscipline of Earth Science tackling the interactions and co-evolution between life and environment (Xie et al. 2023).

Geobiology is a dynamic and rapidly expanding field that offers fresh perspectives on understanding the Earth as a complex system. It focuses on the processes and evolution of the intertwined Earth-life system over time, tackling challenges that neither geology nor biology can address alone. At its core, geobiology integrates diverse subfields from both earth and biological sciences, including ecology, microbiology, physiology, molecular biology, evolutionary biology, paleontology, geology, mineralogy, sedimentology, geochemistry, oceanography, and astrobiology. While it may seem like a straightforward fusion of disciplines, geobiology provides a more holistic approach to studying the interactions between life and the physical environment, both in the past and present. This interdisciplinary perspective enhances our understanding of the co-evolution of the geosphere and biosphere, offering profound insights into Earth's history. Furthermore, geobiological research holds significant practical applications, contributing to fields such as environmental science, resource management, and planetary exploration.

Students may already possess some understanding of ecology, including paleoecology, which examines organisms in relation to their environment. However, ecology tends to focus on a narrower perspective. Ecologists primarily aim to determine how an organism fits its surroundings, posing questions such as: Why does this organism thrive here but not there? How does it obtain food? Is a particular nutrient limiting its population growth? How and where does it reproduce? What factors contribute to its absence in certain areas? How do the young disperse, and what ultimately causes the organism's death? A student could easily generate numerous such inquiries, all emphasizing environmental studies from an organism-centered viewpoint.

In contrast, geobiology adopts a much broader approach, exploring the interactions between life and Earth's surface environment across geological time. This relationship functions bidirectionally—life has profoundly influenced Earth's chemical and physical evolution, while changes in Earth's environment have, in turn, shaped ecosystems. Despite its interdisciplinary nature, geobiology does not rely on a single unified theory or a specific methodological framework. Instead, it synthesizes concepts from various earth and life sciences to provide a holistic perspective, allowing us to view Earth as an interconnected system where biological and geological processes have continuously co-evolved. This integrative approach holds significant value in deepening our understanding of both past and present interactions between life and the physical world.

Like many interdisciplinary textbooks, *Introduction to Geobiology* is a compilation of lecture notes from 25 years of bilingual teaching and learning courses in Geobiology at Northwest University of China. The courses are designed for senior and master's students majoring in Geology or Biology and aim to inspire students to engage in emerging fields in Geobiology and cultivate a holistic understanding of the Earth-life system. The textbook presents thousands of foundational knowledge units—referred to as knowledge cells—that span diverse topics, bridging earth and life sciences. Similar to a biological system, these knowledge cells are organized into a hierarchical framework: from cell-grade paragraphs to tissue- and organ-grade divisions, culminating in comprehensive chapters that form a pluralistic unity of the book. A distinctive feature of this textbook is its emphasis on the geological consequences of biological activity. Chapters 1 through 7 explore the diversity and characteristics of geobiological processes, while Chapter 8 provides a global perspective on their evolution through geological time. The final chapter extends the discussion to astrobiology, examining the broader implications of geobiological phenomena beyond Earth.

This book was made possible with the encouragement and support of my colleagues Joachim Reitner and Jan-Peter Duda in Göttingen, as well as my lifelong mentor, Professor Degan Shu here in Xi'an. Notably, Chapter 8 was finalized during my tenure as a Gauss Professor at the Göttingen Academy of Sciences and Humanities in Lower Saxony, a period that greatly enriched my research. I gratefully acknowledge financial support from the Natural Science Foundation of China (NSFC) and Northwest University. Special thanks go to Mr. Qingshan Wang for his invaluable work in preparing numerous figures for this book.

12 May 2025

Xingliang Zhang
Northwest University, Xi'an

Reference

Xie, S. et al. 2023. *Geobiology (in Chinese)*. Beijing, High Education Press. 570 pp.

About the Companion Website

This book is accompanied by a companion website.

www.wiley.com/go/Zhang_1e

This website contains

For Instructor

- Figure PPTs

1

Biological Diversity: A Geobiological Perspective

Zisheng Guo and Xingliang Zhang

1.1 Introduction

Life can be understood as a living form, either cellular or acellular form, and is composed of interconnected components that are organized in a complex hierarchical structure. This hierarchy ranges from the smallest units—atoms and molecules—to cells, tissues, organs, and ultimately whole organisms. Together with its environment, life forms the biosphere, an intricate and interactive hierarchy consisting of species, populations, communities, and ecosystems (Figure 1.1). Each level in this biological hierarchy, from cells to the biosphere, operates as a self-sustaining and adaptive system that has been constantly co-evolving with physical environment since its origin. Life pervades our planet and potentially extends beyond it. Living organisms thrive in an astonishing variety of environments, from the icy poles to the scorching equator and from the depths of the oceans to several kilometers above the Earth's surface. They can be found in extreme conditions, such as freezing waters, dry valleys, undersea thermal vents, and even deep within the Earth's crust.

Over the past ~4.0 billion years, life on Earth has diversified and adapted to a myriad of environments. Despite this incredible diversity, all living organisms share common traits: they grow, adapt, reproduce, and interact with their surroundings. From microscopic organisms to expansive ecosystems, life takes many forms but is united by these shared characteristics. Every living being has the ability to replicate, driven by DNA as the molecular basis of replication. Moreover, all organisms possess mechanisms to translate the genetic information in DNA into functional products like proteins, fats, and carbohydrates, which are essential for building and maintaining cellular machinery.

Virtually every month, new discoveries are made about surprising occurrences and modes of life forms on Earth, ranging from proteorhodopsin-based phototrophy in the open ocean to methanogenesis driven by geochemical reactions in Earth's interior. Niches once considered to be uninhabitable (such as those with pH of 0 in extremely rich metal solutions) have been found to harbor thriving microbial communities; compounds once thought to be refractory (such as kerogen, or long-chain alkanes under anaerobic conditions) are now known to serve as microbial growth substrates; organisms previously believed to be unculturable (such as anaerobic benzene oxidizers or soil bacteria that produce medically relevant natural products) have been brought into culture and/or their genetic content has been expressed in recombinant strains; and enigmatic geochemical

Introduction to Geobiology, First Edition. Xingliang Zhang, Zisheng Guo, Wei Liu, Ruliang He, Luoyang Li, and Weiduo Hao.

Companion website: www.wiley.com/go/Zhang_1e

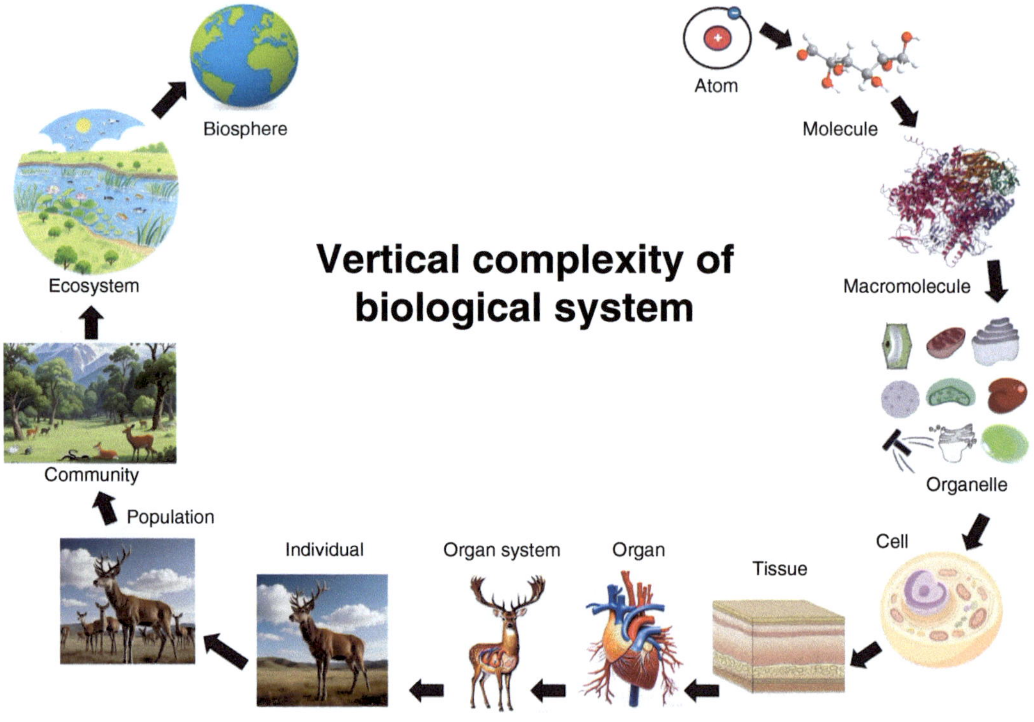

Figure 1.1 Biological hierarchy ranging from the smallest units—atoms and molecules—to cells, tissues, organs, whole organisms, and ultimately the biosphere.

transformations (such as the anaerobic oxidation of methane [AOM] or ammonium) are now attributed to the activity of consortia of bacteria and archaea. The importance and extent of biological diversity have captured the attention of life scientists, and now the community of geosciences is helping to define and explore interesting biogeochemical problems.

Up until now, we have been familiar with organisms that can be said to have a "conventional" metabolism. These are organisms that use organic compounds as energy sources, primarily via aerobic (respiratory) metabolism. In this chapter, we aim to explore the vast biological diversity, especially focusing on microbial metabolisms with geological consequences. To address gaps in biological knowledge for geology students, we begin with the current classification of acellular and cellular life forms. This is followed by an examination of metabolic diversity from a geobiological perspective, which is essential for comprehending geobiological processes.

1.2 Classification of Biological Entities

1.2.1 Acellular Life

Cellular and acellular life represent two fundamental categories of biological entities. Cellular life forms, encompassing prokaryotes and eukaryotes, are living organisms that are composed of one or more cells. These organisms are characterized by the presence of a cell membrane enclosing cytoplasm and genetic material (DNA or RNA), along with their ability to metabolize, grow, reproduce, and respond to stimuli. In contrast, acellular life forms are biological entities that lack cellular structure and are generally not considered fully alive in a traditional sense. Acellular life forms include viruses that consist of genetic material (DNA or RNA) encased in a protein coat, viroids that are small RNA molecules infecting plants, and prions that are misfolded proteins causing diseases in animals and humans (Figure 1.2). Unlike cellular life forms that are autonomous and self-sustaining, acellular entities rely on other organisms to replicate and function. They display biological activity and functions only within living host cells, while existing as inert chemical particles in external environments, which are replicated only within a host organism (parasitic nature) and do not exhibit independent metabolism or growth.

Viruses and their allies may have influenced Earth's geology by infecting bacterial, archaeal, and eukaryotic hosts and altering their physiology. Given that bacteria and archaea have dominated Earth's biosphere for billions of years, these simple, non-self-sustaining, acellular entities—particularly viruses—may be involved in geobiological processes since the Archean Eon in multiple ways. Firstly, viruses contribute to nutrient cycling in marine environments by infecting and lysing microbial cells, such as bacteria and phytoplankton. This process releases organic matter and nutrients into the surroundings, thereby impacting sediment composition over geological time. Secondly, viruses drive microbial evolution through horizontal gene transfer (HGT), also known as lateral gene transfer, the process by which genetic material is transferred between organisms, bypassing traditional parent-to-offspring inheritance, which plays a major role in the evolution

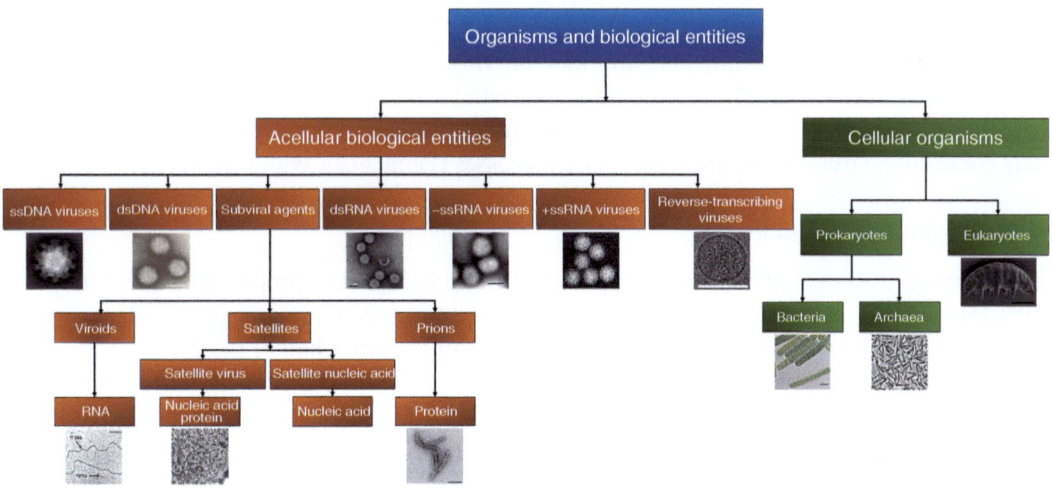

Figure 1.2 Microphotographs of acellular biological entities and cellular organisms. Source: parvovirus B19 (ssDNA viruses), Lee et al. (2024) / Springer Nature / CC BY-NC-ND 4.0; adenovirus (dsDNA viruses), Besson et al. (2020) / MDPI / CC BY 4.0; potato spindle tuber viroid (viroids), from International Committee on Taxonomy of Viruses, https://ictv.global/report_9th/subviral/Viroids; extra small virus (satellite virus), Krupovic et al. (2016) / Springer Nature; prions, Terry et al. (2019) / Springer Nature / CC BY-NC-ND 4.0; rotavirus (dsRNA viruses), Farkas et al. (2013) / Springer Science Business Media New York; influenza virus (-ssRNA viruses), Noda and Kawaoka (2010) / John Wiley & Sons; Aichi virus (+ssRNA viruses), from International Committee on Taxonomy of Viruses, https://ictv.global/report/chapter/picornaviridae/picornaviridae; HTLV-1 (reverse-transcribing viruses), Grigsby et al. (2010) / Springer Nature / CC BY 2.0; cyanobacteria (bacteria), reprinted with permission from Schneider et al. (2025) Schneider et al.; halobacteria (archaea), reprinted with permission from Cui and Dyall-Smith (2021) © 2021, Ocean University of China; tardigrade (eukaryotes), Hashimoto et al. (2016) / Springer Nature / CC BY-NC-SA 4.0.

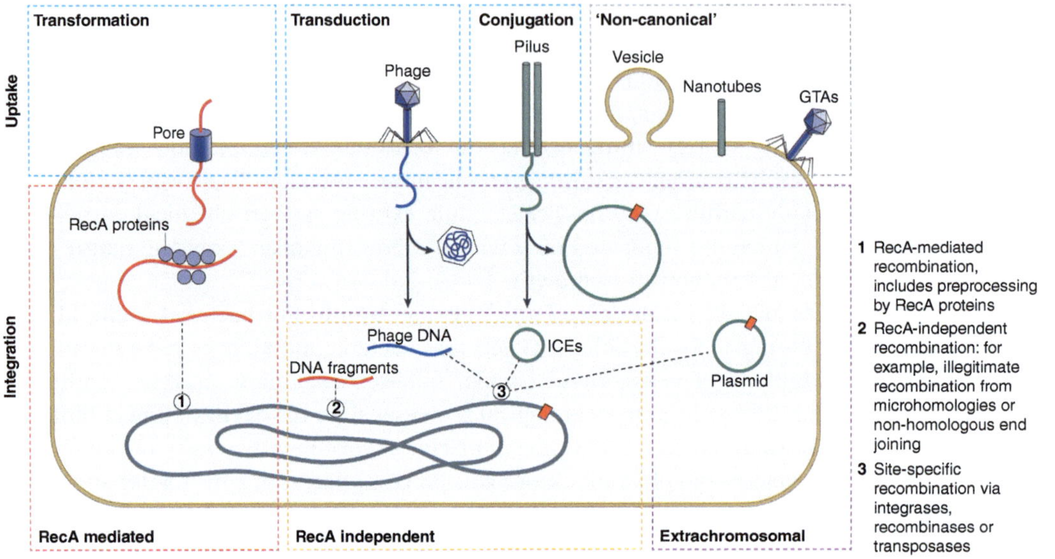

Figure 1.3 Mechanisms of horizontal gene transfer as a driving force of evolution. Transformation: direct uptake of genetic material from the environment; transduction: transfer mediated by viruses; conjugation: exchange through direct cell-to-cell contact. *Source:* Arnold et al. (2022) / with permission of Springer Nature.

of bacteria and archaea (Figure 1.3). HGT can reshape microbial communities and influence the breakdown or formation of mineral deposits by microorganisms. Thirdly, virus-like particles have been discovered in ancient sediments and rocks, offering valuable insights into Earth's early ecosystems and the evolutionary history of life. Lastly, viral activity in oceans affects phytoplankton populations, which play a role in carbon sequestration and regulate atmospheric CO_2 levels, indirectly connecting viruses to geological climate processes (Feng 2020; Wang 2022).

1.2.2 Domains of Cellular Life

The two-kingdom system of organisms, the Animalia and the Plantae, proposed by Carl Linnaeus in the 18th century, laid the groundwork for the systematic study of organisms, creating a structured approach to naming and classifying living forms. In the latter half of the 20th century, evidence began to accumulate that these were insufficient to express the diversity of life, and various schemes were proposed with three, four, or more kingdoms. This coexisted with a scheme dividing life into two main divisions: the prokaryota (bacteria, etc.) and the eukaryota (animals, plants, fungi, and protists).

However, with the advent of molecular taxonomy and phylogeny (sequence comparison of genes coding for 16S ribosomal RNA), our view of life on Earth changed radically. What were once called "prokaryotes" are far more diverse than anyone had suspected. The Prokaryota is now divided into two domains, the bacteria and the archaea. These two groups are as fundamentally different from each other as they are from Eukarya (Table 1.1). Neither of the prokaryotic domains is ancestral to the other, and each exhibits unique characteristics while sharing some common features with the others.

In the last decades, a great deal of additional work has been done to understand the nature of the last universal common ancestor (LUCA) of cellular life, the first eukaryotic common ancestor (FECA), and the last eukaryotic common ancestor (LECA), which is fundamental for

Table 1.1 A comparison of three domains of life.

Characteristics	Bacteria	Archaea	Eukarya
Nuclear envelope	Absent	Absent	Present
Membrane-enclosed organelles	Absent	Absent	Present
Peptidoglycan in cell wall	Present	Absent	Absent
Membrane lipids	Unbranched hydrocarbons	Some branched hydrocarbons	Unbranched hydrocarbons
RNA polymerase	One kind	Several kinds	Several kinds
Initiator amino acid for start of protein synthesis	Formyl-ethionine	Methionine	Methionine
Introns (non-coding parts of genes)	Rare	Present in some genes	Present
Response to the antibiotics streptomycin and chloramphenicol	Growth inhibited	Growth not inhibited	Growth not inhibited
Histones associated with DMA	Absent	Present	Present
Circular chromosome	Present	Present	Absent
Ability to grow at temperatures >100 °C	No	Some species	No

Source: Campbell and Reece (2002) / Benjamin Cummings.

reconstructing the branching pattern of life tree at the base (Figure 1.4). The three-domain system was proposed by Woese et al. (1990) based on genetic and molecular differences, especially in ribosomal RNA sequences, which highlighted the fundamental separation between archaea and bacteria. The two-domain system was proposed based on the eocyte hypothesis positing that eukaryotes evolved from a subgroup of archaea (Lake et al. 1984; Rivera and Lake 1992). Recent work enhances the view that the eukarya arose from well within the archaea and is specifically related to newly discovered archaeal species with eukaryote-like features (Doolittle 2022). Accordingly, archaea and eukaryotes were grouped into one domain due to shared genetic and evolutionary features.

1.2.2.1 The LUCA

The concept of the last universal common ancestor (LUCA) emerged in the late 20th century, significantly enhancing our understanding of the classification of living organisms. LUCA is defined as the ancestral entity from which all cellular life—bacteria, archaea, and eukaryotes—descends, and it represents a pivotal phylogenetic clade. Its genetic composition and biological features have been inferred through comparative genomic analyses on the basis of the conservation of biochemical traits. It was estimated that the LUCA had a genome coding for approximately 355 to 2600 proteins based on various studies (NA n.d.; Moody et al. 2024). These genes likely supported essential functions such as replication, transcription, translation, and basic metabolic pathways. The exact number remains debated due to the challenges of reconstructing LUCA's genome from modern descendants. Key aspects of LUCA remain unclear, such as its membrane structure—raising the possibility of a membrane-less form—and the mechanisms of its replication process. Hydrothermal vent systems are often hypothesized as LUCA's likely habitat. However, this hypothesis is challenged by the complexity of evolution, including phenomena like non-orthologous gene displacement and extensive HGT, which obscure its precise characteristics. Despite these uncertainties,

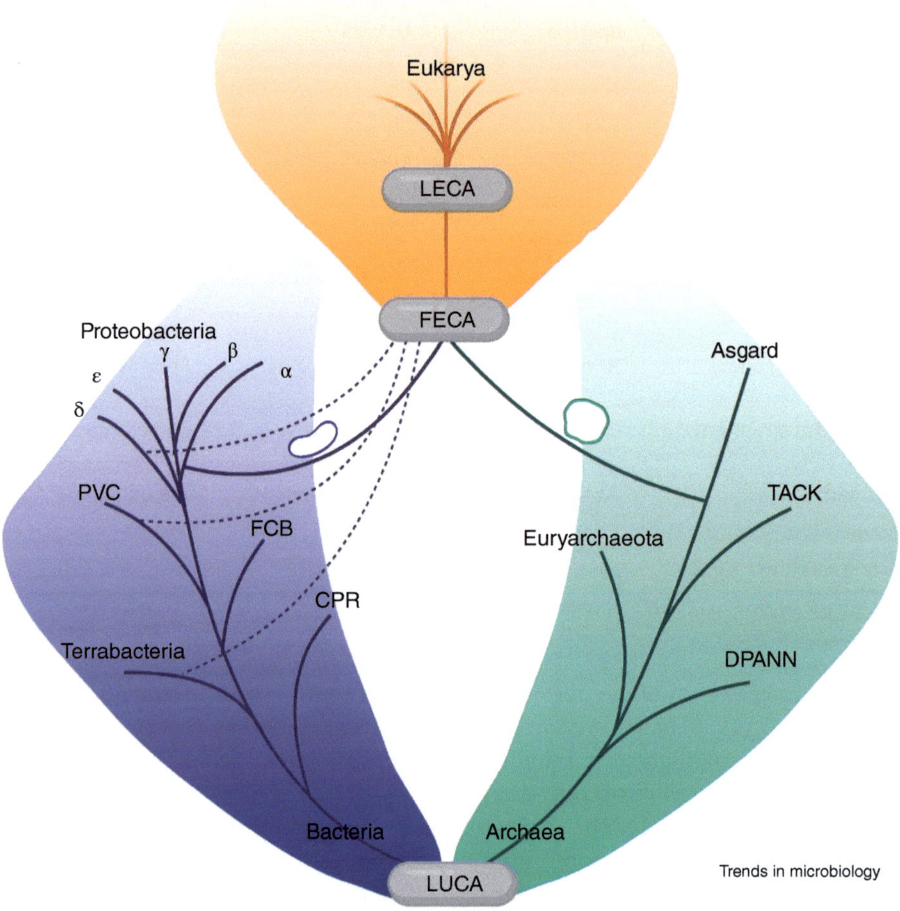

Figure 1.4 Evolution from the LUCA to LECA. A schematic based on the eocyte hypothesis that eukarya emerged from within the archaea. During symbiogenesis, the host cell was an Asgard-like archaeon, while the mitochondria originated from an ancestral alphaproteobacterial endosymbiont, both playing integral roles in shaping the eukaryotic genome. Furthermore, HGT from various bacterial lineages (represented by dotted lines) significantly enriched the genetic repertoire of the LECA, contributing to its evolutionary complexity. Abbreviations: CPR, Candidate Phyla Radiation; DPANN, Diapherotrites, Parv-, Aenigm-, Nano-, and Nanohaloarchaeota; FCB, Fibrobacteres-Chlorobi-Bacteroidetes; FECA, First Eukaryotic Common Ancestor; LUCA, Last Universal Common Ancestor; PVC, Planctomycetes-Verrucomicrobia-Chlamydiae; TACK, Thauma-, Aig-, Cren-, and Korachaeota. *Source:* Nobs et al. (2022) / with permission of Elsevier.

current evidence suggests that the LUCA was a pre-domain organism with a streamlined core genome (a minimal set of essential genes that are crucial for the survival and basic functions of an organism), exhibiting transitional biochemical traits that preceded the evolutionary divergence of bacteria and archaea.

Despite ongoing research, certain controversial questions remain unresolved, such as whether viruses existed before the LUCA or what the nature of interactions between the LUCA and its virome might have been. Recently, a hypothesis was proposed that the LUCA served as a host to viruses, since all known life forms are closely associated with viruses or other mobile genetic elements. Efforts to reconstruct the LUCA virome have been undertaken by examining the distribution of viruses among bacteria and archaea, as well as tracing the evolutionary history of key viral genes (Krupovic et al. 2020). The reconstruction suggested the LUCA had a complex virome that included major groups of viruses currently associated with bacteria and archaea. Furthermore, the LUCA virome was hypothesized to have split into the viromes of the last bacterial common ancestor (LBCA) and the last archaeal common ancestor (LACA). Notably, extensive viral evolution was thought to have occurred even before the existence of the LUCA, adding to the complexities of understanding its role in early life.

1.2.2.2 The FECA and LECA

The FECA is hypothesized as the earliest ancestor of all eukaryotic cells, representing the transition from simpler prokaryotic organisms to more complex eukaryotic life forms. The LECA is considered the most recent shared ancestor of all eukaryotic organisms, representing a critical juncture in evolutionary history that enabled the emergence of complex life. FECA eventually gave rise to LECA that had fully developed eukaryotic features, such as a nucleus, mitochondria, and an endomembrane system. Lynn Margulis famously popularized the hypothesis that eukaryogenesis was driven by endosymbiosis and proposed that mitochondria and chloroplasts originated from free-living bacteria (Sagan 1967). This theory explains the presence of complex internal organelles in eukaryotes, such as mitochondria, which are thought to have evolved from an ancient alphaproteobacterium. However, significant debates in evolutionary biology persist, including the relationship between archaea and eukarya and the identity of archaea as the host cell in eukaryogenesis. These unresolved issues are essential for understanding the basal pattern of the tree of life, particularly whether eukarya are a sister group to archaea or emerged from within the archaeal domain, members of the Asgard archaea or their closely related sister lineage (Nobs et al. 2022).

Eukaryogenesis is understood as an evolutionary process that begins with the microbial cell that diverged from the ancient Asgard lineage, referred to as the FECA that has no extant descendants, and culminates in the LECA, with various transitional stages in between referred to as the proto-eukaryote. Many eukaryogenic theories converge on the idea that an Asgard archaeal host cell and an alphaproteobacterial endosymbiont played fundamental roles in initiating eukaryogenesis. It is suggested that certain eukaryotic cellular features, such as the endomembrane system—comprising the endoplasmic reticulum (ER), Golgi apparatus, lysosomes, vacuoles, vesicles, and plasma membrane—along with a dynamic cytoskeleton, may have predated eukaryogenesis within the host archaeon. However, eukaryotic-specific traits, including genome compartmentalization within the nucleus and phagocytosis, represent distinct innovations that emerged within eukaryotes (Nobs et al. 2022).

The genome of LECA is thought to be a chimeric fusion of genes derived from both archaea and bacteria. Gene duplications, predominantly originating from an Asgard-related host cell, contributed to early eukaryotic complexity. Endosymbiotic gene transfer (EGT) occurred as the ancestral eukaryotic cell engulfed an alphaproteobacterial-related progenitor of the mitochondrion, leading to the integration of genetic material from both the endosymbiont and the host. Additionally, HGT from diverse bacterial groups beyond alphaproteobacterial lineages significantly enriched the proto-eukaryotic genome, further expanding its repertoire of eukaryotic genes. However, only a small fraction of eukaryotic genes appears to have originated directly from alphaproteobacteria, implying that mitochondrial acquisition was a relatively late event in eukaryotic evolution. Consequently, the LECA can be depicted as a cellular and genomic chimera, incorporating innovations from both its diverse archaeal and bacterial predecessors (Figure 1.5).

Many microbial syntrophies, forming a "microbial village," could have nurtured the LECA. The symbiotic relationship could provide proximal space and sources for HGT. López-García and

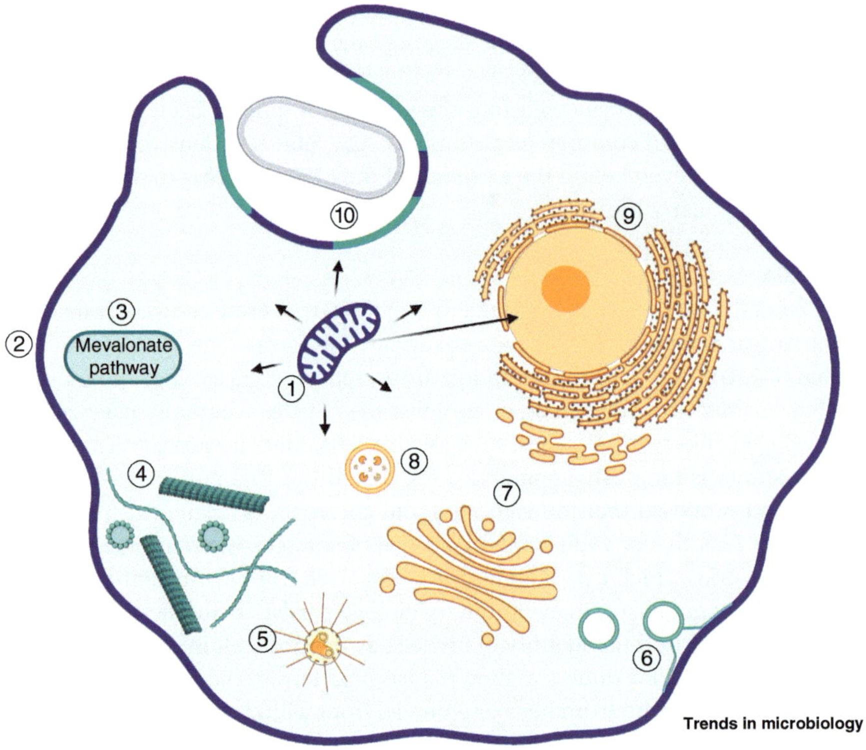

Figure 1.5 Schematic of the chimeric LECA with bacterial contributions in purple, archaeal contributions in aqua green, and eukaryotic innovations in yellow. (1) Mitochondria from an alphaproteobacterial endosymbiont which underwent gene loss and gene transfer to the host nuclear DNA; (2) bacterial-like membrane lipids with ester-bound fatty acids; (3) mevalonate pathway for isoprenoid production. (4) cytoskeletal actin and tubulin proteins; (5) centrosome regulating cytoskeletal proteins for cell shape and chromosome separation; (6) vesicle formation and membrane remodeling; (7) endomembrane system; (8) protein degradation and recycling machinery such as the lysosome; (9) compartmentalization of the nuclear genome; (10) phagocytosis. *Source:* Nobs et al. (2022) / with permission of Elsevier.

Moreira (2020) proposed the revised syntrophy hypothesis, suggesting that Paleoproterozoic microbial mats were most likely the place where eukaryogenesis occurred for dynamic metabolic interactions within the mats. The microbial mats, which are stratified microbial communities, contributed to complex biogeochemical cycling that sustained them. Furthermore, microbial mats offer an ideal ecological niche and taxonomic diversity among microbial communities, facilitating alternative sources of HGT and diverse symbiotic relationships.

1.2.3 Kingdoms of Cellular Life

Before 1970, taxonomists classified all forms of life into two kingdoms: Animalia and Plantae. Bacteria, fungi, and photosynthetic protists were considered plants, and the protozoa were classified as animals. Whittaker (1969) proposed a five-kingdom classification scheme which revolutionized biological taxonomy. This system categorized organisms into five distinct kingdoms based on their cellular structure, mode of nutrition, and ecological roles. First, Whittaker identified two kingdoms of primarily unicellular microorganisms based on whether they exhibit prokaryotic or eukaryotic cellular organization. The kingdom Monera consists of generally single prokaryotic cells, whereas

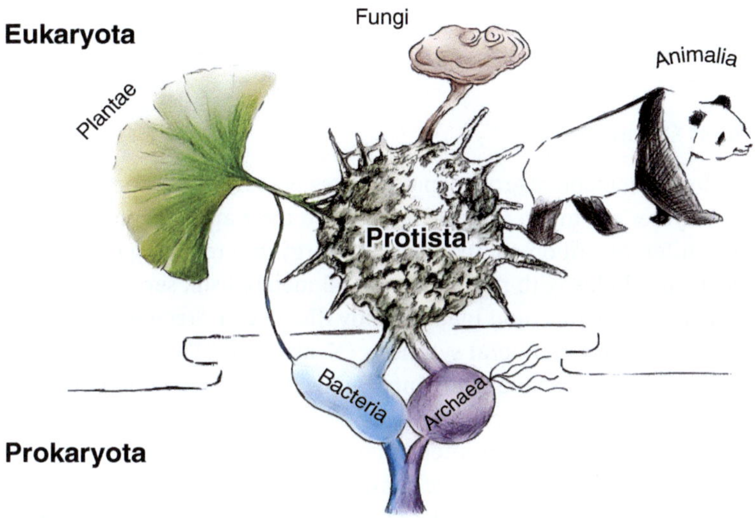

Figure 1.6 Current classification of living cellular organisms, categorized into three domains and six kingdoms.

the kingdom Protista consists of generally single eukaryotic cells. All of the organisms in the remaining three kingdoms (Plantae, Fungi, and Animalia) are eukaryotic, and most of them are multicellular. They may be classified further on the basis of their way of acquiring nutrients. Members of the kingdom Plantae photosynthesize, and members of the kingdom Fungi secrete enzymes outside their bodies and then absorb the externally digested nutrients. In contrast, members of the kingdom Animalia ingest their food and then digest it, either within an internal cavity (many animals) or within individual cells (e.g. sponge cells).

Recently, a six-kingdom scheme has been proposed, which incorporates molecular and evolutionary discoveries, and expands upon older classification methods. Many microbiologists argue that the cell structure of the archaea is so different from that of prokaryotic and eukaryotic cells that they should occupy their own kingdom (Table 1.1). Therefore, the kingdom Monera is split into the bacteria and the archaea. Following the fashion, this textbook adopts the updated six-kingdom classification system (Figure 1.6).

1.3 Diversity of Biological Metabolism

Organisms, especially microbes, show an impressive metabolic diversity in the ways in which they obtain the energy needed for growth. The process involves two basic kinds of transformations, building up or biosynthetic processes mostly via **Calvin Cycle** (the biochemical route of CO_2 fixation in many autotrophic organisms), called **anabolism**, and breaking down or degradative processes, called **catabolism**. In the process of anabolism, microbial cells use energy to convert nutrients and simple compounds into more complex structural and functional macromolecules, which include the formation of protein from amino acids, nucleic acids from nucleotides, polysaccharides from simple sugars, and lipids from glycerol and fatty acids. The opposing process, catabolism, oxidizes organic or inorganic compounds, accompanied by the release of chemical energy and excretion of waste products into the environment. Some of the energy is captured and utilized by organisms for movements, transport of nutrients, and anabolic reactions, while the remaining energy is lost to the environment in the form of heat.

1.3.1 Metabolic Classification of Organisms

The living world can be divided into two metabolic groups: prokaryotes possess rather remarkable metabolic plasticity, and most eukaryotes are much more monolithic in their approach to metabolism (i.e. metabolizing only a few different carbon compounds for energy and respiring only oxygen). For the most part, this division also holds with regard to mineral metabolism, separating the prokaryotes, which harvest energy from minerals (and inadvertently dissolve or precipitate them) from the eukaryotes, which invest energy in biomineral synthesis (Nealson and Rye 2005).

On the basis of energy usage, we can divide metabolic life into three groups. Many microorganisms resemble animals and humans, using organic chemicals as energy sources. We call these organisms **chemoorganotrophs (heterotrophs)**. An interesting collection of microorganisms, exclusively prokaryotic (bacteria and archaea), use inorganic compounds as energy sources. These organisms are called **chemolithotrophs**. Another major class of living organisms, called **phototrophs**, use light as an energy source, which include green plants and also many microorganisms, both prokaryotic and eukaryotic. We will see that there are two distinct types of photosynthesis, oxygenic photosynthesis typified by green plants but also found in both eukaryotic and certain prokaryotic microorganisms, and anoxygenic photosynthesis found exclusively in special groups of bacteria, e.g. purple bacteria, green bacteria, and certain cyanobacteria (see Section 1.3.2.3).

Life can also be categorized according to the carbon sources utilized. Organisms that use inorganic chemicals or light as energy sources are frequently able to grow in the complete absence of organic materials, using carbon dioxide as their sole source of carbon. **Autotrophs** (meaning literally, "self-feeding") are able to obtain all the carbon they need from inorganic sources (CO_2 or bicarbonate). Note that autotrophy does not refer to the energy source used but to the carbon source. Autotrophs are of great importance in the functioning of the biosphere because they are able to bring about the synthesis of organic matter from inorganic (non-living) sources; this process is called primary production. **Chemolithoautotrophs** (in general, chemolithotrophs) are organisms that obtain energy by degrading reduced inorganic compounds and use CO_2 as a carbon source. **Photoautotrophs** use light as the energy source and CO_2 as the major carbon source. **Heterotrophs** utilize energetically rich organic carbon as their carbon sources, breaking down and resynthesizing polymers. **Chemoheterotrophs** are organisms that use chemicals as a source of energy and organic compounds as carbon sources. **Chemoorganoheterotrophs** (in general are chemoorganotrophs) obtain both energy and carbon from organic compounds. **Photoheterotrophs** use light as the energy source and organic compounds as the carbon source.

In addition to alternate styles of energy metabolism, we also discuss in this chapter a number of specialized aspects of catabolism. As for respiratory metabolism, when O_2 is used as an electron acceptor (a substance that accepts electrons during an oxidation-reduction reaction), the process is called **aerobic respiration**. There are also a number of other electron acceptors that can participate in the electron transport process in particular groups of microorganisms. The process in which an electron acceptor other than molecular oxygen is used is called **anaerobic respiration** and will be one of the significant features of metabolic diversity considered here.

Finally, although the major thrust of this chapter is carbon metabolism, our discussion would not be complete without a consideration of one of the important metabolic reactions of microorganisms that involves nitrogen. This is the utilization of N_2 as a source of nitrogen for biosynthesis, a process called **nitrogen fixation**. This important process that is unique to prokaryotes contributes to the recycling of nitrogen into living matter and is important not only agriculturally but also in the total function of the biosphere.

1.3.2 Photosynthesis

1.3.2.1 Pigments, Reactions, and Photosystems

Photosynthesis is the utilization of solar energy by plants, algae, and certain bacteria for the synthesis of organic molecules. Photosynthetic organisms have pigment molecules, chlorophyll (that makes plants look green) in higher plants, algae, and cyanobacteria, or bacteriochlorophyll in green and purple bacteria. There are different types of chlorophyll, but the most common are chlorophyll a–d and f. Chlorophyll a is found in all oxygenic photosynthetic organisms and is the primary pigment in photosynthesis. Chlorophyll b is in green plants and helps broaden the spectrum of absorbed light. Chlorophylls c, d, and f are found in specific algae and cyanobacteria. Both chlorophyll and bacteriochlorophyll are able to capture light energy from the sun and use it to make ATP (adenosine triphosphate) and NADPH (abbreviations for the oxidized/reduced forms of nicotinamide adenine dinucleotide, a diffusible electron carrier). In addition, carotenoids are known to absorb blue and green light, contributing yellow, orange, and red hues, and they also protect against oxidative damage; Phycobilins found in cyanobacteria and red algae help capture light in deep-water environments.

Photosynthesis can be considered in terms of two separate sets of reactions: the light reactions, in which light energy is converted to chemical energy, and the dark reactions, in which this chemical energy is used for CO_2 fixation. Photosynthesis also spins off the energetic molecule, NADPH, which is used by green plants to convert nitrate into ammonia (in assimilative nitrate reduction). Photosynthesis contributes to big changes in the global environment, mainly as a producer of oxygen gas. Virtually all of the oxygen gas in our atmosphere came from photosynthesis. Photosynthesis also has played a minor role as a remover of carbon dioxide gas from the atmosphere (calcium carbonate precipitation greatly overshadows it). Photosynthesis occurs in two main pathways: anoxygenic and oxygenic photosynthesis (see Table 1.2 for comparison).

Photosystems are essential components of photosynthesis, responsible for capturing light energy and converting it into chemical energy (Table 1.2). They are protein-pigment complexes found in the thylakoid membranes of cyanobacteria, algae, and plants. Photosystem I (PSI) uses light energy to produce NADPH that is used for the subsequent photosynthetic dark reaction or Calvin cycle and absorbs light mainly at 700 nm (P700). PSI transfers electrons across the thylakoid membrane from plastocyanin to ferredoxin. Photosystem II (PSII) splits water molecules to release oxygen, electrons, and protons, absorbs light mainly at 680 nm (P680), and plays a key role in the oxygenic photosynthesis process. Enzymes in PSII capture photons of light to excite electrons that are transported through a range of coenzymes and cofactors to reduce plastoquinone to plastoquinol. Excited electrons are replaced by the oxidation of water, resulting in the production of molecular oxygen and hydrogen ions. Oxygenic phototrophs have both PSI and PSII while anoxygenic photoautotrophs carry a single photosystem, either PSI or the one similar to PSII (Blankenship 2010).

1.3.2.2 Oxygenic Photosynthesis

Oxygenic photosynthesis is the most familiar and widespread form of photosynthesis in modern biosphere. On the continents, it occurs mostly in green plants. It is also prevalent in aquatic environments, like lakes, rivers, and oceans. There, oxygenic photosynthesis is performed by algae (Figure 1.7), many protists, and cyanobacteria. Also, corals, lichens, giant clams, and at least one kind of tiny marine slug are photosynthetic because they are inhabited by symbiotic,

Table 1.2 Comparison of photosynthetic mechanisms by different organisms.

	Oxygenic photosynthesis		Anoxygenic photosynthesis	
	Plants, algae	**Cyanobacteria**	**Purple photosynthetic bacteria**	**Green photosynthetic bacteria**
Location of the photosystem[a]	In membranes of thylakoids, which are within the stroma of chloroplasts?	In membranes of thylakoids, located within the cell.	Within the cytoplasmic membranes; extensive invaginations in that membrane effectively increase the surface area	Primarily within the cytoplasmic membrane; chlorosomes attached to the inner surface of the membrane contain the accessory pigments
Type of photosystem[a]	Photosystem I and photosystem II		Similar to photosystem II	Similar to photosystem I
Primary light-harvesting pigment	Chlorophyll *a*		Bacteriochlorophylls	
Mechanism for generating reducing power	Non-cyclic photophosphorylation using both photosystems		Reversed electron transport	Non-cyclic use of the photosystem
Source of electrons for reducing power	H_2O		Varies among the organisms in the group; may include H_2S, H_2, or organic compounds	
CO_2 fixation	Calvin-Benson cycle		Calvin-Benson cycle	Reversed TCA cycle
Accessory pigments	Carotenoids	Carotenoids, phycobilins	Carotenoids	Carotenoids

[a] Photosystem: protein complexes within which chlorophyll and other light-gathering pigments are organized. Two types of photosystems are called photosystem I and photosystem II, in order of their discovery; each has a characteristic of reaction center. The reaction center of photosystem I is known as P700 because the pigment is best at absorbing light having a wavelength of 700 nm, and the reaction center of photosystem II is called P680 because its absorption spectrum has a peak at 680 nm. P700 and P680 are actually identical chlorophyll α molecules. However, their association with different proteins affects the electron distribution in chlorophyll molecules and accounts for the slight differences in light-absorbing properties.
Source: Adapted from Nester et al. (2004) / McGraw-Hill Education.

Figure 1.7 Green alga *Cladophora* at Qingdao's rocky beach in China is actively photosynthesizing, releasing oxygen gas bubbles into the water. *Source:* Photo by Xingliang Zhang.

photosynthetic microorganisms. These organisms use water as the electron donor (a compound that donates electrons during oxidation-reduction reaction) to fix carbon.

Oxygenic photosynthesis is a complex process that involves a cascade of biochemical reactions. In essence, however, it is fairly simple. The general reaction equation is:

$$6CO_2 + 6H_2O \xrightarrow{\text{Sunlight}} C_6H_{12}O_6 \text{ (carbon chain)} + 6O_2$$

This reaction equation demonstrates that the process of photosynthesis takes in carbon dioxide gas from the atmosphere, water from the soil (in the case of green plants), and light energy from the sun. Photosynthesis then uses these resources to make carbon chains. The carbon chains can later be used to make a variety of molecules including sugar. Note that water is used as an electron donor and a source of hydrogen, and oxygen gas is given off as a waste product.

Please draw your attention for a moment to the issue of hydrogen. In order to make organic molecules from carbon dioxide, photosynthesizing organisms need a source of hydrogen that plays an essential role here. During the light-dependent reaction, water molecules are split into oxygen, electrons, and hydrogen ions. The electrons help generate ATP and NADPH, which are energy carriers needed for the Calvin cycle. Meanwhile, the hydrogen ions contribute to creating a proton gradient that drives ATP synthesis.

Oxygenic photosynthesis has played a monumental role in shaping Earth's surface environment. Before oxygenic photosynthesis, Earth's atmosphere lacked free oxygen. Cyanobacteria, the earliest organisms to perform oxygenic photosynthesis, began releasing molecular oxygen perhaps early in the Mesoarchean (3.2–2.8 billion years ago), triggering the Great Oxygenation Event (Figure 1.8). The subsequent rise of eukaryotic phytoplankton and land plants further enhanced the primary production and oxygenated the Earth's surface (see Sections 8.3.4 and 8.4.4). This influx of oxygen dramatically altered atmospheric chemistry and paved the way for aerobic life. As oxygen accumulated in Earth's oceans, it reacted with dissolved iron, forming iron oxide sediments—visible today as banded iron formations (BIFs) in ancient rock records, serving as a major iron resource for global industry. These formations are vital for understanding early Earth processes and serve as valuable iron ore deposits. Oxygen from photosynthesis eventually led to the formation of the ozone (O_3) layer, which protects life from harmful UV radiation. This shielding effect together with atmospheric oxygen accumulation made it possible for aerobic, complex, multicellular organisms to thrive in the ocean and on land, and hence dramatically increased biodiversity. Oxygenic photosynthesis drives the carbon cycle, converting atmospheric CO_2 into organic matter. This process helps regulate Earth's climate, influencing carbon sequestration in sedimentary rocks and fossil fuels.

1.3.2.3 Anoxygenic Photosynthesis

Anoxygenic photosynthesis is probably the oldest form of photosynthesis, perhaps originating within 100 million years or so after the appearance of life ~4.0 billion years ago. Its biochemistry is somewhat simpler than oxygenic photosynthesis. Anoxygenic photosynthesis occurs today in purple sulfur/non-sulfur bacteria, green sulfur/non-sulfur bacteria, and heliobacteria that inhabit shallow, oxygen-free waters and mud in lakes, swamps, and oceans. Some cyanobacteria also grow via anoxygenic photosynthesis. These organisms use a wide range of electron donors (H_2S, H_2, Fe^{2+}, S^0, organic carbon), each of which is oxidized during the process, leading to the fixation of reduced carbon and the oxidation of the electron donors.

Experiments and ecological studies provide key insights, demonstrating that anoxygenic photosynthetic bacteria often outcompete cyanobacteria when alternative electron donors (e.g. H_2S) are

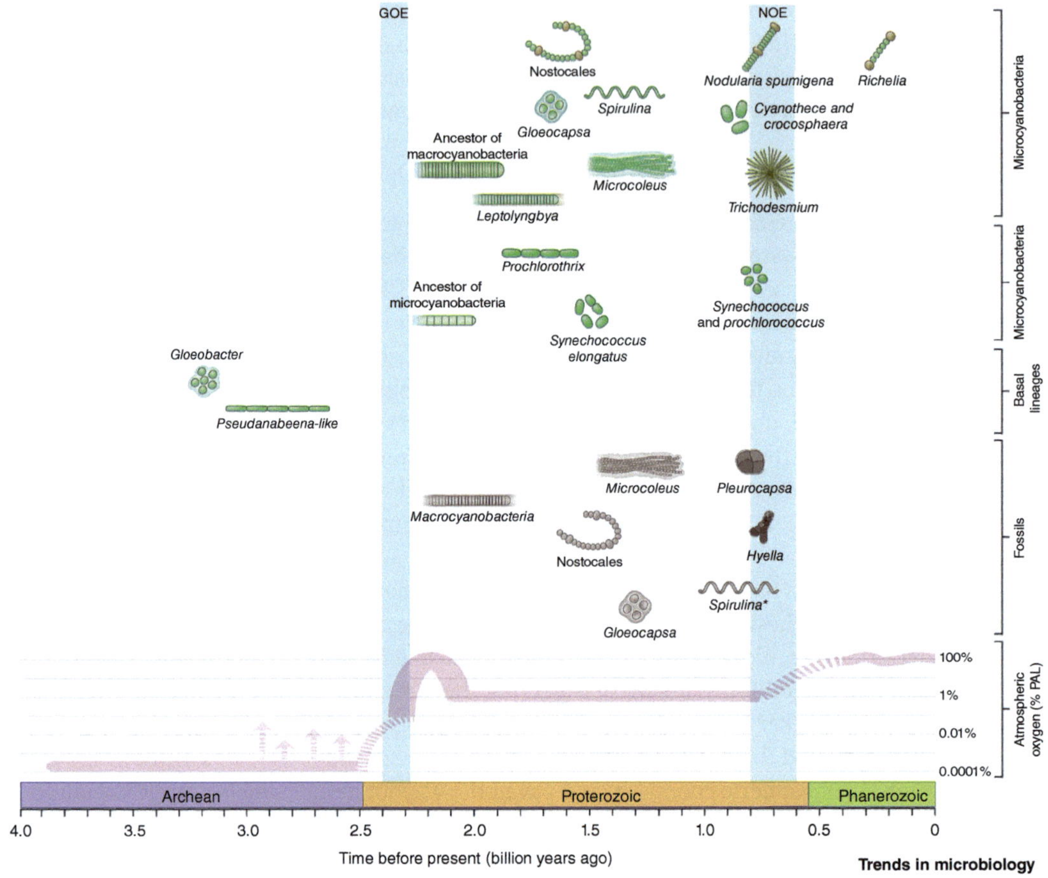

Figure 1.8 Timeline of oxygen concentration in the atmosphere, the antiquity of cyanobacterial lineages based on molecular clock estimates and fossil occurrences. *Source:* Sánchez-Baracaldo et al. (2022) / with permission of Elsevier.

present (Klatt et al. 2015; Ozaki et al. 2019). Additionally, certain modern cyanobacteria (e.g. *Leptolyngbya* sp. strain hensonii) are capable of transitioning to anoxygenic photosynthesis in environments rich in hydrogen sulfide (Hamilton et al. 2018). This metabolic adaptability is fundamentally significant for deciphering Earth's billion-year anoxic history.

Anoxygenic photosynthesis is different from oxygenic photosynthesis in two main ways: (1) a variety of reduced chemicals (e.g. H_2S) can be used instead of water as electron donors and a source of hydrogen; and (2) the bacteria, under strict anaerobic conditions, photosynthetically generate sulfur, sulfate, or iron instead of oxygen gas. The general reaction equation for anoxygenic photosynthesis is:

$$6CO_2 + 12H_2S \xrightarrow{\text{Sunlight}} C_6H_{12}O_6\ \text{(carbon chains)} + 6H_2O + 12S$$

There are major limitations on the ecological distribution of anoxygenic photosynthesis. In order for it to occur, the environment must have a combination of three things: (1) it must be free of oxygen gas; (2) reducing electron donors (e.g. H_2S) must be available; and (3) sunlight must be available. Today, oxygen-less, hydrogen sulfide-rich environment (typical of deep lake waters or deep in muddy environments) is difficult to find with good exposure to sunlight. However, it might be ubiquitous in remote Earth history. During the Archean Eon (~4.031–2.5 billion years ago), the Earth's surface was anoxic and rich in reductants such as H_2S, H_2, and Fe^{2+} that were ready to serve as electron donors.

Anoxygenic photosynthesis may have a profound impact on Earth's early history. Anoxygenic photoautotrophs either predated or coexisted with oxygenic photoautotrophs (Sánchez-Baracaldo et al. 2022), potentially serving as key drivers of geochemical cycles in early ecosystems. They thrived in Earth's surface environments and formed extensive microbial mats that shaped sedimentation patterns. Though their photosynthesis did not release oxygen gas, anoxygenic photoautotrophs (likely cyanobacteria) involved in this process may have played a crucial role in removing excess reductants (H_2S and Fe^{2+}) from ancient oceans, thus creating favorable conditions for the eventual atmospheric accumulation of oxygen. Furthermore, when iron was used as an electron donor, anoxygenic photosynthesis may have contributed to the formation of BIFs (see Sections 1.5 and 6.2.6).

1.3.3 Nitrogen Fixation

Nitrogen fixation is a process in which N_2 is reduced to form ammonia (NH_3), which then can be incorporated into cellular material, by reactions:

$$N_2 + 6H^+ \rightarrow 2NH_3 \text{ and } 2NH_3 + 2R-OH \rightarrow 2R-NH_2 + 2H_2O$$

Nitrogen is required by all living organisms. Most nitrogen on Earth is in the form of dinitrogen (N_2), which is not bioavailable though it makes up 78% of modern atmosphere. Abiotic sources of nitrogen produced through mechanisms such as lightning discharge or mineral-based catalysis are only able to support a small biosphere. Only a small fraction of prokaryotes including bacteria and archaea can fix nitrogen. No eukaryotic organisms have been known to reduce and fix nitrogen. Therefore, just as animals, including humans, depend on other organisms to fix carbon, most organisms rely on prokaryotes to convert atmospheric nitrogen to a form they can assimilate to create biomass.

The reduction process is catalyzed by the enzyme complex nitrogenase, which drives the production of at least half of Earth's fixed nitrogen. This mechanism plays a crucial role in alleviating fixed nitrogen limitations within natural ecosystems and is likely to exert a disproportionate influence on ecosystem functionality compared with inputs from other populations. Consequently, nitrogen-fixing organisms in natural communities are recognized as keystone species (see Boyd and Peters 2013).

Three nitrogenases have been recognized, which occur as either Mo, Fe-only, or V at the active site cofactor. All known nitrogen fixers (or diazotrophs) encode canonical Mo-nitrogenases that are more efficient than Fe-only and V-nitrogenases that are typically viewed as "backup" enzymes, used only when Mo is not available (McRose et al. 2017). These alternative nitrogenases allow microbes to adapt to different environmental conditions, ensuring nitrogen fixation continues even when molybdenum is scarce.

Nitrogenase is readily inactivated by O_2. Therefore, nitrogen-fixing aerobes must have a mechanism to protect them from exposure to O_2. In addition, nitrogen fixation requires a tremendous expenditure of energy, at least 16 molecules of ATP for every molecule of nitrogen fixed, because N_2 has a very stable triple bond. Diazotrophs may be free living, or they may live in symbiotic association with higher organisms, particularly certain plants (Table 1.3).

Symbiotic nitrogen fixers may be localized on the plant root tissues in the form of nodules. In effect, the plant provides nitrogen fixers with a source of energy, in the form of simple sugars, and a sheltered environment in which access to oxygen is controlled so that nitrogenase is not inactivated. Meanwhile, the nitrogen-fixer supplies the plant with a source of nitrogen. Symbiotic nitrogen-fixers are very important agriculturally because they provide a renewable nitrogen source to soils.

Table 1.3 Nitrogen fixers, including free-living and symbiotic prokaryotes (Madigan et al. 1997, 2003, 2018).

Free-living aerobes		
Chemoorganotrophs	**Phototrophs**	**Chemolithotrophs**
Azotobacter *Azomonas* *Azospirillum* *Klebsiella*[a] *Methylomonas*	Cyanobacteria (various, but not all) *Anabaena* *Nostoc* *Gloeothece* *Aphanizomenon*	*Alcaligenes* *Acidithiobacillus* (some species)
Free-living anaerobes		
Chemoorganotrophs	**Phototrophs**	**Chemolithotrophs**
Clostridium *Desulfotomaculum*	Purple bacteria: *Chromatium* *Methanococcus* *Rhodobacter* Green sulfur bacteria *Chlorobium* *Heliobacter*	Archaea: *Methanosarcina* *Methanocaldococcus*
Symbiotic		
Root-nodule-forming bacteria		**Other symbiotic diazotrophs**
Alphaproteobacteria (e.g. *Mesorhizobium, Bradyrhizobium, Sinorhizobium*) and *Betaproteobacteria* (e.g. *Burkholderia*) with leguminous plants such as soybeans, peas, clover; *Actinobacteria* (e.g. *Frankia*) with non-leguminous plants such as alder, bayberry, and autumn olive.		*Teredinibacter* in association with shipworms *Treponema* in termite guts *Glomeribacter* in endomycorrhizal fungi Cyanobacteria in several fungi, algae, and plants

[a] N_2 fixation occurs only under anoxic conditions.

Nitrogen fixers may also live freely in soil and water. In the oceans, planktonic cyanobacteria *Trichodesmium* and *Synechococcus* are primary nitrogen fixers, fueling up to half of the new production, despite making up only 1% of the marine biomass. In subtropical and tropical oceans, nitrogen fixation not only has a major role in the global marine N budget but also impacts the coupling of N-C-P-Fe cycles. Free-living diazotrophs require a means of protecting their nitrogenases from O_2. Some aerobic organisms accomplish this by maintaining a very high rate of O_2 utilization, thereby minimizing the O_2 diffusion into the cell. Many cyanobacteria instead possess specialized cells within their filaments, called heterocysts, which serve as O_2-free microenvironment, while others photosynthesize strictly during daylight and fix nitrogen at night (Madigan et al. 1997).

1.3.4 Chemolithotrophy: Energy from Oxidation of Inorganics

Chemolithotrophs obtain their energy from the oxidation of inorganic compounds, such as H_2, H_2S, S^0, $S_2O_3^{2-}$, Fe(II), Mn(II), NH_4^+, and NO_2^-. These microorganisms are also able to obtain all their carbon from CO_2, generally processed via the Calvin cycle. Therefore, they are also autotrophs. There are many sources of inorganic electron donors for chemolithoautotrophs, which include geological, biological, and anthropogenic sources. Volcanic activity and hydrothermal fluids are major sources of reduced sulfur compounds, as is the microbial sulfate reduction. Agricultural and mining operations add inorganic electron donors to the environment, as does the burning of fossil fuels and input of industrial wastes. The ecological success and metabolic diversity of chemolithoautotrophs indicate that sources and supplies of inorganic electron donors in nature are abundant, which is particularly true on early Earth (see Sections 8.3 and 8.4). In most cases below, O_2 serves as the terminal electron acceptors (TEAs), but different microorganisms can substitute it for less electropositive TEAs, i.e. NO_3^-, Mn(IV), Fe(III), SO_4^{2-}, and S^0 when oxic conditions become diminished.

1.3.4.1 Hydrogen-Oxidation: Using the Chemical Energy in H_2 to Fix Carbon

Hydrogen (H_2) is a common product of microbial metabolism, and a number of chemolithotrophs, called H_2-oxidizing bacteria or archaea, are able to use it to generate ATP. The H_2-oxidizers are the most widespread chemolithoautotrophs, ranging from aerobes to facultative anaerobes that can grow on NO_3^-, Mn(IV), or Fe(III), to obligate anaerobes that use SO_4^{2-}, and finally to methanogens and homoacetogens.

Aerobic H_2-oxidizing bacteria are commonly termed hydrogen bacteria. The reaction:

$$H_2 + 1/2O_2 \rightarrow H_2O$$

is catalyzed by the enzyme hydrogenase and can support the synthesis of at least one ATP. The electrons from H_2 initially being transferred to a quinone acceptor. Some hydrogen bacteria have two hydrogenases. The soluble enzyme takes up H_2 and reduces NAD^+ to NADH, while the membrane-bound hydrogenase is involved in energetics.

Most hydrogen bacteria are facultative chemolithotrophs, e.g. *Ralstonia eutropha*, capable of growing either as chemoorganotrophs on organic compounds or chemolithotrophically by H_2 oxidation via the reaction:

$$6H_2 + 2O_2 + CO_2 \rightarrow (CH_2O) + 5H_2O$$

thus behaving as mixotrophs. When growing autotrophically, hydrogen bacteria fix CO_2 by the Calvin cycle. When organic compounds are present, synthesis of the Calvin cycle is repressed and enzymes involved in utilization of the organic compounds are induced. Therefore, in a natural environment, it is likely that aerobic hydrogen bacteria are able to regulate their catabolic enzymes and shift between chemoorganotrophy and chemolithotrophy (Madigan et al. 2003).

1.3.4.2 Methanogenesis and Acetogenesis from CO_2 Reduction

Carbon dioxide is common in nature and is a major product of energy metabolism of heterotrophs. Several prokaryotic groups are able to use CO_2 as an electron acceptor in anaerobic respiration. The most important CO_2-reducing prokaryotes are methanogens including a large number of Euryarchaeota, a major group of archaea (Madigan et al. 2003). Many of them utilize H_2 as the electron donor (energy source). In addition, formate, carbon monoxide, and other organic compounds such as alcohols provide electrons for CO_2 reduction.

Methanogenesis from CO_2 involves the transfer of eight electrons, with two being added at a time. This stepwise reduction results in intermediate oxidation states of the carbon atom, transitioning from +4 in CO_2 to −4 in CH_4. The process relies on several specialized coenzymes, which can be categorized into two groups: (1) C_1 carriers, responsible for transporting the single-carbon unit through enzymatic reduction, and (2) redox coenzymes, which serve as electron donors (Figure 1.9). The overall reaction is

$$4H_2 + CO_2 \rightarrow CH_4 + 2H_2O$$

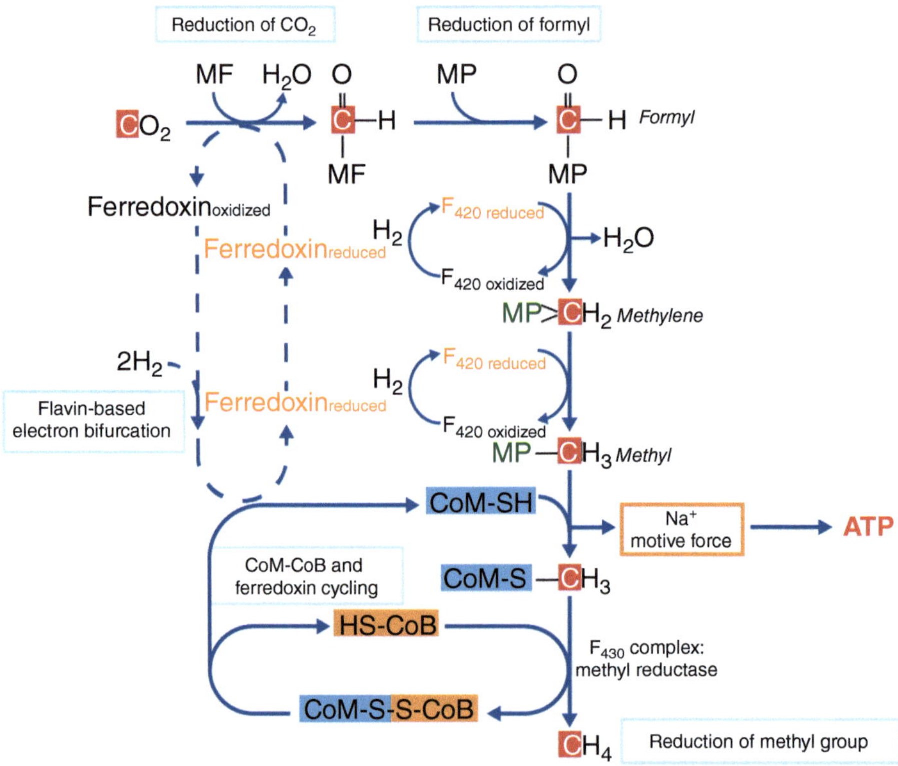

Figure 1.9 Methanogenetic steps in microbial CO_2 reduction by H_2. Carbon atom reduced is highlighted in red, and the source of electrons is highlighted in brown. F_{420} reduced, a decreased form of coenzyme F_{420} that is an unique, flavin-like electron carrier found in archaea (methanogens), bacteria and some fungi; F_{430}, coenzyme F_{430} that is a unique nickel-containing tetrapyrrole cofactor essential for methanogenesis; MF, methanofuran, a coenzyme with a furan ring linked to various side chains and found exclusively in methanogenic archaea; MP, methanopterin, a pterin-based coenzyme that acts as a carrier of one-carbon (C_1) units during the stepwise reduction of CO_2 to methane. CoM (coenzyme M, 2-mercaptoethanesulfonate, $HS\text{-}CH_2\text{-}CH_2\text{-}SO_3^-$) and CoB (coenzyme B, 7-mercaptoheptanoylthreonine phosphate) are central to the final step of methanogenesis.

When growing on $CO_2 + H_2$, the methanogens are chemolithoautotrophs. However, the carbon fixation process is not the Calvin cycle of conventional autotrophs but instead the acetyl-CoA pathway (a pathway of autotrophic CO_2 fixation widespread in obligate anaerobes including methanogens, homoacetogens, and sulfate-reducing bacteria). Quantitatively, this reaction is the most important process involving H_2 in non-sulfate environments, and it accounts for some 40% of the methane produced in sediments (Wolfe 1972).

Another group of CO_2-reducing bacteria is the homoacetogens, which produce acetate rather than CH_4 from CO_2 and H_2 through the acetyl-CoA pathway. The overall reaction is

$$4H_2 + H^+ + 2HCO_3^- \rightarrow CH_3COO^- + 4H_2O$$

In addition to H_2, a variety of C_1 compounds, sugars, amino acids, alcohols, and certain nitrogen bases can be electron donors, depending on the organisms involved. Many homoacetogens are also able to reduce NO_3^- and $S_2O_3^{2-}$, while CO_2 reduction is the major reaction of ecological importance (Madigan et al. 2003). Homoacetogens are not a defined taxonomic group because they include such widely differing organisms as the gram-positive spore-forming bacterium *Clostridium aceticum* and the gram-negative non-spore-forming *Acetobacterium woodii*.

Methanogenesis and acetogenesis through CO_2 reduction play a crucial role in carbon cycling and are found in diverse environments, including soil, sediments, and the gastrointestinal tracts of herbivores. Acetogens compete with methanogens. The ability of methanogens to grow in diverse habitats and their minimal nutrient requirements make them able to survive in extreme conditions. Methanogens generally have a lower half-saturation constant for hydrogen, meaning they can effectively utilize H_2 even at very low concentrations. This gives them a competitive edge over homoacetogens, which may have a higher maximum reaction rate but require higher hydrogen availability to maintain efficient metabolism. In environments where H_2 is limited, methanogens can outcompete homoacetogens simply because they are better adapted to scavenging and using small amounts of hydrogen. This is a key factor in anaerobic processes, where methanogenesis often dominates over acetogenesis, leading to methane production rather than acetate accumulation (Karekar et al. 2022).

1.3.4.3 Oxidation of Reduced Sulfur Compounds

We have discussed sulfur oxidation by bacteriochlorophyll-containing green and purple sulfur bacteria through anoxygenic photosynthesis in Section 1.3.2.3 of this chapter. Here we talk about the colorless sulfur-oxidizing bacteria that are a diverse group of microorganisms capable of growing chemolithoautotrophically on reduced sulfur compounds, such as H_2S, S^0, and $S_2O_3^{2-}$ (thiosulfate). Some sulfur bacteria, such as *Beggiatoa*, *Thiothrix*, *Thiovulum*, *Thiomicrospira*, and *Thioploca*, grow at neutral pH under microaerophilic conditions with O_2 concentrations around 5% atmospheric levels (PAL). Such environments exist at sulfur springs, within microbial mats, low-oxygenated marine waters, and freshwaters polluted by sewage. Similar to Fe(II) oxidation (Section 1.3.4.4), the rates of biological sulfide oxidation are considerably enhanced in comparison to chemical oxidation rates when oxygen concentrations are low. In other words, the bacteria act as catalysts, increasing the reaction kinetics. Others, e.g. *Acidithiobacillus*, *Sulfolobus*, and *Acidianus*, grow in acid environments, characteristic of metal and coal mine drainage.

Hydrogen sulfide is the most common electron donor, SO_4^{2-} and H^+ are the final oxidation products, and oxygen is typically the oxidant. When O_2 is absent, some species can utilize NO_3^- or Fe(III) as their TEAs. Many of the sulfide oxidizers can autotrophically use the Calvin cycle to fix CO_2. The overall oxidation of H_2S to SO_4^{2-} is an $8e^-$ step, necessitating a series of $2e^-$ transfer reactions that successively lead to the intracellular formation of several different sulfur species (Figure 1.10). Less energy is released when any intermediate sulfur oxidation stage is used. There are two ways in which $S_2O_3^{2-}$ is oxidized to SO_4^{2-}. The most common pathway employs the enzyme sulfite oxidase. A few sulfur chemolithotrophs oxidize via a reversal of the activity of adenosine phosphosulfate reductase (Madigan et al. 2003).

The general reaction equation is:

$$6CO_2 + 3H_2S + 6H_2O \rightarrow C_6H_{12}O_6 + 3H_2SO_4$$

This reaction equation, like all shown in this chapter, is a summation and consolidation of a series of reactions. Free O_2 is a necessary ingredient in the initial reactions leading to sulfur oxidation. Therefore, this process does best in oxic environments. Sulfur oxidation produces protons and causes acidification of the environment.

Sulfur oxidizers can tolerate high heat too. One species of *Thermotrix* lives in sulfur hot springs at temperatures up to 185 °F (85 °C). Sulfur-oxidizing bacteria form the ecological base of deep-sea hydrothermal vent communities. The seafloor hot springs spew out sulfur and hydrogen sulfide.

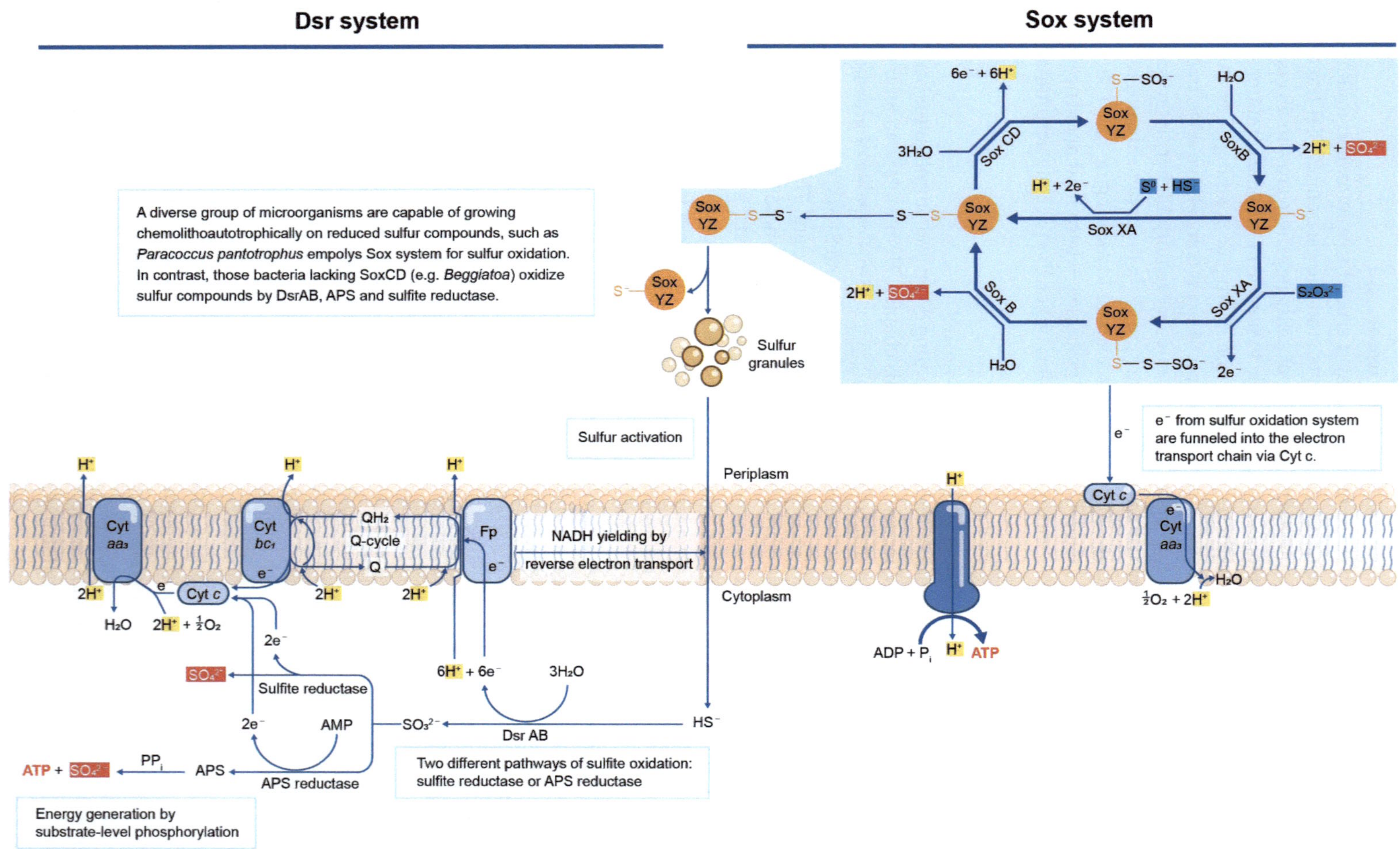

Figure 1.10 Biochemistry of reduced sulfur compound oxidation by sulfur chemolithotrophs. Sox and Dsr are sulfur oxidation and dissimilative sulfite reduction systems, respectively; SoxXA, SoxYZ, SoxB, and SoxCD are four key proteins in the periplasm of the Sox system. The oxidation begins when SoxXA forms a heterodisulfide bond between the sulfur compound to be oxidized (HS^-, S^0, or $S_2O_3^{2-}$) and the carrier protein SoxYZ. The sulfur compound remains to be carried throughout the process and is ultimately released as sulfate through the activity of SoxB. SoxCD (sulfur dehydrogenase) is the enzyme that catalyzes the removal of 6 electrons from the sulfur compound bound to the carrier. Alternatively, bacteria such as *Beggiatoa* that form sulfur granules but lack SoxCD oxidize sulfur compounds using DsrAB (dissimilatory sulfite reductase AB) and APS (adenosine-5′-phosphosulfate) reductase.

Since they occur on ocean bottoms over 6000 feet deep, there is no sunlight whatsoever. Photosynthesis is not even remotely possible. Instead, the sulfur-oxidizing bacteria perform the job normally done by photosynthesis. The sulfur-oxidizing bacteria use the chemical energy in sulfur minerals to make ATP. The ATP can then be used either to support other biosynthetic activities or to make NADPH. Then the ATP and NADPH together feed the carbon fixation cycle. Therefore, sulfur oxidizers may serve as primary producers in chemosynthesis-based hydrothermal vent ecosystems.

1.3.4.4 Iron Oxidation

The aerobic oxidation of Fe^{2+} to Fe^{3+} is able to yield a small amount of energy, and for this reason iron oxidizers must oxidize large amounts of iron in order to grow (Madigan et al. 2003). In modern surficial environments, this reaction is generally limited by the availability of dissolved Fe because at neutral pH and under fully aerated conditions, Fe^{2+} is rapidly oxidized to Fe^{3+} abiologically, which then hydrolyzes to form insoluble ferric hydroxide and precipitate in water. However, Fe^{2+} remains stable under oxic and acidic conditions. Therefore, to overcome the stability limitations, iron oxidizers either grow under acidic conditions (as an acidophile) or under low oxygen concentrations at neutral pH (as a microaerophile). In both cases, the chemical reaction is sufficiently slow so that microbes can harness Fe^{2+} oxidation for growth. In the geological past, ferruginous water bodies may have been prevalent in the bulk ocean and there was sufficient soluble Fe^{2+} for iron oxidizers to grow (see Section 8.4.4).

There are different types of microbial iron oxidation, including aerobic oxidation under acidic conditions, neutrophilic oxidation in microaerobic environments, and nitrate-dependent oxidation under anaerobic conditions (Oshiki et al. 2013). Acidophilic iron-oxidizing bacteria that grow autotrophically on reduced iron, using O_2 as their TEA. The best-known iron-oxidizing acidophiles are *Acidithiobacillus ferrooxidans* and *Leptospirillum ferrooxidans*. They grow well at mine waste disposal sites where reduced sources of iron are continuously regenerated during sulfide mineral oxidation (Liu and Zhang 2015). Another iron-oxidizing microorganism is the archaeon *Sulfolobus acidocaldarius* that lives in hot, acidic springs at temperatures up to the boiling point of water. The oxidation of Fe(II) to Fe(III) is a proton-consuming reaction: $2Fe^{2+} + 0.5O_2 + 2H^+ \rightarrow 2Fe^{3+} + H_2O$; the reaction: $0.5O_2 + 2H^+ \rightarrow H_2O$ occurs on the inner face of the cytoplasmic membrane, whereas the reaction $Fe^{2+} \rightarrow 2Fe^{3+}$ takes place near the outer face of the membrane (Figure 1.11). The overall reaction is

$$26Fe^{2+} + 0.5O_2 + 6CO_2 + 26H^+ \rightarrow C_6H_{12}O_6 + 26Fe^{3+} + 7H_2O$$

There are also iron-oxidizing bacteria that live near neutral pH. As mentioned above, ferrous iron is unstable at neutral conditions in the presence of O_2. While in a neutral pH environment, Fe^{2+} is present at the interface between anoxic and oxic conditions where Fe^{2+} diffuses upward from the anoxic zone. *Gallionella ferruginea* and *Sphaerotilus natans* are known to live at such interfaces. Anammox bacteria *Candidatus* Kuenina stuttgartiensis, *Ca.* Brocadia sinica, and *Ca.* Scalindua sp. are known to oxidize Fe^{2+} using nitrate (Oshiki et al. 2013).

In addition, ferrous iron can be oxidized under anoxic conditions by anoxygenic phototrophs, in which Fe^{2+} is used as an electron donor for photoautotrophic growth through CO_2 reduction or for energy metabolism. Anoxically, Fe^{2+} is stable at neutral pH where these bacteria flourish. Some species of purple bacteria and green sulfur bacteria (e.g. *Chlorobium*) are known to oxidize Fe^{2+} through anoxygenic photosynthesis (Madigan et al. 2003). In the experiment, Fe^{2+} oxidizing anoxygenic phototrophs (purple bacteria) are supplied with $FeCO_3$ and the following reaction takes place:

$$4FeCO_3 + 10H_2O \rightarrow 4Fe(OH)_3 + CH_2O + 3HCO_3^- + 3H^+$$

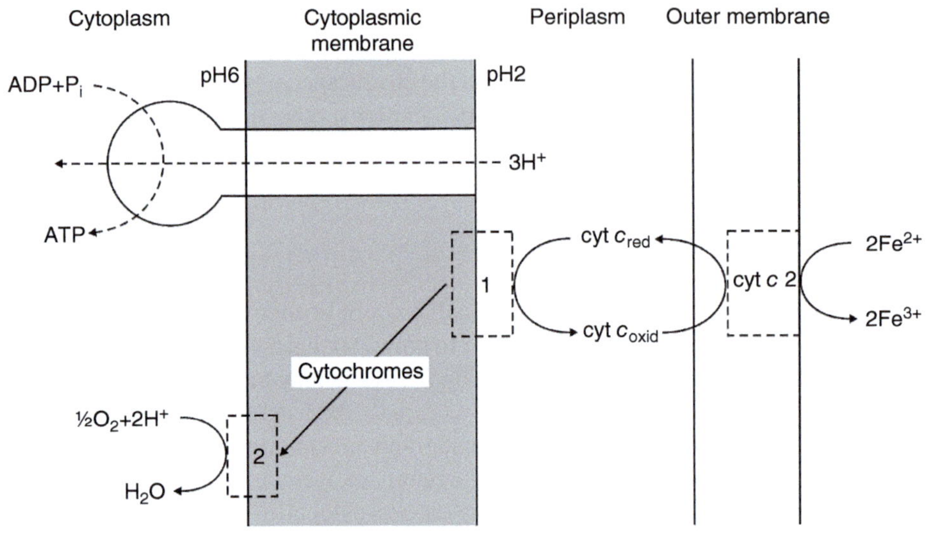

Figure 1.11 ATP synthesis coupled to oxidation of Fe(II) by *Acidithiobacillus ferrooxidans*. The reaction $0.5O_2 + 2H^+ \rightarrow H_2O$ occurs on the inner face of the cytoplasmic membrane, whereas the reaction $Fe^{2+} \rightarrow 2Fe^{3+}$ takes place near the outer face of the membrane. C-type cytochrome transfers electrons from a copper-containing blue protein, rusticyanin (1), to the terminal cytochrome oxidase (2) located at the inner face of the cytoplasmic membrane. *Source:* Adapted from Gottschalk (1986).

The anoxygenic purple bacteria are also able to use FeS as electron donor in which both Fe^{2+} and S^{2-} are oxidized (Ehrenreich and Widdel 1994).

The discovery of Fe(II)-oxidizing phototrophs has important implications in understanding the evolution of photosynthesis and explaining the large deposits of ferric iron found in ancient rocks (Ehrenreich and Widdel 1994; Madigan et al. 2003). Such ferric iron deposits were previously thought to have been formed from the oxidation of ferrous iron by O_2 produced by oxygenic phototrophs. However, these deposits are predominantly in Archean and Early Proterozoic in age (see Section 6.2.4). Therefore, it was proposed that anoxygenic phototrophic iron-oxidizing bacteria contributed to the deposition of BIFs in the Precambrian (Kappler et al. 2005).

1.3.4.5 Manganese Oxidation

Microbial manganese oxidation is a crucial process in biogeochemical cycles. At the Earth's surface, the activation energy of Mn(II) to Mn(III) or Mn(IV) is much larger than that of Fe(II) to Fe(III). While under oxic conditions, the non-biological oxidation reaction proceeds rather slowly when the pH is below 8. A diverse group of bacteria, fungi, and algae can catalyze the oxidation of manganese from its soluble Mn(II) form to insoluble Mn(III) and Mn(IV) oxides through enzyme or non-enzyme–mediated processes (Zhou and Fu 2020). The overall reaction is

$$13Mn^{2+} + 6CO_2 + 0.5O_2 + 19H_2O \rightarrow C_6H_{12}O_6 + 13MnO_2 + 26H^+$$

The end products of biological Mn-oxidation, manganese oxides (e.g. manganite and hausmannite) attract much attention not only because they serve as the sink and source of bioavailable Mn but also because of their tight coupling with the geochemical cycles of other elements (e.g. Cr, S, Fe, Co, C, N, and Se). It is argued that this process is responsible for the majority of Mn(II) oxidation that occurs in natural waters and sediments (Konhauser 2007).

Even though the oxidation of Mn^{2+} to Mn^{4+} is also energetically favorable for growth, and many microorganisms catalyze Mn^{2+} oxidation, as yet no organism has been conclusively shown to derive energy from the oxidation of reduced manganese (Madigan et al. 2018). *Leptothrix* and *Sphaerotilus* are common in freshwater environments and catalyze the oxidation of both iron and manganese but do not appear to conserve energy from these reactions. Instead, they conserve energy from the oxidation of organic matter. In these habitats, their sheaths may become coated with ferric hydroxide [$Fe(OH)_3$] or manganese oxides from the oxidation of these metals.

Experiments demonstrated that the inorganic manganese oxidation was a two-step process for MnO_2 formation. Initially, Mn^{2+} is oxidized to either a metastable Mn(III) oxyhydroxide (manganite) or a mixed valence mineral hausmannite, depending on the initial Mn^{2+} concentration, pH, ionic strength, temperature, and redox conditions of the medium. Manganite subsequently transforms into MnO_2, while hausmannite transforms very slowly to Mn^{2+} and MnO_2 by a reaction called disproportionation (the splitting of a chemical compound into two new compounds, one more oxidized and one more reduced than the original compound). Those steps become rate limiting in the abiological formation of MnO_2. In contrast, microbial catalysis accelerates manganese oxidation by 4–5 orders of magnitude, significantly outpacing abiotic processes. Those organisms may instead utilize a completely different pathway of Mn(II) oxidation. It is possible that the bacteria first enzymatically oxidize Mn^{2+} to Mn^{3+}, and that this intermediate form remains bound to a manganese-oxidizing enzyme. The same enzyme then catalyzes the oxidation of bound Mn^{3+} to MnO_2, rather than a disproportionation to Mn^{2+} and MnO_2. Therefore, a single bacterium can directly oxidize Mn(II) to Mn(IV).

1.3.4.6 Ammonia and Nitrite Oxidation: Nitrification

Ammonia (NH_3) and nitrite (NO_2^-) can be used as electron donors (energy source) by nitrifying microbes to fix carbon. Aerobic ammonium oxidation is a key process in the nitrogen cycle. Ammonia-oxidizing microbes convert ammonium (NH_4^+) into nitrite (NO_2^-), which is then oxidized by the nitrite-oxidizing bacteria to nitrate (NO_3^-). The complete oxidation of ammonia to nitrate, termed nitrification, is an eight-electron transfer process consisting of two different sets of reactions (ammonia oxidation and nitrite oxidation), which are carried out by two separate groups of microbes acting in sequence (Figure 1.12). Nitrifying microbes, including bacteria and archaea, are widely distributed in terrestrial and marine environments. Most nitrifying microbes are only able to catalyze one set of these reactions. For example, *Nitrosomonas* from bacteria and *Nitrosopumilus* from archaea are only able to oxidize NH_3 to nitrite and thus are called ammonia oxidizers. The process is continued by other bacteria, called nitrite oxidizers, such as *Nitrobacter,* that oxidize NO_2^- to NO_3^-. Only certain bacteria in the genus *Nitrospira* can catalyze both sets of reactions, oxidizing NH_3 all the way to NO_3^- or directly into nitrogen gas (N_2) under aerobic or anaerobic conditions. *Nitrospira* species possess genomes encoding enzyme systems for both ammonia and nitrite oxidation, enabling a single organism to catalyze these two processes—a capability termed comammox (complete ammonia oxidation) (Madigan et al. 2018). These bacteria, nicknamed comammox bacteria, derive their name from their unique dual metabolic activity, which streamlines the conversion of ammonia to nitrate within a single species.

Ammonia oxidizers such as *Nitrosomonas* use the enzyme *ammonia monooxygenase* to produce hydroxylamine (NH_2OH) and H_2O. The enzyme, *hydroxylamine oxidoreductase*, then oxidizes NH_2OH to NO_2^-, removing four electrons in the process. In the reaction carried out by ammonia monooxygenase, two electrons and protons are needed to reduce one molecule of O_2 to H_2O. These electrons originate from the oxidation of hydroxylamine and are supplied to ammonia monooxygenase from hydroxylamine oxidoreductase via cytochrome c and ubiquinone (Figure 1.12).

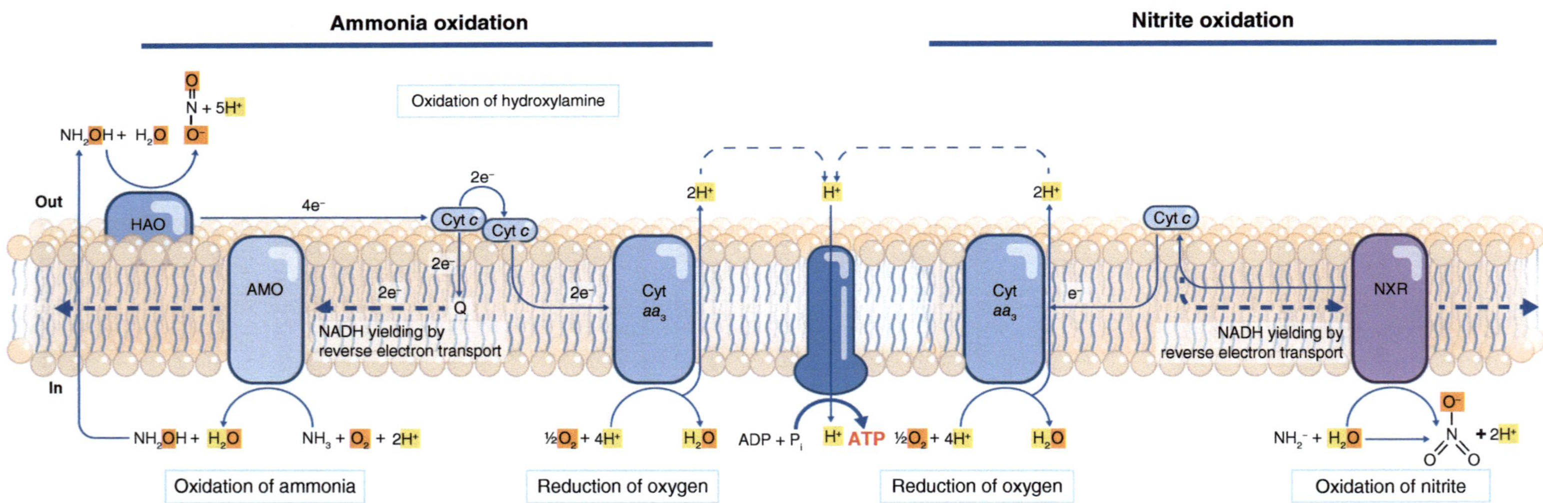

Figure 1.12 Two sets of reactions in nitrification. Left, ammonia oxidation and electron flow in ammonia-oxidizing bacteria. Right, nitrite oxidation by nitrifying bacteria. AMO, ammonia monooxygenase; HAO, hydroxylamine oxidoreductase; Q, ubiquinone; NXR, nitrite oxidoreductase. Cyt *c*, cytochrome *c*, a small, soluble protein located in the intermembrane space of mitochondria and functions as an electron carrier in the mitochondrial electron transport chain; Cyt aa_3 cytochrome aa_3, cytochrome c oxidase that accept and transfer electrons. Cyt *c* and Cyt aa_3 work together in the final steps of oxidative phosphorylation; Cyt *c* delivers electrons to Cyt aa_3, which completes the chain by reducing oxygen.

Nitrite oxidizers such as *Nitrobacter* use the enzyme *nitrite oxidoreductase* to oxidize NO_2^- to NO_3^-. The electrons travel a short transport chain because of the high potential of the NO_3^-/NO_2^- couple to the terminal oxidase. As is the case of iron oxidation (see Section 1.3.4.4), only small amounts of energy are available from nitrite oxidation (Figure 1.12). Hence, minimal amounts of cell material are obtained even though large amounts of nitrite have been oxidized.

Aerobic nitrifying microbes fix carbon via the Calvin cycle. Nitrification is important ecologically since it replenishes supplies of nitrate in the soil which then can be used by denitrifying bacteria or by plants in their biosynthesis. Nitrate is a more usable fixed nitrogen source for algae and cyanobacteria. Globally, nitrification is another process that requires molecular oxygen. As a result, nitrification assists in the ultimate return of nitrogen gas to the atmosphere by supplying nitrate to the denitrifying microbes and reducing the quantity of oxygen gas in the atmosphere.

1.3.4.7 Anammox

Ammonia can also be oxidized under anoxic conditions. This process is called anammox (for anaerobic ammonia oxidation) and is catalyzed by an unusual group of obligately anaerobic bacteria, using NO_2^- as the electron acceptor to yield N_2 by the reaction

$$NH_4^+ + NO_2^- \rightarrow N_2 + 2H_2O$$

Five different anammox genera have been described. *Brocadia anammoxidans* from the Planctomycetes phylum of bacteria is a well-known anammox species. Its cytoplasm has various membrane-enclosed compartments, named anammoxosome organelle (Claret Fernández et al. 2020), where the anammox reaction and energy conversion take place (Figure 1.13). Other anammox bacteria include *Kuenenia*, *Anammoxoglobus*, *Jettenia*, and *Scalindua*, all of which are related to *Brocadia* and also contain anammoxosomes (Madigan et al. 2018).

Anammox bacteria are also autotrophs, but they fix CO_2 by the reductive acetyl-CoA pathway, an autotrophic pathway widespread among some obligately anaerobic autotrophic bacteria and archaea. In the anammox reaction, NO_2^- is first reduced to NO by nitrite reductase, NO reacts with ammonium (NH_4^+) to yield N_2H_4 by the enzyme hydrazine synthase, and then N_2H_4 is

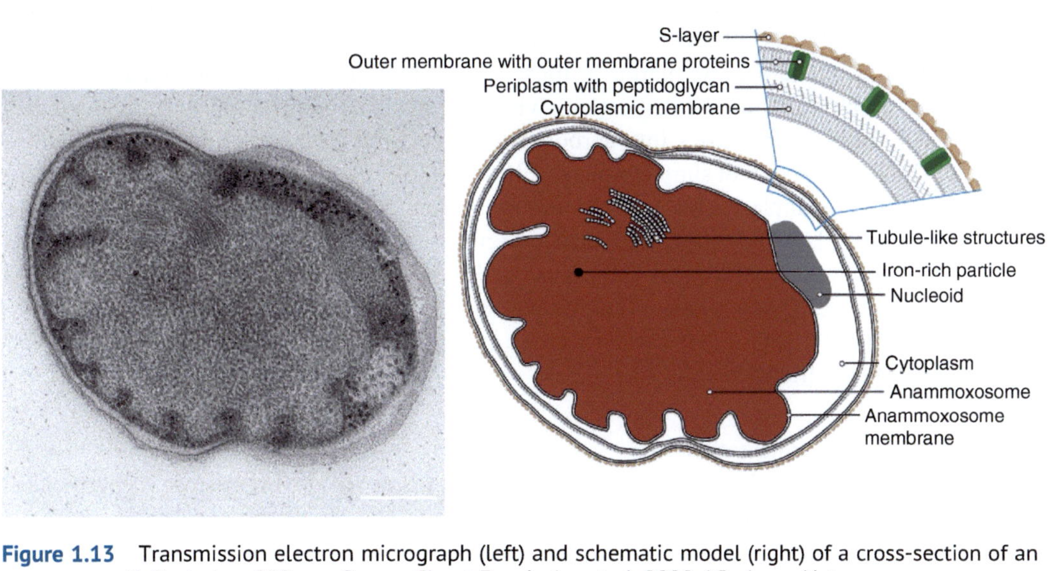

Figure 1.13 Transmission electron micrograph (left) and schematic model (right) of a cross-section of an anammox cell. Scale bar, 200 nm. *Source:* Claret Fernández et al. 2020 / Springer Nature.

oxidized to N_2 plus electrons by the enzyme hydrazine dehydrogenase. The electrons are funneled into the electron transport chain where they are used to reduce nitrite and nitric oxide earlier in the pathway.

The chemolithoautotrophic anammox bacteria are found naturally in anoxic or oxygen-limited terrestrial, marine, and freshwater environments. The anammox has been shown to play a crucial role in the biological nitrogen cycle where anammox bacteria are estimated to produce approximately half of N_2 present in the atmosphere. Besides their presence in nature, all genera of anammox bacteria have been detected in conventional wastewater treatment plants, where they remove ammonium from contaminated waters (Claret Fernández et al. 2020).

1.3.5 Catabolism

Catabolism is a combination of cellular processes that harvest the energy released during the breakdown of preexisting organic compounds such as glucose and this energy is used to synthesize ATP, the energy currency of all cells. In most heterotrophic microorganisms, the oxidation of carbohydrates provides the bulk of the cell's energy. The primary catabolism proceeds through a series of three coupled pathways: glycolysis, tricarboxylic acid cycle (TCA, or citric acid cycle), and respiration chain (electron transport and oxidative phosphorylation). To degrade sugars, cells must first secrete specific extracellular hydrolytic enzymes that break down the glycosidic bonds in polysaccharides into mono- and disaccharides that can then be transported through the plasma membrane into the cytoplasm. For optimal efficiency, organisms benefit from close proximity to the substrate, minimizing the diffusion distance for hydrolysis products to reach the cell surface. Their sugar breakdown follows two pathways: (1) partial degradation begins with glycolysis and transitions directly into fermentation; (2) complete catabolism proceeds through glycolysis, an intermediate reaction forming acetyl CoA, the citric acid cycle, and ultimately, the electron transport chain.

The pathways for the breakdown of organic compounds and energy conservation in ATP can be divided into two major groups: (1) fermentation, in which the degradation occurs in the absence of any added TEAs; and (2) respiration, subdivided into aerobic and anaerobic respirations, in which O_2 or some other oxidants serve as TEAs, respectively. Electron acceptors used in anaerobic respiration include NO_3^-, Mn^{4+}, Fe^{3+}, SO_4^{2-}, etc., each of which is discussed separately in this section. The oxidation pathways of organic matter in most anaerobic processes differ significantly from those in aerobic respiration and denitrification. While aerobes and denitrifiers can fully degrade a wide range of organic compounds into carbon dioxide on their own, anaerobes are generally restricted to a limited selection of simple organic substrates. Consequently, organic matter derived from the hydrolysis of complex polymers must first be broken down by fermentative microorganisms, which transform large polymers into smaller degradation products—such as hydrogen (H_2), pyruvate, lactate, acetate, and formate. These intermediates are then further metabolized by acetogenic bacteria and anaerobic respirers that function syntrophically to achieve the complete oxidation of organic matter into either carbon dioxide (CO_2) or methane (CH_4) within suboxic to anoxic sedimentary environments.

1.3.5.1 Glycolysis and Fermentation

Glycolysis is the primary pathway used by nearly all organisms to convert glucose to pyruvate. A six-carbon glucose molecule is partially broken down into two molecules of pyruvate, with a net gain of two ATP through substrate-level phosphorylation, and two NADH. The decomposition includes a number of complex steps, each catalyzed by its own unique enzyme. Glycolysis does not require oxygen and thus can occur under both oxic and anoxic conditions. The overall reaction is:

$$C_6H_{12}O_6 + 2NAD^+ + 2ADP + 2P_i \rightarrow 2C_3H_3O_3^- + 2NADH + 4H^+ + 2ATP$$

Beyond glycolysis, numerous bacteria have the capability to metabolize five-carbon sugars through the pentose phosphate pathway. This pathway plays a crucial role in generating key intermediate molecules essential for the synthesis of nucleic acids and certain amino acids, while also serving as a primary source of NADPH from $NADP^+$. Unlike glycolysis, the pentose phosphate pathway yields only a single ATP molecule per glucose molecule. Bacteria such as *Escherichia coli*, *Bacillus subtilis*, and *Enterococcus faecalis* utilize this metabolic route. Once pyruvate is formed, it can proceed either through fermentation—essentially the partial oxidation of glucose and other carbohydrates—or respiration.

Fermentation differs from respiration in that it lacks an electron transport chain and operates without externally supplied electron acceptors. In aerobic bacteria, protozoans, algae, fungi (including yeasts), plants, and animals, fermentation serves as a backup energy extraction process. However, for many anaerobic bacteria, it is the primary method of deriving energy from sugars, excluding methanogenic bacteria, sulfate-reducing bacteria, and acetogenic bacteria.

Fermentation occurs when there is a buildup of waste hydrogen at the end of respiration. Normally, oxygen or nitrate clears out these hydrogens, allowing respiration to continue smoothly. However, in environments like deep mud, stagnant water, or certain soils, where oxygen and nitrate are absent, the system gets clogged, causing respiration to halt. To keep ATP production going, cells reroute glycolysis products (pyruvate) through fermentation, which continues the breakdown of carbon chains. Fermentation generates about 20 times less ATP per sugar molecule than respiration, meaning it is far less efficient for sustaining high-energy activities.

Fermentation generates a diverse array of waste products, including lactic acid, ethanol, and acetic acid. When muscles are fatigued, fermentation produces lactic acid, which can contribute to soreness. However, in bacteria, lactic acid plays a valuable role in creating yogurt and cheese. Yeast and certain bacteria produce ethanol and carbon dioxide through fermentation—giving champagne its signature bubbles and alcoholic punch. Bakers also take advantage of fermentation, as the carbon dioxide released by yeast causes dough to rise, transforming it from a dense lump into a fluffy loaf.

Fermentation waste products sometimes serve as carbon chains that contribute to the biosynthesis of larger molecules. More often, these byproducts are released into the environment, where they can lead to contamination—such as yeast dying from ethanol buildup in a sealed bottle of grape juice or wine. Alternatively, other microorganisms may take advantage of these waste products as an energy source, turning what could be harmful into something beneficial.

1.3.5.2 Aerobic Respiration

Prokaryotic and eukaryotic aerobic respirers are ubiquitous in environments where dissolved or free oxygen is available. Aerobic respiration is able to completely oxidize a wide variety of natural and synthetic organic compounds (electron donors) to CO_2 and H_2O, with much greater energy released than during fermentation. The overall reaction is:

$$C_6H_{12}O_6 + 6O_2 \rightarrow 6CO_2 + 6H_2O$$

This capability arises because aerobes possess a suite of hydrolytic enzymes that help break down large macromolecules like proteins, carbohydrates, and fats into smaller, more easily absorbed components. For instance, proteases break down proteins into amino acids, lipases are responsible for breaking down fats into fatty acids and glycerol, amylases catalyze the breakdown of starch and complex carbohydrates into simpler sugars, and cellulases break down cellulose into glucose for energy production. The smaller molecules are further processed by the Link Reaction, Krebs Cycle, and Electron Transport Chain, which are more efficient in energy release.

During aerobic respiration, the total possible yield of ATP is 40: 4 from glycolysis, 2 from the TCA cycle, and 34 from electron transport. However, since 2 ATPs were expended in early glycolysis, this leaves a maximum of 38 ATPs per glucose molecule, though the actual amount can vary based on the organism and specific cellular conditions. Eukaryotic cells often lose a bit of ATP when transporting NADH generated in glycolysis into the mitochondria, reducing the net yield slightly. Meanwhile, aerobic bacteria, which lack mitochondria, can maximize their ATP production since they don't need to expend energy shuttling molecules across membranes. The electron transport chain plays a huge role in ATP production, using oxidative phosphorylation to generate most of the ATP.

A potential side reaction of aerobic respiration is the incomplete reduction of oxygen to superoxide anion (O_2^-) and hydrogen peroxide (H_2O_2), both generated as byproducts of the biochemical reduction of O_2. On the one hand, these toxic oxygen products can be very damaging to cells; aerobes have various neutralizing enzymes (e.g. superoxide dismutase and catalase). One exception is the genus *Streptococcus* (a genus of spherical gram-positive bacteria), which can grow well in oxygen, yet lacks both cytochromes (catalyzing the electron acceptance) and catalase. However, certain species of *Streptococcus* use a special peroxidase that helps them survive oxidative stress. The lack of cytochromes, catalase, and peroxidase in anaerobes as a rule limits their ability to process O_2 and contributes to its toxic effects on them. On the other hand, these highly reactive oxygen-containing radicals (H_2O_2 and O_2^-) can be used to break down refractory compounds, such as lignin.

1.3.5.3 Nitrate Reduction and Denitrification

Anaerobic respiration occurs in environments devoid of oxygen, where alternative electron acceptors replace oxygen in energy production. One of the most common substitutes is nitrate, which plays a key role in the nitrogen cycle. Nitrate reduction can be categorized into two distinct pathways: assimilative and dissimilative, each serving unique biological functions (Figure 1.14).

Both pathways are essential for maintaining nitrogen balance in natural ecosystems, influencing factors like soil fertility, wastewater treatment, and atmospheric nitrogen levels. The assimilative

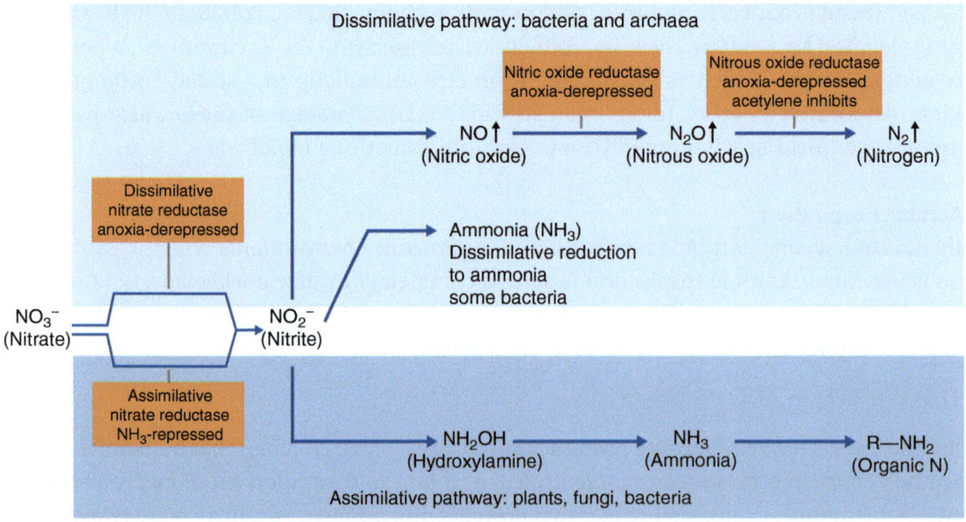

Figure 1.14 Comparison of assimilative and dissimilative processes for nitrate reduction.

pathway is prevalent in plants, fungi, and bacteria that require nitrogen for amino acid and nucleotide synthesis. In this process, nitrate is converted into ammonia (NH_3), which is then incorporated into cellular biomass. Here, nitrate is utilized as a nutrient for growth rather than as an energy source. Conversely, the dissimilative pathway involves the stepwise reduction of nitrate into gaseous nitrogen compounds—N_2O, NO, and ultimately N_2. Since these intermediate and final products are gaseous, they escape from the environment. This entire process is known as denitrification, a crucial process in nitrogen cycling. The overall reaction is:

$$2.5C_6H_{12}O_6 + 12NO_3^- \rightarrow 15CO_2 + 12OH^- + 9H_2O + 6N_2$$

Denitrification is a metabolic process carried out by anaerobic bacteria and archaea, where nitrate serves as an alternative electron acceptor in the absence of oxygen. Remarkably, the eukaryotic species *Globobulimina pseudospinescens*, a shelled amoeba found in anoxic marine sediments, has been identified as a denitrifier (Madigan et al. 2018). The first step of denitrification is catalyzed by nitrate reductase, a membrane-integrated enzyme containing molybdenum. Its synthesis is inhibited by oxygen, as are all subsequent enzymes involved in the pathway, which are coordinately regulated to function only under anoxic conditions. However, certain denitrifying bacteria, such as *Pseudomonas*, exhibit metabolic flexibility, switching between aerobic respiration when oxygen is available and anaerobic respiration using nitrate once oxygen is depleted. The biochemistry of denitrification has been studied in detail in *E. coli*, in which NO_3^- is reduced only to NO_2^- by the enzyme reductase, and *Paracoccus denitrificans* and *Pseudomonas stutzeri*, in which nitrogen oxides are formed from NO_2^- by enzymes nitrite reductase, nitric oxide reductase, and nitrous oxide reductase.

Denitrification plays a crucial role in ecological and environmental systems. Besides producing nitrogen gas (N_2), it also generates nitrous oxide (N_2O) and nitric oxide (NO), which pose environmental concerns. N_2O is a potent greenhouse gas that contributes to climate change and can be transformed into NO through sunlight exposure. NO, in turn, interacts with ozone (O_3) in the upper atmosphere, leading to the formation of nitrite (NO_2^-), which falls to Earth as nitrous acid (HNO_2) in acid rain. Despite these negative effects, denitrification serves as a beneficial function in wastewater treatment by eliminating fixed nitrogen—a key factor in algal blooms that could otherwise occur if nitrate-rich effluent is discharged into environments like rivers or lakes (Madigan et al. 2018).

1.3.5.4 Microbial Manganese Reduction

Microbial manganese reduction is a process in which certain bacteria and archaea facilitate the conversion of manganese oxides into soluble manganese ions. This process plays a crucial role in biogeochemical cycles, influencing metal mobility, nutrient availability, and environmental remediation efforts. It occurs in various natural settings, including aquatic sediments and soil environments, where microbes utilize manganese as an electron acceptor in anaerobic respiration. There are a number of Mn(IV) oxides in soils and sediment, including those with an average composition of MnO_2, such as todorokite, vernadite, pyrolusite, birnessite, and also some mixed valence and trivalent forms, such as hausmannite (Mn_3O_4) and manganite (MnOOH). They are not equal with regard to their propensity to be reduced, amorphous phases with high surface areas being more reactive than highly crystalline oxides with low surface area, and Mn(III) oxyhydroxides being more reactive than Mn(IV) oxides.

Similar to those in Fe(III)-reduction and sulfate reduction (discussed below), the electron transport in the Mn(IV) reduction begins with the transfer of H_2 to a hydrogenase. The H_2 can come from the external environment, whereby the bacteria behave either chemolithoautotrophically as hydrogen-oxidizers or mixotrophically, whereby they assimilate organic compounds as their carbon source. More commonly, H_2 is removed from organic electron donors such as acetate. Under these conditions, microorganisms behave heterotrophically. The sum reaction is:

$$CH_3COO^- + 4MnO_2 + 3H_2O \rightarrow 4Mn^{2+} + 2HCO_3^- + 7OH^-$$

Microbes involved in manganese reduction belong to various taxa, including *Pelobacter*, Colwelliaceae, Arcobacteraceae, Shewanellaceae, and Oceanospirillaceae. Many iron-reducing bacteria can also reduce manganese oxides. Microbial manganese reduction plays a significant role in carbon mineralization, contributing up to 45% in manganese oxide-rich sediments (Wunder et al. 2024). Among the most notable genera are *Pseudomonas*, *Bacillus*, *Geobacter*, and *Shewanella*. While some species function as facultative anaerobes, others primarily act as Fe(III) reducers, shifting to MnO_2 reduction when necessary. Given the high reduction potential of the Mn(IV)/Mn(II) couple, Mn-reducers demonstrate remarkable flexibility in electron donor utilization. For instance, *Geobacter metallireducens* can metabolize butyrate, propionate, lactate, succinate, and acetate, completely oxidizing them into CO_2. Meanwhile, *Shewanella putrefaciens* primarily utilizes lactate and pyruvate, but only oxidizes them to acetate. Notably, *S. putrefaciens* is also capable of reducing Mn(III) hydroxides to Mn^{2+} (see Konhauser 2007).

1.3.5.5 Ferric Iron Reduction

Ferric iron can be reduced to a ferrous state by a variety of microorganisms. The reduction can be a major form of anaerobic respiration because Fe(III) is abundant in many microbial habitats. The reduction potential of the Fe(III)/Fe(II) couple is very electropositive, and thus Fe(III) reduction can be coupled to the oxidation of a wide variety of both organic and inorganic electron donors. Ferric iron minerals occur in nature in a wide variety of forms, ranging from amorphous to crystalline phases. The amorphous to poorly ordered iron oxyhydroxides, such as ferric hydroxide or goethite (FeOOH), are the preferred sources of solid-phase ferric iron for Fe(III)-reducers. More crystalline Fe(III) oxides, such as hematite (Fe_2O_3) and magnetite (Fe_3O_4) are also microbially reducible. The variations in reductive rates are related to a number of factors, including the amount of surface area exposure, crystal morphology, particle aggregation, the composition of the aqueous solution in which the microorganisms grow, and the amount of Fe^{2+} sorbed to the oxide surface.

Microbial Fe(III) reduction has been observed across a diverse array of microorganisms within the domains archaea and bacteria, spanning various chemical and physical conditions. This widespread metabolic capability highlights its ubiquity in microbial life. Extremophilic Fe(III)-reducing microorganisms, including hyperthermophilic, thermophilic, psychrophilic, acidophilic, and alkaliphilic species, have been successfully isolated and cultivated in pure cultures. Among these, one hydrothermal vent archaeal species, *Geogemma barossii* has redefined the upper temperature threshold for life, surviving above 121 °C (Kashefi and Lovley 2003). Some fermentative, sulfate-reducing, and methanogenic microorganisms are also capable of Fe(III) reduction (Weber et al. 2006). Additionally, ferric iron can serve as a TEA in fungal energy metabolism (Li et al. 2024). For instance, *Aspergillus niger* has been reported to biologically leach iron from clay minerals within a pH range of 1.7–3.0, while *Penicillium* sp. can mobilize Fe(III) from jarosite minerals by reducing it to Fe(II) (Li et al. 2024).

Fe(III) oxide minerals are insoluble, preventing their diffusion into microbial cells. As a result, electron transport to the solid-phase TEA cannot take place within the periplasm, as it does with soluble electron acceptors like nitrate or fumarate. Instead, Fe(III) reduction must occur externally to the cell, necessitating close microbial contact with the mineral surface. Notable iron-reducing bacteria include *Geobacter* and *Shewanella* (Figure 1.15). *G. metallireducens* is known to produce surface appendages, such as pili, which facilitate adherence to oxide surfaces and act as nanowires for electron transfer from the cell to the surface of Fe(III) oxides (Figure 1.16). In contrast,

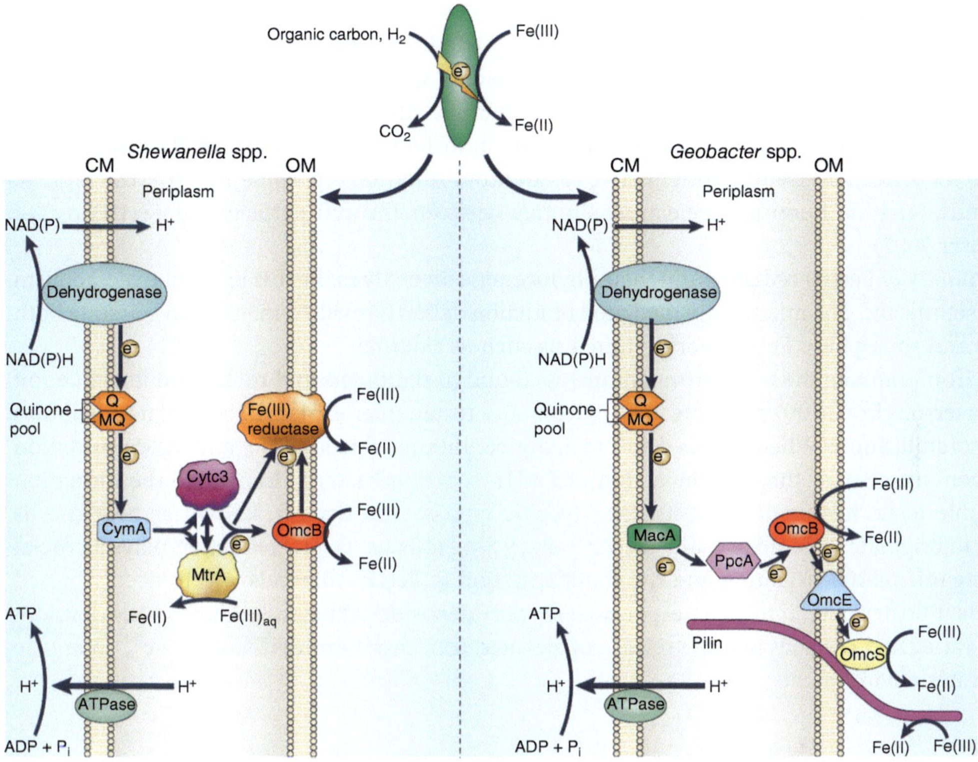

Figure 1.15 Physiological model of the biochemistry in microbial Fe(III) reduction by *Shewanella* and *Geobacter* spp. CM, cytoplasmic membrane; CymA, cytoplasmic membrane-bound tetraheme c-type cytochrome; Cytc3, periplasmic c-type cytochrome; e^-, electrons; MacA, cytoplasmic membrane-bound cytochrome; MQ, menaquinone; MtrA, periplasmic decaheme c-type cytochrome; OM, outer membrane; Omc, outer-membrane-bound cytochromes B, E, and S partially exposed on the cell surface; PpcA, a tri-heme periplasmic c-type cytochrome; Q, quinone. *Source:* Weber et al. (2006) / with permission of Springer Nature.

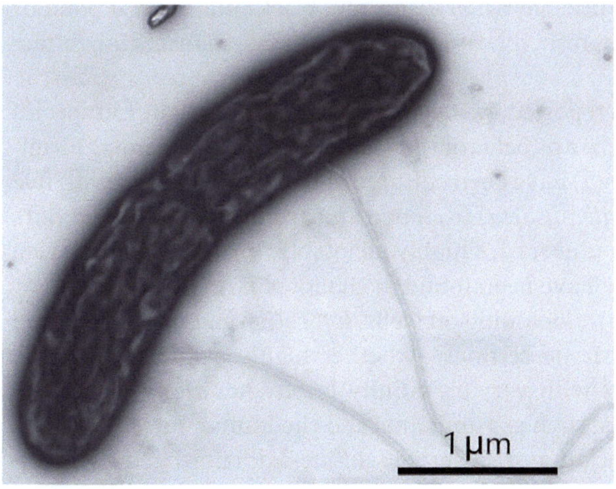

Figure 1.16 A species of *Geobacter* developing pili that helps the bacterium adhere to the oxide surfaces and serves as biological "nanowirer" to transfer electrons from the cell to the surface of Fe(III) oxides. *Source:* Photo by Lucy Yin / The Regents of the University of California / http://genome.jgi.doe.gov/geome/geome.home.html / last accessed on July 15, 2025.

Shewanella algae employs hydrophobic surface proteins to enhance cell adhesion. Fe(III)-reducing bacteria can accumulate nanometer-scale iron oxide particles on their surfaces for later reduction. Upon attachment to the mineral surface, they initiate electron transfer from a reduced source within the cytoplasm, moving across the plasma membrane and periplasm to the outer membrane. Here, iron reductase enzymes facilitate direct electron transfer to the Fe(III) mineral surface, weakening Fe—O bonds and leading to reductive dissolution. Additionally, some bacteria use organic compounds, such as humics, as electron shuttles between the cell surface and Fe(III) oxides (Konhauser 2007).

The ubiquity of Fe(III)-reducers and their phylogenetic diversity make this microbial metabolism globally significant. The microbially mediated reduction of Fe(III) oxide minerals can generate both aqueous and solid-phase Fe(II)-bearing minerals such as siderite.

Ferric iron is among the most prevalent metals found in sediments and rocks, and its reduction produces ferrous iron, a more soluble form of iron. As a result, microbial iron reduction contributes to iron solubilization. When exposed to oxidizing conditions, ferrous iron undergoes oxidation either spontaneously or through the action of Fe(II)-oxidizing bacteria, leading to the formation of insoluble ferric hydroxide deposits. One specific type of iron deposit, known as bog iron, is believed to originate from microbially driven iron cycling in bogs. Fe(III) reduction plays a crucial role in organic matter oxidation and the transformation of Fe(III) minerals within deeply buried sediments. Furthermore, it may be responsible for post-depositional processes such as the development of variegated red beds and the release of elevated concentrations of dissolved Fe^{2+} in anoxic groundwater environments.

1.3.5.6 Sulfate Reduction

In anaerobic respiration, several inorganic sulfur compounds serve as vital electron acceptors. The Earth's crust primarily contains sulfur in the form of sulfate minerals, such as gypsum, and sulfide minerals like pyrite. In marine environments, sulfur is predominantly found as dissolved sulfate. The sulfate ion (SO_4^{2-}), which represents sulfur in its most oxidized state, is a major anion in seawater. It can be transformed into hydrogen sulfide (H_2S) through the activity of sulfate-reducing microbes, which thrive exclusively in oxygen-depleted conditions. A common sign of their presence is the distinct odor of rotten eggs—often noticeable when disturbing black mud from a pond or stagnant water. This characteristic smell arises from the H_2S released by sulfate reducers inhabiting the sediment.

Since sulfate-reducing microbes were first reported by Beijerinck (1895), more than 150 species from at least 50 genera belonging to a number of bacterial phyla (e.g. Desulfobacterota, Firmicutes, Thermodesulfobacteria and Nitrospirota) have been cultured and described, along with five species of two genera (*Archaeoglobus* and *Vulcanisaeta*) from two archaeal phyla, Halobacteriota and Thermoproteota (Figure 1.17). Sulfate-reducers are highly adaptable, thriving across a wide spectrum of environmental conditions. They have been found in shallow marine and freshwater sediments, as well as deep subsurface habitats, including oil wells, hydrothermal vents, and mud volcanoes. Some are able to survive in extreme settings—whether highly acidic or alkaline, intensely hot or cold, and in environments with varying salinity levels. Additionally, sulfate-reducers are present within living organisms, such as ruminants and the human intestinal tract. In marine worms, they establish a close association with aerobic sulfide-oxidizing bacteria, forming a symbiotic relationship essential to their survival (Muyzer and Stams 2008). Besides sulfate, most sulfate-reducing bacteria can also conserve energy from the reduction of elemental sulfur to sulfide ($S^0 + 2H \rightarrow H_2S$).

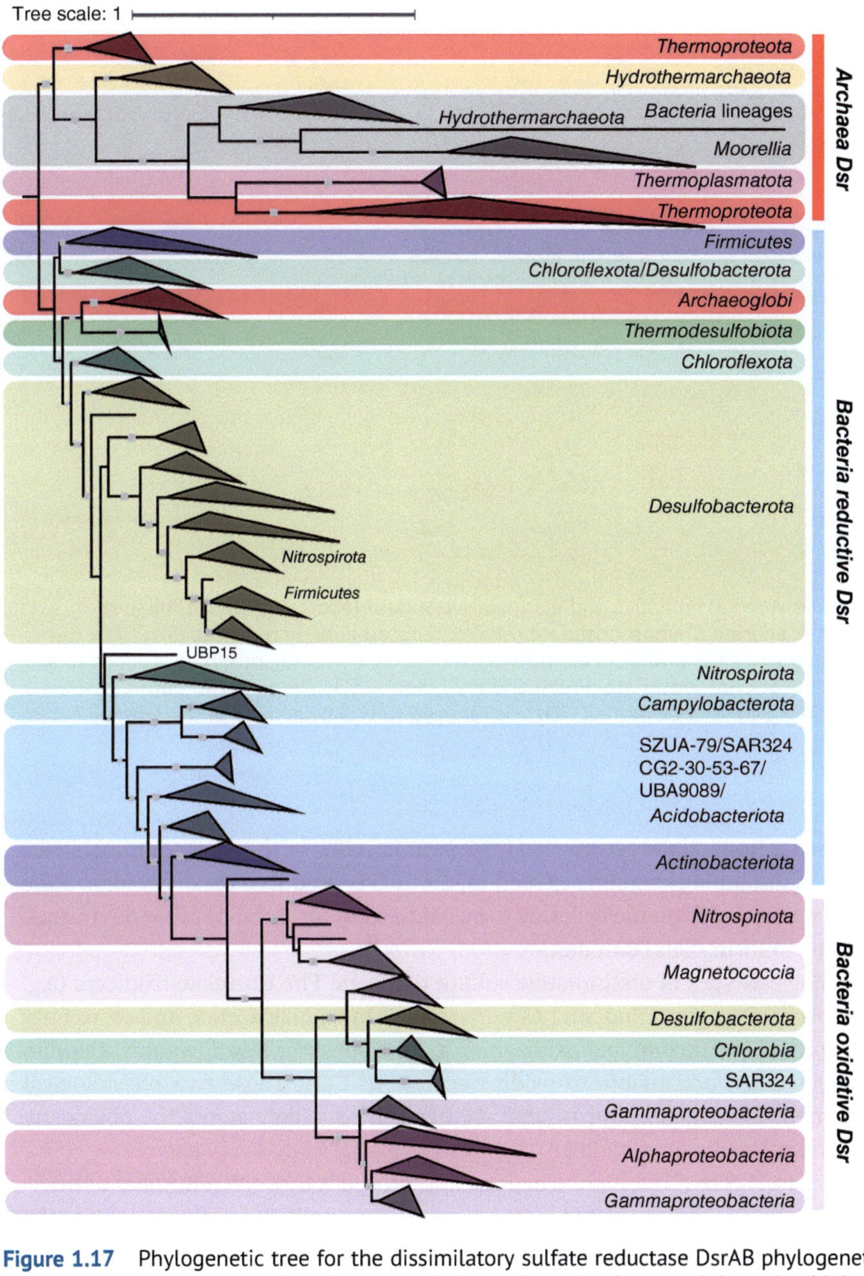

Figure 1.17 Phylogenetic tree for the dissimilatory sulfate reductase DsrAB phylogenetic tree, indicating that sulfate reducers are widely distributed in bacteria and archaea and that microbial sulfate reduction represents an ancient metabolism. *Source:* Cui et al. 2024 / with permission of Elsevier.

Similar to nitrate reduction, microbial sulfate reduction also has assimilative and dissimilative pathways (Figure 1.18). In the assimilative sulfate metabolism, many organisms, including higher plants, algae, fungi, and most prokaryotes, use sulfate as a sulfur source for biosynthetic purposes to make cysteine, methionine, and many other organosulfur compounds. In the dissimilative process, sulfate is used as an electron acceptor for energy conservation, which requires a large-scale reduction of SO_4^{2-} and is restricted to sulfate-reducing bacteria and archaea. In assimilative sulfate

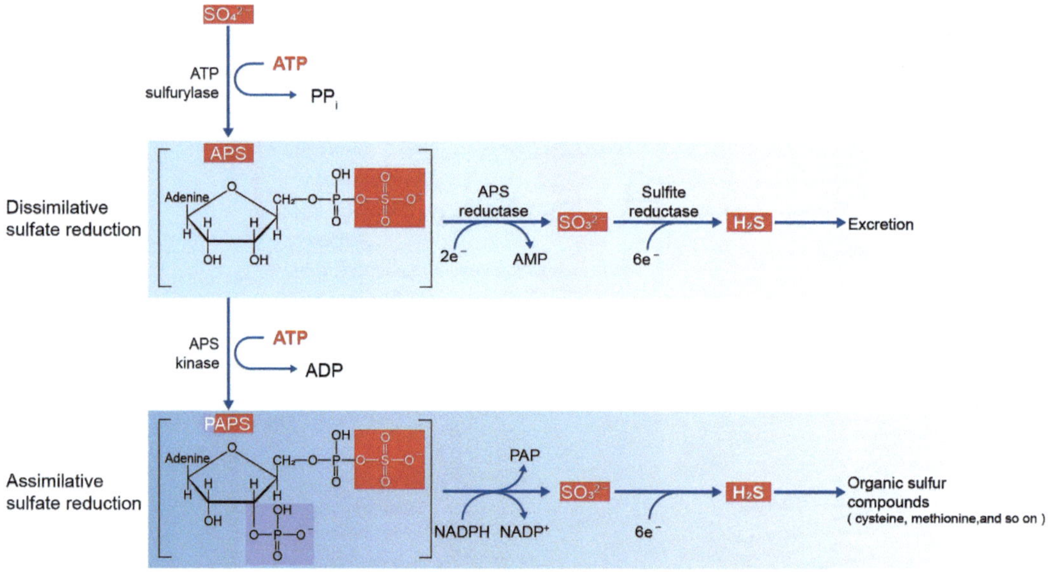

Figure 1.18 Comparison between assimilative and dissimilative sulfate reductions. In dissimilative sulfate reduction, SO_4^{2-} in adenosine 5′-phosphosulfate (APS) is reduced directly to sulfite (SO_3^{2-}) by the enzyme APS reductase with the release of AMP. In assimilative reduction, a second phosphate is added to APS to form 3′-phosphoadenosine 5′-phosphosulfate (PAPS) and only then is SO_4^{2-} reduced. In both cases, the product of sulfate reduction is SO_3^{2-}; once formed, SO_3^{2-} is reduced to H_2S by activity of the enzyme sulfite reductase.

reduction, the H_2S formed is immediately converted into organic sulfur in the form of amino acids, proteins, and so on, but in dissimilative sulfate reduction, the H_2S is produced on a very large scale by sulfate reducers and is excreted from the cell, free to be oxidized by air, used by other organisms, or combined with metals to form metal sulfides.

There are two physiological types of dissimilative sulfate reducers. The complete oxidizers (e.g. *Desulfobacter*, *Desulfococcus*, *Desulfosarcina*, and *Desulfonema*) can oxidize acetate and other fatty acids completely to CO_2, and the incomplete oxidizers (e.g. *Desulfovibrio*, *Desulfomonas*, *Desulfotomaculum*, and *Desulfobulbus*) are unable to oxidize acetate to CO_2. These two physiological groups are not phylogenetically coherent but instead are distributed widely across the phylogeny of sulfate-reducing bacteria (Madigan et al. 2018).

The reduction of SO_4^{2-} to H_2S requires eight electrons and proceeds through a number of intermediate stages (Figure 1.17). It is first activated by the enzyme ATP sulfurylase that catalyzes the attachment of SO_4^{2-} to a phosphate of ATP, forming adenosine phosphosulfate (APS). In dissimilative sulfate reduction, the SO_4^{2-} in APS is reduced directly to sulfite (SO_3^{2-}) by the enzyme APS reductase with the release of adenosine monophosphate (AMP). In assimilative reduction, by contrast, a second phosphate is added to APS to form phosphoadenosine phosphosulfate (PAPS), and only then is the SO_4^{2-} reduced. However, in both cases, the product of sulfate reduction is sulfite, and once SO_3^{2-} is formed, it is reduced to H_2S by activity of the enzyme sulfite reductase.

Microbial sulfate reduction includes specialized variants. Some sulfate-reducing bacteria can generate energy through disproportionation metabolism, a unique process in which a single compound undergoes simultaneous oxidation and reduction. As a result, one fraction becomes more oxidized, while the other is reduced, allowing these microbes to efficiently harness energy in anaerobic environments. For example, *Desulfovibrio sulfodismutans* can disproportionate sulfite, thiosulfate, and sulfur into both sulfide and sulfate by the following reactions:

Sulfite disproportionation: $4SO_3^{2-} + 2H^+ \rightarrow 3SO_4^{2-} + H_2S$

Thiosulfate disproportionation: $S_2O_3^{2-} + H_2O \rightarrow SO_4^{2-} + H_2S$

Sulfur disproportionation: $4S^0 + 4H_2O \rightarrow SO_4^{2-} + 3H_2S + 2H^+$

In addition, the autotrophic, strict anaerobic sulfate-reducing bacterium *Desulfotignum phosphitoxidans* can couple phosphite (HPO_3^-) oxidation to SO_4^{2-} reduction by the reaction:

$$4HPO_3^- + SO_4^{2-} + H^+ \rightarrow 4\,HPO_4^{2-} + HS^-$$

The natural sources of phosphite are likely to be organophosphorus compounds called phosphonates that are generated from the anoxic degradation of nucleic acids, phospholipids, and other cellular sources (Madigan et al. 2018).

Microbial sulfate reduction is a fundamental geobiological process, significantly shaping Earth's sulfur and carbon cycles. Sulfate-reducing bacteria and archaea transform sulfate into hydrogen sulfide (H_2S), which can interact with metal ions to form intracellular or extracellular minerals such as pyrite (Cui et al. 2024). Alternatively, H_2S is released into the environment, where it becomes available for anoxygenic photosynthetic and sulfur-oxidizing bacteria, sustaining a continuous sulfur cycle. Additionally, sulfate-reducing microorganisms facilitate the anoxic oxidation of methane, inducing the precipitation of seep carbonate. In oxygen-deprived settings, these microbes contribute to the decomposition of organic matter, playing a key role in fossil fuel formation and petroleum reservoir dynamics. On a global scale, sulfate reduction accounts for the oxidation of 12–29% of the total carbon flux reaching the ocean floor (Bowles et al. 2014). Furthermore, sulfate reduction competes with methanogenesis, ultimately affecting methane emissions from marine sediments and contributing to greenhouse gas regulation.

1.3.5.7 Other Electron Acceptors

Beyond the above-discussed electron acceptors for anaerobic respiration, various metals, metalloids, and both halogenated and unhalogenated organic compounds are important electron acceptors for microbial anaerobic respiration in nature (Figure 1.19). Additionally, a few strict anaerobes can even utilize protons as electron acceptors. Among these are compounds containing arsenic (As), selenium (Se), chromium (Cr), uranium (U), technetium (Tc), and even plutonium (Pu). Although typically found in low concentrations in natural environments, these substances occasionally appear as pollutants and can sustain the anoxic growth of certain microorganisms. Their microbial reduction serves as a vital mechanism for pollutant removal, contributing to environmental cleanup efforts. Examples are U(VI) reduction and As (V) reduction that are often utilized to clean up the contaminated groundwater.

$$CH_3COO^- + 4UO_2^{2+} + 4H_2O \rightarrow 4UO_2 + 2\,HCO_3^- + 9H^+$$

$$H_2 + UO_2^{2+} \rightarrow 2H^+ + UO_2$$

$$CH_3COO^- + 4HAsO_4^{2-} + 3H^+ \rightarrow 2HCO_3^- + 4H_2AsO_3^-$$

1.3.5.8 Methanogenesis

Methanogens are microorganisms that produce methane (CH_4) as an integral part of their energy metabolism. The process of methane production is called methanogenesis. Methanogens are strict

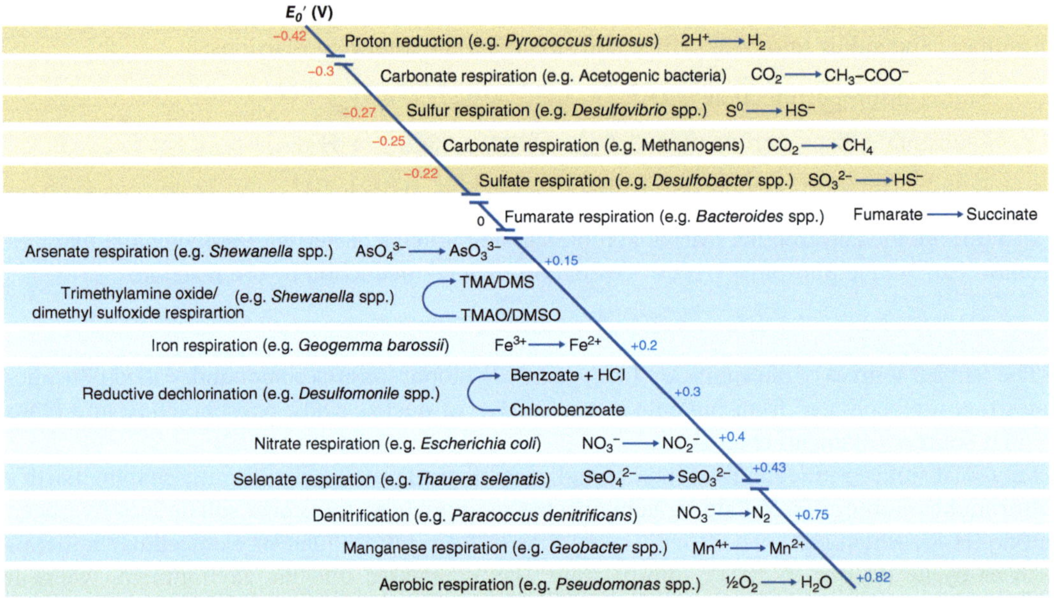

Figure 1.19 Major electron acceptors of microbial respirations. The redox couples are arranged in order from the most electronegative E_0' (top left) to most electropositive E_0' (bottom right) assuming neutral pH.

anaerobic archaea and occur in at least seven orders, including Methanobacteriales, Methanococcales, Methanopyrales, Methanomassiliicoccales, Methanomicrobiales, Methanocellales, and Methanosarcinales. They are spread widely within the Euryarchaeota and do not represent a single coherent phylogenetic group (Madigan et al. 2018). All methanogens use NH_4^+ as nitrogen source, and a few species are known to fix nitrogen. The trace metal nickel is required by all methanogens because it is a component of an important methanogenic coenzyme, *Factor*$_{430}$. Iron and cobalt are also important trace metals for methanogens and, along with nickel, have been the only trace metals shown to be absolutely required for growth of these organisms (Madigan et al. 1997).

Methanogens are important in a wide range of terrestrial and marine environments. They commonly occupy an important niche in anoxic, sulfate-deficient environments such as swamps, wetlands, water-logged soils, rice paddies, tundra, marine sediment, geothermal systems, the subsurface of the Earth's crust, landfill sites, wastewater treatment plants, and within the digestive tracts of many animals. Some can even be found in aerated environments that contain anoxic microenvironments, such as the interior of soil particles.

Methanogenesis occurs through three distinct pathways: CO_2 reduction, methylotrophic methanogenesis, and acetoclastic methanogenesis (Table 1.4). All these pathways depend on coenzyme M, with methane ultimately generated by the reduction of methyl-CoM. We described the methanogenesis by CO_2 reduction in Section 1.3.4.2, a major pathway of methanogenesis, in which microbes grow on $H_2 + CO_2$. This process is viewed as an autotrophy, with CO_2 serving as both carbon source and electron acceptor. In addition to CO_2, however, certain methylated compounds such as methanol and acetate can be converted to methane by methanogens in the absence of O_2.

Methanogenesis represents the final stage in the anaerobic breakdown of organic matter in many cases. Estimates suggest that nearly half of the organic carbon degraded globally by anaerobically respiring microbes ultimately becomes methane. Despite its widespread formation, only a few organic compounds serve as direct precursors to methanogenesis (Table 1.4). Methanogens cannot use long-chain fatty acids (with more than two carbons) or aromatic compounds, so they depend on

Table 1.4 Reactions and substrates in three methanogenic pathways.

Pathways	Reactions	Substrates
CO_2 reduction pathway	$4H_2 + CO_2 \rightarrow CH_4 + 2H_2O$	Carbon dioxide, CO_2 (with electrons derived from H_2, certain alcohols, or pyruvate) Formate, $HCOO^-$ Carbon monoxide, CO
Methylotrophic pathway	$CH_3OH + H_2 \rightarrow CH_4 + CO_2$	Methanol, CH_3OH Methylamine, $CH_3NH_3^+$ Dimethylamine, $(CH_3)_2NH_2^+$ Trimethylamine, $(CH_3)_3NH^+$ Methyl mercaptan, CH_3SH Dimethyl sulfide, $(CH_3)_2S$
Acetoclastic pathway	$CH_3COO^- + H_2 \rightarrow CH_4 + CO_2$	Acetate, CH_3COO^- Pyruvate, CH_3COCOO^-

Table 1.5 Reactions in fermentation and acetogenesis.

Reaction types	Reactions
Fermentation of glucose to acetate, H_2, and CO_2	Glucose $+4H_2O \rightarrow 2$ acetate$^-$ $+ 2HCO_3^- + 4H^+ + 4H_2$
Fermentation of glucose to butyrate, CO_2, and H_2	Glucose $+2H_2O \rightarrow$ butyrate$^-$ $+ 2HCO_3^- + 2H_2 + 3H^+$
Fermentation of butyrate to acetate and H_2	Butyrate$^-$ $+ 2H_2O \rightarrow 2$ acetate$^-$ $+ H^+ + 2H_2$
Fermentation of propionate to acetate, CO_2, and H_2	Propionate- $+ 3H_2O \rightarrow$ acetate$^-$ $+ HCO_3^- + H^+ + H_2$
Fermentation of ethanol to acetate and H_2	2 Ethanol $+2H_2O \rightarrow 2$ acetate$^-$ $+ 4H_2 + 2H^+$
Fermentation of benzoate to acetate, CO_2, and H_2	Benzoate$^-$ $+ 7H_2O \rightarrow 3$ acetate$^-$ $+ 3H^+ + HCO_3^- + 3H_2$
Acetogenesis from $CO_2 + H_2$	$4H_2 + 2HCO_3^- + H^+ \rightarrow$ acetate$^-$ $+ 4H_2O$

Source: Adapted from Madigan et al. (2018).

the activity of other microorganisms to break down complex organic compounds into more suitable and simple substrates. To access a broader array of organic carbon molecules, methanogens often establish syntrophic relationships with fermentative anaerobes, enabling the breakdown of diverse compounds that eventually contribute to methane production (Table 1.5; Figure 1.20). The process is named syntrophy, in which two or more organisms cooperate in the anaerobic degradation of organic compounds. Syntrophs function to supply them with precursors for methanogenesis.

High-molecular-weight substances such as polysaccharides, proteins, and fats, are converted to CH_4 by the cooperative interaction of several physiological groups of prokaryotes. In many anoxic environments, the immediate precursors of CH_4 are H_2 and CO_2, which are generated by the activities of fermentative anaerobes. For conversion of a typical polysaccharide such as cellulose into methane, as many as five major physiological groups of prokaryotes may be involved in the overall process (Figure 1.20). Cellulolytic bacteria cleave the high-molecular-weight cellulose molecule into cellobiose and into free glucose. Glucose is then fermented by primary fermenters to a variety of fermentation products, acetate, propionate, butyrate, succinate, alcohols, H_2, and CO_2 (Table 1.5). These substrates are then consumed by methanogens. Any H_2 produced in primary

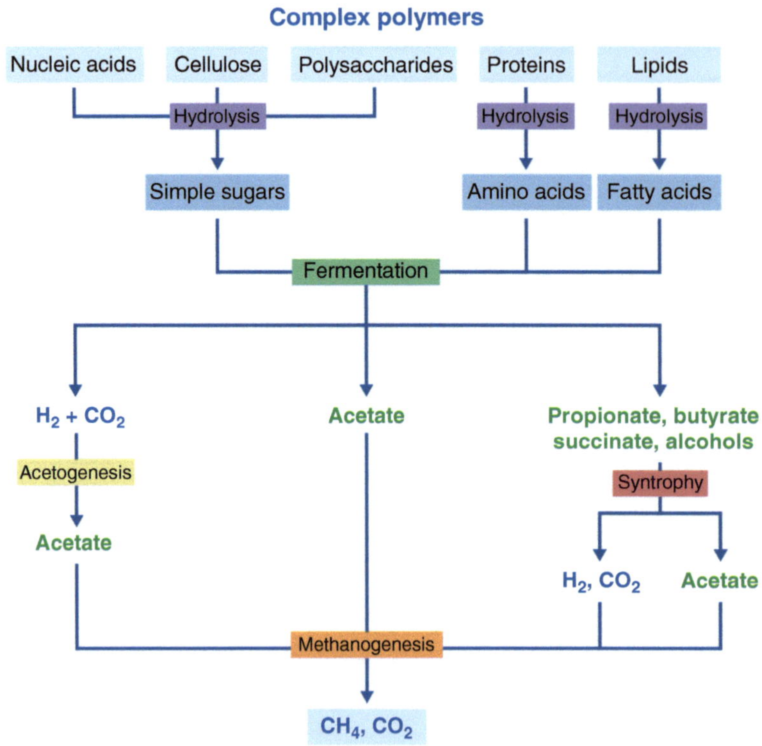

Figure 1.20 The decomposition of complex organic matter in absence of terminal electron acceptors. The breakdown involves a series of microbially mediated steps that differ significantly from the oxidizing decomposition. Hydrolysis breaks down complex polymers into simpler monomers; fermentation converts monomers into volatile fatty acids (VFAs, like propionate and butyrate), alcohols, acetate, hydrogen (H_2), and carbon dioxide (CO_2); Syntrophic oxidation of fermentative products supplies precursors for methanogenesis. Methanogenesis converts acetate or $H_2 + CO_2$ into methane, an end-product of organic matter degradation in the absence of electron acceptors.

fermentative processes is immediately consumed by methanogens, homoacetogens, or sulfate-reducing microorganisms (in environments where sulfate content is significant).

Since methane is an efficient greenhouse gas, methanogenesis has often been linked to early Earth conditions, potentially influencing atmospheric evolution and climate dynamics. In deep-sea sediments, methanogenesis is a key process in methane seeps, supporting unique ecosystems that rely on methane as an energy source (Section 3.3.1). Additionally, the interaction between methanogenesis and anoxic oxidation of methane leads to the formation of specific buildups that serve as geological markers of past methane emissions (Section 6.2.1).

1.3.5.9 Methanotrophs and Methylotrophs

Methylotrophs are microorganisms that can grow only on one-carbon compounds without a carbon–carbon bond, from methane (C^{-4}) to progressively more oxidized forms like methanol (C^{-2}) and methylamine (C^{-2}), through to formaldehyde (C^{0}) and formate (C^{+2}). Methane (CH_4) and many other C_1 organic compounds can be catabolized both aerobically and anaerobically. Many, but not all, methylotrophs are also methanotrophs (microbes that can grow on CH_4). Methanotrophs are also able to grow on more oxidized one-carbon compounds and some can even switch to ammonia oxidation if the supply of methane is temporarily diminished.

Methane can be used by methanotrophs in two ways: the dissimilative manner of oxidation of methane to CO_2 and the assimilative manner in which one-carbon compounds are assimilated into cell material. Two distinct assimilative pathways have been known in methanotrophs. Some methanotrophs (e.g. *Methylosinus* and *Methlocystis*) use the serine pathway, in which acetyl-CoA is synthesized from one molecule of CH_2O (produced from the oxidation of CH_3OH) and one molecule of CO_2. Other methanotrophs (e.g. *Methylomonas*, *Methylomicrobium*, *Methylobacter*, and *Methylococcus*) utilize the ribulose monophosphate pathway for carbon fixation. The ribulose monophosphate pathway is more energy efficient than the serine pathway because all of the carbon for cell material is derived from formaldehyde (CH_2O) at the same oxidation state as cell material, no reducing power is needed (Madigan et al. 1997, 2018).

Methanotrophs are phylogenetically diverse and include bacteria and archaea. Aerobic methane oxidation is catalyzed by particulate and soluble methane monooxygenases (pMMO and sMMO, respectively). These enzymes oxidize methane to methanol, and subsequently, methanol dehydrogenases (MDHs) further oxidize methanol to formaldehyde. After formaldehyde, two more oxidation steps are involved and the intermediates are used for carbon assimilation. Aerobic methanotrophs belong to the gammaproteobacteria (Type I, with families Methylococcaceae and Methylothermaceae), alphaproteobacteria (Type II, with families Methylocystaceae and Beijerinckiaceae), and the Verrucomicrobia phyla (family Methylacidiphilaceae) (Guerrero-Cruz et al. 2021).

Anaerobic methanotrophs (ANME) oxidize methane in the absence of oxygen, using alternative electron acceptors such as sulfate, nitrate, nitrite, iron, manganese, and organic compounds like humic substances. They play a crucial role in mitigating methane emissions, particularly in environments where oxygen is scarce, such as deep-sea sediments, wetlands, and wastewater treatment systems. These methanotrophs include both bacteria and archaea, with archaeal ANME being particularly significant in marine ecosystems. ANME archaea are often found in symbiotic relationships with sulfate-reducing bacteria, facilitating methane oxidation coupled with sulfate reduction (Bhattarai et al. 2019). Meanwhile, bacterial ANME, such as those from the NC10 phylum, can oxidize methane using nitrite, producing oxygen intracellularly to fuel methane oxidation. Interestingly, it was proposed that some methanogens have limited capacity to reverse their normal metabolism, thereby oxidizing methane under anoxic conditions (changing to methanotrophs). Evidence for such syntrophy comes from methane-rich marine sediments, where methanotrophs have been found growing in tight clusters, closely surrounded by their sulfate-reducing bacterial partners (see Konhauser 2007).

AOM has received much attention from geobiologists for a number of reasons: (1) AOM serves as a major methane sink in marine sediments, preventing large-scale methane release into the atmosphere, which could otherwise contribute to climate change; (2) the process leads to the precipitation of carbonate minerals, forming methane-derived authigenic carbonates (see Section 6.2.1), which are often found in cold seep environments and serve as geological records of past methane fluxes; (3) in sulfate-poor environments, AOM can be coupled with the reduction of metal oxides such as manganese and iron, influencing sediment geochemistry and contributing to the global carbon cycle (Xiao et al. 2023); (4) AOM helps maintain the stability of methane hydrates in deep-sea sediments by consuming methane before it escapes, thereby influencing the dynamics of gas hydrate reservoirs.

References

Arnold, B.J., Huang, I., and Hanage, W.P. 2022. Horizontal gene transfer and adaptive evolution in bacteria. *Nature Reviews Microbiology*, 20(4), pp. 206–218.

Beijerinck, M.W. 1895. Über *Spirillum desulphuricans* als Ursache von Sulfatreduktion. *Zentralb Bakteriol Parasitk Infekt Abt*, I, pp. 104–114.

Besson, S., Vragniau, C., Vassal-Stermann, E., Dagher, M.C., and Fender, P. 2020. The adenovirus dodecahedron: beyond the platonic story. *Viruses*, 12(7), pp. 718.

Bhattarai, S., Cassarini, C., and Lens, P.N.L. 2019. Physiology and distribution of archaeal methanotrophs that couple anaerobic oxidation of methane with sulfate reduction. *Microbiology and Molecular Biology Reviews*, 83. 10.1128/mmbr.00074-18.

Blankenship, R.E. 2010. Early evolution of photosynthesis. *Plant Physiology*, 154, pp. 434–438.

Bowles, M.W., Mogollón, J.M., Kasten, S., Zabel, M., and Hinrichs, K.U. 2014. Global rates of marine sulfate reduction and implications for sub-sea-floor metabolic activities. *Science*, 344, pp. 889–891.

Boyd, E., and Peters, J.W. 2013. New insights into the evolutionary history of biological nitrogen fixation. *Frontiers in Microbiology*, 4, pp. 201.

Campbell, N.A., and Reece, J.B. 2002. *Biology* (Sixth Edition), pp. 1247. San Francisco, Pearson Education.

Claret Fernández, L., Mesman, R., and van Niftrik, L. 2020. The anammoxosome organelle: the power plant of anaerobic ammonium-oxidizing (Anammox) bacteria. In: *Bacterial organelles and organelle-like inclusions*. Microbiology Monographs (Vol. 34) (ed. D. Jendrossek), pp. 107–123. Cham, Springer.

Cui, H.L., and Dyall-Smith, M.L. 2021. Cultivation of halophilic archaea (class *Halobacteria*) from thalassohaline and athalassohaline environments. *Marine Life Science and Technology*, 3(2), pp. 243–251.

Cui, L.H., Zhu, K.L., Li, R.Y., Chang, C., Wu, L.Y., Liu, W., Fu, D.J., Liu, P.Y., Qiu, H., Tang, G.Q., Li, Q. L., Gaines, R.R., Tao, Y.X., Wang, Y.Z., Li, J.H., and Zhang, X.L. 2024. The Cambrian microfossil *Qingjiangonema* reveals the co-evolution of sulfate-reducing bacteria and the oxygenation of Earth's surface. *Science Bulletin*, 69, pp. 1486–1494. 10.1016/j.scib.2024.03.001.

Doolittle, W.F. 2022. Evolution: two domains of life or three? *Current Biology*, 30, pp. R177–R179.

Ehrenreich, A., and Widdel, F. 1994. Anaerobic oxidation of ferrous iron by purple bacteria, a new type of phototrophic metabolism. *Applied and Environmental Microbiology*, 60, pp. 4517–4526.

Feng, W. 2020. The role of viruses and bacteria and other microorganisms in the history of the earth biological evolution. *Geology in China*, 47, pp. 655–663 (in Chinese with English abstract).

Farkas, K., Pang, L., Lin, S., Williamson, W., Easingwood, R., Fredericks, R., Jaffer, M.A., and Varsani, A. 2013. A gel filtration-based method for the purification of infectious rotavirus particles for environmental research applications. *Food and Environmental Virology*, 2013(5), pp. 231–235.

Gottschalk, G. 1986. *Bacterial metabolism* (2nd Edition), pp. 359. New York, Springer.

Grigsby, I.F., Zhang, W., Johnson, J.L., Fogarty, K.H., Chen, Y., Rawson, J.M., Crosby, A.J., Mueller, J.D., and Mansky, L.M. 2010. Biophysical analysis of HTLV-1 particles reveals novel insights into particle morphology and Gag stoichiometry. *Retrovirology*, 7, pp. 75. 10.1186/1742-4690-7-75.

Guerrero-Cruz, S., Vaksmaa, A., Horn, M.A., Niemann, H., Pijuan, M., and Ho, A. 2021. Methanotrophs: discoveries, environmental relevance, and a perspective on current and future applications. *Frontiers in Microbiology*, 12, pp. 678057. 10.3389/fmicb.2021.678057.

Hamilton, T., Klatt, J.M., de Beer, D., and Macalady, J.L. 2018. Cyanobacterial photosynthesis under sulfidic conditions: insights from the isolate *Leptolyngbya* sp. strain hensonii. *The ISME Journal*, 12, pp. 568–584.

Hashimoto, T., Horikawa, D.D., Saito, Y. et al. 2016. Extremotolerant tardigrade genome and improved radiotolerance of human cultured cells by tardigrade-unique protein. *Nature Communications*, 7(1), pp. 12808.

Kappler, A., Pasquero, C., Konhauser, K.O., and Newman, D.K. 2005. Deposition of banded iron formations by anoxygenic phototrophic Fe(II)-oxidizing bacteria. *Geology*, 33(11), pp. 865–868.

Karekar, S., Stefanini, R., and Ahring, B. 2022. Homo-acetogens: their metabolism and competitive relationship with hydrogenotrophic methanogens. *Microorganisms*, 10, pp. 397.

Kashefi, K., and Lovley, D.R. 2003. Extending the upper temperature limit for life. *Science*, 301, pp. 934–934.

Klatt, J.M., Al-Najjar, M.A.A., Yilmaz, P., Lavik, G., de Beer, D., and Polerecky, L. 2015. Anoxygenic photosynthesis controls oxygenic photosynthesis in a cyanobacterium from a sulfidic spring. *Applied and Environmental Microbiology*, 81, pp. 2025–2031.

Konhauser, K.O. 2007. *Introduction to geomicrobiology*, pp. 425. Oxford, Blackwell Science.

Krupovic, M., Kuhn, J.H., and Fischer, M.G. 2016. A classification system for virophages and satellite viruses. *Archives of Virology*, 161(1), pp. 233–247.

Krupovic, M., Dolja, V., and Koonin, E. 2020. The LUCA and its complex virome. *Nature Reviews Microbiology*, 18, pp. 661–670.

Lake, J.A., Henderson, E., Oakes, M., and Clark, M.W. 1984. Eocytes: a new ribosome structure indicates a kingdom with a close relationship to eukaryotes. *Proceedings of the National Academy of Sciences of the United States of America*, 81, pp. 3786–3790.

Lee, H., Assaraf, R., Subramanian, S., Goetschius, D., Bieri, J., DiNunno, N.M., Leisi, R., Bator, C.M., Hafenstein, S.L., and Ros, C. 2024. Infectious parvovirus B19 circulates in the blood coated with active host protease inhibitors. *Nature Communications*, 15(1), pp. 9543.

Li, M., Ye, X., Da, Y., Sun, Q., and Zhou, G. 2024. Unveil of the role of fungal taxa in iron(III) reduction in paddy soil. *Frontiers in Microbiology*, 14, pp. 1334051.

Liu, W., and Zhang, X. 2015. Experimental study of microbial pyrite oxidation: a suggestion for geologically useful biosignatures. *Geomicrobiology Journal*, 32, pp. 466–471.

López-García, P., and Moreira, D. 2020. Cultured Asgard archaea shed light on eukaryogenesis. *Cell*, 181, pp. 232–235.

Madigan, M.T., Martinko, J.M., and Parker, J. 1997. *Biology of microorganisms* (Eighth Edition), pp. 986. Upper Saddle River, Prentice Hall.

Madigan, M.T., Martinko, J.M., and Parker, J. 2003. *Brock biology of microorganisms* (Tenth Edition), pp. 1019. Pearson Education, Prentice Hall.

Madigan, M., Sattley, W., Aiyer, J., Stahl, D., and Buckley, D. 2018. *Brock biology of microorganisms* (Fifteenth Edition), pp. 1058. New York, Pearson Education.

McRose, D.L., Zhang, X., Kraepiel, A.M.L., and Morel, F.M.M. 2017. Diversity and activity of alternative nitrogenases in sequenced genomes and coastal environments. *Frontiers in Microbiology*, 8, pp. 267.

Moody, E.R.R., Álvarez-Carretero, S., Mahendrarajah, T.A. et al. 2024. The nature of the last universal common ancestor and its impact on the early Earth system. *Nature Ecology and Evolution*, 8, pp. 1654–1666.

Muyzer, G., and Stams, A.J.M. 2008. The ecology and biotechnology of sulphate-reducing bacteria. *Nature Reviews Microbiology*, 6, pp. 441–454.

NA. n.d. Metabolism, genome and age of the last universal common ancestor. *Nature Ecology and Evolution*, 8, pp. 1573–1574. 10.1038/s41559-024-02474-w.

Nealson, K.H., and Rye, R. 2005. Evolution of metabolism: biomineralization. In: *Biogeochemistry 8*, Treatise on Geochemistry (eds. H.D. Holland and K.K. Turekian) (ed. W.H. Schlesinger), pp. 41–62. Oxford, Elsevier-Pergamon.

Nester, E.W., Anderson, D.G., Robers, C.E., Jr., Pearsall, N.N., Nester, M.T., and Hurley, D. 2004. *Microbiology—A human perspective*, pp. 817. New York, McGraw-Hill.

Nobs, S.-J., MacLeod, F.I., Wong, H.L., and Burns, B.P. 2022. Eukarya the chimera: eukaryotes, a secondary innovation of the two domains of life? *Trends in Microbiology*, 30, pp. 421–431.

Noda, T., and Kawaoka, Y. 2010. Structure of influenza virus ribonucleoprotein complexes and their packaging into virions. *Reviews in Medical Virology*, 20(6), pp. 380–391.

Oshiki, M., Ishii, S., Yoshida, K., Fujii, N., Ishiguro, M., Satoh, H., and Okabe, S. 2013. Nitrate-dependent ferrous iron oxidation by anaerobic ammonium oxidation (Anammox) bacteria. *Applied and Environmental Microbiology*, 79, pp. 13.

Ozaki, K., Thompson, K.J., Simister, R.L., Crowe, S.A., and Reinhard, C.T. 2019. Anoxygenic photosynthesis and the delayed oxygenation of Earth's atmosphere. *Nature Communications*, 10, pp. 3026.

Rivera, M.C., and Lake, J.A. 1992. Evidence that eukaryotes and eocyte prokaryotes are immediate relatives. *Science*, 257, pp. 74–76.

Sagan, L. 1967. On the origin of mitosing cells. *Journal of Theoretical Biology*, 14, pp. 225–274.

Sánchez-Baracaldo, P., Bianchini, G., Wilson, J.D., and Knoll, A.H. 2022. Cyanobacteria and biogeochemical cycles through Earth history. *Trends in Microbiology*, 30, pp. 143–157.

Schneider, M.P.C., Cunha, E., Silva, L., Leão, J., Tavares, V.C., de Sousa, E.B., and Faustino, S. 2025. Cyanobacterial blooms and the presence of cyanotoxins in the Brazilian Amazon. *Toxins (Basel)*, 17(6), pp. 296.

Terry, C., Harniman, R.L., Sells, J., Wenborn, A., Joiner, S., Saibil, H.R., Miles, M.J., Collinge, J., and Wadsworth, J.D.F.2019. Structural features distinguishing infectious ex vivo mammalian prions from non-infectious fibrillar assemblies generated in vitro. *Scientific Reports*, 9(1), pp. 376.

Wang, H. 2022. How do viruses affect the elemental cycle and energy flow in the geological environment? *Journal of Earth Science*, 47, pp. 3840–3841 (in Chinese with an English abstract).

Weber, K., Achenbach, L., and Coates, J. 2006. Microorganisms pumping iron: anaerobic microbial iron oxidation and reduction. *Nature Reviews Microbiology*, 4, pp. 752–764.

Whittaker, R.H. 1969. New Concepts of kingdoms of organisms. *Science*, 163, pp. 150–160.

Woese, C.R., Kandler, O., and Wheelis, M.L. 1990. Towards a natural system of organisms: proposal for the domains Archaea, Bacteria, and Eucarya. *Proceedings of the National Academy of Sciences USA*, 87, pp. 4576–4579.

Wolfe, R.S. 1972. Microbial formation of methane. *Advances in Microbial Physiology*, 6, pp. 107–146.

Wunder, L.C., Breuer, I., Willis-Poratti, G., Aromokeye, D.A., Henkel, S., Richter-Heitmann, T., Yin, X., and Friedrich, M.W. 2024. Manganese reduction and associated microbial communities in Antarctic surface sediments. *Frontiers in Microbiology*, 15, pp. 1398021.

Xiao, X., Luo, M., Zhang, C., Zhang, T., Yin, X., Wu, X., Zhao, J., Tao, J., Chen, Z., Liang, Q., and Dong, X. 2023. Metal-driven anaerobic oxidation of methane as an important methane sink in methanic cold seep sediments. *Microbiology Spectrum*, 11, pp. e05337–e05322.

Zhou, H., and Fu, C. 2020. Manganese-oxidizing microbes and biogenic manganese oxides: characterization, Mn(II) oxidation mechanism and environmental relevance. *Reviews in Environmental Science and Bio/Technology*, 19, pp. 489–507.

2

Life in Extreme Environments

Zisheng Guo and Xingliang Zhang

2.1 Introduction

Most living organisms on Earth have specific requirements for their habitats. When environmental conditions exceed the upper or lower limits of the appropriate conditions, the growth and reproduction of the organisms can be adversely affected, or their survival may even become impossible. However, some exceptional organisms, spanning various branches of the tree of life, have adapted to endure, prefer, or even flourish in one or multiple extreme conditions simultaneously. These unique life forms captivate multidisciplinary scientists, including geobiologists and astrobiologists, because they push the boundaries of what life can endure. Scientists study them to understand the limits of life on Earth and to explore possibilities for life beyond our planet. Their unique adaptations also have practical applications, such as biotechnology and environmental cleanup.

An organism that thrives in an extreme environment is an extremophile. In many cases, extremophiles can thrive under conditions that would kill other creatures. The term "extreme condition" is a relative concept that is fundamentally defined from an anthropocentric perspective. "Extremes" include physical extremes (e.g. temperature, radiation, or pressure) and geochemical extremes (e.g. desiccation, salinity, pH value, oxygen species, or redox potential). These conditions are often inhospitable or even lethal to humans and most other organisms. However, certain organisms can adapt well to the harsh conditions and even live in such environments as their optimal habitats. If an organism can survive in more than one extreme, it is a polyextremophile (Caro-Astorga et al. 2024). For example, *Sulfolobus acidocaldarius*, an archaeon, can flourish at a pH value of 3 and a temperature of 80 °C.

Extremophiles are an important component of Earth's biosphere and constitute a valuable biological resource. As pioneers of harsh environments, extremophiles (e.g. thermophiles, acidophiles, halophiles) colonize extreme habitats and participate in critical processes like sulfur, nitrogen, and carbon cycling, which impact ecosystem stability and global biogeochemical dynamics. Extremophiles, particularly hyperthermophiles, provide clues about early Earth conditions, and their unique biochemical strategies (e.g. DNA repair and protein stability) help scientists understand life's limits and resilience. In addition, they have biotechnological and industrial applications (Coker 2019). Examples are various: extremozymes such as heat-stable, salt-tolerant, or acid-resistant enzymes are used in biotechnology, medicine, and industry; some extremophiles (e.g. methanogens) aid in biofuel production; some (e.g. radiation-resistant *Deinococcus radiodurans*) can be used to clean up toxic waste or radioactive sites (Caro-Astorga et al. 2024).

Introduction to Geobiology, First Edition. Xingliang Zhang, Zisheng Guo, Wei Liu, Ruliang He, Luoyang Li, and Weiduo Hao.

Companion website: www.wiley.com/go/Zhang_1e

The discovery of extremophiles highlights the remarkable adaptability of life and strengthens the possibility of microbial life existing elsewhere in the Solar System and beyond. Growing evidence suggests that extremophiles were among Earth's earliest organisms, with the harsh conditions of the Hadean-Archean Earth likely permitting only such resilient life forms to thrive (see Sections 8.2 and 8.6). As an interdisciplinary field bridging microbiology and geochemistry, extremophile biology has emerged as a central focus of geobiological research (Chen et al. 2006; Dong and Jiang 2023; Xie et al. 2023). Studying their adaptive mechanisms holds significant promise for advancing our understanding of life's origin and evolution, aiding the search for extraterrestrial life, enabling environmental bioremediation, and unlocking applications in industry, agriculture, and scientific innovation.

2.2 Diversity of Extreme Environments and Extremophiles

2.2.1 Extreme Environments

An extreme environment is a habitat characterized by physical or chemical conditions that are close to the perceived limits for life, making survival challenging for most known organisms. These environments are typically inhospitable to humans and "mainstream" life forms (mesophiles) due to factors such as extreme temperatures, pH levels, pressure, radiation, salinity, or toxicity. Conventionally, extreme ecological conditions are classified into habitat-based or parameter-based frameworks.

2.2.1.1 Extreme Ecosystems

Extreme ecosystems are far more widespread than traditionally assumed, occupying a significant portion of Earth's biosphere from mountains to deep sea, from equatorial to polar regions, and from surface into the crust (Figure 2.1; Merino et al. 2019; Shu and Huang 2022). While they represent a small fraction of the planet's surface area, their collective biomass and ecological influence are substantial—particularly in subsurface and oceanic environments (Bar-On et al. 2018). Extreme ecosystems are not randomly distributed but instead follow predictable patterns tied to geological, chemical, and climatic factors. Their occurrence depends on the availability of three key requirements: liquid water at least in trace amounts, electron donors and acceptors, energy sources from sunlight, chemical bonds or radiation, and nutrients such as nitrogen, phosphorus, and trace metals.

Extreme ecosystems are diverse, encompassing a broad spectrum of natural habitats defined by at least one physical or chemical extreme. Crucially, these environments are shaped by the synergistic interplay of multiple extreme factors. For example: soda lakes combine high alkalinity (pH ~9–12) with elevated salinity; deserts exhibit extreme aridity alongside intense ultraviolet radiation; terrestrial hot springs feature high temperatures (often >45 °C), extreme pH (acidic or alkaline), and hypoxic conditions; marine hydrothermal vents are characterized by high temperatures (>350 °C), extreme pressure, and low nutrient availability; hypersaline environments (e.g. salt flats) possess not only high salt concentrations but also frequently high pH and low oxygen/nutrient levels; cryospheric habitats (e.g. polar ice, permafrost) are defined by subzero temperatures, minimal thermal energy, and scarce liquid water; acid mine drainages sustain extreme acidity (pH < 3), toxic metal concentrations, and limited energy-yielding metabolic pathways (Shu and Huang 2022). These ecosystems demonstrate how life adapts to concurrent stressors, often pushing the boundaries of known physiological limits.

Meanwhile, it is essential to emphasize that extreme ecosystems exhibit dynamic adaptations in response to persistent geological and climatic shifts. Periodic geological activity and climate variability can induce rhythmic responses in these ecosystems. Moreover, the definition of "extreme conditions" is both temporally and spatially relative. In Earth's geological past, such conditions

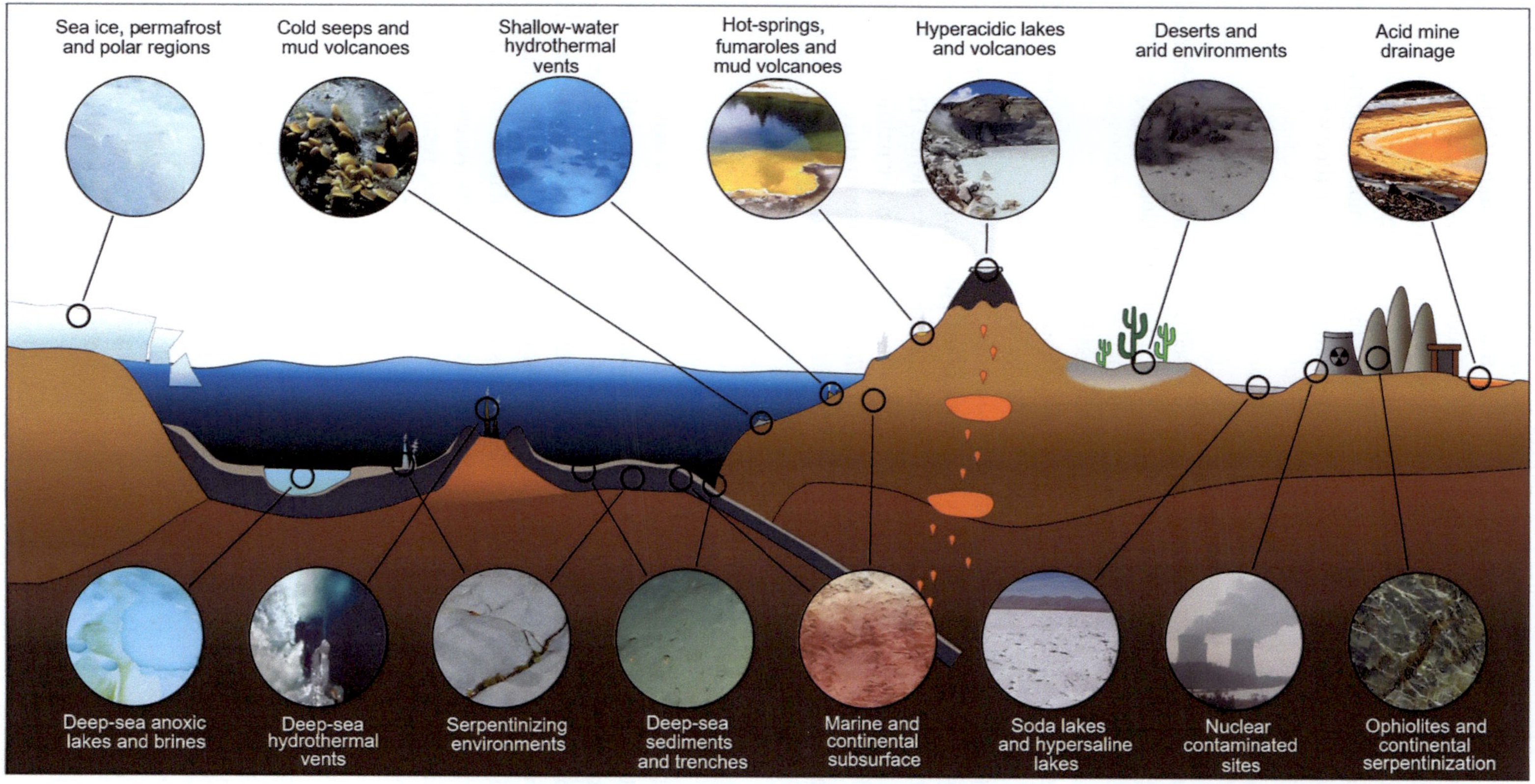

Figure 2.1 Representative idealized cross-section of Earth's surface showing the diversity of extreme ecosystems and their approximate location. *Source:* Merino et al. (2019) / Frontiers Media S.A / CC BY 4.0.

may have been far more widespread. For instance, during the Archean Eon, the planet's surface featured a highly reduced, high-temperature ocean–atmosphere system (Catling and Zahnle 2020), with average temperatures exceeding ~50 °C (see Section 8.3). Despite these harsh conditions, early Earth supported diverse microbial life—organisms that would now be classified as extremophiles. Today, while much of Earth is hospitable to a wide range of life, numerous extreme environments still host thriving ecosystems. This suggests that some modern extremophiles may represent evolutionary relics, preserving ancient survival strategies from a time when extreme conditions were the norm.

2.2.1.2 Extreme Environmental Parameters

Extreme environments can be systematically classified based on measurable physicochemical parameters that exceed the tolerances of most life forms. This framework identifies key environmental stressors, their ranges of extremity, and synergistic interactions that define ecosystem boundaries. A comprehensive summary of core parameters is provided in Table 2.1.

Key advantages of the framework are in multiple folds (Shu and Huang 2022): (1) standardization facilitates direct comparisons of extremophile tolerances across studies; (2) predictive capacity helps to identify potential life-supporting conditions in unexplored environments, including extraterrestrial analogs; (3) dynamic adaptability incorporates temporal and spatial variability (e.g. climate-driven fluctuations or geological cycles) for modeling ecosystem resilience. This framework is widely adopted in scientific communities and provides the foundation for naming extremophile groups based on their dominant stressor (e.g. *thermophiles* for high-temperature specialists, *halophiles* for high-salinity adapters, and *alkaliphiles* for alkaline-tolerant organisms). Its robustness and utility make it the preferred classification system for this book.

2.2.2 Phylogenic Diversity of Extremophiles

Prokaryotes including bacteria and archaea are the most commonly cited extremophiles, while organisms with the capacity to thrive in extreme environments extend across all three domains of life—including eukaryotes (Table 2.2). As illustrated in Figure 2.2, certain eukaryotic organisms exhibit remarkable tolerance to extremes of pH, temperature, and other physicochemical stressors. Furthermore, viruses have also been identified in these harsh habitats, highlighting the broad ecological adaptability of life. Extreme ecosystems typically support microbial communities of low to intermediate complexity, with a high abundance of a few dominant species. As environmental conditions become increasingly extreme, overall microbial diversity declines, resulting in the dominance of a limited number of highly specialized taxa (Shu and Huang 2022).

2.2.2.1 Extremophilic Prokaryotes

The widespread distribution of prokaryotes in extreme environments arises from synergistic adaptations spanning metabolic versatility, physiological characteristics, and protective mechanisms (Figure 2.3). The core driving factor of niche adaptation is metabolic plasticity, enabling diverse substrate utilization (e.g. H_2S/CH_4/glucose co-metabolism), multiple pathways (alternating between aerobic/anaerobic respiration, autotrophy/heterotrophy, or photosynthesis), and extremozymes (e.g. thermo-, acido-, and halostable enzymes). Coupled with low metabolic activity and dormancy strategies, these traits collectively alleviate energy and substrate constraints under physical and geochemical extremes.

Firstly, bacteria and archaea are typically small with dimensions in the micrometer range. For instance, *Pseudomonas* and *Acinetobacter* generally exhibit rod-shaped morphologies (1 μm in width and 2–3 μm in length), while *Azotobacter chroococcum* (a species of nitrogen-fixing bacteria) forms spherical cells of 1–2 μm in diameter. In addition, bacteria reproduce rapidly and can drive

Table 2.1 Classification and examples of extremophiles.

Environmental parameter	Type	Definition	Examples
Temperature	Hyperthermophile	Growth > 80 °C	*Pyrolobus fumarii*, 113 °C
	Thermophile	Growth 60–80 °C	Strain 121, 121 °C
	Mesophile	15–60 °C	*Methanopyrus kandleri*, 122 °C
	Psychrophile	<15 °C	*Synechococcus lividus*
			Homo sapiens
			Psychrobacter, some insects,
			Planococcus halocryophilus OR1, −25 °C
Radiation		Radiation resistant	*Deinococcus radiodurans*
Pressure	Barophile (weight loving)	10–50 MPa	Unknown
			For microbe, 130 Mpa
	Piezophile (pressure loving)	>50 MPa	
	Hyperbarophile		
	Hyperpiezophile		
Desiccation	Xerophiles	$a_w < 0.7$	*Artemia salina*; nematodes, microbes, fungi, lichens, tardigrades
Salinity	Halophile	>8.8% NaCl	Halobacteriaceae
	Hyperhalophile	>14.6% NaCl	*Dunaliella salina*
pH	Alkaliphile	pH > 9	*Natronobacterium*, *Bacillus*
	Hyperalkaliphile	pH > 11	*firmus* OF4, *Spirulina* spp.
	Acidophile	pH < 5	(all pH 10.5),
	Hyperacidophile	pH < 3	*Serpentinomonas sp. B1* (pH 12.5),
			Cyanidium caldarium,
			Ferroplasma sp. (both pH 0), *Picrophilus oshimae* (pH_{opt} 0.7)
Oxygen tension	Anaerobe	Cannot tolerate O_2	*Methanococcus jannaschii*
	Microaerophile	Tolerates some O_2	*Clostridium*
	Aerobe	Requires O_2	*H. sapiens*
Chemical extremes	Gases	Can tolerate high	*C. caldarium* (pure CO_2)
	Metals	concentrations	*Ferroplasma acidarmanus*
		of metal (metallotolerant)	(Cu, As, Cd, Zn); *Ralstonia* sp. CH34 (Zn, Co, Cd, Hg, Pb)
		toxitolerant	

Source: Rothschild and Mancinelli (2001) / Springer Nature.

exponential biomass accumulation in a short time. Under optimal conditions (e.g. LB broth at 37 °C), the model organism *Escherichia coli*, with a doubling time of ~20 minutes and a single-cell wet mass of $\sim 1 \times 10^{-12}$ g, theoretically yields 4.72×10^{21} cells within 24 hours, resulting in a total biomass of 4.72×10^{9} g. Being small allows them to efficiently utilize available resources, while fast reproduction enables them to quickly adapt to changing conditions through genetic mutations and horizontal gene transfer (HGT).

Secondly, prokaryotic organisms benefit from a high surface-area-to-volume ratio, which enhances their interaction with the environment. The lack of complex organelles allows direct

Table 2.2 Overview of the most notable examples of extremophilic microbes and viruses.

	Phylum	Family	Species (or Genus)	Tolerance	Known maximum limit
Algae	Chlorophyta	Chlamydomonadaceae	*Chlamydomonas nivalis*	Cold stress; solar and UV radiation	4 °C; UV 315 nm
		Dunaliellaceae	*Dunaliella salina*	Acidic conditions; saline conditions; contaminated Sites	0.5 pH; 4.5 M NaCl
		Cyanidiaceae	*Cyanidium caldarium*	Hot stress; acidic conditions; saline conditions; contaminated sites	55 °C; 0.5 pH
		Trebouxiophyceae	*Crucigenia*	Cold stress; saline conditions	n.d.
		Trebouxiaceae	*Trebouxia*	Cold stress; low water availability; irradiated environments	113 kGy
		Dolichomastigaceae	*Dolichomastix*	Cold stress; saline conditions; low water availability	n.d.
	Rhodophyta	Galdieraceae	*Galdieria sulphuraria*	Hot stress; acidic conditions	57 °C; pH 0.2; 2 M HCI
Fungi	Ascomycota	*Incertae sedis*	*Cryomyces antarcticus*	Cold stress; low water availability; irradiated environments	117 kGy
		Trichocomaceae	*Aspergillus penicillioides*	Low water availability;	0.585 a_w
		Teratosphaeriaceae	*Hortaea acidophila*	Acidic conditions	0.6 pH
			Fodinomyces uranophilus	Acidic conditions; infected sites	1.7 pH
		Coniochaetaceae	*Coniochaeta fodinicola*	Acidic conditions; infected sites	1.7 pH
		Hypocreaceae	*Cephalosporium*	Acidic conditions	0 pH
		Megasporaceae	*Circinaria gyrosa*	Low water availability; irradiated environments	117 kGy
		Monascaceae	*Monoascus bisporus*	Cold stress; low water availability	0.6 a_w
		Cladosporiaceae	*Cladosporium sphaerospermum*	Cold stress; irradiated environments	5 kGy

	Basidiomycota	Tremellaceae	*Cryptococcus neoformans*	Cold stress; low water availability; irradiated environments	10 kGy
		Sporidiobolaceae	*Rhodotorula*	Cold stress; acidic conditions; low water availability	0 °C; 3 pH
		Leucosporidiaceae	*Glaciozyma antarctica*	Cold stress	0 °C
Archaea	Euryarchaeota	Halobacteriaceae	*Haloarcula marismortui*	Saline conditions; environments without molecular oxygen	25% NaCl
			Haloferax volcanii	Saline conditions; environments without molecular oxygen	25% NaCl
			Haloquadratum walsby	Saline conditions	36% NaCl
		Picrophilaceae	*Picrophilus torridus*	Hot stress; acidic conditions; saline conditions	60 °C; 0–6 pH; 20% NaCl
		Thermococcaceae	*Thermococcus gammatolerans*	Hot stress; irradiated environments	95 °C; 30 kGy
			Thermococcus piezophilus	Hot stress; environments characterized by high or low pH values	95 °C; 125 MPa
		Methanopyraceae	*Methanopyrus kandleri*	Hot stress; environments characterized by high or low pH values	122 °C; 125 MPa
		Methanococcaceae	*Methanothermococcus okinawensis*	Hot stress; environments characterized by high or low pH values	85 °C; 5 MPa
		Methanobacteriaceae	*Methanobacter wolfeii*	Hot stress; environments characterized by high or low pH values	80 °C; 122 MPa
	Crenarchaeota	Sulfolobaceae	*Sulfolobus solfataricus*	Hot stress; acidic conditions	92 °C; 1 pH
		Sulfolobaceae	*Acidianus infernus*	Hot stress; acidic conditions	96 °C; 1 pH

(Continued)

Table 2.2 (Continued)

	Phylum	Family	Species (or Genus)	Tolerance	Known maximum limit
Bacteria	Cyanobacteria	Cyanophyceae	*Chroccoccidiopsis*	Low water availability; irradiated environments	113.25 kGy
	Deinococcus-Thermus	Trueperaceae	*Truepera radiovictrix*	Hot stress; irradiated environments	50 °C; 5 kGy
		Deinococcaceae	*Deinococcus radiodurans*	Low water availability; solar and UV radiation; irradiated environments	15 kGy; UV 295 nm
	Bacteroidetes	Salinibacteraceae	*Salinibacter ruber*	Saline conditions	35% NaCl
	Firmicutes	Halanaerobiaceae	*Halarsenatibacter silvermanii*	Saline conditions	33% NaCl
		Planococcaceae	*Planococcus halocryophilus*	Cold stress; saline conditions	15 °C; 19% NaCl
		Bacillaceae	*Oceanobacillus iheyensis*	Hot stress; acidic conditions; environments characterized by high or low pH values	42 °C; 10 pH; 3 MPa
	Proteobacteria	Acidithiobacillaceae	*Acidithiobacillus ferrooxidans*	Acidic conditions	1.3 pH
		Colwelliaceae	*Colwellia piezophila*	Cold stress; environments characterized by high or low pH values	4 °C; 80 MPa
	Thermodesulfobacteria	Thermodesulfobacteriaceae	*Geothermobacterium ferrireducens*	Hot stress	100 °C
Virus	Uroviricota	Myoviridae	*Halovirus*	Cold stress; low water availability	Not determined
		Siphoviridae	Not determined	Cold stress; environments characterized by high or low pH values	Not determined
		Podoviridae	Not determined	Cold stress; environments characterized by high or low pH values	Not determined
	Saleviricota	Pleolipoviridae	Not determined	Low water availability	Not determined
	Dividoviricota	Sphaerolipoviridae	Not determined	Low water availability	Not determined

Source: Data from Coleine and Delgado-Baquerizo (2022).

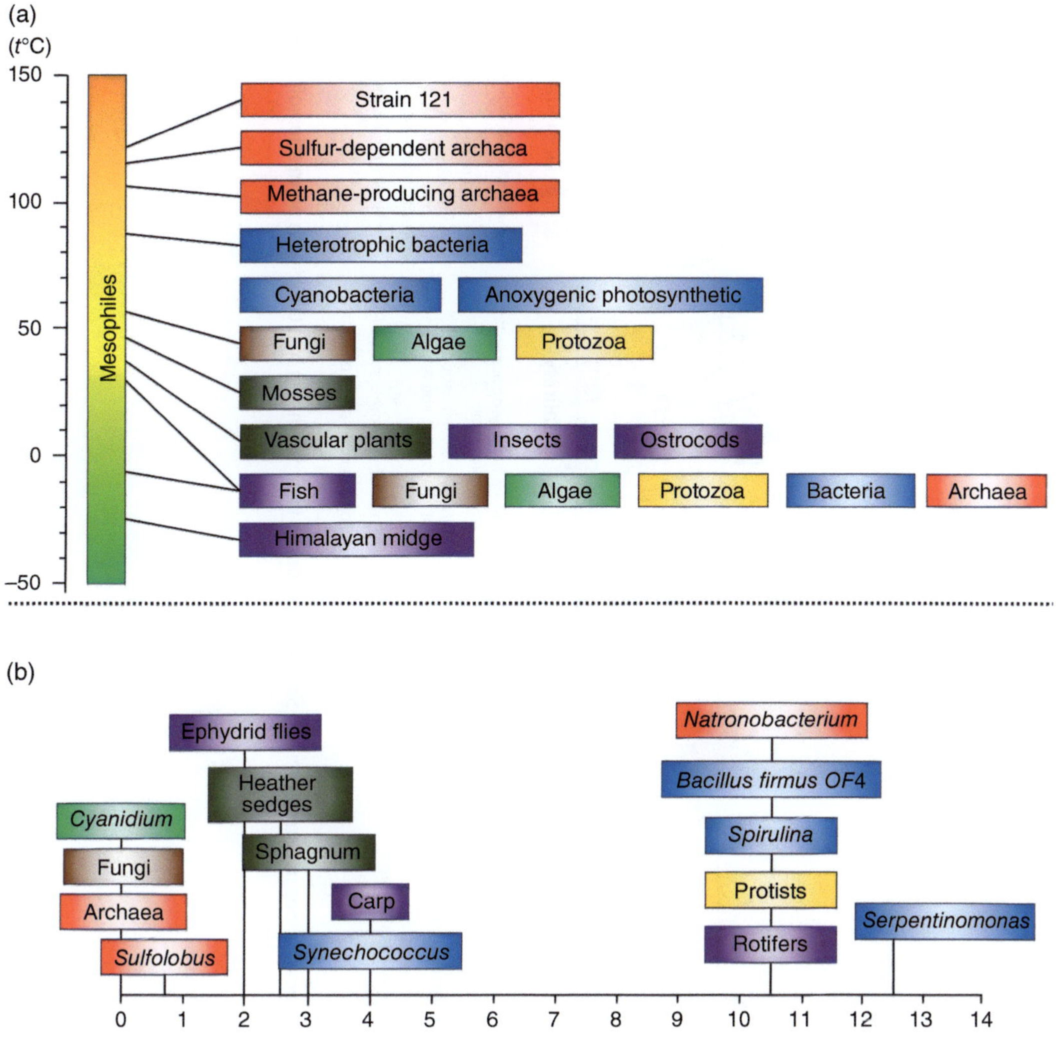

Figure 2.2 Temperature limits (a) and pH limits (b) for life. Archaea are in red, bacteria in blue, algae in light green, assorted protists in yellow, fungi in brown, plants in dark green, and animals in purple. *Source:* Adapted from Rothschild and Mancinelli (2001), Kashefi and Lovley (2003), and Suzuki et al. (2014).

cytoplasmic access to external substrates, improving transport efficiency and environmental responsiveness. Certain cyanobacteria, for instance, adopt filamentous morphologies with increased surface area, optimizing sunlight and carbon dioxide absorption for photosynthesis. This high surface-area-to-volume ratio provides bacteria and archaea with efficient exchange of matter, energy, and signals with their surroundings. It supports rapid substrate uptake, metabolic waste excretion, and cellular homeostasis. Additionally, it enables prokaryotes to swiftly detect and respond to environmental changes—such as fluctuations in temperature, pH, or nutrient availability—by adjusting their physiological activities. In nutrient-limited conditions, some bacteria elongate or form hyphae to maximize surface area, enhancing resource acquisition. Furthermore, a larger surface area aids in resource allocation, accelerating the growth of daughter cells post-division and promoting rapid population expansion.

Thirdly, colonies of prokaryotes, occurring as biofilms and microbial mats in natural ecosystems, are capable of producing extracellular polymeric substances (EPS) and give rise to synergistic interactions. EPS provides protective barriers for cells in extreme environments and enhances the

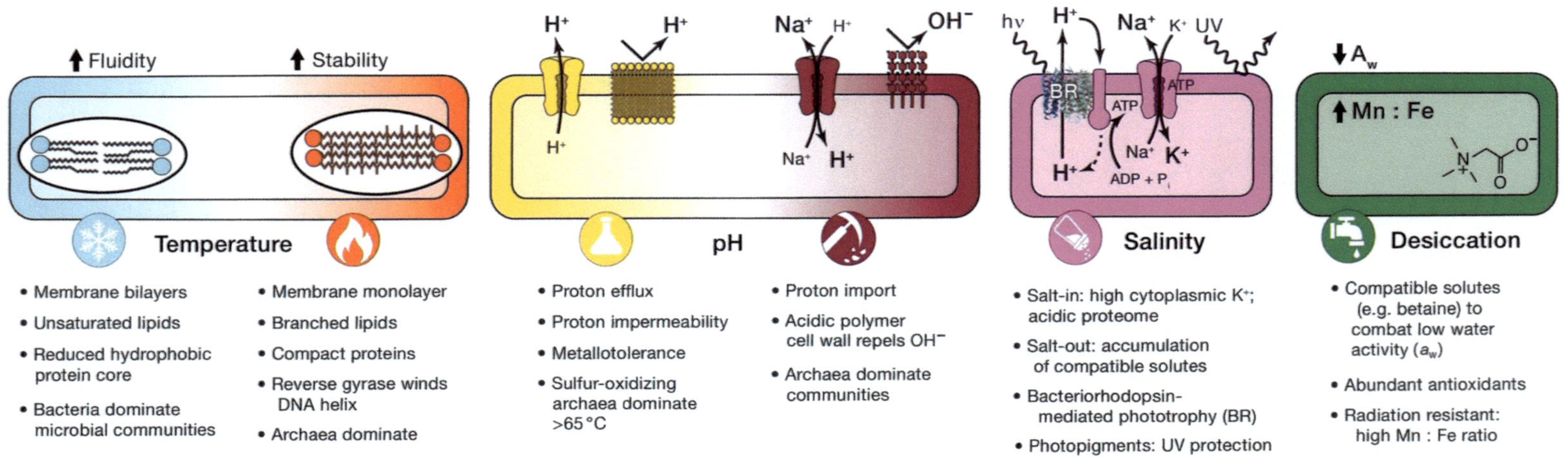

Figure 2.3 Schematic diagram of molecular and cellular adaptations in extremophiles. *Source:* Reproduced from Schmid et al. 2020 / with permission of Elsevier.

adsorption of organic matter, inorganic ions (e.g. metal compounds), and mineral particles from the surroundings, thereby facilitating energy substrate acquisition and electron transfer efficiency. For instance, *Synechocystis* sp. and *Oscillatoria* sp., isolated from marine stromatolites in the Bahamas, are capable of producing EPS (Kawaguchi and Decho 2000); *Oscillatoria* sp., *Leptolyngbya* sp., and *Chroococcus* sp., isolated from a mangrove forest in Lothian Island, can form biofilms during the cultivation process (Veerabadhran et al. 2018). Within microbial mats, cyanobacteria (with EPS secretion ability) can mediate carbonate precipitation in aquatic environments and contribute to microbialite formation through processes of sedimentation and lithification. Furthermore, community regulation (e.g. quorum sensing) and the metabolic collaboration among subgroups can further strengthen the stability of microbial communities, providing advantages such as resource storage, genetic exchange, and resistance to environmental stressors.

Fourthly, prokaryotes generate adaptive potential through HGT and mutation (mutation rates: 10^{-5}–10^{-10} per generation) in response to environmental changes. The adaptive potential facilitates the acquisition of novel genetic functions, phenotypes, and physiological traits critical for environmental adaptation (Averhoff et al. 2021). For example, HGT enables bacteria resistant to certain inhibitory small molecules (e.g. antibiotics) produced by competing microorganisms in the same habitat. Moreover, HGT is considered more prevalent in extreme environments that resemble early Earth conditions, such as hot springs, marine sediments, and other niches characterized by high temperatures, low oxygen, extreme pH, and high salinity.

In addition to their physiological traits, prokaryotes generally have minimal nutrient and energy requirements. However, these factors alone are not the primary drivers of their survival in extreme environments. The key to their persistence lies in their exceptional metabolic diversity and versatility. Bacteria and archaea can exploit a wide array of organic compounds as carbon sources and utilize various inorganic compounds as terminal electron acceptors (Section 1.3). Their ability to process both simple carbon substrates and complex organic matter enables efficient energy generation, supporting their adaptation to diverse and challenging conditions. More importantly, prokaryotes demonstrate remarkable flexibility in alternating between aerobic/anaerobic respiration, autotrophy/heterotrophy, or photosynthesis. Furthermore, autotrophs possess the most varied repertoire of carbon fixation mechanisms, utilizing six known pathways for CO_2 assimilation. Consequently, they display significant metabolic versatility in transitioning between different pathways. For instance, *Shewanella oneidensis*, in addition to aerobic respiration, can perform anaerobic respiration using a wide array of organic and inorganic electron acceptors, including nitrate, nitrite, fumarate, sulfite, thiosulfate, Fe(III), and Cr(VI) (Fredrickson et al. 2008). Purple nonsulfur bacteria (e.g. *Rhodospirillum* and *Rhodopseudomonas*) are metabolically flexible. These organisms can thrive under aerobic, anaerobic, light, or dark conditions. Green nonsulfur bacteria, such as *Chloroflexus*, predominantly employ phototrophy. Under anaerobic conditions with light, they fix CO_2 using H_2S as an electron donor. They can also achieve photoheterotrophic growth using simple organic compounds and be chemoorganotrophic in the absence of light. Recent studies have further revealed that both *Chloroflexus aggregans* and *Chloroflexus aurantiacus* can utilize H_2 as an electron donor for autotrophic growth (Kawai et al. 2019).

Other characteristics of prokaryotic metabolism are metabolic coupling and extremozyme. For example, sulfate reduction is often coupled with the cycles of C, N, P, and Fe in bacterial and archaeal biogeochemical processes. Also, sulfate reduction can couple with anaerobic oxidation of methane (AOM) through syntrophic interactions, forming consortia where sulfate-reducing bacteria (SRB) envelop anaerobic methanotrophic archaea (ANME). The SRB and ANME can be co-localized within the sulfate–methane transition zone. In addition, AOM can be coupled with denitrification processes. ANME-2d (a clade of ANME) is capable of oxidizing methane while reducing nitrate to nitrite and AOM can also be coupled with the reduction of Fe(III) and manganese(IV). Moreover, nitrate reduction can be coupled with iron oxidation and sulfur oxidation processes. For example, *Thauera* sp. AutoDN2, a novel autotrophic denitrifier, is able to couple nitrate reduction with sulfide oxidation while minimizing the sulfate production (Zhang et al. 2025). Extremozymes play a crucial role in extremophile physiology, not only by maintaining structural stability and catalytic activity under extreme conditions but also by reducing the activation energy of biochemical reactions and optimizing substrate conversion efficiency.

2.2.2.2 Extremophilic Eukaryotes

In extreme environments, the abundance and taxonomic diversity of eukaryotic organisms are significantly lower than those of prokaryotes. Eukaryotes exhibit a much broader and complex array of taxa and comprise more intricate evolutionary lineages across protists, fungi, plants, and animals. Generally, macroscopic eukaryotic taxa are sparsely distributed in extreme habitats, and relevant research remains limited. The disadvantages of eukaryotes in extreme environments primarily stem from intrinsic biological limitations, including complexity of structure and gene regulation, metabolic characteristics, process of response to environmental stress, genomic characteristics, evolutionary trajectories, reproductive strategies, and ecological strategies. The complex architecture of eukaryotic cells, characterized by membrane-bound organelles (e.g. nucleus, mitochondria, and endoplasmic reticulum), imposes higher energetic demands for maintenance under extreme conditions. Furthermore, sustaining the stability of these complex structures presents significant challenges to eukaryotes. For instance, sterols contribute to maintaining eukaryotic cell membrane stability, and the membranes become unstable under high temperatures or extreme pH conditions.

The intricate metabolic pathways and regulatory mechanisms of eukaryotes require multiple substrates and coordinated enzyme activities. Under extreme conditions, these processes are frequently disrupted due to substrate scarcity or reduced/inactivated enzyme activity. The functional specialization of eukaryotic organelles (e.g. mitochondria and chloroplasts) restricts metabolic flexibility. Compared to prokaryotes, eukaryotes exhibit limited metabolic versatility, primarily relying on photic or organic energy and aerobic respiration. However, extreme environments frequently lack both light and oxygen, thereby worsening these limitations.

Concurrently, eukaryotic genomes are typically large and complex, containing abundant non-coding regions (ncRNAs) and repetitive sequences. The regulation of eukaryotic gene expression involves multi-level and multi-factorial coordination (Krebs et al. 2017), including transcriptional regulation (e.g. cis-acting elements, trans-acting factors), DNA and histone modifications (e.g. methylation and acetylation), post-transcriptional regulation (e.g. non-coding RNAs and RNA-binding protein), translational regulation (e.g. ribosome binding sites and miRNA), post-translational modifications (e.g. ubiquitination), and signaling pathways (e.g. hormones and receptors). These complex regulatory networks result in delayed adaptive responses to environmental challenges compared to the adaptive strengths and strategies of prokaryotes. In addition, eukaryotes primarily depend on genetic inheritance, with relatively low frequencies of HGT. Moreover, their more complex cell cycle (e.g. mitosis and meiosis) coupled with sexual reproduction restricts population expansion and genetic diversity under extreme conditions.

Prokaryotes established ecological dominance in extreme environments on Earth during the Archean Eon, whereas eukaryotes diversified approximately 2.0 billion years later (Section 8.5.3). This significantly shorter evolutionary timeline may have limited adaptive radiation and insufficient development of specialized survival mechanisms. Consequently, many eukaryotic lineages lack effective adaptive strategies for long-term persistence in extreme conditions. While eukaryotes may have ecological advantages in stable ecosystems, these advantages can be lost in extreme habitats. The survival of eukaryotes in extreme environments frequently depends on stable symbiotic associations, such as free-living amoebae and ciliates are known to host diverse assortments of bacterial symbionts (Gomaa et al. 2022). However, the establishment and maintenance of these symbiotic systems are highly influenced by environmental fluctuations. Furthermore, it is difficult for eukaryotes to challenge the stable ecological networks constructed by prokaryotes in extreme habitats. For example, in saline microbial mats, prokaryotes thrive through photosynthesis and chemolithoautotrophy, while eukaryotes (e.g. diatoms) frequently persist as minor contributors in communities (Saghaï et al. 2017).

Although eukaryotes possess several physiological constraints in responding to harsh conditions, this does not preclude their adaptive strategies to thrive in extreme environments (Figure 2.3). Indeed, eukaryotic microorganisms such as protists and fungi demonstrate strong adaptive capacity and widespread distribution across multiple extreme habitats, including low temperature, hypersalinity, and extreme pH (Coleine et al. 2022; Rappaport and Oliverio 2023). Additionally, eukaryotes inhabit high-temperature environments, heavy metal-contaminated areas, and display tolerance to desiccation and radiation. Algae are capable of thriving in polar environments, with diatoms and chlorophytes additionally adapting to hypersaline and extreme pH conditions. Furthermore, certain plants and animals also inhabit harsh environments (e.g. polar environments). For instance, mosses, Antarctic hair grass, small multicellular animals (including rotifers and Antarctic nematodes), and arthropods (e.g. *Belgica antarctica* and *Polypedilum vanderplanki*) demonstrate resilience in polar ecosystems (Kozeretska et al. 2022; Shaikhutdinov and Gusev 2022). Krill and fish species (e.g. icefish and polar cod) also exhibit physiological adaptations enabling survival in icy aquatic environments (Clark et al. 2023).

2.2.2.3 Extremophilic Viruses

Viruses, in theory, have the potential to infect all life forms, including extremophiles. While extensive research has focused on viruses infecting non-extremophilic hosts, particularly disease-causing pathogens, studies on viruses associated with extremophilic hosts remain relatively scarce. However, recent advancements in metagenomic sequencing and meta-omics approaches have greatly expanded our understanding of virus-prokaryote interactions in extreme environments.

Recent studies highlight the remarkable diversity and ecological roles of viruses in extreme environments. For instance, in the Alpine lake Nam Co. on the Tibetan Plateau—an environment characterized by geographic isolation, extreme temperatures, nutrient scarcity, and intense UV exposure—scientists identified 742 RNA virus species, including numerous novel genus- and family-level groups (Wu et al. 2025). Research on viral diversity and auxiliary metabolic genes (AMGs) in rare earth elements (REEs) mine drainage has shown that viruses encode a higher number of carbon- and sulfur-related AMGs, suggesting their potential influence on microbial carbon and sulfur cycles under REE stress conditions (Hu et al. 2025). Further studies have explored viral diversity and virus-host interactions in various extreme habitats, such as mountain and polar supraglacial zones, deep-sea brine pools, acid mine drainage sites, Earth's subsurface, and Atacama Desert endolithic communities (Busse et al. 2022; Gao et al. 2022; Liu et al. 2023; Minch et al. 2024; Rahlff et al. 2021). A prevailing perspective suggests that viruses play a crucial role in shaping microbial community structures and regulating metabolic processes, thereby contributing to biogeochemical cycles in these extreme environments.

Viruses employ a range of adaptive strategies to thrive in extreme ecological niches (Figure 2.4). These adaptations include specialized morphological features, genomic structures, and protein characteristics, such as enveloped capsids, glycosylated major capsid proteins, enhanced adsorption rates, and optimized virion aspect ratios. Together, these traits improve viral survival and facilitate efficient replication, even in environments with low host density and harsh conditions.

Many archaeal viruses exhibit unique structural features and dynamic adaptations that enhance their replication and survival in extreme environments. These viruses facilitate biophysical adaptations, primarily in thermophilic (>80 °C) or halophilic (>2.0 M) conditions, with many thermophiles also exhibiting acidophilic traits (pH < 4) (Hartman et al. 2019). Due to low host cell

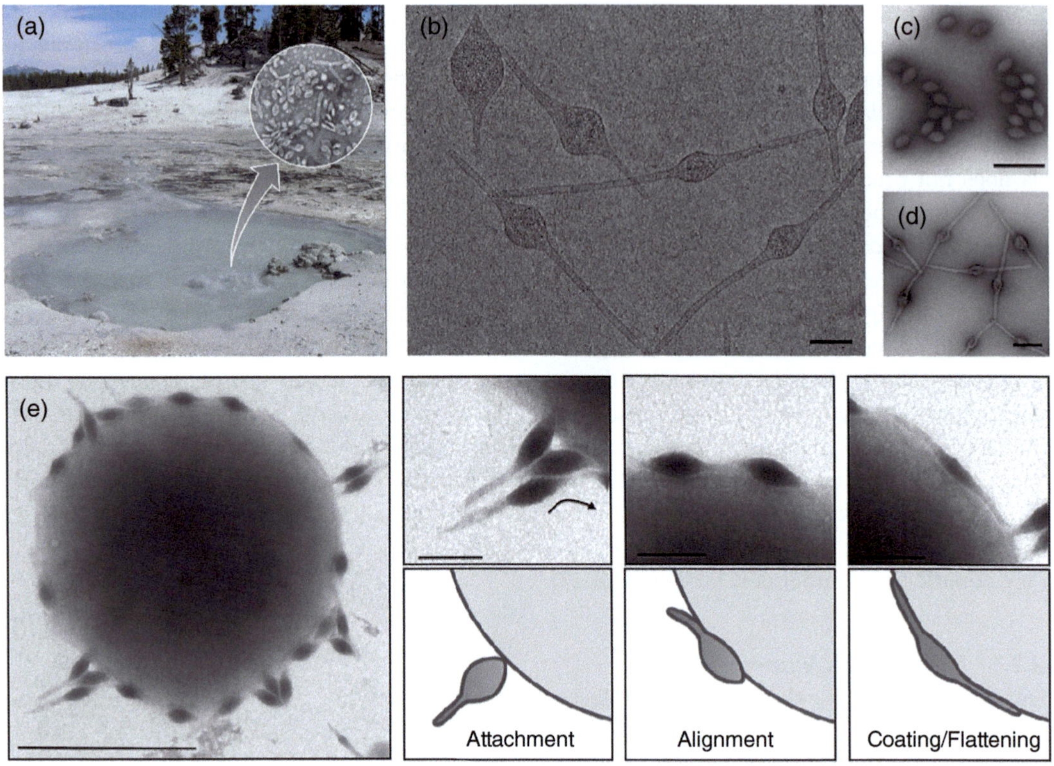

Figure 2.4 Archaeal viruses thrive in extreme acidic environments and electron micrographs of ATV and SMV1 virions. (a) Aarchaeal viruses persist in an acidic Yellowstone hot spring (~pH 2) dominated by hyperthermophilic and acidophilic Crenarchaeota. Inset: Transmission electron micrograph (TEM) of a mixture of archaeal viruses from the boiling mudpot. (b) Cryoelectron microscopy of ATV virions at different stages of tail development, scale bar, 100 nm. (c) TEM of negatively stained tailless ATV virions, scale bar, 100 nm. (d) TEM of negatively stained two tailed ATV virions, scale bar, 200 nm. (e) TEM of SMV1 interaction with host cells, samples were observed after 10 minutes postinfection and negatively stained, left scale bar represents 1 μm, right scale bar represents 200 nm, the three different virus-host association modes are indicated: attachment, alignment, and coating/flattening. *Source:* (a) Madigan et al. (2018) / Pearson Education; (b–d) Prangishvili et al. (2006) / with permission of Elsevier; (e) Uldahl et al. (2016) / American Society For Microbiology.

density and harsh environmental pressures, archaeal viruses face strong selective forces driving rapid host binding and infection. To overcome these challenges, they employ strategies such as increasing adsorption rates and optimizing virion aspect ratios. High-aspect-ratio particles, which exhibit tumbling motion, maximize effective surface area and binding probability.

For instance, in acidic hot springs, virus-to-cell ratios are up to a thousand times lower than those in marine environments. The thermo-acidophilic *Sulfolobus islandicus* rod-shaped virus 2 (SIRV2) demonstrates exceptionally high adsorption rates—ranging from 1×10^3 to 1×10^5 times faster than halophilic archaeal viruses and even 10 times faster than bacteriophages (Hartman et al. 2019). Similarly, *Sulfolobus monocaudavirus* 1 (SMV1) achieves efficient host attachment through three distinct modes: nosecone association, body association, and body/tail association (Uldahl et al. 2016). High-aspect-ratio virions enhance binding probability with host cells while expanding effective surface area.

The Acidianus two-tailed virus (ATV), initially discovered in acidic hot springs (pH 1.5, 85–93 °C) in Italy, is released as a lemon-shaped virion from the host. After host lysis, ATV develops elongated tails at both ends, reaching a length comparable to the capsid's minimum size. Unlike most archaeal viruses in acidic springs that rely on lytic cycles, ATV exhibits lysogenic behavior, which is rare in this group. Notably, tail elongation, coupled with virion thinning and volume reduction, occurs exclusively at high temperatures (>75 °C). Few reports describe archaeal viruses like ATV and SMV1, where viral development occurs independently of host interaction (Prangishvili et al. 2006). The extracellular tail formation of ATV may serve as an adaptive strategy to cope with low host cell density and extreme environmental conditions. Thus, viral adaptations—particularly in virion morphology and dynamics—must be understood within the context of these extreme habitats.

2.3 Thermophiles

Thermophiles are organisms that thrive and reproduce within a temperature range of 60–80 °C, classified according to their optimal growth conditions. In contrast, hyperthermophiles flourish in environments where temperatures exceed 80 °C (Coleine and Delgado-Baquerizo 2022). Geothermal ecosystems, spread across the planet, provide rich habitats for thermophilic organisms, with deep-sea hydrothermal vents and terrestrial geothermal springs being the most prominent. Certain hyperthermophiles depend on high-pressure conditions for growth, as elevated pressure maintains water in its liquid phase at extreme temperatures, pushing the theoretical upper limit to 407 °C under 29.8 MPa (Merino et al. 2019).

The archaeon *Geogemma barossii* strain 121 is capable of cell division at 121 °C in laboratory culture, while *Methanopyrus kandleri* strain 116 is able to grow at 122 °C under 20 MPa of pressure. *Geovibrio ferrireducens* can grow at 100 °C, representing the highest temperature limit for bacteria (Cario et al. 2015; Kashefi and Lovley 2003; Takai et al. 2008). In contrast, eukaryotic cellular activity generally diminishes beyond 60 °C (Figure 2.2), though some protozoa, algae, and fungi may tolerate such conditions (Salinas-García et al. 2025). The upper temperature threshold for mosses is approximately 50 °C, while vascular plants can withstand up to 48 °C, and fish around 40 °C (Figure 2.2), likely due to the reduced solubility of oxygen at higher temperatures. Current environmental and theoretical estimates suggest that the absolute temperature boundary for life is near 150 °C, primarily constrained by the thermal instability of macromolecules at extreme heat (Merino et al. 2019).

2.3.1 Phylogenic Diversity

Thermophiles predominantly belong to the domains Bacteria and Archaea. Recent investigations on geothermal ecosystems demonstrate that elevated temperatures play a crucial role in shaping thermophilic community structure (Merino et al. 2019; Shu and Huang 2022).

Thermophilic bacteria have been found in a variety of geothermal environments, displaying remarkable adaptability to temperatures exceeding 60 °C. They exhibit significant diversity in both phylogeny and metabolism, encompassing autotrophs, phototrophs, heterotrophs, and chemolithotrophs. Among phototrophic bacteria, thermophiles can be found within cyanobacteria, as well as purple and green bacteria. Additionally, various heterotrophic bacteria, including *Bacillus, Clostridium, Thiobacillus, Desulfotomaculum, Thermus*, lactic acid bacteria, actinomycetes, spirochetes, and numerous other genera, contribute to the extensive diversity of thermophilic communities.

More than 146 hyperthermophiles in the temperature range from 80 to 121 °C have been identified (DiGiacomo et al. 2022). These heat-resistant hyperthermophiles are all anaerobic archaea, including species of the genera *Pyrolobus, Pyrodictium*, and *Pyrococcus*. For example, *Pyrolobus fumarii* of the Crenarchaeota, a nitrate-reducing chemolithotroph, grows on the walls of marine hydrothermal vents. It multiplies best at about 105 °C, can reproduce at up to 113 °C, and stops growing in "cooler" environments below 90 °C (Blöchl et al. 1997). Certain hyperthermophile that lives in deep-sea vents, *Methanopyrus*, is of special interest because of its ancient genetic makeup. Analysis of its genes suggests that it may have been among the earliest organisms on Earth.

Geogemma barossii strain 121, isolated from hydrothermal vents in the Pacific Ocean, represents the most thermotolerant organism discovered to date. It is an archaeon coccoid, ~1.0 μm in diameter, with a lophotrichous-like flagellation (Kashefi and Lovley 2003). Strain 121 grows at temperatures ranging from 85 to 121 °C. Notably, it takes approximately 24 hours for strain 121 to multiply at 121 °C (resistant to autoclaving) and the cell division ceases if the temperature drops below 85 °C. Remarkably, strain 121 can even subsist after transient exposure to 130 °C for two hours, and the culture will be resilient when transferred to fresh medium of 103 °C (Kashefi and Lovley 2003). Its unique thermotolerant mechanism may provide new insights into biology and astrobiology.

An exceptional example of eukaryotes with extraordinary resilience is tardigrades, also known as water bear or moss piglet. Closely related to arthropods, tardigrades exhibit remarkable survival capabilities, particularly in their dehydrated state, allowing them to endure extreme physical conditions that would typically be fatal for most life forms, including high temperatures. *Milnesium tardigradum*, for instance, demonstrated over 90% survival after exposure to 100 °C for one hour (Schill and Hengherr 2018). Furthermore, in its cryptobiotic state, a tardigrade can withstand temperatures as high as 151 °C for a few minutes (Glime 2013).

2.3.2 Adaptive Strategies

Thermophilic organisms inhabiting geothermal environments must overcome multiple challenges imposed by thermal stress. Elevated temperatures increase the fluidity of the cytoplasmic membrane, denature proteins, compromise the structural integrity of DNA, and destabilize essential metabolites. To mitigate these challenges, thermophiles and hyperthermophiles have employed multiple survival strategies for the maintenance of cellular homeostasis, normal metabolism, and protein function.

In general, at high temperatures, both the fluidity and permeability of bacterial cell membranes increase rapidly, leading to disordered lipid packing essential for maintaining the liquid crystalline state. As temperatures exceed the optimal range, membrane phospholipids transition from a crystalline liquid phase to a fluid state and subsequently become a nonlamellar form. Thermophilic bacteria maintain membrane stability by increasing the composition ratio of saturated fatty acids, long-chain fatty acids, and branched-chain fatty acids. In addition, thermophilic bacteria with optimal growth above 70 °C generally utilize phospholipids for membrane construction that are analogues of tetraethers and diethers, which are components of archaeal membranes (Pati et al. 2023).

Among thermophilic archaea, tetraether lipids are abundant and predominantly constitute the core membrane lipids. These lipids form exceptionally thermostable monolayer membranes due to restricted hydrocarbon chain mobility. Furthermore, the inclusion of pentacyclic compounds enhances the stability of tetraether-containing membranes by elevating the phase transition temperature. For instance, *Thermococci* exhibit an increasing tetraether/diether lipid ratio with rising temperatures (Pati et al. 2023).

A key thermophilic adaptation of an organism is the presence of heat-stable proteins. These proteins are more densely packed to exclude internal water, more hydrophobic, and have more salt bridges. The thermostable proteins and high activity extremozymes collectively enable thermophiles to exhibit adaptive capacity under thermal stress conditions. For instance, in *Pyrobaculum aerophilum*, over 47% of cysteine residues form disulfide bridges under optimal growth conditions (high temperature). This finding is consistent with the process that thermophilic protein complexes tend to form compact structures, which enhance their stability at high temperatures (Boutz et al. 2007; Pati et al. 2023). Additionally, a critical class of HSP family plays essential roles in adaptation to multiple adverse conditions, such as high temperature, freezing stress, salinity, and radiation. High-molecular-weight HSPs (60–100 KDa) and small HSPs are induced by heat-stress transcription factors (HSFs). HSPs assist in proper protein folding, maintaining protein stability and activity, and can also bind to DNA or RNA to stabilize nucleic acid integrity (Ezemaduka et al. 2017).

DNA exhibits instability under elevated temperatures. Thermophiles have evolved stringent DNA repair mechanisms to maintain genomic stability as a high-temperature adaptation strategy (Wang et al. 2015). For instance, reverse gyrase serves as a thermophilic adaptation trait potentially transferred from thermophilic archaea to bacteria via HGT. This enzyme, classified as a DNA topoisomerase, induces positive supercoiling in double-stranded DNA. Genomic deletion of reverse gyrase coding gene in *Thermococcus kodakarensis* resulted in significantly retarded growth under elevated temperatures, indicating its important role in thermophilic adaptation (Perugino et al. 2009).

The thermophilic chromosome is exposed to elevated levels of damage, including depurination, oxidative stress-induced lesions, deamination, and single- and double-strand breaks. All these damages are exacerbated with increasing temperatures. For example, thermophiles contain polyamines that exhibit enhanced efficacy in stabilizing double-stranded DNA while branched polyamines demonstrate stronger stabilization effects on single-stranded DNA and tRNA (Terui et al. 2005).

2.4 Psychrophiles

Psychrophiles, also known as cryophiles, are organisms that thrive and reproduce in cold environments, typically within a temperature range of −10–20 °C (14–68 °F). They can survive at or below 0 °C, with optimal growth occurring at temperatures of ≤15 °C and a maximum limit of ≤20 °C. In contrast, psychrotolerant organisms are mesophiles that can endure low temperatures but exhibit slower growth. These organisms can grow near 0 °C and exhibit optimal growth with temperatures exceeding 15 °C and a maximum threshold above 20 °C (Coleine and Delgado-Baquerizo. 2022). Beyond their ability to withstand cold, psychrophiles often inhabit extreme environments characterized by factors such as ultraviolet radiation, high osmotic pressure, nutrient scarcity, and low water availability.

Cold habitats include glaciers, ice sheets, snow covers, permafrost, and artificial cryogenic environments. The polar ecosystem serves as one of the quintessential examples. Psychrophiles have been found to thrive in the Arctic and Antarctic oceans, which remain frozen for much of the year. They have also been recovered from core samples taken from the Greenland ice sheet at depths of about 3 km. Many microbes can be preserved successfully at −196 °C in liquid nitrogen, but the lowest recorded temperature for active microbial communities and animals is much higher at −18 °C. Freezing water within a cell is almost invariably lethal. One of the few exceptions to this rule known from nature is the nematode *Panagrolaimus davidi*, which can withstand freezing of all of its body water (Wharton 2003).

2.4.1 Phylogenic Diversity

Among prokaryotes, the bacterium *Planococcus halocryophilus* Or1 exhibits growth and cell division at temperatures as low as −15 °C, while maintaining metabolic activity down to at least −25 °C (Mykytczuk et al. 2013). Additionally, the bacterium *Deinococcus geothermalis* DSM 11300 also demonstrates survivability at −25 °C (Merino et al. 2019). Within the temperature range of −26 to −10 °C, microbial cells are likely to undergo vitrification (without intracellular ice formation), which facilitates low-temperature survival (Clarke et al. 2013). Currently, the coldest temperature at which microbial metabolic activity has been documented is −20 °C, which is observed in enrichment cultures derived from Siberian permafrost soils (Rivkina et al. 2000). The extreme lower limit for microbial life survival currently stands at −25 °C (Tmin, *D. geothermalis* DSM 11300) (Frösler et al. 2017). In deep-ocean bacterial assemblages, Proteobacteria (including Alphaproteobacteria, Gammaproteobacteria, and Deltaproteobacteria), Chloroflexi, and Actinobacteria typically dominate the bacterial communities, while archaeal communities are predominantly composed of Marine Group I Crenarchaeota (Shu and Huang 2022).

In extremely cold environments (the cryosphere), protists exhibit remarkable adaptability (Figure 2.5). The polar diatom *Fragilariopsis cylindrus* demonstrates growth capability at temperatures as low as −20 °C (Bayer-Giraldi et al. 2010). Within fungi, *Cladosporium herbarum* can withstand temperatures of −8 °C (Coleine et al. 2022). Algae also exhibit significant psychrophilic adaptations, for instance, *Chlamydomonas nivalis* produces red pigments and maintains photosynthetic activity in snow environments down to −3 °C (Rappaport and Oliverio 2023). Other psychrophilic algal species, including *Chloromonas tughillensis* (optimal growth at 2.5–5 °C with a minimum of 2.5 °C), *Coccomyxa subellipsoidea* (minimum growth temperature: 0 °C), and *Hemichloris antarctica*, have been successfully cultivated from psychrophilic environments (Hoham et al. 2008; Kania et al. 2022). Algae dominate as primary phototrophs in certain extreme cold ecosystems: diatoms are predominant in sea ice communities, while green algae of the genera *Chloromonas* and *Chlamydomonas* dominate alpine snowfields (Morgan-Kiss et al. 2006; Zhang et al. 2020).

2.4.2 Adaptive Strategies

Perpetual cold conditions are able to cause physical and physiological damage, including enzymes inactivation and disruption of cellular integrity (e.g. piercing of cell membranes by ice crystals). In order to survive and thrive in low-temperature environments, psychrophiles employ multiple mechanisms to solve the chronically cold challenges. The adaptations include maintaining membrane fluidity, keeping enzyme activation, continuing transcription and translation machinery (Sharma et al. 2022).

Low temperatures reduce membrane fluidity. Psychrophiles maintain membrane fluidity by increasing the ratio of polyunsaturated fatty acids to saturated fatty acids. Polyunsaturated fatty acids contain multiple double bonds, making them more flexible than saturated fats. Psychrophiles incorporate a higher proportion of PUFAs into their membranes, preventing them from becoming rigid in cold temperatures. This flexibility ensures proper membrane function, allowing essential biological processes like nutrient transport and enzyme activity to continue even in icy environments. In Antarctic strains of *Pseudomonas syringae* and *Pseudomonas extremaustralis*, elevated levels of hydroxy fatty acids were observed, accompanied by modifications in the fluidity and composition of lipopolysaccharides (Benforte et al. 2018).

Psychrophiles employ multiple survival strategies against freezing, including the accumulation of compatible solutes (e.g. glycine, betaine, sucrose, and mannitol) to depress the intracellular freezing point and stabilize cellular structures. A study on *Pseudoalteromonas haloplanktis* revealed that the uptake of compounds conferring low-temperature tolerance, such as spermine, glutathione, and ornithine, was enhanced under cold stress (Mocali et al. 2017). In addition, psychrophiles produce EPS to protect cells under extreme conditions. In polar marine ecosystems, sea-ice diatoms are known to secrete large amounts of soluble organic matter, with EPS as the main form (Thomas and Dieckmann 2002).

Antifreeze proteins (AFPs) exhibit the capacity to reduce the size and modulate ice crystals. These proteins are localized at the solid–liquid interface, thereby preventing ice crystal growth. Notably, ice recrystallization inhibition occurs at low AFP concentrations, which is critical for minimizing

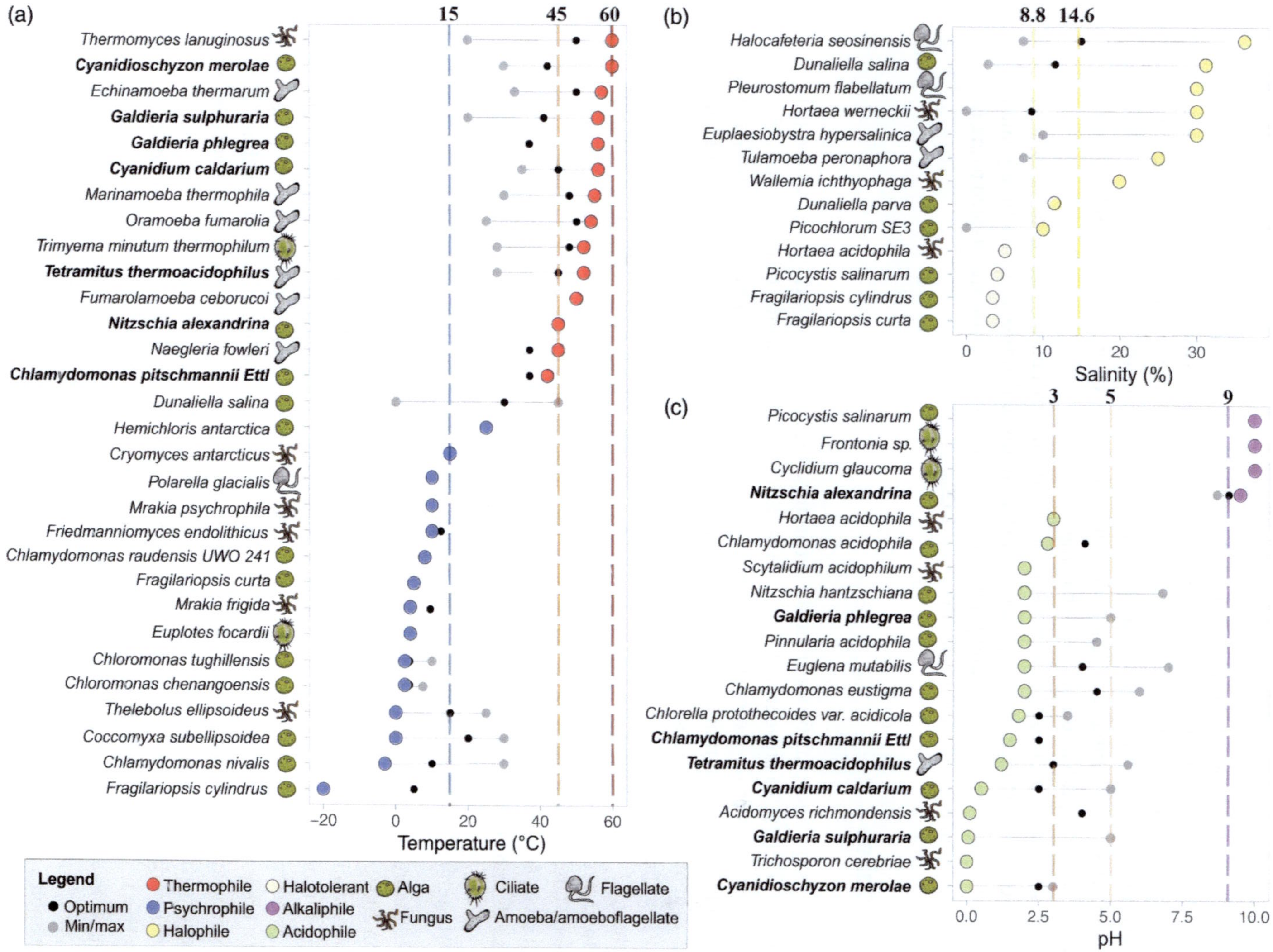

Figure 2.5 Temperature, salinity, and pH limits tolerated by isolated protists. Dumbbell charts highlight the average of optimal ranges when available, or sampling conditions when culturing conditions were not available, and growth ranges of representative isolates by temperature (a), salinity (b), and pH (c). The extreme maximum/minimum condition values are colored by extremophile type. Isolates are also labeled by morphological form, including amoeba, alga, ciliate, flagellate, amoeboflagellate, and fungi for comparison. Algae include diatoms, fungi include yeasts, and flagellates include dinoflagellates; isolate names in bold are polyextremophiles. *Source:* Rappaport and Oliverio (2023) / Springer Nature / CC BY 4.0.

cellular damage during freezing–thawing cycles. Binding sites for AFPs present as complementary structures present on ice crystals and result in the formation of thermal hysteresis, thus increasing the survival of psychrophiles in low-temperature environments (Jia and Davies 2002). For example, antifreeze and ice nucleating activities of AFPs have been identified in Antarctica Lake bacteria *Marinomonas primoryensis* (Antarctica inhabit) and Arctic plant growth-promoting rhizobacterium *Pseudomonas putida* GR12–2 (Gilbert et al. 2004; Muryoi et al. 2004).

Cold shock proteins (CSPs) mediate nucleic acid protection via binding with binding motifs RNP1 and RNP2. When temperature decreases, CSPs are activated for cold stress tolerance in bacteria. For example, CSPs reestablish the membrane fluidity by incorporation of unsaturated fatty acids into their membranes, along with restoring of ribosome function and inducing proper protein folding (Sharma et al. 2022). Jung et al. (2010) demonstrated CSP expression has been induced in *Escherichia coli* under repeated freezing and thawing.

Psychrophiles produce specialized psychrophilic enzymes that exhibit high catalytic activity at low temperatures. For instance, key enzymes in transcription and translation machinery such as elongation factor, RNA polymerase, and peptidyl prolyl cis-trans isomerase are adapted to maintain optimal activity under cold stress (Sharma et al. 2022).

2.5 Acidophiles

Acidophiles are organisms that thrive in acidic environments, such as sulfuric pools, where pH values range from 1 to 5. To prevent high intracellular acidity from damaging essential molecules like DNA, these organisms have evolved mechanisms to continuously pump hydrogen ions out of their cells at a high rate, maintaining a mildly internal pH of approximately 6.5 compared to an external pH of about 2. Extreme acidic environments, both naturally occurring and anthropogenic, are globally distributed. Typical habitats for acidophiles include acidic hot springs, acid lakes, deep-sea hydrothermal vents, acid rock or mine drainage, and industrial acidic wastewater. Notable examples are the acidic hot springs in Yellowstone National Park and the Iron Mountain AMD site in the United States. Sulfide oxidation in these environments is closely linked to acidification, with microbial activity further lowering pH through metabolic products such as sulfate and protons as byproducts (Merino et al. 2019).

2.5.1 Phylogenic Diversity

Acidophiles are predominantly distributed across the Bacteria and Archaea domains (Baker-Austin and Dopson 2007). For example, archaea have been found flourishing under extreme acidity. The aerobic heterotrophs *Picrophilus oshimae* and *Picrophilus torridus* were isolated from Japanese soils permeated with solfataric gases and had optimal growth at pH = 0.7 and 60 °C. *Ferroplasma acidarmanus* has been found growing at pH 0 in acid mine drainage in Iron Mountain in California, thriving in a brew of sulfuric acid and high levels of copper, arsenic, cadmium, and zinc with only a cell membrane and no cell wall. Thermoacidophiles within the genus *Sulfolobus* are also notable for their genetic tractability in research of adaptation mechanisms (Merino et al. 2019).

Among eukaryotes, acidophilic fungi exhibit growth optima below pH = 3, with some species tolerating pH = 1. *Acontium cylindatum*, *Cephalosporium* sp., and *Trichosporon cerebriae* grow near pH 0, while *Neohortaea acidophila* and *Acidomyces acidophilus* tolerate pH 0.6. However, no obligate acidophilic fungus has been described to date (Coleine et al. 2022; Gostinčar et al. 2023).

Protists exhibit a pH tolerance similar to that of bacteria, archaea, and fungi. Cyanobacteria, which cannot survive at pH levels below 4, create conditions where algae become dominant primary producers in highly acidic environments (Weber et al. 2007). Acidophilic protists thrive within a pH range of 4.5 to 2, with the red alga *Cyanidioschyzon merolae* capable of growing at a pH of 0.

Meanwhile, the unicellular red alga *Cyanidium caldarium* flourishes at pH levels below 1 (Rappaport and Oliverio 2023). Beyond red algae, other protists adapt to acidic hot environments. The green alga *Chlamydomonas pitschmannii* inhabits high-temperature, low-pH geothermal spring soils (Pollio et al. 2005), while *Dunaliella acidophila* and the ameboflagellate *Tetramitus thermacidophilus* inhabit high-temperature and acidic geothermal springs and fumaroles (Baumgartner et al. 2009).

2.5.2 Adaptive Strategies

The pH is a crucial parameter in regulating the metabolism of organisms across various scales, acting as the key factor in determining community structure. To sustain cellular functions and metabolism, all organisms must regulate their cytoplasmic pH to remain near-neutral (Merino et al. 2019; Shu and Huang 2022).

Acidophiles employ diverse strategies for pH homeostasis in acidic conditions. Firstly, the protons are blocked by the highly impermeable cell membrane and chemiosmotic gradient created by a Donnan potential. Excess protons are expelled via organic acid degradation or predominance of secondary transporters. Secondly, intracellular pH is buffered by cytoplasmic molecules (e.g. lysine, histidine, arginine), while enzymes are stabilized by "iron rivets" and molecular chaperones repair DNA and protein damage caused by low pH (Baker-Austin and Dopson 2007; Merino et al. 2019).

Proton influx is critical for survival, as excessive intracellular protons disrupt protein and nucleic acid functions, impairing DNA transcription, protein synthesis, and enzyme activity. The archaeal-specific structures composed of tetraether lipids (e.g. *Thermoplasma acidophilum*, *Ferroplasma acidiphilum*, *Sulfolobus solfataricus, and Picrophilus. oshimae*) exhibit enhanced acid resistance due to ether linkages less susceptible to hydrolysis than bacterial ester linkages (Baker-Austin & Dopson 2007; Lloyd et al. 2022).

Membrane channel pore sizes and permeability are also regulated for pH control. For example, *Acidithiobacillus ferrooxidans* upregulates outer membrane porin Omp40 under acidic stress (from pH 3.5 to 1.5), with a large L3 loop controlling the size of the entrance to the pore and ion selectivity at the porin entrance (Baker-Austin and Dopson 2007).

2.6 Alkaliphiles

Alkaliphiles are organisms that prefer highly alkaline environments, such as soda lakes and carbonate-rich soils, with pH values typically ranging from 9 to 11 (Coleine and Delgado-Baquerizo 2022), which are further classified into alkaliphiles (pH > 9) and hyperalkaliphiles (pH > 11). Obligate alkaliphiles cannot grow at pH ≤ 8, whereas facultative alkaliphiles retain this capability. These organisms are predominantly distributed in soda lakes, marine environments, and alkaline wastewater. *Bacillus halodurans* C-125 and *Bacillus pseudofirmus* OF4 serve as typical organisms for alkaliphilic research. Recently, highly alkaliphilic hydrogen-utilizing Betaproteobacteria were isolated from continental serpentinizing sites, leading to the proposal of the novel genus *Serpentinomonas* (Suzuki et al. 2014).

2.6.1 Phylogenic Diversity

Prokaryotes (e.g. the cyanobacterium *Spirulina*) and eukaryotes across diverse taxa, including many species of heterotrophic protists, and some species of metazoan rotifers, exhibit tolerance to pH levels up to $\sim$11.

Among prokaryotes, *Bacillus* species are extensively studied for their adaptations to extreme environments. *B. halodurans* C-125 demonstrates optimal growth at pH of 10–10.5, while *B.*

pseudofirmus OF4 exhibits a broader pH growth range between 7.5 and 10.5 (Hicks et al. 2019; Kim et al. 2023). Notably, *Serpentinomonas* sp. B1, isolated from a terrestrial serpentinizing system, holds the record for the highest documented pH tolerance up to 12.5 and its optimal growth condition is at pH of 11 (Merino et al. 2019). Three *Serpentinomonas* strains (including *Serpentinomonas* sp. B1) are obligate alkaliphiles with optimal growth at pH 11 and utilize hydrogen, calcium carbonate, and oxygen for autotrophic growth (Suzuki et al. 2014).

Gram-negative alkaliphilic bacteria are less characterized than Gram-positive bacteria. For instance, *Pseudomonas alcaliphila* AL15-21^{T} exhibits rough cell surfaces at pH 10 but smooth surfaces at pH 7. Phylogenetic analyses based on 16S rRNA genes reveal clustering of *Pseudomonas mendocina* and *Pseudomonas oleovorans* with alkaliphilic species like *Pseudomonas pseudoalcaligenes*, *P. alcaliphila*, and *Pseudomonas toyotomiensis*, both of *P. mendocina* and *P. oleovorans* are able to grow at pH of 10 (Matsuno and Yumoto 2015). Additionally, the Actinobacteria *Kocuria rosea* strain MG2 demonstrates a wide pH growth range (5–11), with optimal growth at pH of 9.2 (Gholami et al. 2015).

Among eukaryotes, alkalitolerant and alkaliphilic fungi thrive above pH = 8, with optimal growth at pH of 8–9. Species such as *Aspergillus glaucus*, *Emericellopsis alkalina*, *Exophiala dermatitidis*, and *Sodiomyces* spp. have been reported to grow above pH = 11 (Gostinčar et al. 2023). Similarly, *Acremonium* and *Sodiomyces* species isolated from soil and water samples exhibit tolerance to pH 10–11, with *Sodiomyces* spp. strictly restricted to habitats above pH 9 due to their obligate alkaliphilicity (Fernández-López et al. 2023).

Alkaliphilic protists also inhabit extreme pH environments, with documented survival up to pH = 10.48. Alkaliphilic protists are primarily represented by ciliates, though eukaryotic phytoplankton, including *Cyclidium glaucoma* as the most abundant ciliate recovered in alkaline-saline lakes of Kenya and *Frontonia* spp., have been identified in soda lakes in Kenya and Turkey (Rappaport and Oliverio 2023).

2.6.2 Adaptive Strategies

To maintain cytoplasmic pH homeostasis (~pH 8) in much higher pH surroundings, alkaliphiles continuously pump hydrogen ions across their cell membranes into their cytoplasm. However, ATP synthesis becomes energetically challenging under high pH.

Obligate alkaliphilic *Bacillaceae* utilize Na^{+}/H^{+} antiporters and ATPase-driven proton pump to generate ATP in an H^{+} diluted environment. The bacteria retain scarce H^{+} extruded by the respiratory chain on the outer surface of the membrane and increase the potential of the H^{+} for ATP production on the outer surface of the membrane (via specific mechanisms of ATP production). The Mrp (Sha) Na^{+}/H^{+} antiporter is critical for Na^{+} cycling in the cells of both *B. halodurans* C-125 and *B. pseudofirmus* OF4 (Goto et al. 2022).

Additional strategies against alkaline stress include the formation of acidic secondary cell walls, which lower the pH at the outer membrane surface relative to the extracellular environment. *B. halodurans* C-125 forms the teichurono-peptide and teichuronic acid secondary cell wall, whereas *B. pseudofirmus* OF4 produces cell surface protein SlpA and polyglutamic acid. These structures attract H^{+} and repel OH^{-}, protecting intracellular biochemical reactions from extracellular stresses (Matsuno et al. 2018). In addition, *Bacillus* spp. enhances motility through specialized flagellar motors. For example, alkaliphilic *Bacillus clausii* KSM-K16 employs a dual-function MotAB-type flagellar motor, which is identified as the first bacterial species with a single stator–rotor that can utilize both H^{+} and Na^{+} for ion coupling at different pH ranges. Furthermore, MotPS-type flagellar motor of alkaliphilic *Bacillus alcalophilus* AV1934 is able to couple Na^{+}, K^{+}, and Rb^{+} for flagellar rotation (Ito and Takahashi 2017).

2.7 Halophiles

Halophiles are organisms that thrive, grow, and proliferate in highly saturated saline environments, encompassing diverse lineages of archaea, bacteria (e.g. cyanobacteria), fungi, and algae (e.g. some diatoms and green algae). According to the tolerance to salinity ranges, halophiles are categorized into, halotolerant organisms (1.2–2.9% NaCl), halophiles (>8.8% NaCl), and extreme halophiles (>14.6% NaCl) (Coleine and Delgado-Baquerizo 2022). These organisms inhabit both thalassohaline (marine-derived) and athalassohaline (non-marine, including natural and anthropogenic) hypersaline environments, such as saline seas, deep-sea brines, salt lakes, salterns, salt mines, and certain rock sediments. Iconic inland hypersaline systems include the Great Salt Lake, Dead Sea, and Qarhan Salt Lake (Figure 2.6). Hypersaline environments exhibit marked variability in salt concentration, chemical composition, and pH, driving corresponding diversity in resident halophilic species and community structures (Oren 2015). For instance, in the Great Salt Lake, cyanobacteria dominate lower salinity zones, while *Dunaliella* spp. prevails at the highest salinities. In a Spanish saltern pond with 19% salinity, novel microbial consortia, including Actinobacteria and nanohaloarchaea, were prominently detected (Oren 2014).

Figure 2.6 Saline pond of Qarhan Salt Lake. Qarhan Salt Lake is one of the largest salt lakes in China and one of the most significant potash deposits globally. *Source:* Courtesy of Ruimin Wang.

2.7.1 Phylogenic Diversity

2.7.1.1 Halophilic Archaea

In saline environments, halophilic bacteria and various Eukarya dominate, whereas hypersaline habitats are dominated by halophilic archaea, with minor components of bacteria and certain microbial eukaryotes, including certain protists (e.g. *Dunaliella* species) and fungi. The halophilic archaea, which belong to the phylum Halobacteriota and the class Halobacteria, are the most salt-dependent and salt-tolerant microorganisms, requiring more than 100–150 g/L NaCl for both growth and structural stability. This class consists of 71 genera and 275 species, with 49.8% isolated from marine salt environments and 50.2% from inland salt habitats, highlighting the widespread presence of halophilic archaea in both thalassohaline and athalassohaline ecosystems (Cui and Dyall-Smith 2021).

Haloarchaea thrive in extreme saline conditions, such as brines, salt lakes, and saline soils, often reaching densities high enough to impart a pink-red hue to water bodies due to accumulated pigments (Figure 2.6). Their optimal growth occurs at NaCl concentrations of at least 15% (2.5 mol/L), while many species flourish even in saturated salt environments (~35%, 6 mol/L). Notably, nanohaloarchaea (Candidatus Nanohaloarchaeota) inhabit the water column, distinguished by their small cell size (<0.8 μm), compact genomes (<1.2 Mb), and halophilic nature. In contrast, sediments support anaerobic halophilic methanogens, e.g. *Methanosarcinales* and *Methanomicrobiales* (Cui and Dyall-Smith 2021).

Halophilic archaea serve as archaeal model system for investigating archaeal biology and extreme environmental adaptation. For example, *Haloquadratum* has significantly contributed to our understanding of extremophile survival strategies, particularly in hypersaline environments. Its unique square-shaped morphology and ability to thrive in saturated salt conditions make it a fascinating subject for research. Similarly, *Haloferax volcanii*, isolated from the Dead Sea in 1975, thrives in hypersaline conditions and has become a key model archaeon due to its genetic tractability and ease of cultivation. Notably, it was the first archaeon shown to transfer genes via a natural mating system, providing crucial insights into archaeal genetics and HGT mechanisms.

2.7.1.2 Halophilic Bacteria

Halophilic bacteria predominantly belong to the phyla Bacteroidetes, Firmicutes, Proteobacteria, Cyanobacteria, Actinobacteria, and Spirochaetes. Among these, the Firmicutes order Halanaerobiales and the Proteobacteria family Halomonadaceae are almost entirely composed of halophilic taxa. One particularly interesting bacterium, *Salinibacter*, exhibits remarkable competitive fitness at maximal salinities, making it a key player in hypersaline ecosystems. Its discovery in the early 2000s significantly advanced our understanding of saltern crystallizer ecosystems. Thanks to modern culture-independent techniques, researchers can now conduct targeted sampling and ecological assessments before isolating new species. This approach has led to the identification of *Spiribacter salinus* and *Salinibacter ruber*, further enriching our knowledge of microbial life in extreme environments (Oren 2008).

2.7.1.3 Halophilic Eukaryotes

Hypersaline habitats also support diverse eukaryotes, including fungi and protists. Notable examples include *Halocafeteria seosinensis* (stramenopile), initially isolated from a 30% saltern, and now detected in global hypersaline waters. *H. seosinensis* strain EHF34 exhibits optimal growth at 15% NaCl and can tolerate up to 36.3% salinity, but it fails to grow below 7.5% salinity, suggesting it is a borderline extreme halophile (Harding et al. 2017). Its ability to actively ingest prokaryotes in high-salinity waters highlights its ecological role in hypersaline microbial food webs. The unicellular green alga *Dunaliella salina* exhibits remarkable salinity fluctuation tolerance and inhabits environments with salinities up to 30% (approximately 10-fold seawater levels).

Halophilic fungi require >8.8% NaCl for survival in saline environments. *Aspergillus penicillioides*, a ture halophile, inhabits the Dead Sea and salterns. Growth could only be observed after supplementation of the regular medium with 10% salt, even when the microorganisms were cultured in various nutrient-rich media. When added salt absent, fungal conidia either did not germinate, or when germinated, distortions and lysis were seen in the short mycelial forms (Nazareth and Gonsalves 2013). Obligate halophilic fungi, including *Wallemia muriae*, *Phialosimplex salinarum*, *Aspergillus baarnensis*, *Aspergillus salisburgensis*, and *Aspergillus atacamensis*, demand 5–10% NaCl. While the black yeast *Hortaea werneckii* grows across almost the whole salinity range. The extreme halophilic fungus *Wallemia ichthyophaga* thrives optimally at 2.6–3.4 M NaCl, even growing in saturated KCl and $MgSO_4$ solutions and ~2 M $MgCl_2$, representing the most halophilic eukaryote described to date (Coleine et al. 2022).

In addition to microorganisms, halophytic plants are remarkable for their ability to withstand high salinity, enduring NaCl concentrations of 0.5–1.0 M. These plants have evolved specialized mechanisms such as ion compartmentalization, osmotic adjustment, and salt excretion to thrive in saline environments. Beyond plants, halophilic insects like *Ephydra* sp. and *Trichocorixa verticalis* are commonly found around hypersaline lakes, where they have adapted to extreme salt conditions (Kırtel et al. 2018). These insects play a crucial role in hypersaline ecosystems, contributing to nutrient cycling and serving as prey for other organisms.

2.7.2 Adaptive Strategies

Halophiles must withstand osmotic stress caused by elevated salt concentrations, along with challenges to membrane stability and protein functionality, often exacerbated by UV exposure. As a result, ion homeostasis regulation is crucial for their survival. These organisms employ two fundamental strategies to maintain intracellular solute balance, yet the specific mechanisms vary across different halophilic lineages.

2.7.2.1 Salt-in and -out Strategies

The first is salt-in strategy, which involves the accumulation of high intracellular K^+ concentrations, while actively excluding Na^+. It renders organisms obligate halophiles incapable of surviving in low-salt environments. Through this strategy, halophiles maintain cytoplasmic KCl saturation while minimizing Na^+ influx. For instance, *Halobacteriaceae* and *Halanaerobiales* achieve cytoplasmic K^+ saturation through active uptake while minimizing Na^+ influx. Similarly, halophilic bacteria such as *Salinibacter ruber* and *Salisaeta longa* employ comparable methods to maintain cellular homeostasis (Kırtel et al. 2018). Under this K^+ accumulation process, intracellular proteins (including enzymes) have to adapt to high potassium concentrations for proper stability and enzymatic activity. As a result, halophiles using the salt-in strategy typically exhibit acidic proteomes, which facilitate protein function in high-salinity environments.

The second strategy is to accumulate compatible solutes in the cytoplasm, which is employed by moderately halophilic Bacteria and Eukarya across broader salinity ranges through synthesizing or importing organic osmolytes (compatible solutes) to balance osmotic pressure without disrupting the protein and enzyme. Compatible solutes include ectoine, glycine betaine, trehalose, sucrose, glycerol, and tetrahydropyrimidine. These neutral, low-molecular-weight compounds enable flexible responses to fluctuating salinity (Kırtel et al. 2018). This strategy is sometimes described as salt-out strategy, because these hydrophilic organic solutes increase the cytoplasmic osmotic pressure while keeping salt out, thus ensuring water uptake. For example, *Haladaptatus paucihalophilus* synthesizes trehalose for adaptation to high-salt conditions. Also, protists *Halocafeteria seosinensis* and *Dunaliella salina* accumulate high concentrations of compatible solutes in the cell to balance osmotic pressure from the surroundings (Rappaport and Oliverio 2023), allowing them to survive in hypersaline habitats.

2.7.2.2 Other Strategies

Halophiles modulate membrane lipid composition to maintain fluidity under osmotic stress. For example, *Hortaea werneckii* maintains a fluid membrane over a wide range of salinities by keeping a low sterol-to-phospholipid ratio and by decreasing both fatty acyl length and the saturation level of phospholipids (Harding et al. 2017). Other high-salinity adaptations such as evolving higher numbers of acidic residues on protein surfaces to prevent their aggregation and decrease protein pI preventing misfolding (Kırtel et al. 2018). In addition, certain halophiles are able to utilize bacteriorhodopsin to generate ATP and maintain the transmembrane proton gradients under low-oxygen or anaerobic conditions. Furthermore, some halophilic fungi improve halotolerance by altering morphological characteristics. For example, *Wallemia ichthyophaga* withstands high salt concentrations by increasing the intracellular concentration of polyols, using high-affinity K^+ transporters, and increasing the thickness of its cell wall (Zajc et al. 2013).

Extreme halophiles are mostly aerobic and utilize specialized pigmentations, including bacteriorhodopsin for light-driven energy production and carotenoids for ultraviolet protection. The carotenoids give them a pinkish color. To prevent an exodus of water from the cell, halophiles offset the high salt in the environment by accumulating such compounds as potassium and glycine-betaine. This allows a balance of salt inside and outside of the cell preventing water from flowing outward as would be the case if lower salt levels existed within the cells. Saltwater can evaporate leaving evaporite deposits consisting of salts such as sodium chloride (halite) and calcium sulfate (gypsum). Within evaporates are fluid inclusions (small, trapped pockets of water) that can provide a refuge for microbes for at least six months. Cyanobacteria trapped within dry evaporite crusts can continue to have low levels of metabolic function such as photosynthesis (Rothschild et al. 1994).

2.8 Xerophiles

Xerophiles are organisms capable of enduring extreme desiccation through anhydrobiosis—a state in which intracellular water is minimal, and metabolic activity ceases entirely. These organisms are defined by a water activity (a_w) threshold below 0.7 (Coleine and Delgado-Baquerizo 2022). Anhydrobiotic capacity is observed across a broad range of life forms, including bacteria, yeasts, fungi, plants, insects, tardigrades, mycophagous nematodes, and the shrimp *Artemia salina*.

2.8.1 Low Water Activity Stress

Water activity (a_w) is a measure of the availability of water in a substance, rather than just its moisture content. It is defined as the ratio of the vapor pressure of water in a sample to the vapor pressure of pure water at the same temperature. The scale ranges from 0 (completely dry) to 1 (pure water). Low water activity stress stems not only from desiccation-induced water scarcity but also from elevated osmotic pressures. Extremophiles in hypersaline environments and certain polar regions, such as the Antarctic McMurdo Dry Valleys, encounter similar water activity challenges (Sections 2.4 and 2.7). Arid ecosystems remain the dominant terrestrial environment, covering approximately 41% of Earth's land surface (Shu and Huang 2022). Hyper-arid deserts—such as the Sahara, Arabian, Atacama, and Taklimakan—represent extreme desiccation conditions, often characterized by high temperatures, nutrient scarcity, low water activity, and intense ultraviolet radiation (Mohammadipanah and Wink 2016).

Water availability critically regulates cellular physiology, with distinct a_w lower limits observed across biological taxa. Most organisms require a_w ranging from 0.99 to 0.97 for active life processes;

certain bacteria tolerate a_w down to 0.9, while haloarchaea proliferate in saturated NaCl at a_w 0.75 (Coleine et al. 2022). Metabolic cessation generally occurs at the $a_w = 0.605$ threshold for most organisms (Stevenson et al. 2015). Xerotolerant and xerophilic fungi grow in dry environments with $a_w < 0.85$. Xerophiles exhibit growth inhibition at higher a_w values (0.96–0.97). *Xeromyces bisporus* represents one of the most osmophilic microorganisms characterized to date, demonstrating growth at $a_w = 0.61$ and optimal growth at $a_w = 0.85$, with an absolute requirement for reduced water availability (Gostinčar et al. 2023).

2.8.2 Low Water Activity Habitats

High-sugar food environments serve as microbial habitats predominantly dominated by xerophilic (or osmophilic) filamentous fungi and yeasts (Grant 2004). Among microorganisms capable of proliferation below 0.720 a_w, the majority—unlike the obligate osmophile *X. bisporus*—are not restricted to sugar-rich substrates. These include halophilic prokaryotes and xerophilic fungi such as *Aspergillus penicillioides* and *Eurotium herbariorum* (Stevenson et al. 2015). The growth limit for *Eurotium amstelodami* (xerophilic, halotolerant) is 0.7 a_w, while borderline examples like *X. bisporus* and the yeast *Zygosaccharomyces rouxii* exhibit growth limits at 0.61–0.62 a_w (Coleine et al. 2022). Among microbial systems, more than 20–30 microbial taxa have been known to be active at $a_w \leq 0.690$, representing the most extreme life forms adapted to low-water-activity environments (Stevenson et al. 2015). The fungi subclass *Eurotiomycetidae* includes numerous xerophiles, with the genus *Aspergillus* encompassing the majority of xerophilic species. Outside subgenus *Aspergillus* and *Polypaecilum*, *Aspergillus candidus* (sect. Candidi), *A. ochraceus* (sect. Circumdati), and *A. wentii* (sect. Cremei) are documented to grow at or below 0.75 a_w (Pócsi et al. 2024).

Despite polyextreme conditions, arid soils host diverse bacterial lineages, including Actinobacteria, Chloroflexi, Firmicutes, Gemmatimonadetes, and Proteobacteria (Shu and Huang 2022). Notably, Actinobacteria dominate as one of the key taxonomic groups in extreme environments, with elevated relative abundance in hyper-arid desert regions. Consequently, arid and desert habitats serve as unique ecosystems for isolating extremophile or extremotrophic Actinobacteria strains, represented by *Streptomyces*, *Micromonospora*, *Saccharothrix*, *Streptosporangium*, *Cellulomonas*, *Amycolatopsis*, *Geodermatophilus*, *Lechevalieria*, *Nocardia*, and *Actinomadura* (Mohammadipanah and Wink 2016). The family *Geodermatophilaceae* includes only three genera: *Geodermatophilus*, *Blastococcus*, and *Modestobacter*, which thrive under low-a_w and nutrient-limited conditions. *Blastococcus* and *Modestobacter* colonize rock surfaces, while *Geodermatophilus* prefers arid soils.

Xerophilic Actinobacteria such as *Geodermatophilus arenarius* and *G. siccatus* were isolated from Saharan Desert sands in Chad. Additional *Geodermatophilus* species, along with *Actinoplanes* and *Streptomyces* strains, have been gained from Negev Desert and Mojave Desert soils (California–Nevada border). *Streptomyces deserti*, isolated from the hyper-arid Atacama Desert, and *Streptomyces bullii*, from the same region, exemplify desert-adapted *Streptomyces* species (Mohammadipanah and Wink 2016).

Furthermore, in arid environments, certain cyanobacteria, algae, lichens, bryophytes, bacteria, and fungi formed biological soil crusts (Figures 7.15 and 7.16) and facilitate adaptation to harsh survival conditions (Barrón-Sandoval et al. 2023; Hagemann et al. 2017).

2.8.3 Adaptive Strategies

Under low a_w conditions, xerophiles typically maintain cellular homeostasis by accumulating compatible solutes. Additionally, they enhance reactive oxygen species (ROS) scavenging efficiency to mitigate oxidative damage and employ drought-resistant genes and molecular regulatory mechanisms for adaptation. Some organisms also synthesize melanin to reduce UV radiation-induced injury. For example, most xerotolerant fungi, such as *Aspergillus* and *Penicillium* species, accumulate intracellular glycerol as a compatible solute to counteract ambient low a_w (Coleine et al. 2022).

Water can exist as thin aqueous films on various surfaces and interfaces, including minerals, organic matter, and biological structures, or within minerals as fluid inclusions. These films persist in the liquid phase under diverse conditions and serve as critical water resources for sustaining life in arid environments (Stevenson et al. 2015). Bacteria thriving in arid habitats optimize water access for physiological requirements. Most of them are adjacent to mineral soils (e.g. quartz, halite, gypsum), utilizing trapped water from these minerals for growth. Notably, Actinobacteria exhibit high germination and growth rates at $a_w = 0.5$, though non-halophilic species typically remain metabolically inactive when a_w is below 0.80 (Mohammadipanah and Wink 2016).

As a typical eukaryotic example for desiccation tolerance, tardigrades employ multiple adaptive strategies. During metabolically active states, they live in water films on land. When exposed to extreme desiccation, they enter anhydrobiosis by forming a tun state and drastically suppress metabolic activity, concomitantly accumulating high concentrations of trehalose to induce vitrification while upregulating DNA repair proteins expression and ROS scavenging systems (Arakawa 2022).

2.9 Radiodurans

2.9.1 Discovery

Radiodurans are organisms that are able to withstand high doses of ionizing radiation without serious injury. The bacterium *Deinococcus radiodurans* was initially isolated from radiation-sterilized canned food in 1956. This organism is famous for its ability to withstand ionizing radiation (up to 20 kGy of gamma radiation) and UV radiation (doses up to 1000 J/m^2). *D. radiodurans* exhibits significant resilience to withstand other extreme stresses, enabling its widespread distribution from desert to North Pole. To date, over 60 species within the genus *Deinococcus* have been discovered, with the majority of complete sequenced strains exhibiting radiation-resistant phenotypes.

D. radiodurans can withstand 1.5 million rads, a thousand times more than any other life form on Earth and 3000 times more than a human can survive. Its extraordinary tolerance makes it useful at nuclear waste sites, where it consumes nuclear waste and transforms it into more disposable derivatives. Its survival skills have been the source of much curiosity worldwide, even including speculation that it may have originated on Mars, where radiation levels are higher. For example, *D. radiodurans* survived at least 3 years outside the International Space Station (ISS) in a space experiment (Coleine and Delgado-Baqueriz 2022).

2.9.2 Radiotolerant Organisms

Apart from *Deinococcus*, more recent investigations have uncovered a remarkable variety of radioresistant and radiotolerant organisms, spanning both prokaryotic and eukaryotic life, in a range of extreme environments. For instance, the Actinobacteria *Knollia* sp., isolated from the moraine of Mount Everest, can withstand a UV intensity of 1000 J/m^2 (Wang et al. 2024). Additionally, the green alga *Dunaliella bardawil* and two species of *Rubrobacter* have demonstrated strong resistance to radiation. Surprisingly, over 2000 fungal isolates have been identified from Chernobyl and cooling water from nuclear reactors, with the majority being melanized fungi (e.g. *Cladosporium* spp.), which exhibit directional growth in response to ionizing radiation (Lim et al. 2021; Shuryak 2020). Another extremophile, the cryptoendolithic black fungus *Cryomyces antarcticus*, found in sandstone from the Antarctic Dry Valleys (the coldest hyperarid desert and a prime terrestrial analog for Mars), survives under UV-B irradiation doses five to eight times higher than Antarctic UV-B levels (Selbmann et al. 2018). Furthermore, the radiation tolerance of the desert moss *Syntrichia caninervis* was tested under simulated Martian conditions, revealing that this species ranks among the most radiation-resistant plants known. It has been shown to endure exposure to 500 Gy, a lethal dose for most plants (Li et al. 2024b).

2.9.3 Radiotolerant Mechanisms

The molecular mechanisms of radiation resistance in extremophiles remain incompletely characterized. However, several consensus mechanisms have been established (Figure 2.7). The lethal damage to organisms caused by radiation is primarily due to DNA damage induced by ROS. Radiation-generated ROS directly accumulate on DNA, leading to base mutations, single-strand breaks (SSBs), and double-strand breaks (DSBs). Additionally, ROS causes structural protein denaturation, enzyme inactivation, and lipid membrane destabilization. When DNA damage exceeds cellular repair capacity, cell cycle arrest and cell death are triggered. Thus, efficient ROS scavenging (e.g. free radical neutralization) and DNA repair are critical strategies for organisms surviving under high-dose radiation. For instance, *D. radiodurans* possesses efficient and specialized DNA repair mechanisms, including interchromosomal single-strand annealing (SSA), extended synthesis-dependent strand annealing (ESDSA), and unique homologous recombination pathways. These mechanisms enable rapid and efficient restoration of numerous DSBs to normal chromosomal states (Timmins and Moe 2016). Furthermore, *D. radiodurans* employs cleansing systems to efficiently remove DNA and protein fragments.

In addition, physical barriers, e.g. cell membrane properties and pigment synthesis, represent other survival strategies to mitigate or prevent radiation damage. *Deinococcus* species exhibit thicker multilayered cell walls compared to other Gram-negative bacteria, which together with the cell envelope, provide structural support and protection (Farci et al. 2022). Most *Deinococcus* members maintain high intracellular Mn(II) levels (Mn:Fe ratio ≈ 0.24), and Mn(II) deprivation significantly increases radiation sensitivity (Daly et al. 2004). Meanwhile, the high expression levels of superoxide dismutase (SOD), peroxidase, and catalase enable *D. radiodurans* to efficiently scavenge intracellular ROS. Additionally, *Deinococcus* species synthesize pigments (e.g. carotenoids), resulting in red or pink colonies. Carotenoids (e.g. deinoxanthin) contribute to ROS elimination, thereby reducing cellular damage (Tian and Hua 2010). Fungal pigments also play important roles

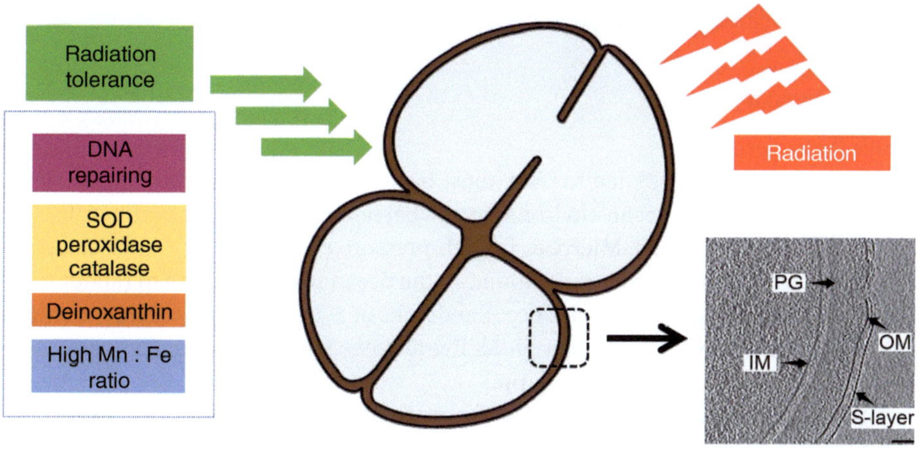

Figure 2.7 Schematic diagram of molecular and cellular adaptations in radiation-tolerant *Deinococcus radiodurans* and cryo-ET of its cell envelope architecture. Inset: A tomographic slice shows the expected density layers on the cell surface corresponding to inner membrane (IM), outer membrane (OM), surface layer (S-layer), and peptidoglycan (PG) of a FIB-milled *D. radiodurans* cell, scale bar, 500 Å. *Source:* Adapted from von Kügelgen et al. (2023).

in radiation resistance. For instance, melanins synthesized by some highly pigmented fungi not only reduce cellular radiation damage but may also utilize ionizing radiation to induce growth through radiosynthesis (Dadachova et al. 2007).

Despite substantial progress in understanding radiation resistance mechanisms, further studies are needed to unravel the unique regulatory networks that enable organisms to survive under lethal doses of radiation. Deciphering these mechanisms will advance our understanding of life's limits on Earth and potentially aid human exploration of extraterrestrial environments.

2.10 Barophiles

Barophiles are organisms, both macroscopic and microscopic, that thrive in environments where the pressure is unusually high, such as deep in the ocean or far underground. Based on pressure tolerance, these organisms can be categorized as Barotolerant/Piezotolerant (0.1–10 MPa), barophiles/piezophiles (10–50 MPa), and hyperbarophiles/hyperpiezophiles (>50 MPa) (Coleine and Delgado-Baquerizo 2022; Merino et al. 2019). The ocean, covering 70% of Earth's surface, is characterized by low temperatures, high hydrostatic pressure (HHP), and darkness. In deep-sea trenches, hydrostatic pressure can reach 110 MPa. Despite these extreme conditions, the deep sea remains the largest aquatic habitat for microbial life on Earth for the past billion years (Orcutt et al. 2011).

The discovery of pressure-tolerant organisms dates back to the 1880s, and the concept of *barophile* was first proposed in 1949. However, it was not until 1979 that Yayanos isolated the first barophilic microorganisms from deep-sea samples, and the term *piezophilic* was initially used to describe pressure-adapted organisms (Scheffer and Gieg 2023). Due to geological limitations in sample accessibility, coupled with the need for specialized sampling methods and expensive laboratory equipment, research progress on piezophiles has lagged behind that of other extremophiles. Most known piezophilic organisms have been discovered within the last 40 years, with *Colwellia* sp. MT-41 being one notable example. Currently, the most widely accepted definition of piezophily stipulates that piezophiles exhibit a maximum specific growth rate (μmax) above atmospheric pressure (0.1 MPa). Under this definition, only 86 microbial isolates with a documented high-pressure optimum (HP_{opt}) > 0.1 MPa have been identified (Scoma 2021), a remarkably small number given the vast diversity and abundance of prokaryotic life on Earth.

2.10.1 Phylogenic Diversity

2.10.1.1 Barophilic Prokaryotes

Among the piezophilic prokaryotes isolated to date, most bacteria are isolated from non-thermal (cold to moderate temperature) deep-ocean environments, whereas archaea are predominantly isolated from deep-sea hydrothermal vents. Microbes in high-pressure environments can be categorized into three groups. Barotolerant microbes are found in the ocean to depths of 4 km (about 400 atm) but grow better at 1 atm. Barophilic species occur at depths of 5–6 km and grow better at high pressure (500–600 atm) than at low. Extreme barophiles live at great depths where the pressure is at least 700 atm and they are not able to grow at 1 atm.

Piezophilic microorganisms have been reported from diverse high-pressure environments, including the deep marine subsurface (hydrothermal vents, deep-sea sediments) and deep terrestrial subsurface (oil reservoirs, Deccan Traps, hot brines, and mines) (Scheffer and Gieg 2023). For example, *Marinomonas piezotolerans* YLB-05^T from deep-sea sediments of the Yap Trench demonstrates optimal growth at 0.1 MPa and tolerance up to 50 MPa (Yu et al. 2019). *Thermococcus piezophilus* from the Mid-Cayman Rise hydrothermal vent, a hyperthermophilic and piezophilic archaeon, shows a broad pressure tolerance range (0.1–125 MPa, with an optimum at 50 MPa) (Dalmasso et al. 2016). Investigations of the Challenger Deep (10,918 m) and the Sirena Deep (10,667 m) revealed microbial communities dominated by heterotrophic Gammaproteobacteria, followed by *Nitrosopumilus* and Euryarchaeota, with minor members from Alphaproteobacteria, Betaproteobacteria, Epsilonproteobacteria, Deltaproteobacteria, and anaerobic methanotrophic archaea (Tarn et al. 2016).

Generally, deep-ocean bacterial assemblages are predominantly composed of Proteobacteria (Alphaproteobacteria, Gammaproteobacteria, and Deltaproteobacteria), Chloroflexi, and Actinobacteria (DeLong et al. 2006; Orcutt et al. 2011), while archaeal communities are typically dominated by Marine Group I Crenarchaeota (Karner et al. 2001).

2.10.1.2 Barophilic Eukaryotes

Beyond prokaryotes, piezophilic eukaryotes have also been found. For instance, metazoans such as holothurians (sea cucumbers) are reported to be abundant in the world's deepest location, the Challenger Deep in the West Pacific, 400 km southwest of Guam, approximately 110 MPa.

Recent studies have also explored fungal communities in oceanic samples. A survey of six Mid-Oceanic Ridge sediment samples from the East Pacific and South Indian Oceans, employing both culture-dependent and culture-independent methods, identified 97 fungal isolates belonging to 7 genera and 10 species. Notably, *Penicillium* (two species), *Rhodotorula*, *Meyerozyma*, *Ophiocordyceps*, *Vishniacozyma*, *Aspergillus* (three species), and *Phoma* were detected. *Ophiocordyceps heteropoda*, previously reported only from terrestrial habitats, was found in both oceans (Tang et al. 2020).

Another study conducted a comprehensive fungal identification and phylogenetic diversity analysis (based on morphological characteristics and sequencing data) of 10 sediment samples collected from five locations in the East Pacific Ocean (4545–7068 m depth). This analysis identified 12 fungal genera, including *Aspergillus* (five strains), *Aureobasidium* (three strains), *Candida* (two strains), *Cladosporium* (56 strains), *Cystobasidium* (one strain), *Devriesia* (nine strains), *Knufia* (one strain), *Nigrospora* (three strains), *Penicillium* (18 strains), *Rhodotorula* (four strains), *Sarocladium* (three strains), and an unclassified *Xylariales* (one strain). Notably, the *nirK* gene was detected in *Aspergillus niger* and *Penicillium chrysogenum*, suggesting a potential role for fungal denitrification activity in these environments (Wei et al. 2018).

2.10.2 Adaptive Strategies

High pressure exerts multifaceted impacts on organisms, including reduced cell membrane fluidity, diminished stability of macromolecules, increased difficulty in double-strand DNA unwinding, and altered respiratory chain and metabolic systems, as well as impaired flagellar motility (cellular motility). While adaptive strategies for piezophiles under extreme pressures have been reported through studies on other extreme environments (temperature, pH, and salinity), piezophiles and hyperpiezophiles have employed those mechanisms to thrive under high-pressure conditions (Jebbar et al. 2015).

First, cell membranes maintain membrane fluidity with elevated levels of unsaturated fatty acids at high pressures. For instance, under high-pressure cultivation, *Shewanella* sp., *Moritella* sp. DB21MT-5, and *Photobacterium profundum* SS9 exhibit increased monounsaturated fatty acid content in their membranes. Similarly, the piezotolerant *Shewanella piezotolerans* WP3 shows elevated relative abundance of C20:5 (Tamby et al. 2023). Additional adaptations include upregulation of chaperone-encoding genes, modification of respiratory chain components, expression of different porins, and production of osmolytes. *Shewanella violacea* DSS12 (growth range: 0.1–70 MPa; optimal: 30 MPa) synthesizes pressure-adapted proteins such as respiratory proteins, which demonstrate enhanced stability and activity under high pressure (Ohke et al. 2013). *P. profundum* SS9 possesses a unique aquaporin (AqpSS9) that enhances membrane permeability under HHP conditions (Wei et al. 2017). This species also employs dual flagellar systems: lateral flagella are expressed exclusively under high-pressure/high-viscosity conditions, while polar flagella function in liquid cultures (McCarter 2004). When *Halomonas titanicae* ANRCS81 was incubated at 40 MPa, genes related to antioxidant defense, anaerobic respiration, and fermentation were upregulated, indicating intracellular oxidative stress induced by high pressure (Li et al. 2023).

Genomic analyses reveal that piezophilic organisms exhibit genetic and metabolic signatures for deep-sea pressure adaptation: (1) pressure-regulated outer membrane protein genes, cytochrome-related genes, monounsaturated fatty acid synthases, malate dehydrogenase, isopropylmalate dehydrogenase, and glycosyltransferases; (2) under high pressure, genes associated with translation, ribosome structure, secretion, motility, and starch/cellulose metabolism are upregulated; (3) these organisms tend to encode hydrophobic and smaller molecular, along with increased synthesis of pressure-responsive genes; (4) lateral flagellar systems are highly developed (enhancing motility); (5) genes encoding 3-hydroxybutyrate are present, 3-hydroxybutyrate and its oligomers are piezolytes and help stabilize proteins; (6) rRNA folding, ribosome assembly, and protein translation are optimized for high-pressure conditions (Li et al. 2023; Liu et al. 2020; Oger and Jebbar 2010; Peoples et al. 2020; Scheffer & Gieg 2023).

Despite limited isolation of strict piezophiles, studies demonstrate that oxygen consumption rates in sediments (Mariana Trench) are nearly double those at 6000 m, suggesting active microbial communities in hadal zones that play significant roles in global carbon cycling (Glud et al. 2013). Furthermore, organisms potentially accommodate high pressures relatively easier than other harsh conditions, for example, piezotolerant strains can temporarily withstand 2000 MPa (Sharma et al. 2002; Vanlint et al. 2011). At present, progress in piezophile research lags behind other extremophiles due to multiple limiting factors, prompting the hypothesis that vast, undiscovered "piezophilic kingdoms" thrive in deep ocean or far underground, potentially exhibiting greater diversity than currently recognized.

2.11 Tardigrades as a Model Animal for Astrobiology

Investigating the survival potential of terrestrial life in space environments or simulated space environments by selecting extremotolerant model species and systematically analyzing their adaptive mechanisms provides a robust framework for astrobiological research. This approach not only refines biosignature detection strategies for extraterrestrial life but also advances fundamental understanding of planetary habitability. However, identifying appropriate model organisms for astrobiological research—particularly those capable of surviving extreme space conditions—remains a significant challenge. Key requirements include organisms exhibiting a high tolerance for freezing, radiation, and desiccation. The organisms with the best chance for survival under such conditions are polyextremophiles, such as some species of Bacteria, Archaea, Rotifera, Nematoda, Arthropoda, and Tardigrada (Figure 2.8). For instance, *Deinococcus radiodurans* survived at least 3 years outside the International Space Station (Coleine and Delgado-Baquerizo 2022; Kawaguchi et al. 2020) and tardigrades were exposed to the space environment for 10 days in space experiments (Jönsson et al. 2008; Rebecchi et al. 2009).

2.11.1 Survival Strategies

Tardigrades are exceptionally resilient microscopic animals, renowned as one of the toughest animals on Earth. They are highly unique within the extremophile group due to their complex and efficient stress-resistant adaptations. These adaptations, coupled with their relatively complex

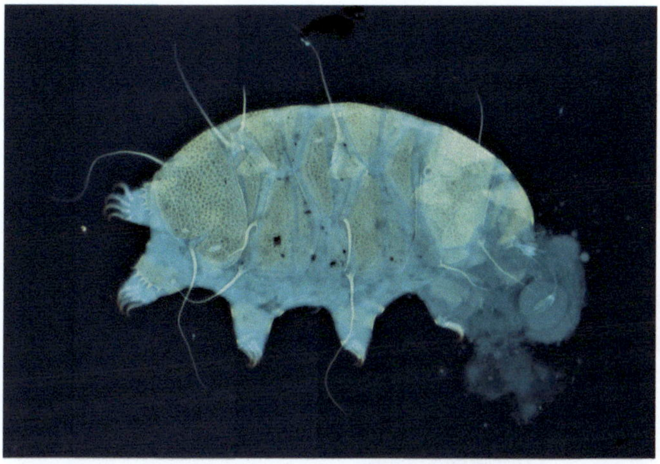

Figure 2.8 Tardigrade *Echiniscus* sp. from mosses in Mount Taibai located in Shaanxi Province of China, 3500 m above the sea level. The animal is ~260 μm long. *Source:* Photo by Xingliang Zhang.

structure (for their size) and their relative ease of observation, have made tardigrades an excellent model animal for space research. They exhibit extraordinary survival capabilities in extreme environments through a combination of unique physiological and molecular adaptations. Survival strategies among extant tardigrades include osmoregulation, as well as cryptobiosis and diapause in the form of encystment, cyclomorphosis, and resting eggs. Cryptobiosis is an extreme form of dormancy in which an organism enters a state where all measurable metabolic activity ceases. Unlike cryptobiosis, diapause does not involve a complete cessation of metabolism but rather a significant reduction in activity.

In tardigrades, both encystment and cyclomorphosis function as diapause mechanisms, involving significant morphological transformations that are not directly tied to growth but rather to atypical molting processes. These diapause stages, marked by reduced metabolic activity, highlight the exceptional phenotypic plasticity of tardigrades, mediated by epigenetic mechanisms to endure harsh environmental conditions. Encysted tardigrades take on a distinct oval, contracted shape, encased in retained exuviae or cuticular layers that are often opaque, sometimes appearing in shades of orange, purple-red, or brown. Cyclomorphosis, observed in *Halobiotus*, features the formation of a mobile cyst, characterized by the preservation of the old cuticle, modifications to the buccal–pharyngeal system and claws, and changes in gonadal development (Guidetti and Møbjerg 2018). Additionally, the production of diapause resting eggs, along with the considerable variation in hatching times, may serve as a bet-hedging strategy to navigate unpredictable environmental conditions (Glime 2013).

Under a sudden environmental change, tardigrades rapidly enter cryptobiosis and reduce its metabolic activity to as low as 0.01% of normal levels (Coleine and Delgado-Baquerizo 2022). Once the stress factor is removed, they quickly revive from their ametabolic state. Cryptobiosis may be induced by a range of extreme conditions leading to different cryptobiotic stages corresponding to anhydrobiosis, osmobiosis, cryobiosis, and possibly anoxybiosis and chemobiosis. Most notably, stress conditions to tardigrade that induce tun forming (Figure 2.9) and the water loss, i.e. desiccation (anhydrobiosis), rise in surrounding osmotic pressure (osmobiosis), or freezing (cryobiosis). The lack of oxygen (anoxybiosis) and environmental toxicants (chemobiosis) may also induce cryptobiosis (Møbjerg and Neves 2021). Additionally, tardigrades generally possess chitin as a key component of their cuticle. In some terrestrial species (e.g. heterotardigrades), the chitinous armor (Figure 2.8) may slow desiccation and protect against damage under desiccation stress (Glime 2013).

Figure 2.9 Micrographs of the limnoterrestrial eutardigrade *Ramazzottius varieornatus* in the active state (dorsal view) and tun state (ventral view). *Source:* Reprinted with permission from Møbjerg and Neves 2021, © 2020 Møbjerg and Neves.

2.11.2 Adaptations to Anhydrobiosis

Tardigrades are able to enter a desiccation-tolerant state known as anhydrobiosis, allowing them to extend their lifespan by up to 20 years without senescence (Schill and Hengherr 2018). During desiccation, intracellular proteins and membranes form hydrogen bonds with other molecules, thereby compensating for the loss of hydrogen bonds with water. These forced molecular interactions, such as protein–protein interactions, can induce irreversible conformational changes and lead to the loss of enzyme activity. Water loss triggers a phase transition from the biologically active liquid crystalline to the gel phase in membranes and may also cause organelle fusion (Schill and Hengherr 2018). Tun typically forms during anhydrobiosis and may also be associated with osmobiosis, cryobiosis, or chemobiosis, all characteristic of the cryptobiotic state. Marine tidal species can enter the tun state within seconds when exposed to desiccation, while limno-terrestrial tardigrades generally take around 30 minutes to complete tun formation (Sørensen-Hygum et al. 2018).

Several mechanisms likely contribute to the protection of tardigrades in the anhydrobiotic state. Heat shock proteins (e.g. Hsp70 and Hsp90) serve as molecular chaperones to preserve or restore protein integrity, while tardigrade-specific intrinsically disordered proteins (TDPs), disaccharides, and other metabolites help prevent the formation of damaging compartment aggregates during desiccation (Schill and Hengherr 2018). Sugars (e.g. trehalose) appear to play an important role in desiccation tolerance across microorganisms, invertebrates, and plants. Tardigrades synthesize protectants such as trehalose and glycerol to facilitate recovery from the tun state, though trehalose levels vary significantly among species (Sadowska-Bartosz and Bartosz 2024).

Certain tardigrades (e.g. *Paramacrobiotus* and *Macrobiotus* spp.) enable the transition into vitrification (glassy state) to maintain cellular homeostasis and immobilize proteins. Not only sugars but also proteins such as late embryogenesis abundant (LEA) proteins may participate in vitrification and cell stabilization. The accumulation of these proteins during dehydration contributes to glass formation and enhances stability. This may explain why minimal or no trehalose is detected in tardigrades during the vitrified state (Schill and Hengherr 2018).

TDPs encompass multiple protein groups. For example, LEA proteins act as molecular chaperones, forming a physical barrier between partially unfolded neighboring proteins to prevent aggregation, which have been identified in the anhydrobiotic state of *Macrobiotus hufelandi* (Schill and Hengherr 2018). Besides LEA proteins, mitochondrial abundant heat soluble (MAHS) proteins are found in *Ramazzottius varieornatus*, while cytoplasmic abundant heat soluble (CAHS) and secretory abundant heat soluble (SAHS) protein families are present in *Hypsibius dujardini* and *Milnesium tardigradum* (Tanaka et al. 2015; Yamaguchi et al. 2012). LEA, MAHS, CAHS, and SAHS families are collectively termed TDPs. All TDPs form amorphous solids upon desiccation, and this vitrified state contributes to protecting tardigrades (Schill and Hengherr 2018).

2.11.3 Adaptations to Cryobiosis

Tardigrades also exhibit remarkable tolerance to low temperatures, with some species even surviving −196 °C during cryobiosis (Hengherr and Schill 2018). *Acutuncus antarcticus* from Antarctica was successfully retrieved from frozen phytobenthos after up to 30.5 years (Tsujimoto et al. 2016). Tardigrades employ antifreeze proteins (AFPs) to prevent intracellular ice formation, while ice-nucleating proteins and other ice-nucleating agents (INAs) regulate ice crystal formation and morphology, thereby mitigating cellular damage (Glime 2013; Hengherr and Schill 2018). Furthermore, the survival rates of adult tardigrades and their embryonic stages under subzero temperatures are influenced by cooling rates. Although studies on cold tolerance of tardigrades are ongoing, the physiological and biochemical adaptations underlying these mechanisms remain poorly understood. Notably, the survival stress posed by low-temperature stress (e.g. membrane damage and oxidative stress) resembles that of desiccation, suggesting a cross-tolerance mechanism between anhydrobiosis and cryobiosis (Hengherr and Schill 2018).

2.11.4 Molecular Mechanisms of Radiation and ROS Tolerance

Tardigrades, renowned as extraordinary extremophiles in astrobiology, exhibit remarkable resistance to radiation and are considered among the most radiation-tolerant animals on Earth. Some species can endure ionizing radiation doses reaching several kGy (Jönsson et al. 2018). In space environments, they face exposure to galactic cosmic radiation and ionizing solar radiation, primarily composed of protons (85–95%) and alpha particles (5–15%), both key components of cosmic radiation affecting their survival. Additionally, UV solar radiation presents the greatest challenge, significantly reducing their viability (Jönsson et al. 2018; Sadowska-Bartosz and Bartosz 2024).

No significant differences in tolerance have been observed between sparsely ionizing radiation (low-LET, such as photons and electrons) and densely ionizing radiation (high-LET, including alpha particles, protons, neutrons, and heavy ions). Both anhydrobiotic and hydrated tardigrades display comparable resilience to ionizing radiation (Jönsson et al. 2018). While molecular mechanisms underlying their radiation tolerance have increasingly been studied, they remain largely unresolved. Due to their similarity to desiccation tolerance pathways, DNA repair is thought to play a crucial role, alongside ROS scavenging, in supporting their remarkable resistance to extreme conditions. Dehydration, high/low temperature, osmotic stress, and radiation are all capable of inducing ROS accumulation and subsequent DNA damage. Tardigrades possess highly sophisticated molecular mechanisms that help mitigate ROS-induced cellular destruction and protect their genetic integrity.

2.11.4.1 ROS Scavenging Mechanisms

Within the antioxidant defense systems of tardigrades, diverse small-molecular antioxidants such as the tripeptide glutathione (γ-Glu-Cys-Gly) are capable of scavenging ROS (Sadowska-Bartosz and Bartosz 2024). *Hypsibius henanensis* may have acquired the dihydroxyphenylalanine (DOPA) dioxygenase gene *DODA1* via HGT from bacteria. DODA1 protein confers radiation resistance in tardigrades by synthesizing betalains with radical scavenging activity (Li et al. 2024a). Most tardigrade species possess a substantial number of proteins associated with antioxidant defense, including critical ROS scavenging enzymes such as superoxide dismutases (SODs) and catalase. Nearly all tardigrade species possess an expanded family of glutathione transferases, constituting a further detoxification system. For instance, *Ramazzottius varieornatus* upregulates over 1300 genes encoding GSTs, peroxiredoxins, and thioredoxins under UV stress (Yoshida et al. 2017). Under extreme stress conditions, tardigrades significantly reduce their metabolic rates, even to the state of nearly halting metabolic activity during cryptobiosis like anhydrobiosis, thereby minimizing the ROS production.

2.11.4.2 DNA Repair Mechanisms

In addition to ROS scavenging mechanisms, tardigrades exhibit remarkable DNA repair capabilities. Nearly all tardigrades possess all major DNA repair pathways, including homologous recombination, non-homologous end joining, nucleotide excision repair, base excision repair, and mismatch repair. A tardigrade-specific nucleosome-binding protein, termed damage suppressor (Dsup), was identified in *R. varieornatus*. Dsup tightly associates with nuclear DNA, protecting chromosomal DNA from hydroxyl radical-induced damage (Hashimoto et al. 2016). Dsup has been identified in two species within the Hypsibioidea superfamily. Anoud et al. (2024) discovered a novel tardigrade-specific gene encoding Tardigrade DNA damage Response 1 (TDR1). The TDR1 protein, widely conserved across tardigrades, interacts with DNA and forms high-concentration aggregates. Notably, TDR1 is among the most highly expressed genes under ionizing radiation, suggesting its potential role in both DNA condensation and preservation of chromosome organization throughout the DNA repair process (Anoud et al. 2024; Sadowska-Bartosz and Bartosz 2024). In *Hypsibius henanensis*, a tardigrade-specific radiation-induced disordered protein (TRID1) mediates phase separation, thereby accelerating the repair of DNA double-strand breaks (DSBs) (Li et al. 2024a).

Furthermore, tardigrades also modulate stress-signaling pathways to regulate cellular responses to radiation, oxidative stress, and other pressures (Sadowska-Bartosz and Bartosz 2024). Hashimoto et al. (2016) hypothesized that tardigrades suppress autophagy induction to avoid excessive cellular damage caused by severe stress, potentially enabling the utilization of damaged biomolecules for resumption after rehydration.

The synergistic integration of efficient antioxidant systems, strong DNA repair pathways, and unique DNA protective proteins (e.g. Dsup, TRID1, and TDR1) reinforces the remarkable resilience of tardigrades. In particular, their resistance to dehydration and ionizing radiation has directly enabled their utilization in space programs and survival during multiple extraterrestrial exposure experiments.

2.11.5 Tardigrades in Space

At least two space missions with tardigrades involved have been reported: the FOTON-M3 mission in 2007 and the Endeavour mission in 2011. The FOTON-M3 mission included three distinct projects: (1) The Tardigrade Resistance to Space Effects (TARSE) Project aimed to observe and analyze the influence of environmental stress, life history traits, and DNA damage in space (on spacecraft). During the experiment, active and anhydrobiotic *Paramacrobiotus richtersi* were exposed to radiation under microgravity conditions. The results indicated no induction of HSPs (70 and 90 kDa), while the antioxidant response was induced. (2) The Tardigrada In Space (TARDIS) Project primarily investigated whether *Milnesium tardigradum* and *Richtersius coronifer* could survive under open space conditions. Desiccated tardigrades survive when exposed to space vacuum, cosmic radiation, and UV radiation at low Earth orbit for 10 days. However, galactic cosmic radiation, ionizing solar radiation, and UV solar radiation significantly reduced their survival rate. (3) The RoTaRad (Rotifers, Tardigrades and Radiation) Project examined the effects of space conditions (primarily cosmic radiation) on the initial survival, long-term survival, and fecundity of experimental organisms. In the subsequent Endeavor mission, the TARDIKISS Project (Tardigrades in Space) aimed to further understand the life history traits and mechanisms of repairing structural DNA damage in *P. richtersi* and *Ramazzottius oberhaeuseri* when exposed to space conditions for 16 days. The results indicated no significant effects of microgravity and cosmic radiation on survival rates (Sadowska-Bartosz and Bartosz 2024; Weronika and Łukasz 2017).

However, another study revealed significant differences between the experimental groups sent into space and control animals on Earth in ROS scavenging enzymes, glutathione levels, and fatty acid composition (Rizzo et al. 2015). Additionally, recent research demonstrated that *Paramacrobiotus experimentalis* exhibits a long-term resistance to high concentrations of magnesium perchlorate—up to 1.5 mM—within levels expected on Mars (Wilanowska et al. 2024). Given their remarkable resilience to extreme conditions and the valuable insights gained from space experiments, tardigrades are expected to continue contributing as key model organisms in astrobiology.

References

Anoud, M., Delagoutte, E., Helleu, Q. et al. 2024. Comparative transcriptomics reveal a novel tardigrade-specific DNA-binding protein induced in response to ionizing radiation. *eLife*, 13, pp. RP92621.

Arakawa, K. 2022. Examples of extreme survival: tardigrade genomics and molecular anhydrobiology. *Annual Review of Animal Biosciences*, 10, pp. 17–37.

Averhoff, B., Kirchner, L., Pfefferle, K. et al. 2021. Natural transformation in Gram-negative bacteria thriving in extreme environments: from genes and genomes to proteins, structures and regulation. *Extremophiles*, 25, pp. 425–436.

Baker-Austin, C., and Dopson, M. 2007. Life in acid: pH homeostasis in acidophiles. *Trends in Microbiology*, 15, pp. 165–171.

Bar-On, Y.M., Phillips, R., and Milo, R. 2018. The biomass distribution on Earth. *Proceedings of the National Academy of Sciences of the United States of America*, 115, pp. 6506–6511.

Barrón-Sandoval, A., Martiny, J.B., H., Pérez-Carbajal, T. et al. 2023. Functional significance of microbial diversity in arid soils: biological soil crusts and nitrogen fixation as a model system. *FEMS Microbiology Ecology*, 99, pp. 1–15.

Baumgartner, M., Eberhardt, S., De Jonckheere, J.F. et al. 2009. *Tetramitus thermacidophilus* n. sp., an amoeboflagellate from acidic hot springs. *Journal of Eukaryotic Microbiology*, 56, pp. 201–206.

Bayer-Giraldi, M., Uhlig, C., John, U. et al. 2010. Antifreeze proteins in polar sea ice diatoms: diversity and gene expression in the genus *Fragilariopsis*. *Environmental Microbiology*, 12, pp. 1041–1052.

Benforte, F.C., Colonnella, M.A., Ricardi, M.M. et al. 2018. Novel role of the LPS core glycosyltransferase WapH for cold adaptation in the Antarctic bacterium *Pseudomonas extremaustralis*. *PLoS One*, 13, pp. e0192559.

Blöchl, E., Rachel, R., Burggraf, S. et al. 1997. *Pyrolobus fumarii*, gen. and sp. nov., represents a novel group of archaea, extending the upper temperature limit for life to 113 °C. *Extremophiles*, 1, pp. 14–21.

Boutz, D.R., Cascio, D., Whitelegge, J. et al. 2007. Discovery of a thermophilic protein complex stabilized by topologically interlinked chains. *The Journal of Molecular Biology*, 368, pp. 1332–1344.

Busse, L., Tisza, M., and DiRuggiero, J. 2022. Viruses ubiquity and diversity in Atacama Desert endolithic communities. *Viruses*, 14, pp. 1983.

Cario, A., Grossi, V., Schaeffer, P. et al. 2015. Membrane homeoviscous adaptation in the piezo-hyperthermophilic archaeon *Thermococcus barophilus. Frontiers in Microbiology*, 6, pp. 1152.

Caro-Astorga, J., Meyerowitz, J.T., Stork, D.A. et al. 2024. Polyextremophile engineering: a review of organisms that push the limits of life. *Frontiers in Microbiology*, 15, pp. 1341701.

Catling, D.C., and Zahnle, K.J. 2020. The Archean atmosphere. *Science Advances*, 6, pp. eaax1420.

Clark, M.S., Hoffman, J.I., Peck, L.S. et al. 2023. Multi-omics for studying and understanding polar life. *Nature Communications*, 14(1), pp. 7451.

Chen, J., Lian, B., Wang, B. et al. 2006. The occurrence and biogeochemistry of microbes in extreme environments. *Earth Science Frontiers*, 13, pp. 199–207.

Clarke, A., Morris, G.J., Fonseca, F. et al. 2013. A low temperature limit for life on Earth. *PLoS One*, 8, pp. e66207.

Coker, J.A. 2019. Extremophiles and biotechnology: current uses and prospects. *F1000Research*, 5, pp. 396.

Coleine, C. and Delgado-Baqueriz, M. 2022. Unearthing terrestrial extreme microbiomes for searching terrestrial-like life in the Solar System. *Trends in Microbiology*, 30, p. 1101–1115.

Coleine, C., Stajich, J.E., and Selbmann, L. 2022. Fungi are key players in extreme ecosystems. *Trends in Ecology and Evolution*, 37(6), pp. 517–528.

Cui, H., and Dyall-Smith, M.L. 2021. Cultivation of halophilic archaea (class Halobacteria) from thalassohaline and athalassohaline environments. *Marine Life Science & Technology*, 3, pp. 243–251.

Dadachova, E., Bryan, R.A., Huang, X. et al. 2007. Ionizing radiation changes the electronic properties of melanin and enhances the growth of melanized fungi. *PLoS One*, 2, pp. e457.

Dalmasso, C., Oger, P., Selva, G. et al. 2016. *Thermococcus piezophilus* sp. nov., a novel hyperthermophilic and piezophilic archaeon with a broad pressure range for growth, isolated from a deepest hydrothermal vent at the Mid-Cayman Rise. *Systematic and Applied Microbiology*, 39, pp. 440–444.

Daly, M.J., Gaidamakova, E.K., Matrosova, V.Y. et al. 2004. Accumulation of Mn(II) in *Deinococcus radiodurans* facilitates gamma-radiation resistance. *Science*, 306, pp. 1025–1028.

DeLong, E.F., Preston, C.M., Mincer, T. et al. 2006. Community genomics among stratified microbial assemblages in the ocean's interior. *Science*, 311, pp. 496–503.

DiGiacomo, J., McKay, C., and Davila, A. 2022. ThermoBase: a database of the phylogeny and physiology of thermophilic and hyperthermophilic. *PLoS One*, 17, pp. e0268253.

Dong, H.L., and Jiang, H.C. 2023. *Geomicrobiology*, pp. 320. Beijing, Higher Education Press [in Chinese].

Ezemaduka, A.N., Wang, Y., and Li, X. 2017. Expression of CeHSP17 protein in response to heat shock and heavy metal ions. *Journal of Nematology*, 49, pp. 334–340.

Farci, D., Haniewicz, P., and Piano, D. 2022. The structured organization of *Deinococcus radiodurans*' cell envelope. *Proceedings of the National Academy of Sciences of the United States of America*, 119, pp. e2209111119.

Fernández-López, M.G., Batista-García, R.A., and Aréchiga-Carvajal, E.T. 2023. Alkaliphilic/alkali-tolerant fungi: molecular, biochemical, and biotechnological aspects. *Journal of Fungi*, 9, pp. 652.

Fredrickson, J.K., Romine, M.F., Beliaev, A.S. et al. 2008. Towards environmental systems biology of *Shewanella. Nature Reviews Microbiology*, 6, pp. 592–603.

Frösler, J., Panitz, C., Wingender, J. et al. 2017. Survival of *Deinococcus geothermalis* in Biofilms under desiccation and simulated space and Martian conditions. *Astrobiology*, 17(5), pp. 431–447.

Gao, S., Paez-Espino, D., Li, J. et al. 2022. Patterns and ecological drivers of viral communities in acid mine drainage sediments across Southern China. *Nature Communications*, 13(1), pp. 2389.

Gholami, M., Etemadifar, Z., and Bouzari, M. 2015. Isolation a new strain of *Kocuria rosea* capable of tolerating extreme conditions. *Journal of Environmental Radioactivity*, 144, pp. 113–119.

Gilbert, J.A., Hill, P.J., Dodd, C. et al. 2004. Demonstration of antifreeze protein activity in Antarctic lake bacteria. *Microbiology*, 150(Pt 1), pp. 171–180.

Glime, J.M. 2013. Tardigrade survival. Chapter 5-1. In: *Bryophyte ecology, 2, bryological interaction* (ed. J. M. Glime), pp. 26. Houghton, Michigan Technological University.

Glud, R.N., Wenzhöfer, F., Middelboe, M. et al. 2013. High rates of microbial carbon turnover in sediments in the deepest oceanic trench on Earth. *Nature Geoscience*, 6, pp. 284–288.

Gomaa, F., Utter, D.R., Loo, W. et al. 2022. Exploring the protist microbiome: the diversity of bacterial communities associated with *Arcella* spp. (Tubulina: Amoebozoa). *European Journal of Protistology*, 82, pp. 125861.

Gostinčar, C., Stajich, J.E., and Gunde-Cimerman, N. 2023. Extremophilic and extremotolerant fungi. *Current Biology*, 33, pp. R752–R756.

Goto, T., Ogami, S., Yoshimume, K. et al. 2022. Differences in bioenergetic metabolism of obligately alkaliphilic *Bacillaceae* under high pH depend on the aeration conditions. *Frontiers in Microbiology*, 13, pp. 842785.

Grant, W.D. 2004. Life at low water activity. *Philosophical Transactions of the Royal Society of London. Series B, Biological Sciences*, 359, pp. 1249–1266.

Guidetti, R., and Møbjerg, N. 2018. Environmental adaptations: encystment and cyclomorphosis. In: *Water bears: the biology of tardigrades, Zoological Monographs* (Vol. vol 2) (ed. R. Schill), pp. 249–271. Cham, Springer.

Hagemann, M., Henneberg, M., Felde, V.J.M.N.L. et al. 2017. Cyanobacterial populations in biological soil crusts of the northwest Negev Desert, Israel-effects of local conditions and disturbance. *FEMS Microbiology Ecology*, 93, pp. 1–9.

Harding, T., Roger, A.J., and Simpson, A.G.B. 2017. Adaptations to high salt in a halophilic protist: differential expression and gene acquisitions through duplications and gene transfers. *Frontiers in Microbiology*, 8, pp. 944.

Hartman, R., Munson-McGee, J., Young, M. et al. 2019. Survey of high-resolution archaeal virus structures. *Current Opinion in Virology*, 36, pp. 74–83.

Hashimoto, T., Horikawa, D.D., Saito, Y. et al. 2016. Extremotolerant tardigrade genome and improved radiotolerance of human cultured cells by tardigrade-unique protein. *Nature Communications*, 7, pp. 12808.

Hengherr, S., and Schill, R.O. 2018. Environmental adaptations: cryobiosis. In: *Water bears: the biology of tardigrades, Zoological Monographs* (Vol. vol 2) (ed. R. Schill), pp. 295–309. Cham, Springer.

Hicks, D.B., Jereen, A., Fackelmayer, O.J. et al. 2019. Mutational loss of carotenoids in alkaliphilic *Bacillus pseudofirmus* OF4 results in sensitivity to oxidative stress and growth at high pH. *Microbiology-sgm*, 165, pp. 1001–1012.

Hoham, R.W., Frey, F.M., Mohn, W.W. et al. 2008. Optimum growth temperatures of three species of green *Chloromonas* snow algae from Upstate New York and the white mountains, Arizona. *Arctic, Antarctic, and Alpine Research*, 40, pp. 355–363.

Hu, C., Lu, J., Chen, Z. et al. 2025. Viral diversity and auxiliary metabolic genes in rare earth element mine drainage in South China. *Water Research*, 281, pp. 123666.

Ito, M., and Takahashi, Y. 2017. Nonconventional cation-coupled flagellar motors derived from the alkaliphilic *Bacillus* and *Paenibacillus* species. *Extremophiles*, 21, pp. 3–14.

Jebbar, M., Franzetti, B., Girard, E. et al. 2015. Microbial diversity and adaptation to high hydrostatic pressure in deep-sea hydrothermal vents prokaryotes. *Extremophiles*, 19, pp. 721–740.

Jia, Z., and Davies, P.L. 2002. Antifreeze proteins: an unusual receptor-ligand interaction. *Trends in Biochemical Sciences*, 27, pp. 101–106.

Jönsson, K.I., Rabbow, E., Schill, R.O. et al. 2008. Tardigrades survive exposure to space in low Earth orbit. *Current Biology*, 18, pp. R2008729–R2008731.

Jönsson, K.I., Levine, E.B., Wojcik, A. et al. 2018. Environmental adaptations: radiation tolerance. In: *Water bears: the biology of tardigrades, Zoological Monographs* (Vol. vol 2) (ed. R. Schill), pp. 311–329. Cham, Springer.

Jung, Y.H., Yi, J.Y., Jung, H.J. et al. 2010. Overexpression of cold shock protein a of *Psychromonas arctica* KOPRI 22215 confers cold resistance. *Protein Journal*, 29, pp. 136–142.

Kania, K., Levytska, A., Drożak, A. et al. 2022. The effect of temperature conditions during growth on the transformation frequency of *Coccomyxa subellipsoidea* C-169 obtained by electroporation. *Biochemical and Biophysical Reports*, 30, pp. 101220.

Karner, M.B., DeLong, E.F., and Karl, D.M. 2001. Archaeal dominance in the mesopelagic zone of the Pacific Ocean. *Nature*, 409, pp. 507–510.

Kashefi, K., and Lovley, D.R. 2003. Extending the upper temperature limit for life. *Science*, 301, pp. 934.

Kawaguchi, T., and Decho, A.W. 2000. Biochemical characterization of cyanobacterial extracellular polymers (EPS) from modern marine stromatolites (Bahamas). *Preparative Biochemistry and Biotechnology*, 30, pp. 321–330.

Kawaguchi, Y., Shibuya, M., Kinoshita, I. et al. 2020. DNA damage and survival time course of deinococcal cell pellets during 3 years of exposure to outer space. *Frontiers in Microbiology*, 11, pp. 2050.

Kawai, S., Nishihara, A., Matsuura, K. et al. 2019. Hydrogen-dependent autotrophic growth in phototrophic and chemolithotrophic cultures of thermophilic bacteria, *Chloroflexus aggregans* and *Chloroflexus aurantiacus*, isolated from Nakabusa hot springs. *FEMS Microbiology Letters*, 366, pp. fnz 122.

Kim, S.K., Min, Y.H., and Jin, H.J. 2023. Characteristics of the ErmK protein of *Bacillus halodurans* C-125. *Microbiology Spectrum*, 11, pp. e0259822.

Kırtel, O., Versluys, M., Van den Ende, W. et al. 2018. Fructans of the saline world. *Biotechnology Advances*, 36, pp. 1524–1539.

Kozeretska, I., Serga, S., Kovalenko, P. et al. 2022. *Belgica antarctica* (Diptera: Chironomidae): A natural model organism for extreme environments. *Insect Science*, 29(1), pp. 2–20.

Krebs, J.E., Goldstein, E.S., and Kilpatrick, S.T. 2017. *Lewin's GENES XII* (twelfth Edition). Burlington, Jones and Bartlett Learning.

von Kügelgen, A., van Dorst, S., Yamashita, K. et al. 2023. Interdigitated immunoglobulin arrays form the hyperstable surface layer of the extremophilic bacterium *Deinococcus radiodurans*. *Proceedings of the National Academy of Sciences of the United States of America*, 120, pp. e2215808120.

Li, J., Xiao, X., and Zhou, M. 2023. Strategy for the adaptation to stressful conditions of the novel isolated conditional piezophilic strain *Halomonas titanicae* ANRCS81. *Applied and Environmental Microbiology*, 89, pp. e0130422.

Li, L., Ge, Z., Liu, S. et al. 2024a. Multi-omics landscape and molecular basis of radiation tolerance in a tardigrade. *Science*, 386, pp. eadl0799.

Li, X., Vai, W., Yang, Q. et al. 2024b. The extremotolerant desert moss *Syntrichia caninervis* is a promising pioneer plant for colonizing extraterrestrial environments. *The Innovation*, 5, pp. 100657.

Lim, S., Bijlani, S., Blachowicz, A. et al. 2021. Identification of the pigment and its role in UV resistance in *Paecilomyces variotii*, a Chernobyl isolate, using genetic manipulation strategies. *Fungal Genetics and Biology*, 152, pp. 103567.

Liu, P., Ding, W., Lai, Q. et al. 2020. Physiological and genomic features of *Paraoceanicella profunda* gen. nov., sp. nov., a novel piezophile isolated from deep seawater of the Mariana Trench. *Microbiology*, 9, pp. 1–16.

Liu, Y., Jiao, N., Zhong, K. et al. 2023. Diversity and function of mountain and polar supraglacial DNA viruses. *Science Bulletin*, 68, pp. 2418–2433.

Lloyd, C.T., Iwig, D.F., Wang, B. et al. 2022. Discovery, structure, and mechanism of a tetraether lipid synthase. *Nature*, 609, pp. 197–203.

Madigan, M.T., Bender, K.S., Buckley, D.H. et al. 2018. *Brock biology of microorganisms* (15th Edition). New York, Pearson Education.

Matsuno, T., and Yumoto, I. 2015. Bioenergetics and the role of soluble cytochromes C for alkaline adaptation in Gram-negative alkaliphilic *Pseudomonas*. *BioMed Research International*, 2015, pp. 847945.

Matsuno, T., Goto, T., Ogami, S. et al. 2018. Formation of proton motive force under low-aeration alkaline conditions in alkaliphilic bacteria. *Frontiers in Microbiology*, 9, pp. 2331.

McCarter, L.L. 2004. Dual flagellar systems enable motility under different circumstances. *Journal of Molecular Microbiology and Biotechnology*, 7, pp. 18–29.

Merino, N., Aronson, H.S., Bojanova, D.P. et al. 2019. Living at the extremes: extremophiles and the limits of life in a planetary context. *Frontiers in Microbiology*, 10, pp. 780.

Minch, B., Chakraborty, M., Purkis, S. et al. 2024. Active prokaryotic and eukaryotic viral ecology across spatial scale in a deep-sea brine pool, *Isme. Communications*, 4, pp. ycae084.

Møbjerg, N., and Neves, R.C. 2021. New insights into survival strategies of tardigrades. *Comparative Biochemistry and Physiology A-Molecular and Integrative Physiology*, 254, pp. 110890.

Mocali, S., Chiellini, C., Fabiani, A. et al. 2017. Ecology of cold environments: new insights of bacterial metabolic adaptation through an integrated genomic-phenomic approach. *Scientific Reports*, 7, pp. 839.

Mohammadipanah, F., and Wink, J. 2016. Actinobacteria from arid and desert habitats: diversity and biological activity. *Frontiers in Microbiology*, 6, pp. 1541.

Morgan-Kiss, R.M., Priscu, J.C., Pocock, T. et al. 2006. Adaptation and acclimation of photosynthetic microorganisms to permanently cold environments. *Microbiology and Molecular Biology Reviews*, 70, pp. 222–252.

Muryoi, N., Sato, M., Kaneko, S. et al. 2004. Cloning and expression of afpA, a gene encoding an antifreeze protein from the arctic plant growth-promoting rhizobacterium *Pseudomonas putida* GR12-2. *Journal of Bacteriology*, 186, pp. 5661–5671.

Mykytczuk, N.C.S., Foote, S.J., Omelon, C.R. et al. 2013. Bacterial growth at −15 °C; molecular insights from the permafrost bacterium *Planococcus halocryophilus* Or1. *ISME Journal*, 7, pp. 1211–1226.

Nazareth, S., and Gonsalves, V. 2013. Aspergillus penicillioides—a true halophile existing in hypersaline and polyhaline econiches. *Annals of Microbiology*, 64, pp. 397–402.

Oger, P.M., and Jebbar, M. 2010. The many ways of coping with pressure. *Research in Microbiology*, 61, pp. 799–809.

Ohke, Y., Sakoda, A., Kato, C. et al. 2013. Regulation of cytochrome c- and quinol oxidases, and piezotolerance of their activities in the deep-sea piezophile *Shewanella violacea* DSS12 in response to growth conditions. *Bioscience Biotechnology and Biochemistry*, 77, pp. 1522–1528.

Orcutt, B.N., Sylvan, J.B., Knab, N.J. et al. 2011. Microbial ecology of the dark ocean above, at, and below the seafloor. *Microbiology and Molecular Biology Reviews*, 75, pp. 361–422.

Oren, A. 2008. Microbial life at high salt concentrations: phylogenetic and metabolic diversity. *Aquatic Biosystems*, 4, pp. 2.

Oren, A. 2014. Taxonomy of halophilic archaea: current status and future challenges. *Extremophiles*, 18, pp. 825–834.

Oren, A. 2015. Halophilic microbial communities and their environments. *Current Opinion in Biotechnology*, 33, pp. 119–124.

Pati, S., Banerjee, S., Sengupta, A. et al. 2023. Adaptation strategies of thermophilic microbes. In: *Bacterial survival in the hostile environment* (eds. A. Kumar and S. Tenguria), pp. 231–249. Cambridge MA, Academic Press.

Peoples, L.M., Kyaw, T.S., Ugalde, J.A. et al. 2020. Distinctive gene and protein characteristics of extremely piezophilic *Colwellia*. *BMC Genomics*, 21, pp. 1–18.

Perugino, G., Valenti, A., D'amaro, A. et al. 2009. Reverse gyrase and genome stability in hyperthermophilic organisms. *Biochemical Society Transactions*, 37, pp. 69–73.

Pócsi, I., Dijksterhuis, J., Houbraken, J. et al. 2024. Biotechnological potential of salt tolerant and xerophilic species of *Aspergillus*. *Applied Microbiology and Biotechnology*, 108, pp. 521.

Pollio, A., Cennamo, P., Ciniglia, C. et al. 2005. *Chlamydomonas pitschmannii* Ettl, a little-known species from thermoacidic environments. *Protist*, 156, pp. 287–302.

Prangishvili, D., Vestergaard, G., Häring, M. et al. 2006. Structural and genomic properties of the hyperthermophilic archaeal virus ATV with an extracellular stage of the reproductive cycle. *Journal of Molecular Biology*, 359(5), pp. 1203–1216.

Rahlff, J., Turzynski, V., Esser, S.P. et al. 2021. Lytic archaeal viruses infect abundant primary producers in Earth's crust. *Nature Communications*, 12, pp. 4642.

Rappaport, H.B., and Oliverio, A.M. 2023. Extreme environments offer an unprecedented opportunity to understand microbial eukaryotic ecology, evolution, and genome biology. *Nature Communications*, 14, pp. 4959.

Rebecchi, L., Altiero, T., Guidetti, R. et al. 2009. Tardigrade resistance to space effects: first results of experiments on the LIFE-TARSE mission on FOTON-M3 (September 2007). *Astrobiology*, 9, pp. 581–591.

Rivkina, E.M., Friedmann, E.I., McKay, C.P. et al. 2000. Metabolic activity of permafrost bacteria below the freezing point. *Applied and Environmental Microbiology*, 66, pp. 3230–3233.

Rizzo, A.M., Altiero, T., Corsetto, P.A. et al. 2015. Space flight effects on antioxidant molecules in dry tardigrades: the TARDIKISS experiment. *BioMed Research International*, 2015, pp. 167642.

Rothschild, L.J., and Mancinelli, R.L. 2001. Life in extreme environments. *Nature*, 409, pp. 1092–1101.

Rothschild, L.J., Giver, L.J., White, M.R. et al. 1994. Metabolic activity of microorganisms in evaporites. *Journal of Phycology*, 30, pp. 431–438.

Sadowska-Bartosz, I., and Bartosz, G. 2024. Antioxidant defense in the toughest animals on Earth: Its contribution to the extreme resistance of tardigrades. *International Journal of Molecular Sciences*, 25, pp. 8393.

Saghaï, A., Gutiérrez-Preciado, A., and Deschamps, P. 2017. Unveiling microbial interactions in stratified mat communities from a warm saline shallow pond. *Environmental Microbiology*, 19(6), pp. 2405–2421.

Salinas-García, M.Á., Fernbach, J., Rinnan, R. et al. 2025. Extreme smells—microbial production of volatile organic compounds at the limits of life. *FEMS Microbiology Reviews*, 49, fuaf004.

Scheffer, G., and Gieg, L.M. 2023. The mystery of piezophiles: understudied microorganisms from the deep, dark subsurface. *Microorganisms*, 11, pp. 1629.

Schill, R.O., and Hengherr, S. 2018. Environmental adaptations: desiccation tolerance. In: *Water bears: the biology of tardigrades* (ed. R. Schill), pp. 273–293. Cham, Springer.

Schmid, A.K., Allers, T., and DiRuggiero, J. 2020. SnapShot: microbial extremophiles. *Cell*, 80, pp. 818–818.e1.

Scoma, A. 2021. Functional groups in microbial ecology: updated definitions of piezophiles as suggested by hydrostatic pressure dependence on temperature. *ISME Journal*, 15, pp. 1871–1878.

Selbmann, L., Pacelli, C., Zucconi, L. et al. 2018. Resistance of an Antarctic cryptoendolithic black fungus to radiation gives new insights of astrobiological relevance. *Fungal Biology*, 122, pp. 546–554.

Shaikhutdinov, N., and Gusev, O. 2022. Chironomid midges (Diptera) provide insights into genome evolution in extreme environments. *Current Opinion in Insect Science*, 49, pp. 101–107.

Sharma, A., Scott, J.H., Cody, G.D. et al. 2002. Microbial activity at gigapascal pressures. *Science*, 295, pp. 1514–1516.

Sharma, S., Chaturvedi, U., Sharma, K. et al. 2022. An overview of survival strategies of psychrophiles and their applications. In: *Survival Strategies in Cold-adapted Microorganisms* (eds. R. Goel, R. Soni, D.C. Suyal, and M. Khan), pp. 133–151. Singapore, Springer.

Shu, W.S., and Huang, L.N. 2022. Microbial diversity in extreme environments. *Nature Reviews Microbiology*, 20, pp. 219–235.

Shuryak, I. 2020. Review of resistance to chronic ionizing radiation exposure under environmental conditions in multicellular organisms. *Journal of Environmental Radioactivity*, 212, pp. 106128.

Sørensen-Hygum, T.L., Stuart, R.M., Jørgensen, A. et al. 2018. Modelling extreme desiccation tolerance in a marine tardigrade. *Scientific Reports*, 8, pp. 11495.

Stevenson, A., Burkhardt, J. et al. 2015. Multiplication of microbes below 0.690 water activity: implications for terrestrial and extraterrestrial life. *Environmental Microbiology*, 17, pp. 257–277.

Suzuki, S., Kuenen, J.G., Schipper, K. et al. 2014. Physiological and genomic features of highly alkaliphilic hydrogen-utilizing Betaproteobacteria from a continental serpentinizing site. *Nature Communications*, 5, pp. 3900.

Takai, K., Nakamura, K., Toki, T. et al. 2008. Cell proliferation at 122 C and isotopically heavy CH_4 production by a hyperthermophilic methanogen under high-pressure cultivation. *Proceedings of The National Academy of Sciences of United States of America*, 105, pp. 10949–10954.

Tamby, A., Sinninghe Damsté, J.S., and Villanueva, L. 2023. Microbial membrane lipid adaptations to high hydrostatic pressure in the marine environment. *Frontiers in Molecular Biosciences*, 9, pp. 1058381.

Tanaka, S., Tanaka, J., Miwa, Y. et al. 2015. Novel mitochondria-targeted heat-soluble proteins identified in the anhydrobiotic tardigrade improve osmotic tolerance of human cells. *PLoS One*, 10, pp. e0118272.

Tang, X., Yu, L., Xu, W. et al. 2020. Fungal diversity of deep-sea sediments in Mid-Oceanic Ridge area of the East Pacific and the South Indian Oceans. *Botanica Marina*, 63, pp. 183–196.

Tarn, J., Peoples, L.M., Hardy, K. et al. 2016. Identification of free-living and particle-associated microbial communities present in hadal regions of the Mariana Trench. *Frontiers in Microbiology*, 7, pp. 665.

Terui, Y., Ohnuma, M., Hiraga, K. et al. 2005. Stabilization of nucleic acids by unusual polyamines produced by an extreme thermophile *Thermus thermophilus*. *Biochemical Journal*, 388, pp. 427–433.

Thomas, D.N., and Dieckmann, G.S. 2002. Antarctic Sea ice-a habitat for extremophiles. *Science*, 295 (5555), pp. 641–644.

Tian, B., and Hua, Y. 2010. Carotenoid biosynthesis in extremophilic *Deinococcus-Thermus* bacteria. *Trends in Microbiology*, 18, pp. 512–520.

Timmins, J., and Moe, E. 2016. A decade of biochemical and structural studies of the DNA repair machinery of *Deinococcus radiodurans*: Major findings, functional and mechanistic insight and challenges. *Computational and Structural Biotechnology Journal*, 14, pp. 168–176.

Tsujimoto, M., Satoschi, I., and Hiroshi, K. 2016. Recovery and reproduction of an Antarctic tardigrade retrieved from a moss sample frozen for over 30 years. *Cryobiology*, 72, pp. 78–81.

Uldahl, K.B., Jensen, S.B., Bhoobalan-Chitty, Y. et al. 2016. Life cycle characterization of *Sulfolobus* monocaudavirus 1, an extremophilic spindle-shaped virus with extracellular tail development. *Journal of Virology*, 90, pp. 5693–5699.

Vanlint, D., Mitchell, R., Bailey, E. et al. 2011. Rapid acquisition of gigapascal-high-pressure resistance by *Escherichia coli*. *MBio*, 2, pp. e00130–e00110.

Veerabadhran, M., Chakraborty, S., Mitra, S. et al. 2018. Effects of flask configuration on biofilm growth and metabolites of intertidal Cyanobacteria isolated from a mangrove forest. *Journal of Applied Microbiology*, 125, pp. 190–202.

Wang, Q., Cen, Z., Zhao, J. et al. 2015. The survival mechanisms of thermophiles at high temperatures: an angle of omics. *Physiology*, 30, pp. 97–106.

Wang, X., Liu, Y., Chen, Z. et al. 2024. Genomic functional analysis of novel radiation-resistant species of *Knollia* sp. nov. S7-12^T from the North Slope of Mount Everest. *Microorganisms*, 12, pp. 1748.

Weber, A.P.M., Horst, R.J., Barbier, G.G. et al. 2007. Metabolism and metabolomics of eukaryotes living under extreme conditions. *International Review of Cytology*, 256, pp. 1–34.

Wei, P., Wang, Q., Hang, B. et al. 2017. High-level cell-free expression and functional characterization of a novel aquaporin from *Photobactetrium profundum* SS9. *Process Biochemistry*, 59, pp. 172–179.

Wei, X., Guo, S., Gong, L. et al. 2018. Cultivable fungal diversity in Deep-Sea sediment of the East Pacific Ocean. *Geomicrobiology Journal*, 35, pp. 790–797.

Weronika, E., and Łukasz, K. 2017. Tardigrades in space research – past and future. *Origins of Life and Evolution of Biospheres*, 47, pp. 545–553.

Wharton, D.A. 2003. The environmental physiology of Antarctic terrestrial nematodes: a review. *Journal of Comparative Physiology B-biochemical Systems and Environmental Physiology*, 173, pp. 621–628.

Wilanowska, P.A., Rzymski, P., and Kaczmarek, Ł. 2024. Long-Term survivability of tardigrade *Paramacrobiotus experimentalis* (Eutardigrada) at increased magnesium perchlorate levels: implications for astrobiological research. *Life (Basel)*, 14, pp. 335.

Wu, L., Liu, Y., Shi, W. et al. 2025. Uncovering the hidden RNA virus diversity in Lake Nam Co: evolutionary insights from an extreme high-altitude environment. *Proceedings of The National Academy of Sciences of United States of America*, 122, pp. e2420162122.

Xie, S.C., Qiu, X., Dong, H.L. et al. 2023. *Geobiology*, pp. 570. Beijing, Higher Education Press [in Chinese].

Yamaguchi, A., Tanaka, S., Yamaguchi, S. et al. 2012. Two novel heat-soluble protein families abundantly expressed in an anhydrobiotic tardigrade. *PLoS One*, 7, pp. e44209.

Yoshida, Y., Koutsovoulos, G., Laetsch, D.R. et al. 2017. Comparative genomics of the tardigrades *Hypsibius dujardini* and *Ramazzottius varieornatus*. *PLoS Biology*, 15, pp. 2002266.

Yu, L., Zhou, Z., Wei, S. et al. 2019. *Marinomonas piezotolerans* sp. nov., isolated from deep-sea sediment of the Yap Trench, Pacific Ocean. *International Journal of Systematic and Evolutionary Microbiology*, 69, pp. 739–744.

Zajc, J., Liu, Y., Dai, W. et al. 2013. Genome and transcriptome sequencing of the halophilic fungus *Wallemia ichthyophaga*: haloadaptations present and absent. *BMC Genomics*, 14, pp. 617.

Zhang, Z., Qu, C., Zhang, K. et al. 2020. Adaptation to extreme Antarctic environments revealed by the genome of a sea ice green alga. *Current Biology*, 30, pp. 3330–3341.e7.

Zhang, Y., Wang, Q., Rogers, M. et al. 2025. Autotrophic denitrification under anoxic conditions by newly discovered mixotrophic sulfide-oxidizing bacterium. *Bioresource Technology*, 18, pp. 132553.

3

Microbial Ecosystems

Wei Liu and Xingliang Zhang

3.1 Introduction

Microbial ecosystem refers to a community of microorganisms, such as bacteria, archaea, fungi, protists as well as acellular life forms such as viruses, that coexist in a specific environment and interact with one another through nutrient cycles and energy flows. These ecosystems can be found in a wide range of habitats, including soil, water, air, and the animal body (e.g. the gut microbiome), as well as in extreme environments like deep-sea vents or hot springs. Microbial ecosystems play a critical role in maintaining ecological balance, promoting nutrient cycling, influencing climate, and supporting the development of biosphere and anthroposphere.

Biofilms and microbial mats, composed of highly structured, multispecies communities of microorganisms, are among the most common microbial ecosystems in the natural world. These communities thrive in a wide variety of modern and ancient environments (Figure 3.1). The study of biofilms and microbial mats has evolved into a multidisciplinary field, shifting its focus from examining individual free-floating microbial cells (plankton) or those growing in isolated liquid and agar cultures, to a more advanced approach centered on surface-attached communities. In recent years, the critical role of biofilms and microbial mats in global processes has gained increasing recognition. As our understanding of their functions, both within organisms and in ecosystems, particularly in sediments and on rock surfaces, continues to grow, it's becoming clear that biofilms and microbial mats are among the most widespread life forms that have shaped our planet since the life origin billions of years ago, and therefore are geobiologically significant.

In this chapter, we will explore the development, distribution, and preservation of biofilms and microbial mats, and highlight their significant implications in exploring extraterrestrial life.

3.1.1 From Cell to Microbial Communities

By the 1940s, ecologists and microbiologists began noticing that a significant number of microorganisms in aquatic environments were living as biofilms on submerged surfaces (Figure 3.2) rather than floating freely in the water. These biofilms are complex communities of microorganisms, including bacteria, archaea, algae, fungi, and protozoans, that adhere to surfaces and are embedded in a matrix of extracellular polymeric substances (EPS) they produce.

Introduction to Geobiology, First Edition. Xingliang Zhang, Zisheng Guo, Wei Liu, Ruliang He, Luoyang Li, and Weiduo Hao.

Companion website: www.wiley.com/go/Zhang_1e

Figure 3.1 An extant microbial mat on the tufa wall at Erasbach, South Germany, where the photosynthesis of mat species induces carbonate precipitation of the tufa. *Source:* Photo by Xingliang Zhang.

Biofilms and microbial mats can be regarded as interactive microbial communities, where metabolic activities are integrated, and developmental processes similar to those of multicellular organisms can be observed. The interactions and structural complexities are made possible by the expression of specific gene sets, causing the phenotypes of cells within biofilms to differ significantly from those of planktonically grown cells of the same species. Molecular and microscopic evidence suggest that biofilms exhibit a succession of distinct phenotypes. It was found that bacterial cells upregulate a large number of genes when forming sessile communities. Modern proteomic techniques have shown that the phenotype of biofilm cells differs from their planktonic counterparts in the expression of up to 70% of their genes. Analysis of gene expression in biofilms, using mRNA-based methods, further confirms that biofilm cells differ from their planktonic counterparts as much as planktonic cells of one species differ from those of another (Stoodley et al. 2002).

The intricate genetic interactions between cells in prokaryotic communities are a defining feature of biofilms and microbial mats. Their development is facilitated by the localization of cells on surfaces, which also provides a protective niche where attached cells can establish a localized homeostatic environment. The development of EPS aids in the intercellular transfer of genetic material.

3.1.2 Extracellular Polymeric Substances (EPS)

The most important and still not clearly settled phenomenon in relation to microbial communities is their capacity to excrete large amounts of extracellular substance which, in many cases, make up more than 99% of the total metabolic products. The extracellular compounds were first named extracellular polysaccharides, later EPS.

Figure 3.2 Pebbles (a) coated with biofilms and carbonate mineral precipitates (b) in the river Moosach, Germany. *Source:* Photos by Xingliang Zhang.

EPS are produced and secreted by microorganisms, and may vary in their physical state, chemical properties, and functional roles to cells. They are produced by many different types of colonized microorganisms, including bacteria, algae, and fungi. Depending on the producers and specific habitats, EPS may comprise of numerous types of polysaccharides, proteins, nucleic acids, and amphiphilic polymers or a mixture of these. Due to their polymeric chemistry, EPS molecules may involve in many functions, some of which are listed in Table 3.1.

The EPS are often produced in association with biofilms and/or microbial mats. The production of EPS influences many processes that have broader biological and geobiological importance. For example, EPS gels alter the physical stability and optical signatures of sediments and seawater, induce mineral precipitation, and influence the global flux of carbon and other elements from and to oceans. Finally, biofilms and their EPS are prominent components of extreme environments ranging from cold Antarctic epipontic communities to anhydrophilic hypersaline ponds and to high-temperature and pressure ecosystems in deep-sea hydrothermal vent systems (Decho 2020).

Table 3.1 Functions of extracellular polymeric substances (EPS) in biofilms and microbial mats.

Adhesive • irreversible adhesion (attachment) • de-adhesion (detachment)	**Structure** • polymer network = matrix • cohesive force within biofilms
Protection • predation • desiccation • environmental changes • heavy metals and organic compounds	**Recognition** • co-aggregation • co-adhesion • microcolony formation • lectin interactions
Sorption • particulate and detritus • nutrients • ionic interactions • hydrophobic interactions	**Physiology** • regulatory or protective role in enzyme reactions • storage and reserve functions • intercellular transfer of genetic material

Source: Neu et al. (2003)/Springer Nature.

3.1.3 Preservation Potential

Microbial communities within microbe-dominated ecosystems are capable of forming both organic and mineral deposits, such as the tufa wall shown in Figure 3.1, which was first described by Alexander von Humboldt in 1793 (Krumbein et al. 2003). As they mature or disappear, these structures leave behind a combination of organic polymers and minerals formed through microbial metabolism and microbially induced sedimentary structures (see Section 6.4). As a result, biofilms can be identified in ancient rocks, such as a fossilized biofilm found in the Ediacaran deposits (Figure 3.3). Dried and detached microbial mats may be carried by the wind, sometimes traveling hundreds of kilometers inland. These were once mistaken for organic meteorites (chondrites) or "paper

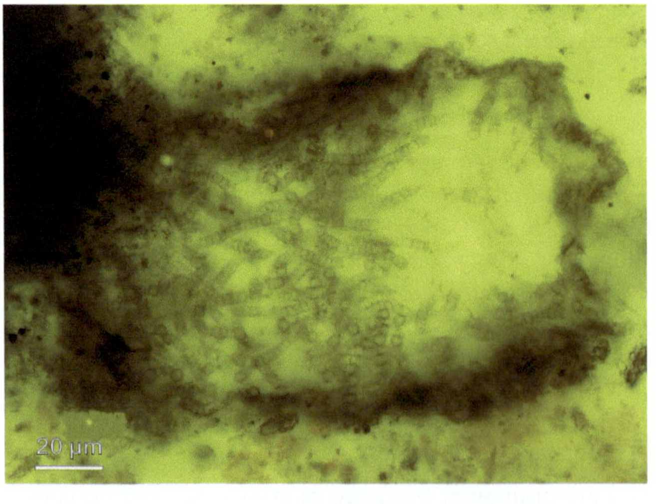

Figure 3.3 Autofluorescent photomicrograph of a silicified biofilm from the chert nodule of the Doushantuo Formation, Ediacaran, South China. *Source:* Photo by Xingliang Zhang.

meteorites," a misconception that persisted for about 250 years, until Ehrenberg (1839) identified them as wind-blown microbial mats or carpets (Krumbein et al. 2003).

3.2 Biofilms and Microbial Mats

In the broadest sense, both biofilms and microbial mats are microbial communities, primarily made up of prokaryotes, that colonize on surfaces. In both cases, microbial colonization depends on the formation of EPS, which trap inorganic and abiotic components and helps retain water. There are close interactions between the microbes, the surface they colonize, and the surrounding environment. Factors like the physical and chemical environment, species composition, and species interactions all play a role in shaping the structure of these communities. These similarities have led to the suggestion that biofilms and microbial mats are essentially the same.

Biofilms are microbial populations that adhere to surfaces within a matrix of EPS, and they form the foundation for microbial mats. Typically, biofilms are thin, organic layers, ranging from tens to a few hundred microns in thickness. They are generally small-scale, often consisting of a single layer of organisms. Biofilms are commonly found in cryptic environments, such as between sedimentary grains, in pore spaces, and within the bodies of animals and plants.

Microbial mats can be seen as more complex biofilms. They are thicker and cover larger surface areas, sometimes spanning several square kilometers. What sets microbial mats apart is their steep chemical gradients, the abundance of phototrophic microorganisms, and the stratification of microbial populations into distinct layers.

3.2.1 Biofilms

3.2.1.1 What Are Biofilms?

The term "biofilm" was first introduced in the early 1970s by microbiologist J. William Costerton and his colleagues, who were investigating bacterial communities that form on surfaces in aquatic environments. Their research revealed that these communities are not random aggregates of individual cells; instead, they are highly organized structures with specialized functions, including nutrient uptake, waste removal, and protection against environmental stressors (Bamford et al. 2023).

Biofilms are now recognized as thin layers of microbial communities, primarily composed of prokaryotes, that are embedded in a matrix of EPS and attached to either inert or living surfaces. The biofilm matrix consists of a variety of biomolecules, including polysaccharides, proteins, extracellular DNA and RNA, lipids, and other compounds. These components collectively form a protective and adhesive layer around the microbial community. The composition of the matrix can vary significantly based on the microbial species present and the surrounding environmental conditions. For example, *Pseudomonas aeruginosa* produces at least three distinct exopolysaccharides in its biofilm. Similarly, matrix production in *Bacillus subtilis* can change depending on the strain and environmental factors, such as temperature (Bamford et al. 2023).

Biofilms are complex, dynamic communities of bacteria encased in a self-produced matrix that holds cells in close proximity. This matrix allows signal molecules to accumulate and enables coordinated behavior through a communication system known as quorum sensing. In biofilms, bacteria exhibit different genetic expression compared to their free-floating, or planktonic, counterparts. Quorum sensing allows bacterial populations to detect when their numbers are sufficient to activate specific genes, such as those for toxin production in pathogenic species. Additionally, cell signaling in biofilms can regulate differentiation into specialized subpopulations that perform distinct roles within the community. Biofilms also afford bacteria increased resistance to antimicrobial agents, a defense attributed to both their physical structure and intrinsic genetic shifts that occur upon attachment to surfaces. Cells can detach from biofilms either individually or in clumps, and clumps retain the biofilm's reduced susceptibility to antimicrobials. Under certain conditions, biofilms are even capable of migrating across surfaces using various mechanisms. The discovery of such coordinated behavior in microbial communities has greatly enhanced our understanding of bacterial survival strategies.

You've probably encountered biofilms in your daily life: the plaque on your teeth, the slippery slime on pebbles in creeks, or the gel-like film inside a vase that once held flowers for a week. Biofilms can also form in the gut linings of animals and on plant roots. They tend to develop on solid surfaces exposed to moisture, soft tissue surfaces in living organisms, or at liquid–air interfaces. Classic natural locations for biofilm development include rocks and other substrates in marine or freshwater environments. In short, biofilms form anywhere there is sufficient moisture and nutrients. In fact, the entire globe, both above and below ground, is "seeded" with the bacteria that form biofilms, and bacterial communities flourish, disperse, or become dormant depending on changing environmental conditions.

3.2.1.2 Why Learn About Biofilms?

Biofilms have become one of the most exciting areas of study in microbiology, medical science, environmental science, and geobiology. Traditionally, microbiologists focused on free-floating bacteria in laboratory cultures, but recent research has shown that in natural environments, most microbes exist as biofilms, which behave very differently. In fact, in any aquatic system, about 99% of bacteria are likely to be found in biofilms attached to surfaces.

These biofilms are a significant source of free-floating microbes in water, some of which can cause infections and diseases in animals. Aquatic environments, especially those impacted by industrial wastewater, are often polluted by the development of biofilms. Both the cellular material and EPS within biofilm systems play a key role in adsorbing charged chemicals, such as heavy metals, as well as in the adsorption and degradation of toxic organic compounds. In this way, biofilms contribute to both the contamination and potential remediation of aquatic ecosystems. In addition to serving as a reservoir for harmful microbes, biofilms can also cause corrosion in stainless steel structures. This adds another layer of concern when dealing with biofilm presence, particularly in industrial or healthcare settings. Both the beneficial and detrimental aspects of biofilms are summarized in Table 3.2.

From a geobiological perspective, microbial communities within biofilms play a crucial role in the cycling of essential elements like carbon, nitrogen, and phosphorus, as well as nutrient recycling. Biofilm systems, including the cells and their EPS, may be involved in mineral formation and can trap very small mineral grains. Furthermore, biofilm growth can significantly alter the processes of transformation in major rock-forming minerals, which are chemically and physically resistant (see Section 7.3).

Biofilms have a significant impact on both geology and the environment. In geology, biofilms grow rapidly on carcasses and can influence soft-tissue preservation by promoting the replication of soft tissues in minerals or by mineralizing themselves to outline the soft tissues (pseudomorphing), e.g. pyritization through microbial sulfate reduction (Figure 3.4). They also play a complex role in sedimentology by enveloping grains and forming organic coatings that stabilize sediments (see Section 6.4). These biofilms can become calcified before internal sediment and cement form, and in some cases, they erode rocks and minerals by dissolving the substrate, creating microbial habitats

Table 3.2 Beneficial and detrimental aspects of biofilms.

Beneficial aspects	Detrimental aspects
In the natural environment • playing a major role in shaping Earth's surface chemistry • key to recycling vital elements • making up most of Earth's living biomass • contributing to the formation of sedimentary minerals and rocks **Waste and wastewater treatment** • developing water-cleaning systems **Remediation of contaminated soil and groundwater** • rendering initially toxic organic material into harmless by-products by bioremediation process **Microbial leaching** • recovery of metals such as Au, Ag, Cu, U, Co, and Mo	**In natural environments** • affect climate by producing and consuming atmospheric gases • mobilizing toxic elements such as Hg, As, and Se • producing toxic algal blooms • creating oxygen depletion zones in lakes, rivers, and coastal environments • inducing to microbial diseases such as plague, cholera, Lyme disease, and West Nile virus **Industrial environments** • biofouling • biocorrosion • equipment damage • product contamination **Public health** • reservoirs for pathogenic organisms and sources of disease outbreaks (such as otitis media, bacterial endocarditis, cystic fibrosis) • nosocomial (hospital-acquired) infections • increasing illnesses associated with pathogenic organisms in the household

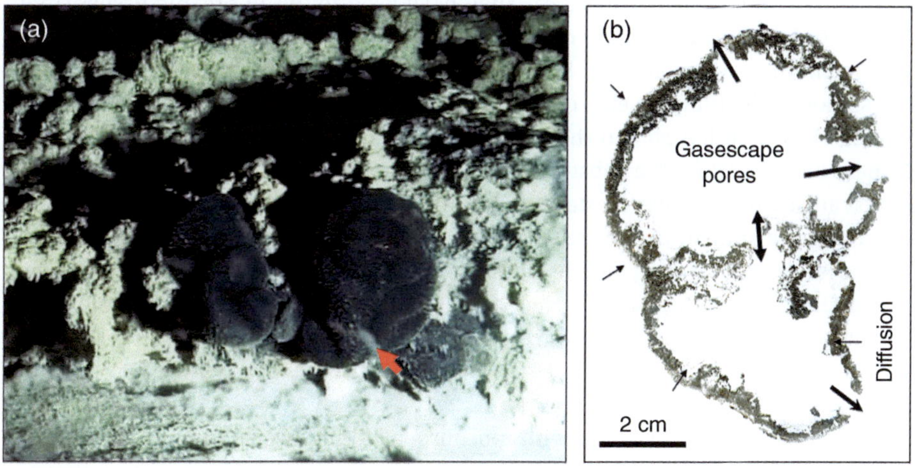

Figure 3.4 Biofilms in the Black Sea: (a) showing stationary biofilms with a "sponge" organization and a free-floating biofilm (see arrow); (b) the internal structures and channels for substances (such as nutrients and metabolites) between the interior and exterior. *Source:* Courtesy of Joachim Reitner.

both on the exterior (epiliths) and interior (endoliths) of solid surfaces (see Section 7.2). In the environment, biofilms attached to contaminated soil particles and aquatic sediments help degrade soil-bound contaminants from chemical spills, while biofilms on plant roots (e.g. *Rhizobium*) assist in nutrient cycling, enhancing agricultural productivity.

To truly appreciate their role in both everyday life and global processes, past and present, it is essential to understand how biofilms develop, their structure, their involvement in geobiological processes, the potential problems they cause, and the strategies we use to control them.

3.2.1.3 Types of Biofilms

Pure cultures and defined mixed culture biofilms can be obtained in laboratories, but the complexity of natural environments introduces additional layers of variability in the structure and function of biofilms. This variability arises from the presence of a wide range of prokaryotic species, the coexistence of eukaryotic organisms, the diverse EPS produced by both prokaryotic and eukaryotic microbes, and interactions with dissolved, colloidal, and particulate compounds in the environment (Neu et al. 2003). In natural settings, three distinct types of biofilms can be identified: (1) subaquatic biofilms, (2) subaerial biofilms, and (3) biodictyon or microbial networks.

Subaquatic biofilms are characterized by a layer of biologically solidified water, where more than 99% of the volume consists of microbial cells from various taxa. These cells modify the chemical gradient between a solid surface and the surrounding water, facilitating metabolic activities that sustain life processes. The internal structure of these biofilms is complex and has been compared to the structure of large cities. In contrast, subaerial biofilms form a thin organic coating on surfaces exposed to minimal water, such as rocks and mineral particles in the atmosphere or within soils, rocks, or dunes. These biofilms can survive for long periods without liquid water, maintaining life processes during intermittent water availability. Almost all surfaces exposed to the atmosphere are covered by some form of biofilm community, visible or invisible. The term "biodictyon" comes from the Greek words "bio" (life) and "diktyon" (net), signifying a living network. The biodictyon is a three-dimensional, often concentric, network of filamentous and unicellular microbes embedded in soil, sediments, or rock fissures. These networks can be as complex as a multicellular organism, resembling the structure of a sponge (Figure 3.4). They have been found in organic tissues, root areas, decaying wood, or in deep sedimentary or rock biospheres where environmental conditions prevent the development of two-dimensional biofilms or multilayered microbial mats.

In lotic systems, stationary biofilms develop on submerged surfaces such as pebbles, rocks, wood, leaves, and sediments. Research has shown the complex interactions between bacterial and algal activities within these biofilms. Both the substrate and the hydrodynamics of a lotic system influence biofilm biomass and activity. However, stationary biofilms play a relatively minor role in large rivers and streams due to the limited surface area available. In these systems, many microorganisms associate with mobile interfaces, forming mobile biofilms. These mobile aggregates, akin to marine and lake snow, are referred to as "river snow" (Neu 2000). Compared to marine and lake biofilms, lotic biofilms are subject to continuous shear forces, making them smaller and more rigid.

3.2.1.4 Development of Biofilms

As soon as a clean surface comes into contact with moisture, a biofilm begins to form. The development of a biofilm occurs in several stages. Initially, individual cells attach to the surface, followed by the formation of larger biofilm areas. At any stage in the biofilm's development, single cells or larger sections of the biofilm may detach from the interface, allowing the cells to colonize new habitats and expand to other areas.

When a clean surface first comes into contact with water, the initial substances that are associated with the surface are not microbial cells but trace organic materials. Almost immediately, an organic layer forms on the water–solid interface, creating a "conditioning layer" (Figure 3.5). This layer helps neutralize excessive surface charge and surface free energy, which can prevent microbial cells from approaching closely enough to initiate attachment. Additionally, the organic molecules

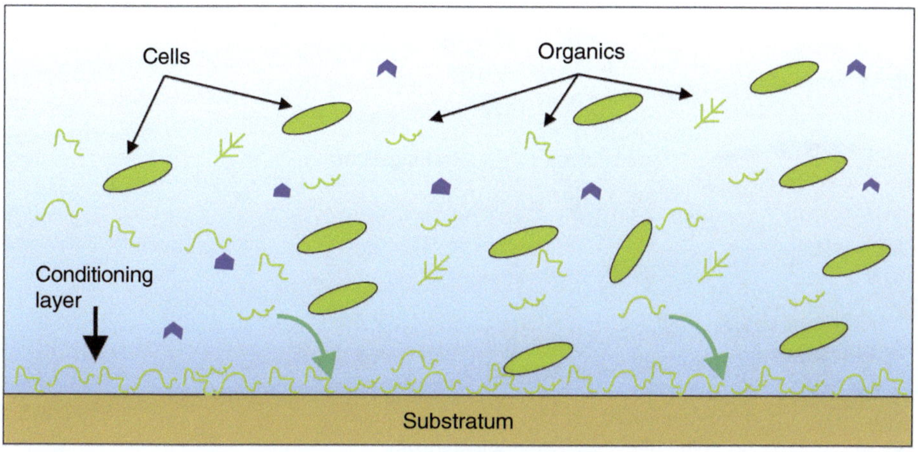

Figure 3.5 Adsorption of organic molecules on a clean surface forms a conditioning layer.

adsorbed onto the surface often serve as a nutrient source for microbial growth, facilitating their subsequent colonization.

Some planktonic (free-floating) microbial cells will approach the interface, where they become entrained within the boundary layer. A portion of these cells will collide with the surface and adsorb for a brief period, only to desorb later. This process is known as reversible adsorption. The initial attachment is driven by electrostatic and physical forces, rather than chemical bonding. Some of the reversibly adsorbed cells, however, begin to prepare for a more permanent attachment by forming structures that can bind them to the surface. These cells then undergo irreversible adsorption, also called permanent attachment (Figure 3.6). Additionally, microbial cells may activate or deactivate specific genes, transitioning from planktonic to sessile states. Various microbial species employ different adhesion mechanisms, which may involve specialized cell surface molecules or structures like fimbriae or pili. In natural environments, pioneer microorganisms are not limited to prokaryotes; they may also include algae, fungi, and protozoans.

Once microbial cells reach the surface, they begin to metabolize available nutrients, grow into microcolonies, and eventually form a biofilm with a distinct structure. Biofilm cells secrete EPS (Figure 3.7), which bind the biofilm together and anchor it to the substrate. In addition to providing

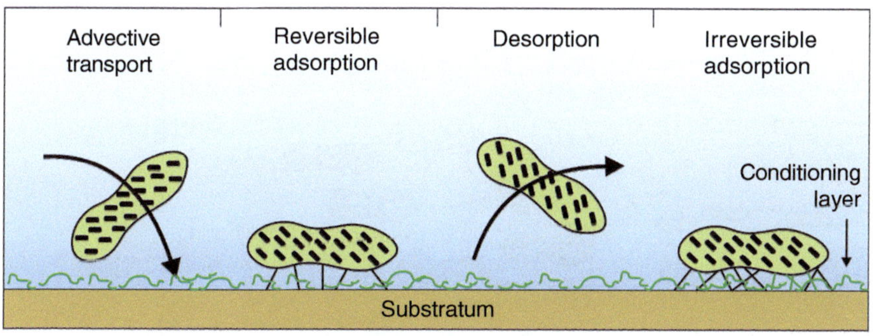

Figure 3.6 Adhesion of bacterial cells to the conditioned surface, reversible adsorption, desorption, and irreversible adsorption occurring at this stage.

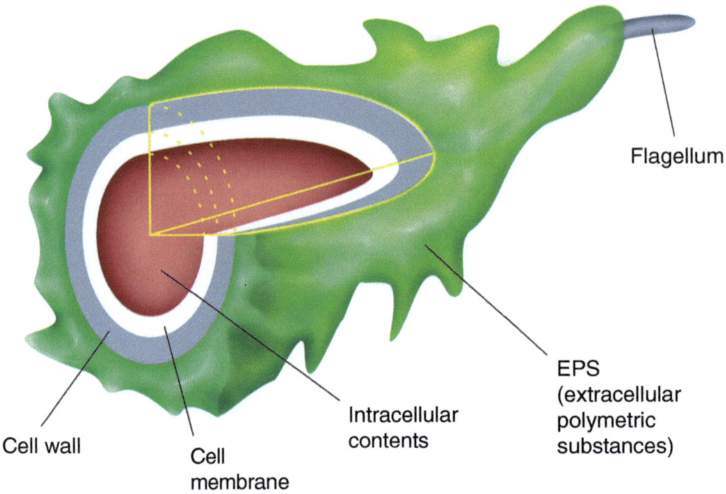

Figure 3.7 A three-dimensional model of bacterial cell to show the extracellular slime layer.

structural support, EPS trap scarce nutrients and protects the cells from environmental stressors. Research shows that attachment is facilitated by EPS extending from the bacterial cell wall. This polymeric material, known as glycocalyx, contains both charged and neutral polysaccharide groups that not only promote attachment but also function as an ion-exchange system to capture and concentrate trace nutrients from the surrounding water. The EPS also acts as a protective shield, helping the bacteria resist biocides and other toxic substances.

As nutrients accumulate, the pioneer cells begin to reproduce, with daughter cells producing their own glycocalyx, which significantly increases the surface area for ion exchange. This leads to the development of a thriving microbial colony (Figure 3.8). In a mature biofilm, the volume is primarily occupied by the loosely organized glycocalyx matrix (75%–95%), while microbial cells make up only a small portion (5%–25%). Since the glycocalyx holds a substantial amount of water, biofilm-covered surfaces become gelatinous and slippery. In essence, biofilm is composed of microbial cells entangled within a dense network of extracellular polymers, resembling a spider's web.

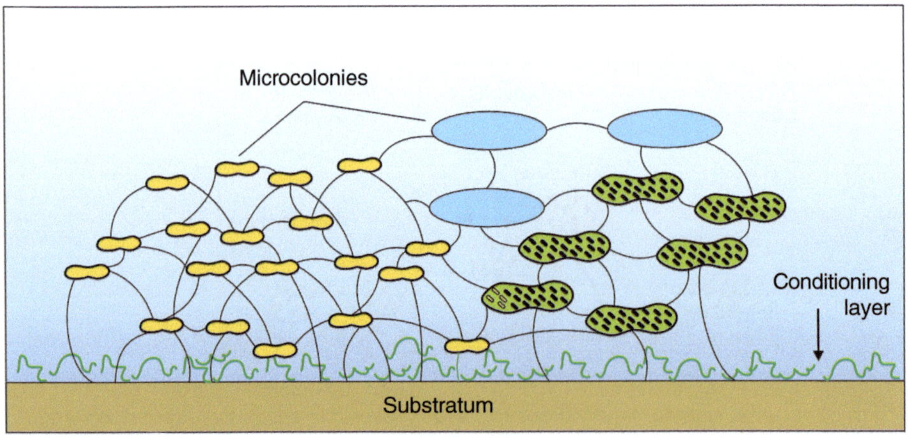

Figure 3.8 A matured biofilm, in which bacteria and other microorganisms develop cooperative microcolonies.

3.2.1.5 Factors that Affect Biofilm Attachment and Growth

The material of a surface has little or no effect on biofilm development. Even on the silver-coated ion exchange resins, designed to inhibit microbial growth, bacteria that are tolerant to silver can still emerge (Flemming and Geesey 1991). While smoother surfaces may slow down the initial attachment of bacteria, they do not seem to have a significant effect on the overall amount of biofilm that will eventually form. Meltzer (1993) investigated *Pseudomonas* attachment to stainless steel and concluded that no surface is entirely resistant to biofouling. While surface structure can affect the initial rate of fouling, this influence is limited to the first few hours of exposure. Generally, smooth surfaces experience a slower initial fouling rate compared to rough ones, but biofilm formation becomes unavoidable over time.

High water flow rates may influence biofilm growth, but they would not prevent bacteria from attaching to surfaces or remove existing biofilm due to the low flow velocity in the boundary layer, also called laminar layer (the area adjacent to the substrate surface where the flow rate is not turbulent) and the strong adhesion of the EPS. Regardless of water velocity, the flow is slowest in the layers closest to the substrate surface. In systems with continuous high-velocity flow, the biofilm tends to accumulate filamentous bacteria, such as *Pseudomonas*, which are particularly well-suited for attachment with their filamentous structures. These bacteria anchor themselves to surfaces using their "sticky" EPS. However, water flow velocity does affect the density and thickness of biofilms. At higher flow rates, biofilms tend to be denser and more tenacious. As a result, surfaces may appear free of foulants, as they do not feel slimy to the touch. The maximum thickness of the biofilm is generally limited by the thickness of the laminar flow layer. Under steady-flow conditions, biofilm thickness stabilizes at an equilibrium dictated by hydrodynamic shear forces and nutrient concentration. Excessive growth beyond the laminar sublayer triggers sloughing of planktonic pioneer cells, which can seed new biofilm formation downstream (Patterson et al. 1991). In systems with fluctuating water flow, such as automated watering systems that include periodic flushing, bacteria are sloughed off during the flush. This results in random "particle showers" of bacteria, which can explain day-to-day fluctuations in total bacterial counts.

Like all living organisms, microbes require specific nutrients for growth and reproduction. They obtain some of these nutrients from both substrates and water systems. When these nutrients are limited, bacterial growth is restricted. Under ideal conditions, a bacterial cell divides into two daughter cells every 20 minutes (Harfst 1992). This means a single cell and its descendants could grow exponentially, reaching over 2 million cells in just eight hours or even 4 million pounds of bacteria in 24 hours! However, such rapid growth is rarely observed, especially in clean drinking water, due to constraints like available space and nutrients. Nonetheless, water systems typically contain enough nutrients to support microbial growth, as a variety of nutrients can be detected even in pure water (Table 3.3).

3.2.1.6 Fully Functioning Biofilm: Cooperate, Grow, and Spread

Biofilm consists of a cooperative "consortium" of species. A mature, fully functional biofilm behaves like a living tissue on its substrate. It is a complex, metabolically cooperative community made up of different species, each occupying a customized microniche. For instance, an anaerobic biofilm may develop beneath the aerobic layer (Figure 3.9). Some researchers even suggest that biofilms have primitive circulatory systems. The article "Slime City" describes mature biofilms creatively: "*Different species live cheek-by-jowl in slime cities, helping each other to exploit food supplies and to resist antibiotics through neighborly interactions. Toxic waste produced by one species might be hungrily devoured by its neighbor. And by pooling their biochemical resources to build a communal slime city, several species of bacteria, each armed with different enzymes, can break down food supplies that no single species could digest alone. The biofilms are permeated at all levels by a network of*

Table 3.3 Nutrients for bacterial growth found in pure water systems.

Nutrient	Sources
Organic carbon	Humic and fulvic acids (source water)
	Pipe plasticizers and solvents
	Fiberglass-reinforced plastics (FRPs)
	Pump and gage lubricants
	Microbial byproducts
	Personnel
	Airborne dust
Nitrogen	Humic and fulvic acids (source water)
	Nitrates and nitrites (source water)
	Microbial byproducts
	Airborne dust
Phosphorus	Phosphates (source water)
	Microbial byproducts
	Airborne dust
Sulfur	Sulfates (source water)
	Sulfuric acid (reverse osmosis pretreatment)
	Membrane surfactants
	Airborne dust
Trace metals and salts	Source waters
	Process piping-reinforced plastics (FRPs)
	Stainless steel system components
	Reverse osmosis pretreatment chemicals
	Personnel
	Airborne dust

Source: After Mittelman (1985).

channels through which water, bacterial garbage, nutrients, enzymes, metabolites and oxygen travel to and fro. Gradients of chemicals and ions between microzones provide the power to shunt the substances around the biofilm." (Coghlan 1996).

Biofilms spread at their own rate through ordinary cell division, but they also periodically release new pioneer cells to colonize new habitats (Figure 3.9). As the biofilm grows and thickens, extending through the boundary layer into areas with higher flow velocities and more turbulence, some cells are sloughed off. These new pioneer cells have an easier time than their predecessors, as the parent biofilm will release waste products into the surrounding environment, which can serve as an initial organic coating for uncolonized surfaces or as nutrients for other cell types.

Mature biofilm development ranges from several hours to several weeks, depending on the specific system. *P. aeruginosa*, a common pioneer bacterium in biofilm research, has been shown to adhere to surfaces rapidly. It was once reported that *Pseudomonas* cells attached to stainless steel, including electropolished surfaces, within just 30 seconds of exposure (Vanhaecke et al. 1990).

3.2.1.7 Biofilm Architecture

Due to limitations in technology, microbiologists once believed that biofilms were simply disordered clumps of bacteria with no specific structure or organization, as traditional light microscopy

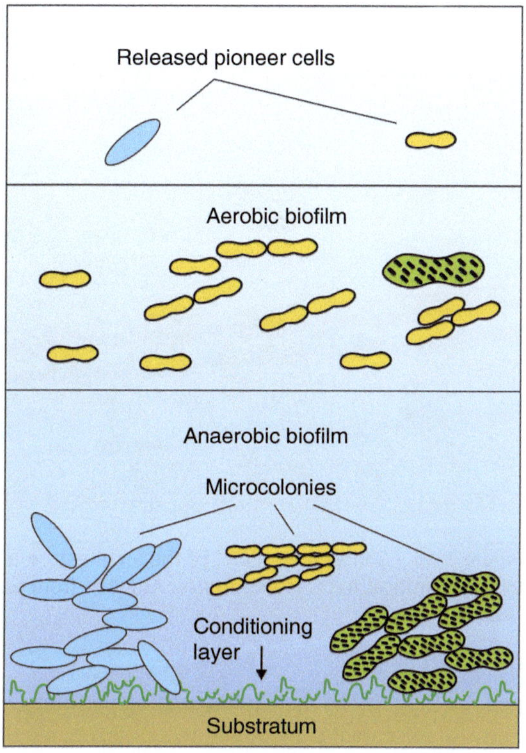

Figure 3.9 A fully functioned biofilm periodically releases pioneer cell. Metabolically cooperative micro-communities of different species are living in customized microniches, with an anaerobic biofilm developing underneath the aerobic layer.

could only provide two-dimensional images, limiting understanding. However, with the advent of laser scanning techniques like confocal laser scanning microscopy (CLSM) and 2-photon laser scanning microscopy (2-PLSM), researchers are now able to observe biofilms in three dimensions, providing a much clearer and more detailed view of living, fully hydrated biological samples.

When observed with conventional light or electron microscopy, the EPS matrix of a biofilm appears as an amorphous structure primarily composed of polysaccharide macromolecules. These polymeric molecules are secreted around the cells, which seem randomly distributed within the matrix. However, new techniques that allow researchers to magnify biofilms without disrupting their gel-like structure have revealed the complex organization of biofilms, almost like viewing a city from a satellite. The article "*Slime City*" (Coghlan 1996) describes this structure: "*In most cases, the base of the biofilm is a bed of dense, opaque slime 5 to 10 μm thick. It is a sticky mix of polysaccharides, other polymeric substances and water, all produced by the bacteria. Soaring 100 to 200 μm upwards are colonies of bacteria shaped like mushrooms or cones. Above street level comes more slime, this time of a more watery makeup and variable consistency with a network of channels through which water, bacterial garbage, nutrients, enzymes, metabolites, and oxygen travel* (Figure 3.10)." CLSM observations have shown that biofilm cells are not randomly arranged, but rather often form aggregated clusters within the EPS matrix (Korber et al. 1993). Areas with high cell concentrations and high microbial activity are typically adjacent to regions with more EPS and fewer cells. Microbial cells within the EPS matrix are motile and can adjust their positions relative to one another within the biofilm.

3.2.1.8 Biochemistry of Biofilm Bacteria

Earlier researchers assumed that biofilm microorganisms behaved similarly to solitary or free-floating microbial cells. However, recent findings reveal that while biofilm cells share the same

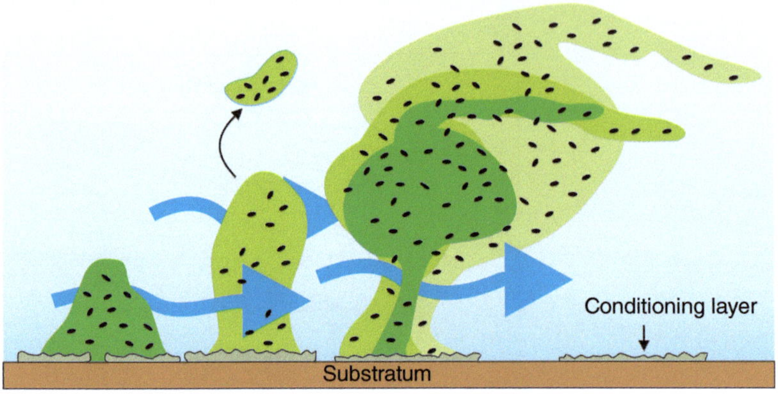

Figure 3.10 Conceptual model of the architecture of a single-species biofilm based on direct observations using a confocal microscope. *Source:* Adapted from Costerton et al. (1995).

genetic makeup as their free-roaming counterparts, their biochemistry differs significantly. This is because microbial cells switch to a different set of genes once they form a biofilm. For instance, the Center for Biofilm Engineering at Montana State University studied *P. aeruginosa* and how it forms biofilms. The result demonstrated that, as soon as these bacteria attached to a surface, they activated genes involved in producing alginate, a particularly sticky form of slime, and then switched these genes off once they became enveloped in alginate. There are studies suggest that up to 30%–40% of the proteins in bacterial cell walls are different between sessile (biofilm-associated) and planktonic (free-floating) bacteria, referred to as "city dwellers" and "free-rovers" by Coghlan (1996). Some of the targets for antibiotics are absent in biofilm bacteria, making them much harder to kill. This issue is particularly problematic when biofilms form inside humans and animals.

Scientists have also studied the chemicals, named sigma factors, which allow bacteria to shift their biochemistry to adapt to life in a biofilm. If they can also discover a "reverse sigma factor," on the other hand, that would convert biofilm bacteria back into planktonic cells, it might be possible to break down biofilms by sending them an "evacuation signal" (Coghlan 1996).

3.2.2 Microbial Mats

In favorable conditions, biofilms can develop into thick organic layers that extend laterally, sometimes covering several square kilometers. These layers, which resemble carpets, are referred to as microbial mats (Figure 3.11). Often called algal mats, cyanobacterial mats, or potential stromatolites, microbial mats are named for the predominance of cyanobacteria and their role in forming laminated sedimentary structures (Krumbein 1994). For this discussion, microbial mats are defined as laminated microbial ecosystems that cover large substrate areas. They are vertically stratified and are primarily dominated by prokaryotic microbes often associated with fungi, algae, or protozoans.

Prokaryotic cyanobacteria form the key component of most modern microbial mats in photic zones, producing cohesive, viscous layers that can range in thickness from a few millimeters to

Figure 3.11 A modern cyanobacterial mat in a near-dried-up pond: (a) showing folded structures, gas bubbles, and trapped insects; (b) the photomicrograph of the microbial mat, predominated by filamentous cyanobacteria. *Source:* Photos by Xingliang Zhang.

several meters. These mats typically appear as thin blue-green films of organic matter and can be found in diverse environments, including intertidal flats, thermal springs, Antarctic lakes, hypersaline lagoons, and arid deserts. In remote areas like ephemeral pools and ponds, microbial mats are major contributors to biodiversity (Figure 3.12). However, microbial mats are relatively rare in both Phanerozoic and modern ecosystems, and their limited distribution and fossil record can be attributed to grazing and sediment disturbance by animals. Consequently, microbial mats are most commonly found in environments where metazoan activity is restricted, such as high salinity, high temperature, and unstable environments that experience frequent wetting and drying (Canfield et al. 2005). In contrast, microbial mats thrived in the Archean and Proterozoic oceans, where the lack of bioturbation allowed them to flourish. Fossil records show that microbial mats date back to 3.5 billion years ago, representing some of the earliest forms of life on Earth, as was evidenced by microbialites (e.g. stromatolites) and microbially induced sedimentary structures (Sections 6.4 and 8.2).

Figure 3.12 Modern microbial mats in an ephemeral pool after rain. *Source:* Photo by Wei Liu.

In aquatic environments, microbial mats can form on the bottom (benthic mats), at the chemocline, or even on the surface (floating mats). They can be classified into phototrophic and chemotrophic types based on their energy source. Phototrophic microbial mats are highly self-sustaining ecosystems that thrive in surface environments where light is available. Chemotrophic microbial mats, on the other hand, are found in remote locations such as deep-sea hydrothermal vents, methane seepage areas, and cave ecosystems, where they use chemically reducing substances instead of sunlight to produce organic compounds. For example, large microbial towers are found in the anoxic settings of the Blake Sea, where methane is oxidized in conjunction with sulfate reduction during anaerobic methane oxidation. While chemotrophic microbial mats remain relatively less explored, the emphasis here is on those primarily powered by photosynthesis.

3.2.2.1 Structure of Microbial Mats

The population structure of microbial mats can be observed microscopically or analyzed using various chemical and light microsensors. While species composition and interactions certainly affect the structure of microbial communities, their relative importance compared to physical and chemical factors is not yet fully understood.

Microbial mats are typically vertically stratified, with different populations positioned according to their needs for light and chemical interfaces, such as between oxygen and sulfide (Figure 3.13). The population structure of microbial mats was well studied in the hypersaline seawater evaporation pond at Exportadora de Sal (ESSA), Guerrero Negro, Baja California, Mexico (Canfield et al. 2005). The composite diagram illustrates the vertical structure of the microbial mat at this site (Figure 3.13a). The surface layer is predominantly composed of diatoms, which are highly pigmented and adapted to the intense light exposure at the mat surface during the day. Just beneath this is an assemblage of filamentous cyanobacteria, where most of the oxygenic photosynthesis occurs. Microelectrode profiling has shown that the peak oxygen concentration occurs in this layer (Figure 3.13b). The density of cyanobacteria decreases sharply with depth, and oxygen concentrations also drop due to heterotrophic respiration and the oxidation of chemically reduced metabolic products, such as sulfide and ammonia, that diffuse from below. At the oxygen–sulfide interface,

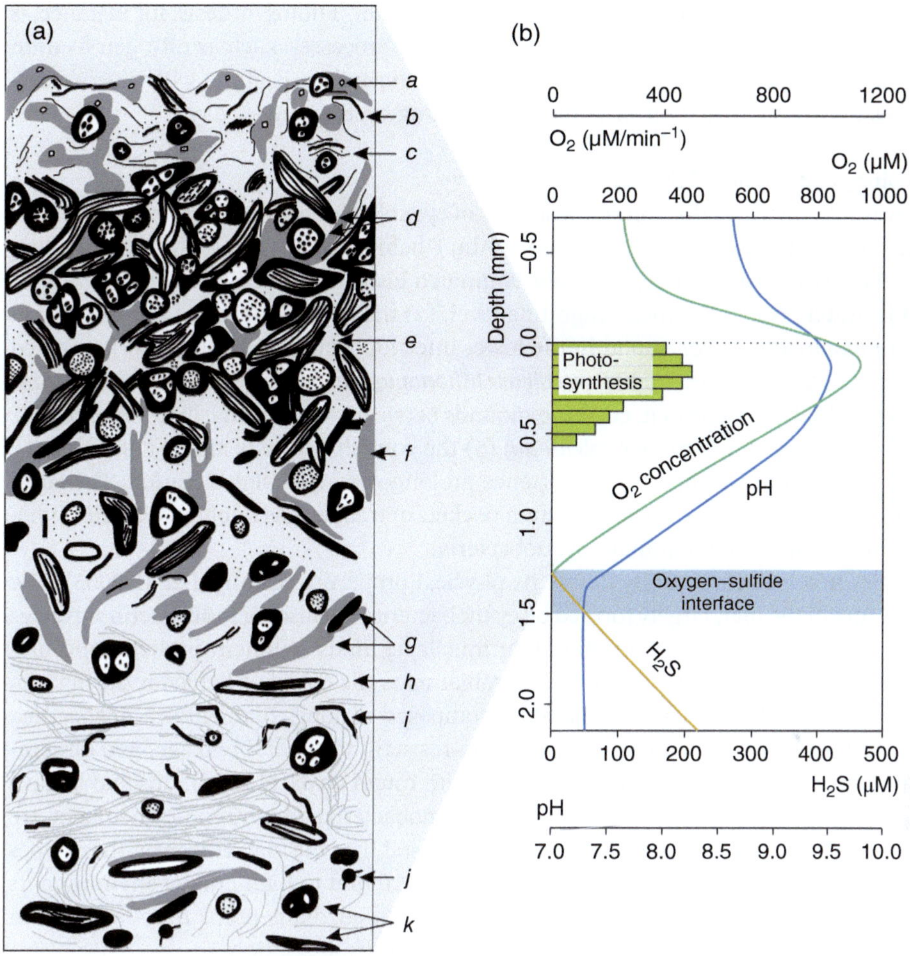

Figure 3.13 Communities and main metabolic processes within the microbial mat: (a) a schematic vertical section of the topmost 2 mm of subtidal mat dominated by *Microcoleus chthonoplastes* cyanobacteria, showing populations of microbiota; (b) daytime depth gradients in O_2, oxygenic photosynthesis, sulfide, and pH. Letters along the right margin in A indicate the following: *a*, diatoms; *b*, *Spirulina* spp. cyanobacteria; *c*, *Oscillatoria* spp. cyanobacteria; *d*, *M. chthonoplastes* cyanobacteria; *e*, non-photosynthetic bacteria; *f*, unicellular cyanobacteria; *g*, fragments of bacterial mucilage; *h*, green non-sulfur bacteria (capable of anoxygenic photosynthesis); *i*, *Beggiatoa* spp. (non-photosynthetic sulfide-oxidizing bacteria); *j*, metazoans (e.g. nematodes); *k*, abandoned cyanobacterial sheaths. *Source:* Des Marais (2003)/with permission of The University of Chicago Press Books.

filamentous green non-sulfur bacteria are found. Below this interface, a heterotrophic community is engaged in organic carbon mineralization and sulfate reduction, processes that produce sulfide for the sulfide-oxidizing populations just above. These population zones, ranging from primary-producing cyanobacteria to sulfate reducers, are often compressed into a depth of just 2–5 mm.

One striking feature of microbial mats is the sharp zonation of pigmentation. For example, the sandy mats in the Great Sippewissett Salt Marsh have five distinct layers, while the flat laminate mat at Laguna Figueroa has four discernible layers (Stolz 2000). Despite these distinct layers, there is still significant heterogeneity within them. Functional units within populations form microcolonies, with some species confined to specific layers, while others are active throughout the entire mat. The effect of diurnal fluctuations is also important to consider. Photosynthesis, for instance, is light-dependent and thus shows a clear diurnal pattern. Other processes, such as nitrogen fixation and methanogenesis, have also been shown to follow a diurnal rhythm. Bacteria with diverse metabolic capabilities may even alter their physiology depending on the time of day.

3.2.2.2 Morphology and Physical Behaviors

Mat morphology is influenced by the dominant species of cyanobacteria and environmental factors (Stolz 2000). Six types of mat in the intertidal zone at Abu Dhabi of the Persian Gulf has been identified (Golubic 1992), which are (1) the gelatinous laminated biscuits, formed by *Phormidium hendersonii*, found in waters below the mean high tide level; (2) the mamillate mat, built by coccoid cyanobacterium *Entophysalis major*, found in the lower intertidal zone; (3) the low flat laminated mats, composed of *Lyngbya aestuarii* and *Microcoleus chthonoplastes*, typically in tidal pool bottoms and channels; (4) the high pinnacle mat, covering mounds between pools, primarily composed of *L. aestuarii* and an undermat of *Schizothrix splendida*; (5) the convoluted mats and (6) folded mats, located in the upper intertidal zone, which experience prolonged periods of desiccation. The latter two types of mats are leathery in texture and contain pockets of trapped gas bubbles, with *M. chthonoplastes* and *S. splendida* as the dominant cyanobacteria.

The main species in a mat community define its physical properties (Canfield et al. 2005). The filamentous structure of the mat, largely formed by cyanobacteria, but also including actinomycetes and sulfate-reducing bacteria, along with abundant mucilage, binds sediment particles and prevents their mobilization by currents. This gives microbial mats the cohesive behavior during erosion, transport, and deposition (Figure 3.11). Mats composed of long, filamentous, ensheathed cyanobacteria, like those formed by *Lyngbya*, are tough, resistant to shear, breakage, and desiccation. These mats are typically found in intertidal areas with rough physical conditions and behave like a supple tissue. In mats formed by unicellular cyanobacteria, phototrophic members are embedded in thick EPS, mostly polysaccharides, which help stabilize the mat. However, these types of mats are less resistant to physical stress and are usually not found in high-energy environments (Stal 1994). Microbial mats are susceptible to disruption, but their filamentous framework often leads to fragmentation into polygonal pieces, controlling their disintegration.

3.2.2.3 Microenvironment Within the Microbial Mats

The environment within the mat is quite different from that in the overlying water column (Figure 3.14). Just beneath the mat surface, the community experiences steep vertical gradients of light intensity and redox conditions that fluctuate significantly over the course of the day. Gradients of oxygen and sulfide create contrasting chemical environments, forming microenvironments that are separated by just a few millimeters. At night, when oxygenic photosynthesis ceases, the upper layers of the mat become highly reduced and sulfidic, and the oxygen–sulfide interface moves closer to the mat surface.

The light flux penetrating the mat can be measured as a downward irradiance (the total downwelling light passing through a horizontal plane) or as scalar irradiance (the sum of all light converging on a given point within the mat). Since scalar irradiance reflects the total light available at a specific location, it provides the most meaningful description of the environment within the mat. The intensity and spectral composition of light that penetrates the mat are altered by absorption and scattering. Motile photosynthetic organisms adjust their position according to the resulting light gradient, and some species harvest light in the infrared range. Jørgensen and Des Marais (1988) measured the scalar irradiance of two microbial mats: one dominated by the filamentous cyanobacterium *M. chthonoplastes*, and the other, at higher salinity, dominated by unicellular

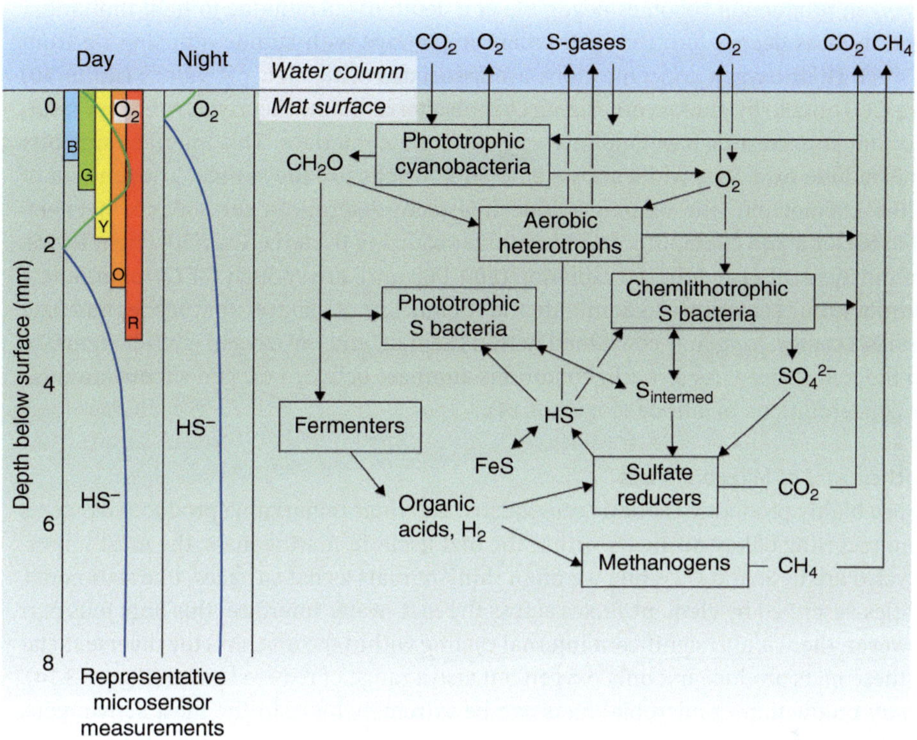

Figure 3.14 Schematic of a cyanobacterial microbial mat with associated depth-related light and chemical gradients. Boxes denote functional groups of microorganisms, and arrows denote flows of chemical species into or out of microorganisms. $S_{intermed}$ indicates sulfur in intermediate oxidation states. Schematic at left depicts vertical gradients of O_2 and sulfide during the day and at night. Oxygen concentrations are shown decreasing to zero at a depth of 2 mm during the day, and just below the mat surface at night. The vertical bars at upper left represent the relative depths of penetration of blue (B), green (G), yellow (Y), orange (O), and red (R) light. Such chemical gradients and light penetration profiles of both filamentous and unicellular mats are qualitatively similar, although the depth scale (mm) of such profiles tends to be greater for unicellular mats. *Source:* Des Marais (2003) / with permission of The University of Chicago Press Books.

cyanobacteria. The dark olive mat dominated by *Microcoleus* shows a sharp decline in light intensity and a marked change in spectral composition with depth. The minimum in the spectra corresponds to the absorption maxima of photosynthetic pigments in cyanobacteria. Chlorophyll-a absorbs light at wavelengths of about 430 and 670 nm, phycocyanin at around 620 nm, and various carotenoids between 450 and 500 nm. In contrast, the unicellular cyanobacteria mat, with a lower cell density and a more gelatinous texture, is light orange-tan in color. Light penetrated deeper into this mat, although blue light is strongly attenuated, with carotenoids absorbing most of the light. In both mats, longer-wavelength light, especially wavelengths longer than 900 nm, penetrated the farthest (Figure 3.14).

The high rate of oxygenic photosynthesis in microbial mats creates steep and fluctuating gradients in pH, dissolved inorganic carbon (DIC), and oxygen, as seen in the depth profiles (Figures 3.13 and 3.14). Studies have shown that dissolved oxygen levels can be nearly five times higher than those in air-saturated brine, but this oxygen has a very short residence time of only 2 minutes (Jørgensen and Des Marais 1986). Oxygen concentration peaks where cyanobacteria are most abundant, and remarkably, oxygen can accumulate to high levels and then drop to zero within just 2 mm of mat depth. Oxygen production becomes negligible at a depth of 0.5 mm due to light limitation. However, oxygen diffuses deeper into the mat, where it overlaps with sulfide diffusing up from below (Figure 3.13). High oxygen concentrations are associated with high pH values (up to 10) because of active CO_2 uptake by photosynthesizing cyanobacteria. Below the oxygenated layer, sulfide begins to accumulate, creating a well-defined oxygen–sulfide interface. This interface supports non-phototrophic sulfide oxidizers, while anoxygenic phototrophs use the sulfide. The amount of light reaching the chemocline (the oxygen–sulfide interface) determines the balance between photoautotrophic bacteria and chemolithotrophic sulfide-oxidizing bacteria. Even low light levels, as little as 1% of the incident near-infrared radiation (800–900 nm), are enough for *Chromatium*, a phototrophic purple sulfur bacterium, to dominate the chemocline. As sunset approaches and oxygenic photosynthesis ceases, oxygen is consumed within the mat, and the oxygen–sulfide interface moves closer to the mat surface. As a result, conditions alternate between oxygen supersaturation and millimolar concentrations of sulfide (Figure 3.14).

3.2.2.4 Biogeochemistry of Microbial Mats

Microbial mats are highly productive aquatic ecosystems, and their remarkable productivity stems from the efficient recycling of key nutrients within the mat itself. In marine mats, the most important elements cycled are C, S, and O. While we often think of mats as flat surfaces, like sediments, with their activities described by element fluxes across the mat–water interface, this only tells part of the story. However, there is also significant internal cycling within the mat, and the diverse metabolisms within these mats produce not only oxygen but also a range of reduced gases (Figure 3.14).

Rates of primary production in microbial mats can be extremely high. In the most active mats, even millimeter-thin layers of cyanobacteria have gross photosynthesis rates comparable to those in hardwood forests and tropical rainforests (Canfield et al. 2005). During the day, DIC from various sources is fixed through photosynthesis. Anoxygenic photosynthesis contributes less than 10% of the total carbon fixation rate. The sources of DIC include both external DIC diffusing into the mat from the overlying water column and internal DIC, which is produced within the mat. The internal DIC originates from sulfate reduction in both the oxic zone (where sulfate reducers co-exist with facultative aerobic heterotrophs) and deeper layers, as well as from aerobic respiration (aerobic heterotrophy, photorespiration, and dark respiration by cyanobacteria). Measurements show that most of the DIC used in primary production comes from internal carbon sources, with microbial sulfate reduction being the primary source of DIC at night.

The cycling of oxygen during the day mirrors that of DIC. Oxygen is produced solely through primary production, while its sinks include various forms of O_2 respiration within the mat, sulfide oxidation, and the efflux of O_2 from the mat. At night, in the absence of the phototrophic community to sequester nutrients, the mat community actively sequesters ammonia. During the day, most of the O_2 produced is recycled within the mat through O_2 respiration, with sulfide oxidation occurring as well. At night, O_2 diffuses into the mat and is consumed primarily by sulfide oxidation near the mat–water interface. A comparison of the O_2 and DIC fluxes across the mat–water interface shows that during the day, more DIC diffuses into the mat than O_2 diffuses out, while at night, more DIC diffuses out of the mat than O_2 diffuses in. However, both net O_2 and DIC fluxes are balanced over the full 24-hour cycle. This indicates that during the day, carbon with an oxidation state greater than zero is incorporated into the mat, and similarly high oxidation state carbon is released at night (Des Marais 2003).

During the day, photosynthetic microbial mats produce substantial amounts of O_2. At night, O_2 is consumed by sulfide oxidation at the mat surface and in the lowermost water column (Canfield and Des Marais 1993), making the entire mat anoxic (Figure 3.14). Consequently, cyanobacteria in the mat must ferment to obtain energy at night, likely producing a variety of reduced low-molecular-weight compounds. Hoehler et al. (2001) observed that subtidal *Microcoleus* mats produced CO, CH_4, and significant quantities of H_2. The rates of CO emission correlated with photosynthesis rates, suggesting that cyanobacteria, diatoms, or both may be the sources. H_2 emission rates were highest at night, consistent with fermentation under anoxic conditions. Methane emission rates remained constant throughout the diel cycle, suggesting that the source lies beneath the zone of the mat that becomes oxygenated during the day. Organic photosynthates in the mat also interact with sulfides in the pore waters to produce dimethyl sulfide and other organosulfur gases, some of which escape the mat (Visscher et al. 2003).

3.3 Microbial Ecosystems on Earth

Different habitats on Earth generally have distinct geological conditions, leading to the development of specific microbially dominated ecosystems. For example, glacier–rock interfaces host chemoautotrophic microorganisms that use hydrogen as an energy source, while marine and lake sediments rich in organics are home to heterotrophic bacteria like sulfate reducers and methanogens. Mid-ocean ridges, rich in reduced gases (e.g. hydrogen, hydrogen sulfide) and hydrocarbons, support sulfur-oxidizing and hydrogen-oxidizing bacteria. Basaltic crusts, with active water-rock reactions, produce hydrogen and methane, promoting autotrophic growth. Continental shelf sediments host both heterotrophic and autotrophic communities. Terrestrial oil reservoirs and coal seams provide substrates for heterotrophs, including sulfate reducers and hydrocarbon degraders. In deep subsurface environments, geological hydrogen fuels microbial processes like sulfate reduction and methanogenesis. Subterranean caves are rich in microorganisms that fix carbon and oxidize methane and hydrogen sulfide, while terrestrial groundwater, often anaerobic, supports a wide variety of bacteria, including sulfate reducers, nitrate reducers, and methanogens (Dong and Jiang 2023).

3.3.1 Microbial Ecosystems in Marine Settings

The ocean covers approximately 71% of the Earth's surface and contains 97% of the Earth's water volume, making it the largest ecosystem on the planet. Marine sediments are a crucial component of the marine ecosystem and host ecosystems formed by unique geological structures such as mud volcanoes, hot springs, and cold seeps, providing suitable environments for the growth of various organisms (Figure 3.15), particularly microorganisms.

Through the 16S rRNA gene, the most widely targeted taxonomic marker in environmental microbiology studies, we have gained an initial understanding of the diversity of microbial communities in marine sediments. For example, Orcutt et al. (2011) summarized the microbial diversity in marine sediments: bacterial groups such as Alphaproteobacteria, Gammaproteobacteria, Deltaproteobacteria, Actinobacteria, and Chloroflexi are widely present in deep-sea surface sediments; meanwhile, in continental shelf edge sediments, cold seeps, hydrothermal vents, and deep sediments, mainly inhabited by archaea including Bathyarchaeota, Lokiarchaeota, Thaumarchaeota, Euryarchaeota, and Woesarchaeota.

3.3.1.1 Basaltic Ocean Crust Ecosystems

Basalt is the main rock component of the upper oceanic crust, rich in reducible minerals such as iron-titanium oxides, pyroxenes, olivine, and metal sulfides (e.g. Fe^{2+} accounts for about 9%, while

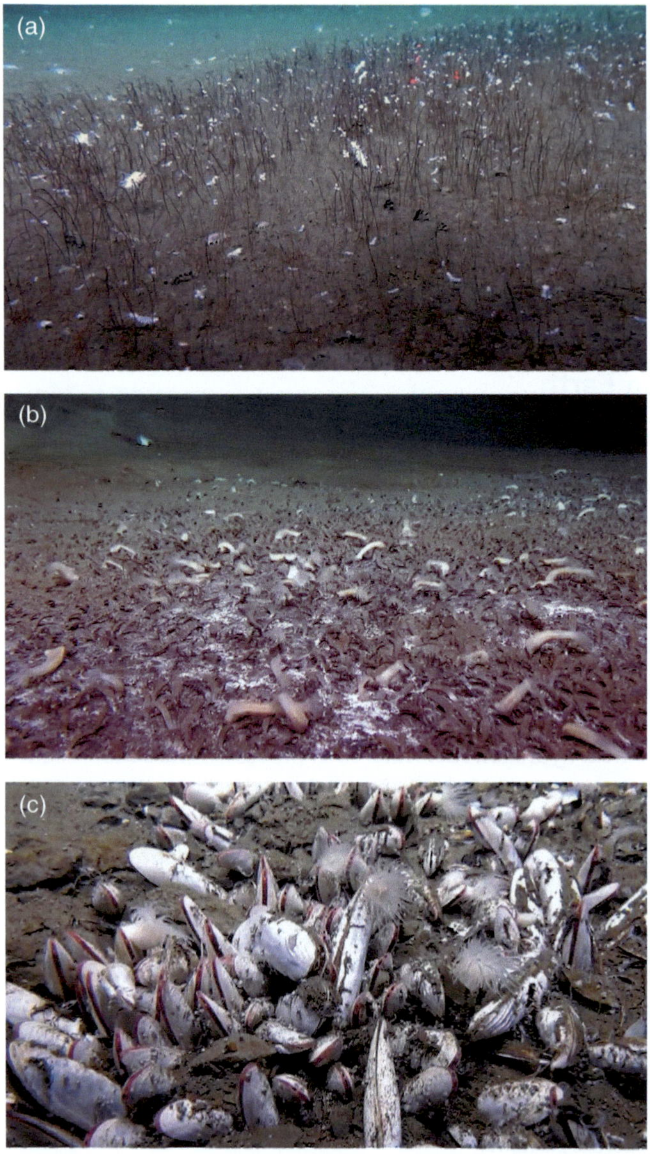

Figure 3.15 Photographs of representative fauna in the dark ocean (the Kuril–Kamchatka Trench and western Aleutian Trench): (a) abundant free-moving polychaetes *Macellicephaloides grandicirra* (white) in the tightly packed frenulate siboglinids at the Cotton Field (9332 m below the sea level); (b) tube-dwelling polychaetes *Anobothrus* sp. and Actiniaria with spots of white microbial mats at the Aleutian Deepest (6870 m below the sea level); (c) aggregation of vesicomyid bivalves *A. phaseoliformis* with Actiniaria at the Clam Bed (5743 m below the sea level). *Source:* Peng et al. (2025) / Springer Nature / CC BY 4.0.

Mn^{2+} and sulfides each make up approximately 0.1%), providing significant energy and nutrient sources for microorganisms. Fluid within the basalt of oceanic crust is relatively active, and fluid exchange between the crust and the overlying seawater creates an imbalance in the chemical substances (electron acceptors) within the crust; additionally, the trace dissolved organic carbon (DOC) carried by fluids flowing through the basaltic ocean crust can serve as electron donors for microorganisms (Lin et al. 2012), promoting redox reactions within the crust and sustaining the microbial ecosystem.

Compared to the overlying bottom seawater, the microbial community in the basaltic ocean crust exhibits high diversity, with bacteria being the dominant group in terms of abundance. The predominant bacterial phyla include Proteobacteria (especially Alphaproteobacteria and Gammaproteobacteria), Actinobacteria, Bacteroidetes, Chloroflexi, Firmicutes, and Planctomycetes (Orcutt et al. 2011; Santelli et al. 2008).

In the basaltic rocks in contact with seawater, reduced sulfur- and iron-rich minerals are oxidized by oxygen and nitrate, with microorganisms deriving energy from these chemical reactions. Hydrogen gas, a byproduct of serpentinization when bottom seawater interacts with basalt, serves as a substrate for methane production—an important chemoautotrophic process (Santelli et al. 2008). Nitrogen fixation also leads to higher concentrations of ammonium ions in the crust fluids. When these ammonium ions enter the seawater, they are taken up by ammonia-oxidizing bacteria, which convert them into nitrite. This nitrite is further oxidized by nitrite-oxidizing bacteria to form nitrate, completing the full nitrification process (Mason et al. 2009).

In addition to the chemoautotrophic microorganisms mentioned above, the oceanic crust also hosts a large number of heterotrophic microorganisms that play a key role in breaking down organic matter in the crustal fluids. For instance, an analysis of the amino acid composition in these fluids revealed highly active organic matter degradation (Lin et al. 2015). Furthermore, an examination of aliphatic compounds in the fluids showed that long-chain n-alkanes (C_{20} to C_{33}) were much more abundant than short-chain n-alkanes, suggesting a preference among heterotrophic microorganisms for utilizing short-chain n-alkanes (Cowen et al. 2003).

3.3.1.2 Hydrothermal Vent Ecosystems

The deep-sea thermal ecosystem is a unique and ancient chemoautotrophic system, fueled by energy from the deep beneath the seabed, several kilometers below the surface. This makes it a classic example of a dark energy ecosystem. Hydrothermal zones in the deep sea are typically found at mid-ocean ridges, subduction zones, back-arc basins, and hotspot volcanoes, where tectonic plates diverge or converge at depths ranging from 700 to 5000 m. By far, over 600 hydrothermal vent fields have been identified in these regions (source: InterRidge Global Database). Of these 304 confirmed active vent sites listed in the Database, 44% occur in the Western Pacific Ocean (Tunnicliffe et al. 2023) and more than 80 hydrothermal vents have been found in the Pacific Ocean at sites of rapid seafloor spreading. A few vents have also been found in the Atlantic Ocean, at sites of slower spreading. Hydrothermal vents have also been found at sites of seafloor spreading in the Red Sea. In 1995, scientists started looking for hydrothermal vents in the Indian Ocean and quickly discovered one that had previously been active. Scientists have also found hydrothermal vents at hot-spots where magma rises in the middle of lithospheric plates to form volcanic islands, such as Hawaii.

The genesis of hydrothermal fluid is well understood. Taking the mid-ocean ridge as an example (Figure 3.16a), seawater seeps down through cracks reaching the magma chamber above the mantle plume. There, it is heated and reacts with the surrounding rock, generating a variety of reducing substances. These substances then dissolve in the heated seawater, forming a hydrothermal fluid. The heated fluid then forces its way back up through the crust and emerges in hydrothermal vents, or deep-sea hot springs. Two main types of hydrothermal vents have been identified. Warm vents release hydrothermal fluids at temperatures ranging from 6 to 23 °C, significantly warmer than the surrounding seawater, which is around 2 °C. On the other hand, hot vents, often called "black smokers" due to the dark cloud of mineral-rich precipitates formed when the hot water mixes with seawater, discharge hydrothermal fluids at temperatures between 270 and 380 °C (Figure 3.16b).

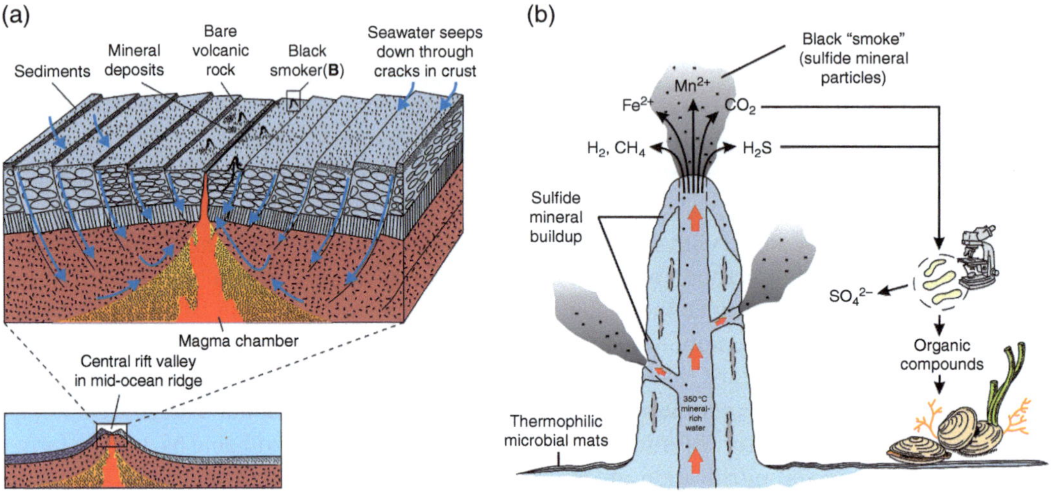

Figure 3.16 Cross-section of a mid-ocean ridge and black smoker: (a) seawater seeps down through cracks in the crust, becomes heated, and then reemerges as underwater hot springs; (b) as the hot, mineral-rich water from the rift zone meets the cold ocean water, minerals precipitate and accumulate, gradually building the chimney of the black smoker. *Source:* Adapted from Castro and Huber (2008).

The flow rates of these vent types are also different: warm vents emit fluids at 0.5–2 cm per second, while hot vents have much higher flow rates, ranging from 1 to 2 m per second.

The hydrothermal fluid in black smokers is typically acidic (pH 2–3). It contains a variety of substances produced by water-rock interactions, including H_2, CH_4, NH_3, H_2S, short-chain alkanes (CnH_{2n+2}), and metal elements such as Fe, Mn, Cu, Zn, Ca, and Ba (Figure 3.17). Si is also present, while O and Mg elements are relatively scarce. The fluid has a wide range of salinity, a high temperature, and a broad distribution. In contrast, white smokers are much rarer and are primarily

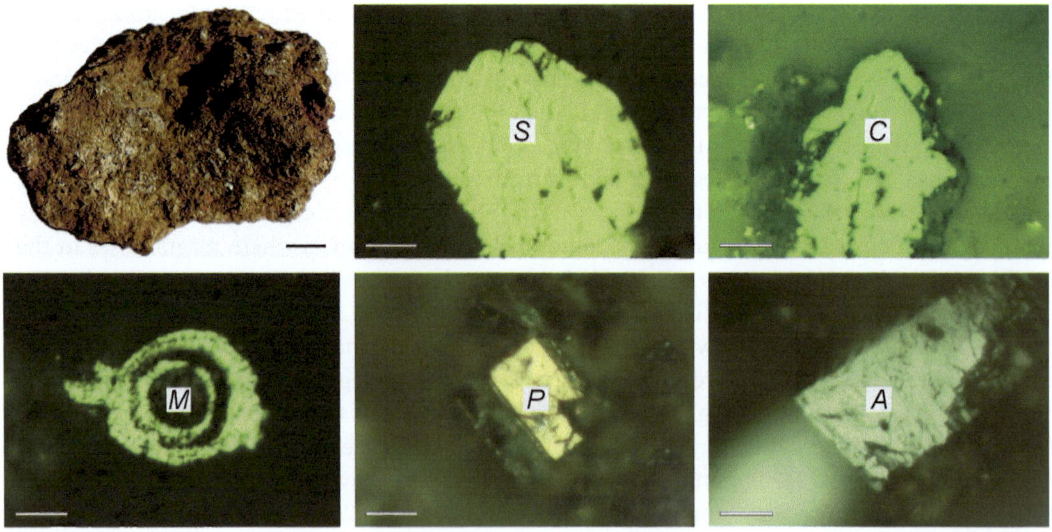

Figure 3.17 The sulfide mineral debris from the submarine hydrothermal deposits of the East Pacific Rise. Optical microphotographs and Raman analyses of the sulfide mineral debris identifying four types of sulfide minerals such as *S*, sphalerite; *C*, chalcopyrite; *M*, marcasite; and *P*, pyrite; as well as one type of sulfate mineral: *A*, anhydrite. *Source:* Photos by Wei Liu.

found in areas like the Lost City in the North Atlantic. The hydrothermal fluid here is alkaline (pH >7), with temperatures around 90 °C. It is rich in light mineral elements such as Ba, Ca, and Si, but lacks H_2S and CO_2. Instead, white smokers emit prominent levels of H_2 and CH_4, which serve as the primary energy source for the white smoker ecosystems. Scientists believe that environments like Lost City could resemble conditions on early Earth and even on other planets, such as Enceladus or Europa, where hydrothermal activity might support life. As such, it has generated scientific interest as a prime location for investigating the origin of life on Earth and other planets similar to it (Levinson 2020).

Deep-sea hydrothermal vents support vibrant and diverse microbial communities. The primary production for these communities comes from microbial chemosynthesis (chemoautotrophy), rather than photosynthesis. The water around hydrothermal vents is teeming with bacteria and archaea that thrive on the toxic emissions from the vents, often clouding the water. While hyperthermophiles do not live directly in the superheated fluids, they thrive along a gradient where the hot water mixes with the cold bottom seawater. A large number of sulfur-oxidizing bacteria, such as *Thiobacillus*, *Thiomicrospira*, *Thiothrix*, and *Beggiatoa*, are found in and around the vents, particularly within the walls and adjacent sediments of black smokers (Figure 3.18). Samples

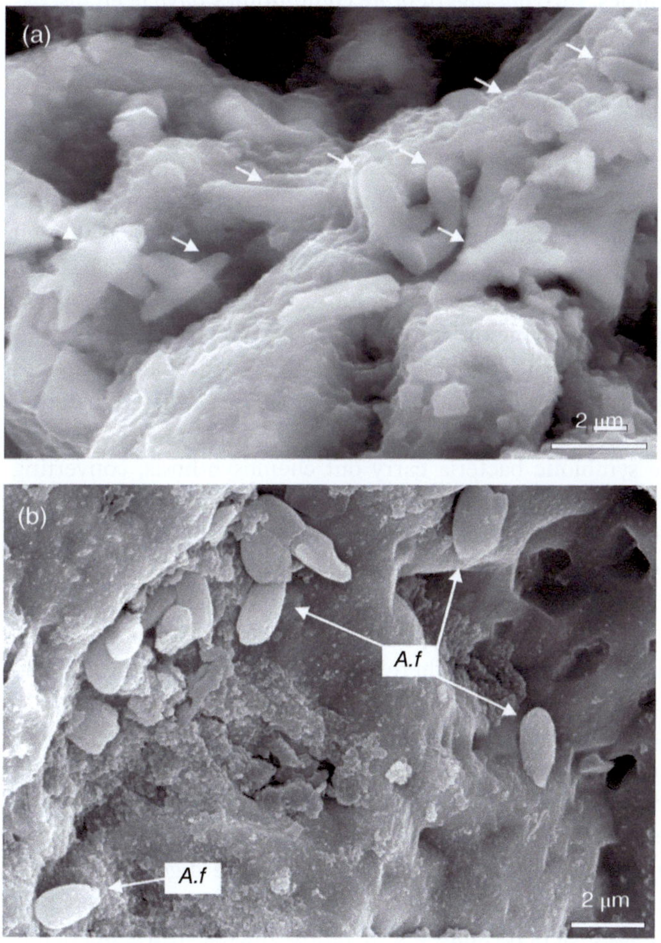

Figure 3.18 (a) A curious morphology from the hydrothermal deposits of the East Pacific Rise; (b) *Acidithiobacillus ferrooxidans* in laboratory. Both are of the same size and are adsorbed onto the pyrite surface. *Source:* Photos by Wei Liu.

collected near these vents have yielded cultures of these bacteria, and in situ experiments have demonstrated their ability to fix CO_2 and oxidize H_2S and $S_2O_3^{2-}$. Other types of bacteria found in these environments include nitrifying bacteria, hydrogen-oxidizing bacteria, iron- and manganese-oxidizing bacteria, and methylotrophic bacteria, which likely feed on methane and carbon monoxide emitted from the vents.

One of the most intriguing insights from studies of microbial ecology in hydrothermal systems is the hypothesis that life may have originated in a hot aqueous environment. The strongest evidence for this idea comes from the observation that deeply rooted lineages within the universal phylogenetic tree are represented by archaea and bacteria that thrive in hot water. Additionally, the relatively short evolutionary distances between contemporary hyperthermophilic archaea and other chemolithoautotrophic organisms suggest that these archaea may have evolved in hot environments and have remained there, requiring few major evolutionary adaptations. In contrast, moderately thermophilic and mesophilic bacteria exhibit longer evolutionary lineages, which are marked by higher rates of rRNA sequence evolution. However, recent studies have raised questions about the validity of the "hot origin" hypothesis. Some researchers suggest that the branching of these deeply rooted genera might not be as straightforward as once thought, challenging the idea that life originated solely in a high-temperature environment.

In addition to the microbial communities, thriving invertebrate communities have been discovered within a few meters of hydrothermal vents. While some vent animals feed by filtering bacteria from the surrounding water, this does not seem to be the primary feeding method for most vent animals. One of the dominant species in many vent ecosystems, the giant tube worm (*Riftia*) (Figure 3.19), which can grow up to 2 m in length, does not filter bacteria. Remarkably, the giant tube worm lacks a mouth, gut, or anus, and instead has a specialized gastrointestinal structure called the trophosome (also known as the feeding body), which makes up about 50% of the worm's body weight. This tissue is packed with sulfur granules, and microscopic analysis reveals a high concentration of prokaryotic cells—around 3.7×10^9 cells per gram of trophosome tissue. The large spherical cells in the trophosome closely resemble the marine sulfur-oxidizing bacterium *Thiovulum*. The trophosome tissue also contains the enzyme rhodanese, which plays a key role in the conversion of sulfite ($S_2O_3^{2-}$) into elemental sulfur (S_0) and sulfate (SO_3^{2-}). Additionally, enzymes from the Calvin cycle, which is used by autotrophic organisms to fix CO_2 into organic material, are present in the trophosome.

Inside the tube worm's body, these symbiotic bacteria carry out chemosynthesis, converting hydrogen sulfide and carbon dioxide into organic matter. The worm, in turn, benefits by absorbing much of the organic material produced by the bacteria. The worm's bright-red plume functions similarly to a gill, exchanging carbon dioxide, oxygen, and hydrogen sulfide with the surrounding water. Hydrogen sulfide is toxic to most organisms, but the tube worm's blood contains a specialized hemoglobin that binds to hydrogen sulfide, preventing poisoning. This hemoglobin also transports hydrogen sulfide to the bacteria in the trophosome, where it is used for chemosynthesis. Most other vent animals must chemically process hydrogen sulfide to avoid its toxic effects.

Other vent animals, such as mussels (*Bathymodiolus*) and large clams (*Calyptogena*), also harbor symbiotic bacteria, although they are capable of filter feeding as well. These animals rely on both their symbiotic bacteria and non-symbiotic bacteria as food sources. For example, a shrimp species (*Rimicaris*) that dominates in some vent communities feeds by scraping off and consuming bacteria-covered minerals from the chimneys around the vents. The bacteria are digested and any remaining mineral matter is excreted. Like the tube worms, these giant clams and mussels host sulfur-oxidizing bacteria in their gill tissues. Genetic analyses, including nucleic acid sequencing, have revealed that each vent animal typically hosts only one major species of bacterial symbiont, and the species varies depending on the type of animal.

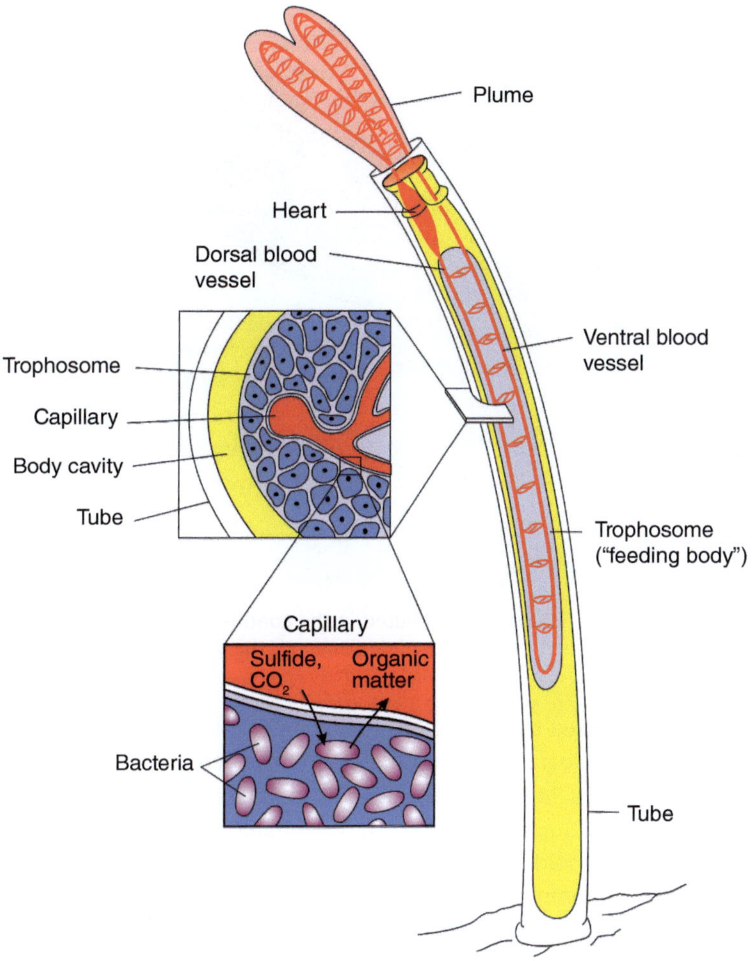

Figure 3.19 The anatomy of the giant hydrothermal-vent tube *Riftia pachyptila*. The plume at the end acts like a gill, except that it exchanges H_2S as well as CO_2 and O_2. H_2S and CO_2 are carried in the blood to the trophosome, where symbiotic bacteria use them to make organic matter by chemosynthesis. *Source:* Adapted from Castro and Huber (2008).

3.3.1.3 Cold Seep Ecosystems

A cold seep, also known as a cold vent, is usually distributed under the submarine sedimentary surface of passive and/or active continental margin slopes, where hydrocarbon-rich fluids, especially methane and hydrogen sulfide, leak into the water, often forming a brine pool (Figure 3.20). Despite what its name might suggest, a "cold" seep does not mean that the fluid seeping through to the surface is colder in temperature than the surrounding water. In fact, the temperature of these fluids tends to be slightly higher than the surrounding water. The term "cold" refers to the fluids being cooler in comparison to the very hot conditions found at hydrothermal vents.

Cold seeps systems provide a unique environment that supports chemosynthetic communities, including species that are often endemic to these areas. Over time, these seeps can contribute to the development of distinctive topographies. The reaction between seawater and methane at cold seeps can also lead to the formation of carbonate rocks and chemoherms (Section 6.2). Cold seep ecosystems were first discovered in 1979 in the San Clemente Fault Zone off California and later confirmed as cold spring habitats in Florida in 1983.

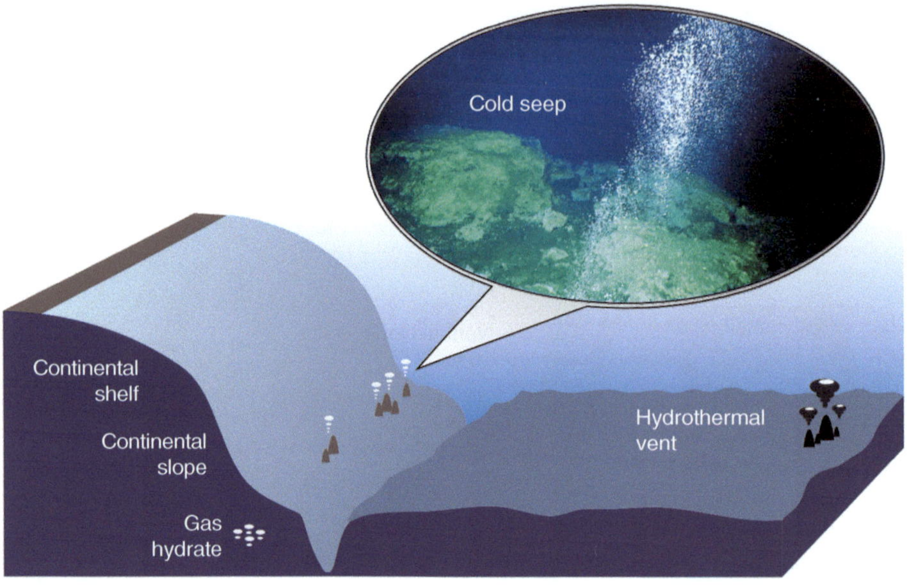

Figure 3.20 Cold seep, also known as a cold vent, is usually distributed in the continental margin slopes, where hydrocarbon-rich fluids, especially methane and hydrogen sulfide, leak into the water.

Methane and other organic materials released by these seeps fuel chemosynthetic microorganisms, forming the foundation of a unique biological community. One of the deepest known cold seep communities was discovered at a depth of 7326 m in the Japan Trench. Unlike hydrothermal vents, which are unstable, cold seeps release fluids at a slower, more reliable rate. This stability, along with cooler temperatures, allows many cold-seep species to live longer than those in hydrothermal vent ecosystems. For instance, siphonophores, some of the longest-living non-colonial invertebrates, can live between 170 and 250 years. These ecosystems are remarkable in their ability to support life in such extreme conditions, relying on chemosynthesis rather than photosynthesis to sustain the organisms within them.

Besides the larger organisms like mussels and tube worms, cold seep ecosystems host a variety of microorganisms, including methanogens, anaerobic methane-oxidizing archaea, sulfate-reducing bacteria, and sulfur-oxidizing bacteria. On the aerobic surface of the sediments, the primary methane-oxidizing microorganisms are aerobic methanotrophic bacteria, such as Methylococcales from the Gammaproteobacteria group, which often coexist with mussels and tube worms. While in the anoxic sediments, methane oxidation is primarily carried out by anaerobic methane-oxidizing archaea and sulfate-reducing bacteria. Sulfur-oxidizing bacteria, like the Beggiatoaceae from the Gammaproteobacteria, can oxidize sulfides from sulfate reduction. Additionally, other archaea (such as Thermoplasmata, Bathyarchaeota, and Lokiarchaeota) with unknown functions, along with bacteria like Planctomycetes and Proteobacteria, also inhabit cold seep sediments. These microorganisms may also play a role in the methane metabolic cycle (Inagaki et al. 2006).

3.3.2 Microbial Ecosystems in Terrestrial Settings

3.3.2.1 Glacier and Frozen Soil Systems

Low-temperature environments refer to habitats that are consistently or periodically below 5 °C, such as glaciers (including the polar ice caps), ice sheets, sea ice, lake ice, snow, and frozen ground

Figure 3.21 Mountain glacier in the Tianshan Mountains, which is affected by the high temperature in summer, causing them to melt and reveal a vast expanse of sediment at their base. *Source:* Photo by Wei Liu.

(Permafrost) (Margesin and Collins 2019). This region is often referred to as the cryosphere in the literature. Mountain glaciers (Figure 3.21) and polar ice caps cover around 10% of the global land area, while snow-covered area fluctuates between 1.3% and 30.6% depending on the season. Frozen areas make up about 9%–12% of the land surface (Dong and Jiang 2023).

Even the most extreme cold and frozen environments harbor diverse and metabolically active microbial populations. In recent years, studies on microorganisms in cold habitats have been mainly focused on the Earth's cryosphere, i.e. frozen ecosystems, which include glaciers, ice sheets, sea ice, lake ice, and frozen ground (permafrost) (Margesin and Collins 2019). These microorganisms, known as psychrophilic, cold-tolerant, or cold-adapted, face a range of challenges due to the harsh conditions. In addition to low temperatures, which slow down physical and biological processes, they must contend with factors such as low nutrient availability, high salinity, dryness, low water activity, and even high UV radiation and oxidative stress at high altitudes or high pressure in the deep sea (D'Amico et al. 2006). Interestingly, the primary limiting factor for life in cold environments is not the temperature itself, but the lack of liquid water at low temperatures (Bakermans 2017). Microbial reproduction has been observed at subzero temperatures as low as −20 °C, and metabolic activity has been detected down to −30 °C. While activities at even lower temperatures have been reported, they still require further verification (Bakermans 2017). Based on their growth temperature, these microorganisms can be classified into two categories: psychrophiles and psychrotolerant. Psychrophiles thrive at temperatures below 15 °C, with a maximum growth temperature below 20 °C. In contrast, psychrotolerant can survive in temperatures below 0 °C but grow better between 20 and 25 °C.

The temperature limit for the completion of a microorganism's life cycle depends on its cellular structure, with unicellular organisms typically having a lower temperature threshold than multicellular ones. For free-living unicellular microorganisms, the low-temperature limit for active metabolism is determined by the vitrification (glass transition) temperature of the cytoplasm, which is influenced by freeze-induced desiccation. As the cytoplasm cools, its viscosity increases, halting the movement of gases and metabolites, thus stopping metabolism. However, unlike intracellular freezing, vitrification is not lethal. Once the cell warms up and rehydrates, metabolic activity can resume (Bakermans 2017; Margesin and Collins 2019).

Most bacteria and archaea are capable of surviving in low-temperature environments, but on a larger scale, microbial communities in these environments differ from those in warmer habitats. For example, Betaproteobacteria are dominant in glacier environments, while Flavobacteria and Gammaproteobacteria are prevalent in sea ice. In frozen soil habitats, Actinobacteria are the dominant group (Dong and Jiang 2023).

A comparative metagenomic study of snow samples from the Arctic and Antarctic revealed notable differences in microbial community structure and ecological drivers. Snow from an Alaskan glacier hosted a highly diverse microbial population, including about 2500 bacterial species, primarily Proteobacteria, Bacteroidetes, and Firmicutes, along with around 30 species of archaea. Many of these microbes overlapped with those found in Arctic soil, suggesting that snow microbes may contribute to soil formation during glacier retreat. In Arctic Spring snow, microbial composition was primarily influenced by photochemical activity and oxidative stress. In contrast, Antarctic snow communities were highly site-specific, dominated by Flavobacterium species unique to the region, likely due to strong environmental selection pressures (Choudhari et al. 2014; Lopatina et al. 2016; Maccario et al. 2014).

Cryoconite holes, which harbor unique and often less diverse microbial communities compared to surrounding ice-marginal environments, are dominated by cyanobacteria and heterotrophic Proteobacteria. These communities exhibit temporal and spatial variation, with seasonality having a greater impact than local environmental factors. They play a significant role in carbon cycling through processes like carbon fixation and the breakdown of organic matter. During melt events, cryoconite holes also export microbes and DOC downstream. Metagenomic studies highlight a high abundance of functional genes related to nutrient uptake, particularly for nitrogen, sulfur, iron, and phosphorus, with lower levels of genes associated with photosynthesis (Franzetti et al. 2016; Stibal et al. 2015).

Subglacial environments are primarily dominated by bacterial taxa such as Proteobacteria, Firmicutes, Actinobacteria, and Bacteroidetes, with archaea and eukaryotes being less common. Key microbial processes in these environments include methane cycling and the chemolithoautotrophic oxidation of reduced iron, sulfur, and nitrogen. Molecular and culture-based studies have confirmed active microbial metabolism in these systems, emphasizing the importance of microbial chemolithotrophy in subglacial biogeochemical cycles (Achberger et al. 2017).

Permafrost, found in both polar and high-altitude regions, supports diverse and active microbial communities even at subzero temperatures. Using advanced techniques, researchers have identified active bacterial, archaeal (especially methanogens), and fungal populations. These organisms adapt through unique mechanisms, such as modifying membrane compositions, producing cryoprotectants, and shifting metabolic pathways. Bacteria dominate permafrost microbiomes, with smaller contributions from archaea (e.g. Euryarchaeota, Crenarchaeota) and fungi (primarily Ascomycota). Active microbial processes at subzero temperatures include carbon degradation, methanogenesis, methane oxidation, nitrification, denitrification, and iron/sulfur cycling (Altshuler et al. 2017).

3.3.2.2 Desert System

Around one-fifth of the Earth's land is covered by deserts (including the *Gobi* Desert). Deserts typically experience extreme continental climates, characterized by low annual precipitation and high

Figure 3.22 Landscape of the Tengger Desert in Ningxia, China (a), and landscape of *Gobi* Desert in Xinjiang, China (b). *Source:* Photos by Wei Liu.

evaporation rates (Figure 3.22). Due to intense sunlight, temperature fluctuations in these areas are drastic, with significant temperature differences between day and night. The soil has low specific heat capacity and is nutrient-poor, which together contribute to the fragile desert ecosystem with low biodiversity, simple community structures, and limited self-regulation.

Deserts host various microbial groups, including *Bacillus*, salt-tolerant bacteria, and radiation-resistant species. Studies show that actinomycetes, such as *Streptomyces* and *Micromonospora*, make up an average of 25.5% of the microbial population in global deserts. In more extreme, arid regions, this proportion can exceed 70%, and in some areas, it may reach more than 94% (Dong and Jiang 2023).

One key component of the desert ecosystem is the biological soil crusts (Figure 3.23), composed of algae, lichen, moss, heterotrophic bacteria, fungi, and soil particles. This microbial community forms on the surface of desert soils and plays a vital role in maintaining species diversity, preventing

Figure 3.23 Biological soil crusts on the surface of sandy soil. The communities are primarily composed of cyanobacterial lichens, which form the thickest crusts, ranging from 5 to 10 mm in thickness. *Source:* Photo by Wei Liu.

wind erosion, stabilizing sand, and transforming soil nutrients and energy. Studies on the diversity and abundance of bacteria and oxygenic phototrophs in biological desert crusts under different salinity conditions showed that cyanobacteria, mainly *Microcoleus vaginatus* of the order Oscillatoriales, were predominant in the low saline crusts, while other phototrophs, such as diatoms, were the main microorganism group responsible for the oxygenic photosynthesis in the high saline crusts (Li et al. 2013). Through photosynthesis, cyanobacteria fix CO_2 into organic matter and other metabolites, which serve as a food source for other heterotrophic microorganisms. Moreover, cyanobacteria, lichens, and heterotrophic microorganisms also play a role in nitrogen fixation by converting N_2 into NH_4^+, which becomes incorporated into the nitrogen cycle in the soil (Belnap 2002).

The higher salt content in crusts may stimulate the growth of bacteria, including Deinococcus-Thermus, Bacteroidetes, and some subdivisions of Proteobacteria. Furthermore, the abundance of oxygenic phototrophs was significantly higher under lower saline conditions than that in the higher saline crusts, suggesting that the higher salinity in desert crusts could suppress the numbers of total bacteria and phototrophic bacteria but did highly improve the diversity of salt-tolerant bacteria (Li et al. 2013).

Desert surfaces, covered by pebbles or rocks (desert pavement), are home to hypolithic microbial communities that thrive on the undersides of translucent rocks like quartz or marble (Figure 3.24). These communities are especially prominent in hyperarid deserts, where they can form either as extensions of biological soil crusts or as distinct communities in extreme habitats. In hyperarid deserts like the Atacama, hypolithic colonization rates are very low (about 3:4000), while in more humid regions like the Namib Desert, rates exceed 95%. Moisture availability, such as fog-derived moisture in the Namib, is crucial for community development (Lacap et al. 2011; Warren-Rhodes et al. 2013). Studies have shown that photosynthetic activity in these communities depends on soil moisture levels of at least 15% (Tracy et al. 2010). The microbial composition of hypolithic communities is dominated by cyanobacteria, particularly *Chroococcidiopsis* and *Nostoc* species, which play a key role in nitrogen fixation and contribute to soil nutrient cycling. Other significant groups include heterotrophic bacteria from phyla like Actinobacteria, Proteobacteria, and Acidobacteria, some of which are radiation-tolerant. The presence of *Agrobacterium*, a nitrogen-fixing genus, and

Figure 3.24 Cyanobacteria-dominated hypolithic biofilm intimately associated with the rocks. *Source:* Photo by Wei Liu.

Methylobacterium, which metabolizes methanol, highlights the specialized adaptations of hypolithic microbes (Makhalanyane et al. 2015). The microbial diversity in the hypolithic community mirrors that of surrounding soils, suggesting selective recruitment from local populations. However, colonization is a slow process, potentially taking decades.

3.3.2.3 Soil System

Soil is a biologically active, porous layer that forms the uppermost part of Earth's crust. It plays a vital role in supporting life on our planet, acting as a reservoir for water and nutrients, filtering and breaking down harmful wastes, and contributing to the cycling of carbon and other elements within the global ecosystem. Over time, soil has developed through weathering processes shaped by biological, climatic, geological, and topographic factors (Section 7.4).

Soil microbial communities are essential to ecosystem functionality, playing a key role in maintaining soil fertility and health. Their diversity is strongly influenced by various land-use practices and environmental conditions. Soils managed under conventional agricultural systems, such as paddy and upland fields, tend to exhibit higher microbial diversity than soils affected by high salinity, contamination, or low agricultural suitability. Prokaryotic communities in these soils are primarily composed of Proteobacteria, Chloroflexi, Acidobacteria, and Bacteroidetes, with their abundance patterns closely tied to soil condition and quality. Similarly, eukaryotic communities are mainly dominated by Opisthokonta, Stramenopiles, Alveolates, Rhizaria, and Amoebozoa, displaying distinct distribution patterns across different soil types (Yoon et al. 2024).

Soil provides a natural habitat for a vast array of microorganisms, with about 98% of them living in biofilms. These biofilms consist of tightly bound clusters of microorganisms attached to surfaces, with the cells often embedded in a self-produced matrix made up of EPS. Soil biofilms follow a series of common steps during their formation (Figure 3.25): (1) initial conditioning film formation and microbial adhesion; (2) irreversible adhesion and biochemical activation; (3) biofilm maturation and stability enhancement; (4) structural development and environmental adaptation in biofilms; and (5) biofilm dispersal and microbial shedding (Zhang et al. 2024).

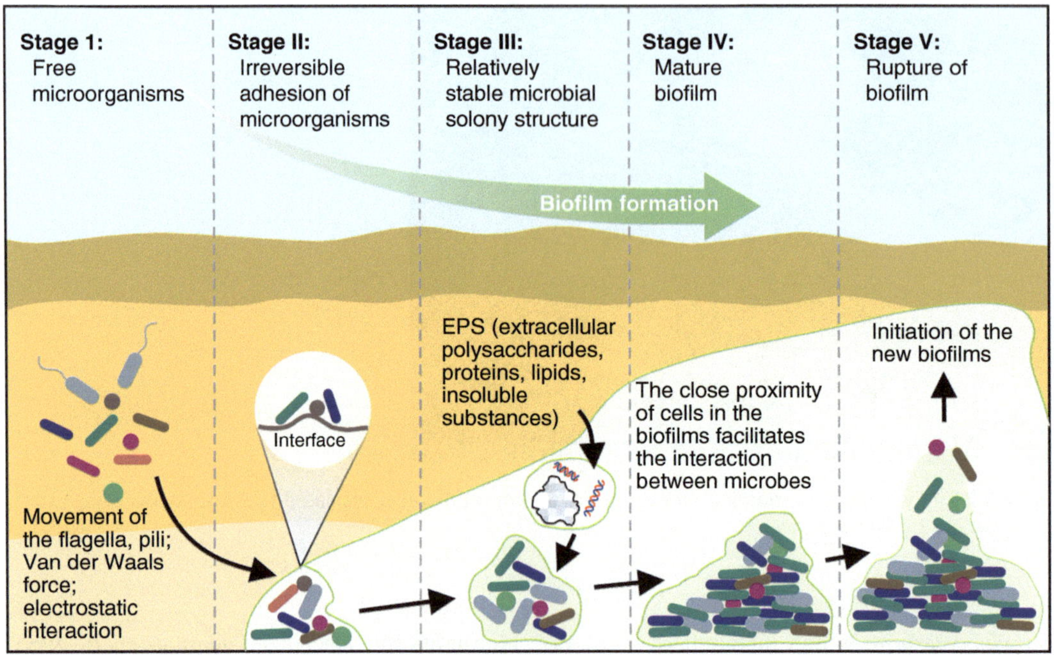

Figure 3.25 The forming process of soil biofilms. Stage I: free microorganisms initiate the biofilm formation process under the influence of flagella, pili, van der Waals force and electrostatic interaction; stage II: microorganisms undergo irreversible attachment on surfaces with characteristics such as cessation of flagella-mediated movement and induction of lipopolysaccharide biosynthesis; stage III: biofilm maturation stages: microorganisms secrete more EPS to form a relatively stable colonization structure; stage IV: ultimately form a mature structure which depends on environmental conditions, signaling molecules, and nutritional competition; and stage V: the rupture of biofilm allows microbial cells to regain motility, marking the beginning of the formation of a new biofilm. *Source:* Zhang et al. (2024) / with permission of Elsevier.

Soil biofilms significantly influence the structure and behavior of microorganisms, fostering both cooperation and competition within the microbial community (Mishra et al. 2022; Petrova et al. 2010). These interactions enhance microbial adaptability, with biofilm-forming microorganisms often dominating. Microorganisms in biofilms communicate through quorum sensing, adjusting their behavior in response to population density and environmental changes, which improves their survival and stress tolerance compared to free-floating cells. The EPS secreted by biofilms also act as a protective barrier against environmental stressors. In addition to aiding microbial adhesion and protection, soil biofilms play a crucial role in pollutant removal (Mishra et al. 2021). Their surface contains functional groups that can bind and immobilize pollutants, while the biofilm's large surface area and high density of binding sites enable efficient pollutant capture (Ayangbenro and Babalola 2017). Furthermore, biofilms can transform complex contaminants into simpler, less toxic compounds through enzymatic processes.

Soil biofilms are involved in complex nutrient cycles, interacting with the soil environment to regulate key elements like carbon, nitrogen, phosphorus, and other minerals. (1) Biofilms help microbes survive external pressures, aiding carbon and nitrogen accumulation. They utilize organic carbon for metabolism and can fix atmospheric carbon dioxide. Nitrogen-fixing bacteria in biofilms convert nitrogen into forms usable by plants, enhancing soil nitrogen fertility (Bhattacharyya and Jha 2012; Odoh 2017). Biofilms also regulate nitrification and denitrification by altering pH and redox conditions, supporting soil nitrogen cycling. (2) Biofilms promote the mineralization and release of phosphorus by secreting enzymes like acid phosphatases (Chauhan et al. 2021). They also support the cycling of sulfur, iron, and potassium through processes like solubilizing sulfur and producing siderophores. This makes biofilms essential for maintaining healthy soil nutrient cycles.

Soil biofilms play a critical role in their formation and stability. (1) Biofilms promote soil aggregate formation by secreting EPS, which bind to soil particles through electrostatic and hydrogen bonding (Jiang et al. 2007). This helps create stable soil structures, enhancing nutrient cycling and microbial activity. (2) Biofilms provide a protective environment for microbes, allowing them to thrive and contribute to soil health. Microbial activities within biofilms help in nutrient exchanges, organic matter transformation, and water retention, all of which are crucial for soil aggregate stability (Lünsdorf et al. 2010). (3) Biofilms increase the stability of soil aggregates by enhancing nutrient cycling and producing humus, which binds soil particles together. EPS plays a key role in this process, as demonstrated by studies showing that biofilms with higher EPS production led to more stable aggregates (Sher et al. 2020). Biofilms also help soil withstand environmental stresses like heat and drought.

Soil biofilms are essential in regulating pollutant migration and promoting soil remediation by acting as barriers to contaminants, adsorbing pollutants, and enhancing biodegradation processes. Through the secretion of EPS, biofilms sequester pollutants like heavy metals (e.g. Cu, Zn, Pb) and organic contaminants (e.g. PAHs, pesticides) via functional groups in EPS that interact with pollutants, reducing their bioavailability and toxicity (Späth et al. 1998; Xing et al. 2022). Biofilms facilitate bioremediation by producing extracellular enzymes that break down complex pollutants, and through co-metabolism, where microbes share metabolic pathways for enhanced pollutant degradation. EPS also acts as a biosurfactant, increasing the bioavailability of organic pollutants (Gauri et al. 2012). Biofilms immobilize heavy metals through adsorption, ion exchange, and complexation, converting toxic metals into less harmful forms. Gene expression related to metal-binding proteins further enhances metal detoxification. These capabilities also help reduce heavy metal uptake in crops, offering potential for agricultural applications in contaminated environments (Zhang et al. 2024).

Rhizosphere biofilms are essential for plant growth and stress resilience, forming through complex interactions between soil microorganisms and plant roots. These biofilms enhance nutrient availability, water retention, and disease resistance, while also promoting root development (Zhang et al. 2022). Microbes colonize plant roots by responding to root exudates, using mechanisms like chemotaxis, quorum sensing, and signaling pathways to form cohesive biofilm communities. Plants contribute by secreting nutrients and signaling molecules that support microbial growth, while also managing immune responses to regulate biofilm formation (Entcheva-Dimitrov and Spormann 2004). Rhizosphere biofilms improve plant resilience to environmental stresses like nutrient deficiency, high salinity, and drought by enhancing nutrient cycling, ion balance, and water retention. Additionally, biofilms act as natural biocontrol agents, suppressing pathogens through the production of antimicrobial compounds, promoting plant health, and improving disease resistance (Brokate et al. 2024; Rudrappa et al. 2008). This symbiotic relationship between plants and rhizosphere biofilms underscores their potential in sustainable agriculture for improving crop yields, resilience, and pathogen control (Zhang et al. 2024).

3.3.2.4 Karst Cave System

A karst cave is a natural underground space primarily formed through the chemical dissolution of groundwater acting on soluble rocks (mainly carbonate rocks), with mechanical processes like water erosion and collapse also playing a role. Karst caves are an important part of the terrestrial ecosystem (Figure 3.26).

Figure 3.26 Structures and biodiversity within the Karst caves: (a) cave plants are distributed in the entrance and weak light areas, mainly consisting of lower plants such as ferns, mosses, lichens, and algae, Guizhou, China; (b) vadose shafts are domepits created by a relatively slow and steady flow formed in the vadose zone, Tanding Cave, Indonesia; (c) cave scallops are formed by the dissolution and erosion of flowing water. The scallops usually appear on the ceiling and/or sidewall of the karst cave with the steep direction indicating the incoming direction of the fluid, Hang Sung Sot, Vietnam; (d) stalactites and stalagmites, elongated forms of various minerals (mainly $CaCO_3$) deposited from solution by slowly dripping water. A stalactite hangs like an icicle from the ceiling or sides of a cavern; while a stalagmite appears like an inverted stalactite, rising from the floor of a cavern, Guizhou, China; (e) the goose pipe stone is the early period of stalactite development. When the CO_2-rich solution drips down from the ceiling, it releases CO_2 due to pressure differences, causing supersaturated calcium carbonate to precipitate and grow downward, eventually forming a crystal clear "goose pipe" stone, Dongge Cave, Guizhou, China; (f) cave bats are the most common mammal in caves, living on the ceiling of caves and nocturnal, Tanding Cave, Indonesia; (g) external clastic deposits refer to the coarse sediments abandoned and deposited in the original passages during the waning stages of cave development, Dongmen Cave, Guizhou, China; (h) fungus gnats, a type of Macrocera insect, live in karst caves. Their larvae produce threads that hang sticky droplets, forming a beautiful "pearl curtain." These threads, however, are actually traps designed to capture and prey on small flying insects, Guangxi, China; and (i) actinomycetes, the yellow mats on the cave wall, are a special group of prokaryotes that can form branching hyphae and conidia. It grows in filamentous form and mainly reproduces through spores, Guizhou, China. *Source:* Courtesy of Yu Liu.

Karst caves typically have the following characteristics: (1) they are usually formed near the water table, and the landscape is often shaped by underground karst formations; (2) except for the entrance where sunlight may penetrate, the interior is dark year-round; (3) due to the long-term isolation from the external environment, the caves maintain a relatively constant temperature with minor fluctuations; (4) the air humidity in the caves is consistently high, often reaching 90% or even 100%; (5) the organic matter is scarce in the caves, mainly transported by wind, water, or other external factors; and (6) the concentration of CO_2 and other reducing gases (such as H_2S, CH_4, and NH_3) is elevated, which supports the growth of chemoautotrophs, developing specialized microbial communities in the caves (Dong and Jiang 2023).

Bacteria are the dominant group of microorganisms in caves. The most commonly detected bacterial groups include Actinobacteria, Proteobacteria, Firmicutes, and Bacteroidetes, among others. Within these groups, there are hundreds of species of Actinomycetes, such as *Nonomuraea cavernae*, *Saccharopolyspora cavernae*, and *Streptomyces canchipurensis*, along with other newly identified species (Dong and Jiang 2023). There are also a large number of fungal groups in the caves, mainly Ascomycota, in addition to Basidiomycota and Zygomycota. The composition of fungi shows a high degree of habitat specificity, such as *Penicillium* in the air, *Capnodiales* in the rock wall, *Sordariales* and *Agaricales* corresponding to sediments and dripping water, respectively. Viruses are also a growing area of interest among cave life forms. According to preliminary data, over 80 viruses have been isolated from cave-dwelling bats. Some of them are linked to major zoonotic diseases, including Ebola, Lyssavirus, SARS-related coronaviruses, and so on, presenting a significant threat to public health (Dong and Jiang 2023).

In addition to soil, caves are considered to be another important atmospheric methane sink in terrestrial ecosystems, and the oxidation of atmospheric methane by microorganisms is considered to be the main reason. Increasingly, it is thought that the consumption of methane by methanotrophic bacteria plays a key role in the low methane concentrations in caves. Studies have shown that the methane-oxidizing bacteria in caves are primarily from upland soil clusters γ, which have a high affinity for methane and can use trace methane (0.0018‰) in the air as a carbon source and energy source for growth (Dong and Jiang 2023).

3.3.3 Microbial Ecosystems in Other Extreme Settings

3.3.3.1 Extremely Acidic System

Extreme acidic environments are found worldwide, and both natural processes and human activities can create strongly acidic conditions. Natural acidic environments comprise hydrothermal sites on land or in the deep sea, cave systems, acid sulfate soils, and acidic fens, as well as naturally exposed ore deposits (gossans). Man-made acidic environments are mostly mining sites including mine waste dumps and tailings, acid mine drainage (Figure 3.27), and biomining operations. The acidification of these environments is primarily driven by the sulfuric acid produced from the oxidation of sulfide minerals, such as pyrite and chalcopyrite. Among these, acid mine drainage is the most common human-made acidic environment and presents a significant challenge for environmental management.

Microorganisms can also contribute to acidic environments through their metabolic processes. Taking *Acidithiobacillus ferrooxidans* as an example, experimental analysis shows that it can release H^+ to generate sulfuric acid while oxidizing pyrite and other metal sulfides (Liu and Zhang 2015), which can further lower the pH of the solution to as low as 2.0 (Figure 3.28). These strong acidic environments often result in the release of heavy metal ions from minerals, which can be toxic to most organisms. However, these extreme conditions still support a variety of microorganisms that

Figure 3.27 Acid mine drainage from the Mike Horse Mine complex enters upper Blackfoot River watershed, USA. *Source:* Earthworks/https://earthworks.org/issues/acid-mine-drainage/last accessed on June 18, 2025.

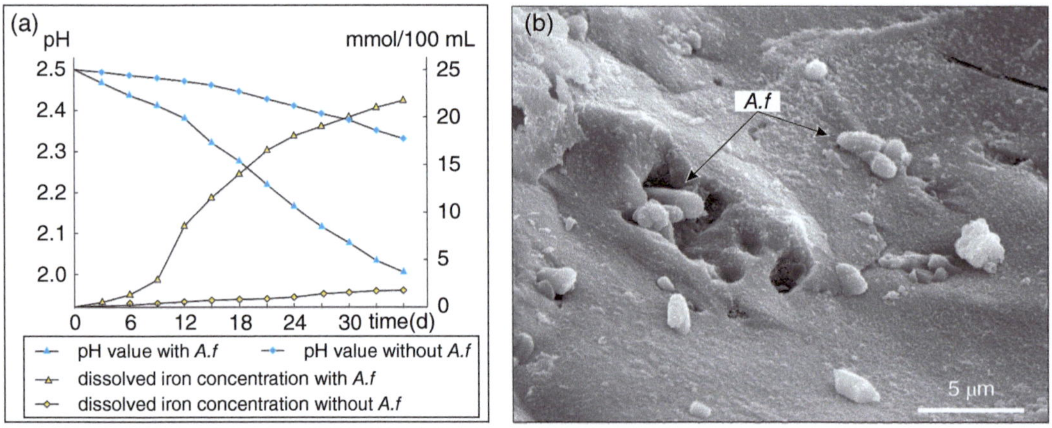

Figure 3.28 An experimental study on microbial pyrite oxidation shows that *Acidithiobacillus ferrooxidans* (*A.f*) accelerates pyrite erosion. *Source:* Liu and Zhang (2015)/Informa UK Limited. (a) In the presence of *A.f*, the concentrations of total dissolved iron and H^+ significantly increased; (b) numerous erosion pits formed on the pyrite surface, consequently.

thrive in low pH environments. These acid-loving microbes, i.e. *acidophiles*, can resist both high acidity and heavy metal toxicity by adopting various survival strategies. These microorganisms often have autotrophic metabolisms, meaning they can use substances like ferrous ions, hydrogen, and reducing sulfides as energy sources. Additionally, they exhibit strong resistance to osmotic pressure and heavy metals and can produce extracellular polymers.

Extremely acidic environments are often oligotrophic, meaning they contain very low concentrations of soluble organic matter, typically less than 20 mg/L. The first microorganisms to thrive in sulfur-rich acidic conditions are chemoautotrophs that oxidize sulfide and ferrous ions. Their growth leads to the accumulation of organic matter, creating favorable conditions for heterotrophic organisms to emerge. For example, *Acidithiobacillus* species, such as *Acidithiobacillus thiooxidans* and *A. ferrooxidans*, are among the most common obligate chemoautotrophs in these environments. *A. thiooxidans* can grow in a pH range of 0.9 ~ 4, with an optimal pH of 2.5, while *A. ferrooxidans* thrives in a pH range of 2 ~ 3 (Dong and Jiang 2023). Both species can fix CO_2 via the Calvin-Benson-Bassham cycle, contributing to the carbon cycle. Additionally, *A. ferrooxidans* can fix N_2 through a nitrogenase complex encoded by the *nif* operons.

3.3.3.2 High Saline-alkaline System

High saline-alkaline environments are primarily found in arid climate regions and are considered as one of the most extreme habitats both in modern times and throughout geological history on Earth (Figure 3.29). A high saline environment generally refers to an extreme setting where soluble salinity is much higher than that of normal seawater (35‰). Such environments typically form due to the intense evaporation of soluble inorganic salts (e.g. NaCl, Na_2SO_4), which accumulate naturally or through human activities, as seen in salt lakes and salt fields. Additionally, evaporation can lead to the concentration of alkaline inorganic salts like Na_2CO_3, making these environments not only saline but also highly alkaline, as in saline-alkaline lands.

Figure 3.29 Salt flats in Badwater Basin, Death Valley National Park, USA. *Source:* Photos by Wei Liu.

While high saline-alkaline environments are inhospitable to most macro-organisms, they provide a thriving habitat for salt- and alkali-tolerant microorganisms. Based on their salinity requirements, these microbes are classified into: extreme halophiles (growing best in media containing 2.5 ~ 5.2 mol/L), moderate halophiles (growing best in media containing 0.5 ~ 2.5 mol/L), and halotolerant microorganisms that do not show an absolute requirement for salt for growth but grow well up to often very high salt concentrations (considered extremely halotolerant if the growth range extends above 2.5 mol/L) (Oren 2008).

Many microbes in high-salinity environments must also withstand alkaline stress (high pH). They are categorized as: Alkaliphiles, which grow best at pH 9 ~ 10, and Alkalitolerants, which can tolerate high pH but prefer environments with pH < 9, sometimes even weakly acidic conditions. Alkaliphiles are diverse and found across both prokaryotic and eukaryotic domains, including Bacteria, Archaea, and Eukaryotes. Notably, certain microbial families, such as Halobacteriaceae (Euryarchaeota), Halanaerobiales (Firmicutes), and Halomonadaceae (Proteobacteria), are almost entirely composed of halophilic species (Dong and Jiang 2023). Research on saline-alkali soils and salt lakes found that halophilic Archaea exhibited a high degree of consistency at the family level, with Haloferacaceae, Halomicrobiaceae, and Nanosalinaceae being the predominant families. This was determined through the optimization of culture-dependent experiments. A total of 98 strains belonging to 26 genera were successfully isolated, among which haloarchaea constituted a significant proportion; notably, 22 potential novel taxa were identified (Ding et al. 2025).

To survive in high-salinity environments, microorganisms employ two main strategies to balance osmotic pressure (Oren 2008): (1) high-salt-in strategy, used primarily by extreme halophiles, this approach involves accumulating high concentrations of K^+ inside the cell to counteract external salinity. Microbes utilizing this strategy adapt their enzymes and proteins to function in high-potassium conditions, maintaining stability and activity even in nearly saturated salt environments; (2) the organic-solutes-in strategy, by which microbes approach synthesize or absorb compatible organic solutes from their surroundings to resist extracellular salt stress. This strategy allows greater adaptability to varying salinity levels and is commonly used by most halophilic bacteria, eukaryotes, and some halophilic methanogenic archaea.

3.3.3.3 Deep Subsurface System

The deep subsurface system refers to environments beneath the terrestrial and seafloor surfaces, where microorganisms thrive without relying on photosynthesis for energy. These extreme subsurface environments have shaped unique physiological and ecological characteristics in microorganisms, including traits such as anaerobic, autotrophic, thermophilic, barophilic, and oligotrophic (Edwards et al. 2012). These microorganisms play crucial roles in mineral and rock weathering, element cycling, the formation and migration of oil, gas, and metal resources, as well as the degradation of organic matter. While deep-sea systems have been extensively discussed (see Section 3.3.1), we focus here on deep terrestrial subsurface systems, including bedrock and fluids within the continental depths, as well as oil-gas reservoirs.

Research on microorganisms in deep continental bedrock primarily relies on ultra-deep continental drilling projects such as the International Continental Scientific Drilling Program (ICDP), the International Ocean Discovery Program (IODP), and the China Continental Scientific Drilling Project (CCSD) (Figure 3.30). Molecular biology studies have confirmed the presence of microorganisms at depths exceeding 4500 m in the CCSD core, where temperatures reach around 120 °C, nearly the upper limit for microbial survival (Dong et al. 2009).

Microbial distribution in the deep subsurface biosphere is heavily influenced by geological conditions and nutrient availability (Figure 3.31), with depth playing a key role. Generally, microbial abundance decreases with increasing depth (Gihring et al. 2006). Shallow, low-temperature,

Figure 3.30 "Shenditake 1," the deepest vertical well in Asia, reached a depth of 10,910 m below the surface on February 20, 2025. *Source:* http://www.cctv.com.

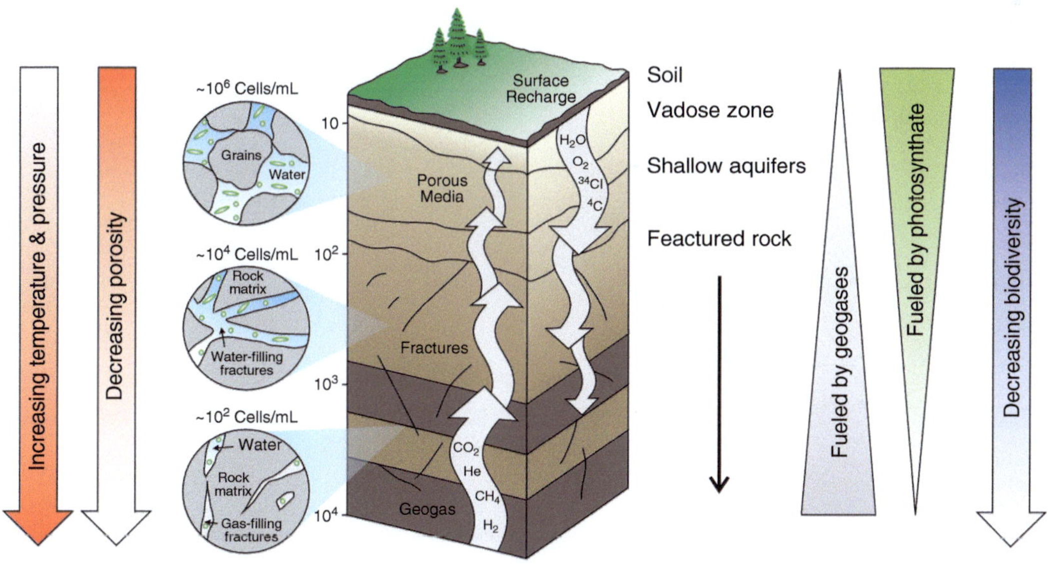

Figure 3.31 Overview of subsurface continental environments. *Source:* Adapted from Kieft (2016). Microbial abundance in pore water generally declines with depth, from shallow aquifers to underlying fractured rock environments. The quantity and quality of photosynthetically derived organic C decline with depth, and the importance of "geogas" energy sources (e.g. H_2, CH_4) generally increases with depth. Temperature increases with depth according to the local geothermal gradient; pressure increases with depth (~10 MPa/km).

micro-aerophilic environments, with relatively young and low-salinity fluids, often feature microbial communities dominated by Proteobacteria. These environments experience frequent exchanges with surface fluids. In contrast, deeper, higher-temperature, higher-salinity, ancient, and anaerobic environments tend to be dominated by Firmicutes and Euryarchaeota (Magnabosco et al. 2014). As depth increases, Firmicutes, capable of producing spores, become more prevalent.

In oil-gas reservoirs, microorganisms primarily inhabit the pore spaces within the reservoir's porous media. At a microscopic level, these microbes thrive in various microenvironments, such as the water phase in fractures, the oil–water interface, or microbial biofilms (Meckenstock et al. 2014). The bacterial composition in these reservoirs includes 15 phyla, including Proteobacteria, Firmicutes, Bacteroidetes, and Actinobacteria. Archaea, primarily methanogenic, are represented by genera like *Methanoculleus*, *Methanocalculus*, and *Methanolinea* (Li et al. 2017).

Hydrogen-driven, sulfur-nitrogen coupling, and syntrophic processes are key metabolic pathways in the deep subsurface biosphere. The diversity of species and functions of deep subsurface microorganisms is closely tied to the energy available within these environments. In sedimentary rock-dominated deep subsurface environments, organic carbon can be brought in from the surface or deeper subsurface, providing a carbon source and energy for microorganisms. However, in other types of rock, organic matter is scarce, and the primary energy source for deep subsurface microbes is geogenic H_2 and CH_4. H_2 serves as both an energy and electron source for microbial growth and metabolism (Sherwood Lollar et al. 2014). H_2 oxidation can couple with various reduction processes

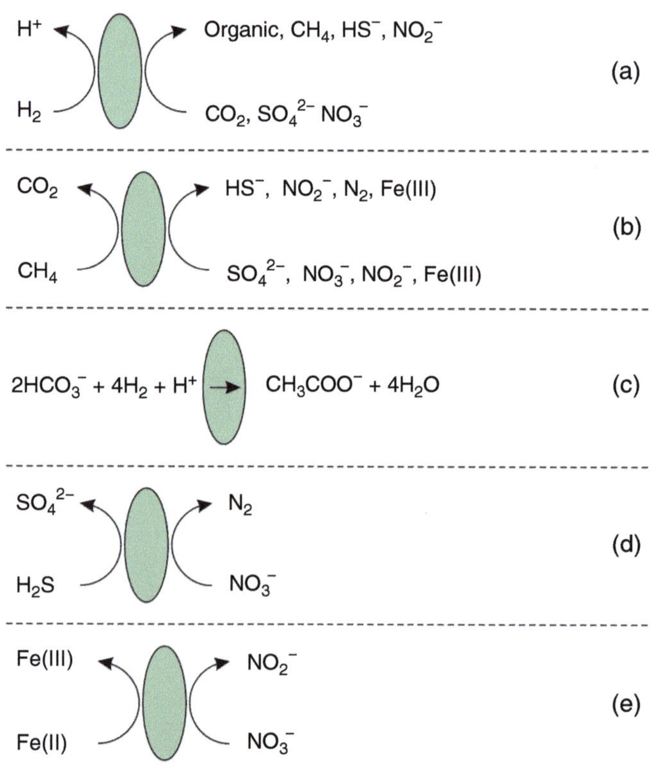

Figure 3.32 Metabolic diversity of Microorganisms in subsurface environments. *Source:* Adapted from Yin and Dong (2020). (a) CO_2, SO_4^{2-}, and NO_3^- reduction coupled with H_2 oxidation; (b) SO_4^{2-}, NO_3^-, NO_2^- and Fe(III) reduction coupled with CH_4 oxidation; (c) acetic acid reaction; (d) NO_3^- reduction coupled with H_2S oxidation; and (e) NO_3^- reduction coupled with Fe(II) oxidation.

(Figure 3.32), such as the conversion of H_2 and CO_2 into acetic acid through sulfate reduction by sulfate-reducing bacteria, or hydrogenotrophic methanogens converting H_2 and CO_2 into CH_4 (Colman et al. 2017). Acetic acid supports acetotrophic methanogens and acetic acid fermentation, while CH_4 sustains methanophilic microorganisms.

3.4 Microbial Ecosystem as a Model for Exploring Life Beyond

Microbial ecosystems on Earth provide valuable analogs for studying extraterrestrial life, as they thrive in extreme, isolated environments that resemble conditions on other planets and their moons.

Microbial mats, which dominated Earth's early biosphere and persist in extreme environments, could also exist in extraterrestrial environments. Much of Earth's life remains hidden beneath the surface, such as in geothermal vents, the deep ocean, Antarctic shelves, and even within the Earth's crust. These subsurface ecosystems hold significant potential for advancing our understanding of evolution, physiology, biogeochemistry, and biodiversity. These ecosystems offer insights into how life could survive in harsh extraterrestrial habitats, such as those on Europa or Mars. For example, microbial mats, found in hot springs, deep-sea vents, and polar regions, mimic conditions that may exist in extraterrestrial environments. They help scientists understand how life can sustain itself without sunlight, relying on chemical energy instead of solar energy in extreme environments with high pressure, low (or high) temperature, and limited resources, much like what could exist beneath the icy surface of Europa.

Furthermore, by studying these Earth-based microbial ecosystems, researchers can refine their methods for detecting life beyond Earth. The tools, techniques, and lessons learned from exploring microbial life in extreme environments on our planet can be applied when searching for extraterrestrial life, helping scientists design more effective missions and experiments. These include the search for chemical signatures, the use of rovers or landers to explore subsurface habitats, and the development of instruments capable of analyzing remote or difficult-to-access environments.

However, we must acknowledge that studying microbial ecosystems in extreme environments on Earth does not provide all the answers. Extraterrestrial ecosystems may present unforeseen challenges, requiring adaptable and flexible methodologies. The possibility of encountering familiar or entirely novel forms of life remains an exciting prospect. These discoveries may reveal life synchronized with the unique rhythms of distant worlds, just as life on Earth is aligned with our planetary rotation. As cosmologist Katie Mack noted, exploring the unknown is like reaching the edge of a map, we do not know what we'll find. But as evolutionary biologist Stephen Jay Gould suggested, the simplest forms of life could be widespread across the cosmos. While Earth is the only known cradle of life, the universe's chemical conditions suggest that life could exist elsewhere, perhaps in places we have not yet explored. Missions to Mars and the outer moons of our solar system offer promising opportunities to uncover new forms of life, whether extinct or currently thriving. Like explorers at the edge of a map, we venture forward, eager to discover what lies beyond (Sielaff and Smith 2019).

References

Achberger, A., Michaud, A., Vick-Majors, T., Christner, B., Skidmore, M., Priscu, J., and Tranter, M. 2017. Microbiology of subglacial environments. In: *Psychrophiles: from Biodiversity to Biotechnology* (Second Edition) (ed. R. Margesin), pp. 83–110. Berlin, Springer.

Altshuler, I., Goordial, J., and Whyte, L.G. 2017. Microbial life in permafrost. In: *Psychrophiles: from Biodiversity to Biotechnology* (Second Edition) (ed. R. Margesin), pp. 153–179. Berlin, Springer.

Ayangbenro, A.S., and Babalola, O.O. 2017. A new strategy for heavy metal polluted environments: a review of microbial biosorbents. *International Journal of Environmental Research and Public Health*, 14, pp. 94.

Bakermans, C. 2017. Determining the limits of microbial life at subzero temperatures. In: *Psychrophiles: from biodiversity to biotechnology* (ed. R. Margesin), pp. 21–38. Berlin, Springer.

Bamford, N.C., MacPhee, C.E., and Stanley-Wall, N.R. 2023. Microbial Primer: an introduction to biofilms - what they are, why they form and their impact on built and natural environments. *Microbiology*, 169, pp. 001338.

Belnap, J. 2002. Nitrogen fixation in biological soil crusts from southeast Utah, USA. *Biology and Fertility of Soils*, 35, pp. 128–135.

Bhattacharyya, P.N., and Jha, D.K. 2012. Plant growth-promoting rhizobacteria (PGPR): emergence in agriculture. *World Journal of Microbiology and Biotechnology*, 28, pp. 1327–1350.

Brokate, O., Papenbrock, J., and Turcios, A.E. 2024. Biofilm-forming microorganisms in the rhizosphere to improve plant growth: coping with abiotic stress and environmental pollution. *Applied Soil Ecology*, 202, pp. 105591.

Canfield, D.E., and Des Marais, D.J. 1993. Biogeochemical cycles of carbon, sulfur, and free oxygen in a microbial mat. *Geochimica et Cosmochimica Acta*, 57, pp. 3971–3984.

Canfield, D.E., Kristensen, E., and Thamdrup, B. 2005. *Aquatic geomicrobiology*, pp. 642. Amsterdam, Elsevier Academic Press.

Castro, P., and Huber, M.E. 2008. *Marine biology*, pp. 459. New York, WCB/McGraw-Hill.

Chauhan, P., Verma, P., Pandey, S., Bhattacharya, A., Tripathi, A., Giri, V.P., Singh, S.P., and Mishra, A. 2021. Endophytic microbial interaction with legume crop for developing resistance against nutrient stress. In: *Microbes in Land Use Change Management* (eds. J.S. Singh, S. Tiwari, C. Singh, and A.K. Singh), pp. 363–387. Netherlands: Elsevier.

Choudhari, S., Lohia, R., and Grigoriev, A. 2014. Comparative metagenome analysis of an Alaskan glacier. *Journal of Bioinformatics and Computational Biology*, 12, pp. 1441003.

Coghlan, A. 1996. Slime city. *New Scientist*, 15, pp. 32–36.

Colman, D.R., Poudel, S., Stamps, B.W., Boyd, E.S., and Spear, J.R. 2017. The deep, hot biosphere: twenty-five years of retrospection. *PNAS*, 114, pp. 6895–6903.

Costerton, J.W., Lewandowski, Z., Caldwell, D.E., Korber, D.R., and Lappin-Scott, H.M. 1995. Microbial biofilms. *Annual Reviews of Microbiology*, 49, pp. 711–745.

Cowen, J.P., Giovannoni, S.J., Kenig, F., Johnson, H.P., Butterfield, D., Rappé, M.S., Hutnak, M., and Lam, P. 2003. Fluids from aging ocean crust that support microbial life. *Science*, 299, pp. 120–123.

D'Amico, S., Collins, T., Marx, J.C., Feller, G., and Gerday, C. 2006. Psychrophilic microorganisms: challenges for life. *EMBO Reports*, 7, pp. 385–389.

Decho, A.W. 2020. Extracellular polymeric substances (EPS). In: *Encyclopedia of geobiology* (eds. J. Reitner and V. Thiel), pp. 359–362. Netherlands, Springer.

Des Marais, J.D. 2003. Biogeochemistry of hypersaline microbial mats illustrates the dynamics of modern microbial ecosystems and the early evolution of the biosphere. *Biological Bulletin*, 204, pp. 160–167.

Ding, Y., Ke, J.T., Hong, T., Zhang, A.D., Wu, X., Jiang, X.R., Shao, S.L., Gong, M., Zhao, S.D., Shen, L., and Chen, S.X. 2025. Microbial diversity and ecological roles of halophilic microorganisms in Dingbian (Shaanxi, China) saline-alkali soils and salt lakes. *BMC Microbiology*, 25, pp. 287.

Dong, H.L., and Jiang, H.C. 2023. *Geomicrobiology*, pp. 321. Beijing, Higher Education Press.

Dong, H.L., Yu, B.S., and Lu, G. 2009. Recent developments in geomicrobiology. *Geological Review*, 55, pp. 552–580.

Edwards, K.J., Becker, K., and Colwell, F. 2012. The deep, dark energy biosphere: intraterrestrial life on earth. *Annual Review of Earth and Planetary Sciences*, 40, pp. 551–568.

Ehrenberg, C.G. 1839. Über das im Jahre 1686 in Curland Vom Himmel gefallene Meteorpapier und über dessen Zusammensetzung aus Conferven und Infusorien. *Annalen derrr Physik und Chemie*, 16, pp. 187–188.

Entcheva-Dimitrov, P., and Spormann, A.M. 2004. Dynamics and control of biofilms of the oligotrophic bacterium Caulobacter crescentus. *Journal of Bacteriology*, 186, pp. 8254–8266.

Flemming, H.C., and Geesey, G.G. 1991. *Biofouling and biocorrosion in industrial water systems*, pp. 220. New York, Springer.

Franzetti, A., Tagliaferri, I., Gandolfi, I., Bestetti, G., Minora, U., Mayer, C., Azzoni, R.S., Diolaiuti, G., Smiraglia, C., and Ambrosini, R. 2016. Light-dependent microbial metabolisms drive carbon fluxes on glacier surfaces. *The ISME Journal*, 10, pp. 2984–2988.

Gauri, S.S., Mandal, S.M., and Pati, B.R. 2012. Impact of Azotobacter exopolysaccharides on sustainable agriculture. *Applied Microbiology and Biotechnology*, 95, pp. 331–338.

Gihring, T.M., Moser, D.P., Lin, L.H., Davidson, M., Onstott, T.C., Morgan, L., Millesson, M., Kieft, T.L., Trimarco, E., Balkwill, D.L., and Dollhopf, M.E. 2006. The distribution of microbial taxa in the subsurface water of the Kalahari Shield, South Africa. *Geomicrobiology Journal*, 23, pp. 415–430.

Golubic, S. 1992. Microbial mats of Abu Dhabi. In: *Environmental evolution* (eds. L. Margulis and L. Olendzenski), pp. 103–130. Cambridge, MIT Press.

Harfst, W.F. 1992. Fundamentals in microbiology for high-purity water treatment. *Ultrapure Water*, 9, pp. 33–35.

Hoehler, T.M., Bebout, B.M., and Des Marais, J.D. 2001. The role of microbial mats in the production of reduced gases on the early Earth. *Nature*, 412, pp. 324–327.

Inagaki, F., Nunoura, T., Nakagawa, S., Teske, A., Lever, M., Lauer, A., Suzuki, M., Takai, K., Delwiche, M., Colwell, F.S., Nealson, K.H., Horikoshi, K., D'Hondt, S., and Jørgensen, B.B. 2006. Biogeographical distribution and diversity of microbes in methane hydrate-bearing deep marine sediments on the Pacific Ocean Margin. *PNAS*, 103, pp. 2815–2820.

Jiang, D., Huang, Q., Cai, P., Rong, X., and Chen, W. 2007. Adsorption of Pseudomonas putida on clay minerals and iron oxide. *Colloids and Surfaces B: Biointerfaces*, 54, pp. 217–221.

Jørgensen, B.B., and Des Marais, J.D. 1986. Competition for sulfide among colorless and purple sulfur bacteria in cyanobacterial mats. *FEMS Microbiology Ecology*, 38, pp. 179–186.

Jørgensen, B.B., and Des Marais, J.D. 1988. Optical properties of benthic photosynthetic communities: fiber optic studies of cyanobacterial mats. *Limnology and Oceanography*, 33, pp. 99–113.

Kieft, T.L. 2016. Microbiology of the Deep Continental Biosphere. In: *Their World: A Diversity of Microbial Environments* (ed. C.J. Hurst), pp. 225–249. Switzerland, Springer.

Korber, D.R., Lawrence, J.R., Hendry, M.J., and Caldwell, D.E. 1993. Analysis of spatial variability within Mot + and Mot-*Pseudomonas fluorescens* biofilm using representative elements. *Biofouling*, 7, pp. 339–358.

Krumbein, W.E. 1994. The year of the slime. In: *Biostabilisation of sediments* (eds. W.E. Krumbein, D.M. Paterson, and L.J. Stal), pp. 1–7. Oldenburg, BIS-Verlag.

Krumbein, W.E., Brehm, U., Gerdes, G., Gorbushina, A.A., Levit, G., and Palinska, K.A. 2003. Biofilms, biodictyon, biomat, microbialites, oolites, stromatolotes, geophysiology, global mechanism, parahistology. In: *Fossils and recent biofilms—a natural history of life on Earth* (eds. W.E. Krumbein, D.M. Paterson, and G.A. Zavarzin), pp. 1–27. Netherlands, Kluwer Academic Publishers.

Lacap, D.C., Warren-Rhodes, K.A., McKay, C.P., and Pointing, S.B. 2011. Cyanobacteria and chloroflexi-dominated hypolithic colonization of quartz at the hyper-arid core of the Atacama Desert, Chile. *Extremophiles*, 15, pp. 31–38.

Levinson, G. 2020. *Rethinking evolution: the revolution that's hiding in plain sight*, pp. 500. Hackensack, World Scientific Publishing.

Li, K., Liu, R.Y., Zhang, H.X., and Yun, J.L. 2013. The diversity and abundance of bacteria and oxygenic phototrophs in saline biological desert crusts in Xinjiang, Northwest China. *Environmental Microbiology*, 66, pp. 40–48.

Li, X.X., Mbadinga, S.M., Liu, J.F., Zhou, L., Yang, S.Z., Gu, J.D., and Mu, B.Z. 2017. Microbiota and their affiliation with physiochemical characteristics of different subsurface petroleum reservoirs. *International Biodeterioration and Biodegradation*, 120, pp. 170–185.

Lin, H.T., Cowen, J.P., Olson, E.J., Amend, J.P., and Lilley, M.D. 2012. Inorganic chemistry, gas compositions and dissolved organic carbon in fluids from sedimented young basaltic crust on the Juan de Fuca Ridge flanks. *Geochimica et Cosmochimica Acta*, 85, pp. 213–227.

Lin, H.T., Amend, J.P., LaRowe, D.G., Bingham, J.P., and Cowen, J.P. 2015. Dissolved amino acids in oceanic basaltic basement fluids. *Geochimica et Cosmochimica Acta*, 164, pp. 175–190.

Liu, W., and Zhang, X.L. 2015. Experimental study of microbial pyrite oxidation: a suggestion for geologically useful biosignatures. *Geomicrobiology Journal*, 32, pp. 466–471.

Lopatina, A., Medvedeva, S., Shmakov, S., Logacheva, M.D., Krylenkov, V., and Severinov, K. 2016. Metagenomic analysis of bacterial communities of Antarctic surface snow. *Frontiers in Microbiology*, 7, pp. 398.

Lünsdorf, H., Erb, R.W., Abraham, W.R., and Timmis, K.N. 2010. 'Clay hutches': a novel interaction between bacteria and clay minerals. *Environmental Microbiology*, 2, pp. 161–168.

Maccario, L., Vogel, T.M., and Larose, C. 2014. Potential drivers of microbial community structure and function in Arctic spring snow. *Frontiers in Microbiology*, 5, pp. 413.

Magnabosco, C., Tekere, M., Lau, M.C.Y., Linage, B., Kuloyo, O., Erasmus, M., Cason, E., Heerden, E.V., Borgonie, G., Kieft, T.L., Olivier, J., and Onstott, T.C. 2014. Comparisons of the composition and biogeographic distribution of the bacterial communities occupying South African thermal springs with those inhabiting deep subsurface fracture water. *Frontiers in Microbiology*, 5, pp. 679–689.

Makhalanyane, T.P., Valverde, A., Gunnigle, E., Frossard, A., Ramond, J.B., and Cowan, D.A. 2015. Microbial ecology of hot desert edaphic systems. *FEMS Microbiology Reviews*, 39, pp. 203–221.

Margesin, R., and Collins, T. 2019. Microbial ecology of the cryosphere (glacial and permafrost habitats): current knowledge. *Applied Microbiology and Biotechnology*, 103, pp. 2537–2549.

Mason, O.U., Di Meo-Savoie, C.A., Van Nostrand, J.D., Zhou, J.Z., Fisk, M.R., and Giovannoni, S.J. 2009. Prokaryotic diversity, distribution, and insights into their role in biogeochemical cycling in marine basalts. *The ISME Journal*, 3, pp. 231–242.

Meckenstock, R.U., Von, N.F., Stumpp, C., Lueders, T., Himmelberg, A.M., Hertkorn, N., Schmitt-Kopplin, P., Harir, M., Hosein, R., Haque, S., and Schulze-Makuch, D. 2014. Water droplets in oil are microhabitats for microbial life. *Science*, 345, pp. 673–676.

Meltzer, T.H. 1993. *High-purity water preparation for the semiconductor, pharmaceutical, and power industries*, pp. 833. Littleton, Tall Oaks Publishing, Inc.

Mishra, S., Lin, Z.Q., Pang, S.M., Zhang, W.P., Bhatt, P., and Chen, S.H. 2021. Recent advanced technologies for the characterization of xenobiotic-degrading microorganisms and microbial communities. *Frontiers in Bioengineering and Biotechnology*, 9, pp. 26.

Mishra, S., Huang, Y.H., Li, J.Y., Wu, X.Z., Zhou, Z., Lei, Q.Q., Bhatt, P., and Chen, S.H. 2022. Biofilm-mediated bioremediation is a powerful tool for the removal of environmental pollutants. *Chemosphere*, 294, pp. 133609.

Mittelman, M.W. 1985. Biological fouling of purified-water systems: Part 1, Bacterial growth and replication. *Microcontamination*, 3, pp. 51–70.

Neu, T.R. 2000. In situ cell and glycoconjugate distribution in river snow studied by confocal laser scanning microscopy biofilms. *Aquatic Microbial Ecology*, 21, pp. 85–95.

Neu, T.R., Eitner, A., and Paje, M.L. 2003. Development and architecture of complex environmental biofilms. In: *Fossils and recent biofilms—a natural history of life on Earth* (eds. W.E. Krumbein, D.M. Paterson, and G.A. Zavarzin), pp. 29–45. Netherlands, Kluwer Academic Publishers.

Odoh, C.K. 2017. Plant growth promoting rhizobacteria (PGPR): a bioprotectant bioinoculant for sustainable agrobiology. A Review. *International Journal of Advanced Research in Biological Sciences*, 4, pp. 123–142.

Orcutt, B.N., Sylvan, J.B., Knab, N.J., and Edwards, K.J. 2011. Microbial ecology of the dark ocean above, at, and below the seafloor. *Microbiology and Molecular Biology Reviews*, 75, pp. 361–422.

Oren, A. 2008. Microbial life at high salt concentrations: phylogenetic and metabolic diversity. *Saline Systems*, 4, pp. 1–13.

Patterson, M.K., Husted, G.R., Rutkowski, A., and Mayette, D.C. 1991. Isolation, identification, and microscopic properties of biofilms in high-purity water distribution systems. *Ultrapure Water*, 8, pp. 18–24.

Peng, X.T., Du, M.R., Gebruk, A. Liu, S.Q., Gao, Z.M., Glud, R.M., Zhou, P., Wang, R.H., Rowden, A.A., Kamenev, G.M., Maiorova, A.S., Papineau, D., Chen, S., Gao, J.W., Liu, H.L., He, Y., Alalykina, I.L., Dolmatov, I.Y., Zhang, H.Y., Li, X.G., Malyutina, M.V., Dasgupta, S., Saulenko, A.A., Shilov, V.A., Liu, S.T., Xie, T.T., Qu, Y.G., Song, X.K., Zhang, H.B., Liu, H., Zhang, W.J., Huang, X.X., Xu, H.Z., Xu, W.J., Mordukhovich, V.V. and Adrianov, A.V. 2025. Flourishing chemosynthetic life at the greatest depths of hadal trenches. *Nature*, 645, pp. 679–685. 10.1038/s41586-025-09317-z

Petrova, L.P., Shelud'ko, A.V., and Katsy, E.I. 2010. Plasmid rearrangements and alterations in Azospirillum brasilense biofilm formation. *Microbiology*, 79, pp. 121–124.

Rudrappa, T., Biedrzycki, M.L., and Bais, H.P. 2008. Causes and consequences of plant-associated biofilms. *FEMS Microbiology Ecology*, 64, pp. 153–166.

Santelli, C.M., Orcutt, B.N., Banning, E., Bach, W., Moyer, C.L., Sogin, M.L., Staudigel, H., and Edwards, K.J. 2008. Abundance and diversity of microbial life in ocean crust. *Nature*, 453, pp. 653–656.

Sher, Y., Baker, N.R., Herman, D., Fossum, C., Hale, L., Zhang, X.X., Nuccio, E., Saha, M., Zhou, J.Z., Pett-Ridge, J., and Firestone, M. 2020. Microbial extracellular polysaccharide production and aggregate stability controlled by switchgrass (*Panicum virgatum*) root biomass and soil water potential. *Soil Biology and Biochemistry*, 143, pp. 107742.

Sherwood Lollar, B., Onstott, T.C., Lacrampe-Couloume, G., and Ballentine, C.J. 2014. The contribution of the Precambrian continental lithosphere to global H_2 production. *Nature*, 516, pp. 379–382.

Sielaff, A.C., and Smith, S.A. 2019. Habitability of mars: how welcoming are the surface and subsurface to life on the red planet? *Geosciences*, 9, pp. 361–384.

Späth, R., Flemming, H.C., and Wuertz, S. 1998. Sorption properties of biofilms. *Water Science and Technology*, 37, pp. 207–210.

Stal, L.J. 1994. Microbial mats in coastal environments. In: *Mirobial mats, structure, development and environmental significance* (eds. L.J. Stal and P. Caumette), pp. 21–32. Heidelberg, Springer.

Stibal, M., Schostag, M., Cameron, K.A., Hansen, L.H., Chandler, D.M., Wadham, J.L., and Jacobsen, C.S. 2015. Different bulk and active bacterial communities in cryoconite from the margin and interior of the Greenland ice sheet. *Environmental Microbiology Reports*, 7, pp. 293–300.

Stolz, J.F. 2000. Structure of microbial mats and biofilms. In: *Microbial sediments* (eds. R.E. Riding and S. M. Awramik), pp. 1–8. Berlin, Springer.

Stoodley, P., Sauer, K., Davies, D.G., and Costerton, J.W. 2002. Biofilms as complex differentiated communities. *Annual Review of Microbiology*, 56, pp. 187–209.

Tracy, C.R., Streten-Joyce, C., Dalton, R., Nussear, K.E., Gibb, K.S., and Christian, K.A. 2010. Microclimate and limits to photosynthesis in a diverse community of hypolithic cyanobacteria in northern Australia. *Environmental Microbiology*, 12, pp. 592–607.

Tunnicliffe, V., Chen, C., Giguère, T., Rowden, A.A., Watanabe, H.K., and Brunner, O.2023. Hydrothermal vent fauna of the western Pacific Ocean:Distribution patterns and biogeographic networks. *Diversity and Distributions*, 30, pp. e13794.

VanHaecke, E., Remon, J.P., Moors, M., Raes, F., De Rudder, D., and Van Peteghem, A. 1990. Kinetics of *Pseudomonas aeruginosa* adhesion to 304 and 316-L stainless steel: role of cell surface hydrophobicity. *Applied and Environmental Microbiology*, 56, pp. 788–795.

Visscher, P.T., Baumgartner, L.K., Buckley, D.H., Rogers, D.R., Hogan, M., Raleigh, C., Turk, K., and Des Marais, J.D. 2003. Dimethyl sulfide and methanethiol as biogenic signatures in laminated microbial ecosystems. *Environmental Microbiology*, 5, pp. 296–308.

Warren-Rhodes, K.A., McKay, C.P., Boyle, L.N., Wing, M.R., Kiekebusch, E.M., Cowan, D.A., Stomeo, F., Pointing, S.B., Kaseke, K.F., Eckardt, F., Henschel, J.R., Anisfeld, A., Seely, M., and Rhodes, K.L. 2013. Physical ecology of hypolithic communities in the central Namib Desert: the role of fog, rain, rock habitat and light. *Journal of Geophysical Research: Biogeosciences*, 118, pp. 1451–1460.

Xing, Y.H., Tan, S.X., Liu, S., Xu, S.Z., Wan, W.J., Huang, Q.Y., and Chen, W.L. 2022. Effective immobilization of heavy metals via reactive barrier by rhizosphere bacteria and their biofilms. *Environmental Research*, 207, pp. 112080.

Yin, H.F., and Dong, H.L. 2020. *The deep subsurface biosphere*, pp. 279. Beijing, Science Press.

Yoon, J.H., Adhikari, M., Jeong, S.S., Lee, S.P., Kim, H.S., Lee, G.S., Park, D.H., Kim, H., and Yang, J.E. 2024. Microbial diversity of soils under different land use and chemical conditions. *Applied Biological Chemistry*, 67, pp. 111.

Zhang, H., Qian, Y.Y., Fan, D.D., Tian, Y.N., and Huang, X. 2022. Biofilm formed by *Hansschlegelia zhihuaiae* S113 on root surface mitigates the toxicity of bensulfuron-methyl residues to maize. *Environmental Pollution*, 292, pp. 118366.

Zhang, B., Hu, X.Y., Zhao, D.L., Wang, Y.P., Qu, J.H., Tao, Y., Kang, Z.H., Yu, H.Q., Zhang, J.Y., and Zhang, Y. 2024. Harnessing microbial biofilms in soil ecosystems: enhancing nutrient cycling, stress resilience, and sustainable agriculture. *Journal of Environmental Management*, 370, pp. 122973.

4

Earth as a System and Biogeochemical Cycles

Weiduo Hao and Xingliang Zhang

4.1 Introduction

The universe consists of matter and energy, both of which are fundamental to understanding Earth's dynamic systems. Our planet, Earth system comprises the solid Earth, the fluid Earth (atmosphere and hydrosphere), and the living Earth (biosphere). The solid Earth comprises the core, mantle, and crust. The upper mantle and crust form Earth's outer rigid part, called lithosphere. These spheres contain matter in gaseous, liquid, or solid states, while energy drives their interactions and matter cycles. The law of conservation of matter and energy states that matter and energy cannot be created or destroyed but can be transformed into different forms. For example, gases condense to become liquid; oxygen reacts with iron to form "rust" (iron oxide). Earth's matter is finite, yet it undergoes continuous recycling, shifting between various forms. Any matter within the Earth system serves merely as a temporary reservoir, existing in a specific location and state before transforming again. These transformations occur through three primary types of processes. Physical processes, such as evaporation, erosion, and transportation, involve changes in the state or movement of matter without altering its chemical structure; chemical processes, e.g. oxidation and combustion, involve reactions that alter the molecular structure of substances; biological processes, like photosynthesis, chemosynthesis, and respiration, are driven by living organisms, making essential nutrients available for life. These processes ultimately cause the matter to be transported or change its chemical state in Earth's system.

Life on Earth is inextricably linked to the environment through a variety of interacting cycles and feedback loops. These cycles ensure that essential elements like carbon, nitrogen, phosphorus, and water are recycled and made available for life processes. In this chapter, we will mainly discuss element cycling among Earth's surface subsystems, i.e. through the atmosphere, hydrosphere, lithosphere, and biosphere.

4.2 Earth System: An Overview

4.2.1 A System Approach

The Earth is composed of four interconnected spheres (lithosphere, hydrosphere, atmosphere, and biosphere) through which matter cycles and energy flows. A systems approach offers a useful

Introduction to Geobiology, First Edition. Xingliang Zhang, Zisheng Guo, Wei Liu, Ruliang He, Luoyang Li, and Weiduo Hao.

Companion website: www.wiley.com/go/Zhang_1e

framework for understanding these complex interactions. Technically, a **system** is defined as a set of components that function together as a unified whole. This perspective allows us to isolate specific parts of the world and focus on the elements that interact more closely with one another than with external factors. For instance, within the human body, individual cells interact more intensively with other cells in the body than with the surrounding environment. Further, we might examine the digestive system, which consists of specialized cells and organs working together to process food. This example illustrates a key principle: most systems are hierarchical, meaning they are comprised of smaller subsystems and those subsystems can further contain their own subsystems, each made up of interacting components.

4.2.2 Energy Flows of the Earth System

4.2.2.1 Law of Thermodynamics

The first law of thermodynamics states that energy cannot be created or destroyed but can only be transformed. The second law of thermodynamics explains that when energy is transformed from one kind to another, it becomes degraded, meaning that the energy becomes less capable of doing work. For instance, in an automobile, only about 30% of the chemical energy in gasoline is converted into the energy of motion. Similarly, photosynthesis transforms solar energy into chemical energy by creating organic compounds from carbon and other atoms, storing this energy in the bonds between atoms. However, this transformation is far from 100% efficient, as most of the incoming solar energy is lost as heat, which is considered low-quality energy and less capable of performing work compared to high-quality energy.

Entropy refers to the amount of low-quality energy in a system. When entropy is high, matter tends to disorganize into simpler states. The second law of thermodynamics is sometimes called the law of entropy because all energy transformations increase the entropy of a system unless new high-quality energy, such as solar energy, is introduced to the system to replenish it.

4.2.2.2 Earth's Energy Budget

Energy transformations exhibit variable efficiencies depending on system design and thermodynamic constraints. Advances in photovoltaic technology have achieved solar energy conversion efficiencies approaching 30% under standard test conditions, while nuclear fission reactors and internal combustion engines demonstrate distinct energy conversion profiles for electricity generation and mechanical work, respectively. Fundamental thermodynamic principles dictate unavoidable energy loss, typically manifesting as thermal losses due to entropy production.

The Earth's energy budget is dominated by solar irradiance (Figure 4.1), with approximately 30% reflected via albedo effects (clouds, aerosols), 51% absorbed and thermally reradiated, and the remainder driving atmospheric and geochemical processes. Photosynthetic efficiency is limited to ~0.06% of incident solar irradiance. Secondary energy contributions, including tidal forces and geothermal flux, constitute a minor fraction of Earth's energy, about 0.002% (Skinner and Murck 2011), while the geothermal energy is sufficient to drive plate tectonics and sustain chemosynthetic ecosystems at hydrothermal vents. The equilibrium between the energy Earth receives from the Sun and the energy it radiates back into space is crucial for maintaining a stable climate. Recent studies indicate that Earth's energy imbalance has increased significantly over the past few decades, primarily due to human activities such as greenhouse gas emissions and changes in cloud cover and ice reflectivity (Loeb et al. 2021).

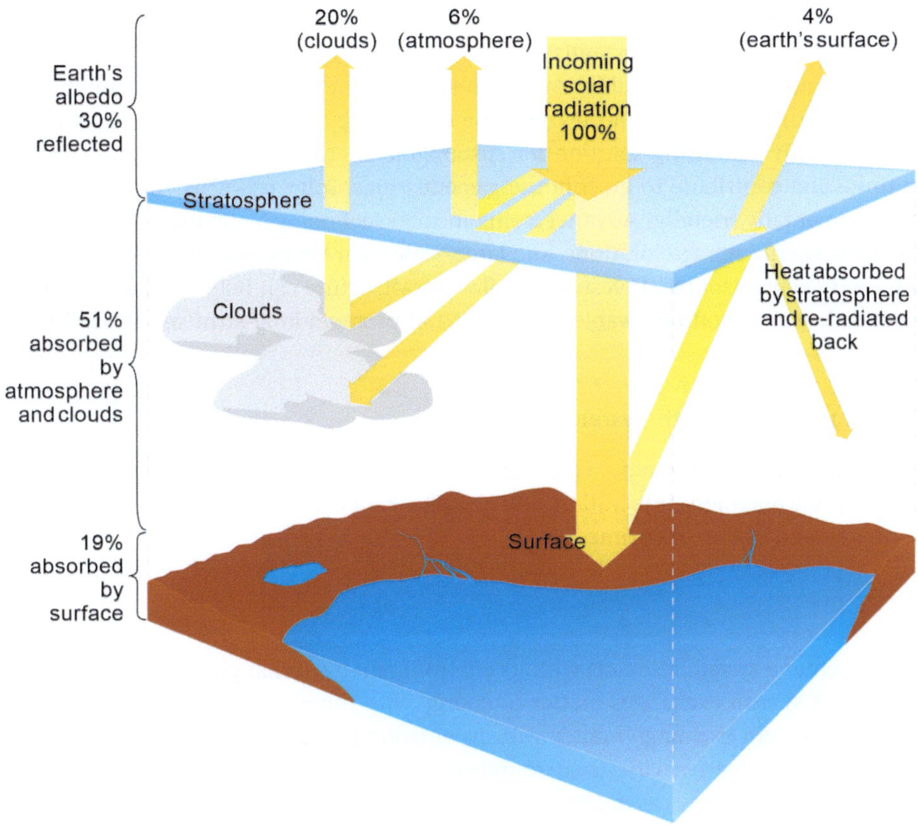

Figure 4.1 Budget for incoming solar energy. Thirty percent of the solar radiation striking the Earth is immediately reflected back into space; fifty-one percent is absorbed and radiated into space as heat. The remaining energy creates wind and drives the water cycle, photosynthesis, and other processes.

4.2.2.3 Human Use of Energy Flows

Modern civilization relies heavily on the utilization of Earth's energy flows, including solar panels, hydropower, wind energy, and fossil fuels. The fossil fuels are currently serving as the cornerstone of our energy infrastructure. These fossilized biological materials store ancient solar energy in the form of chemical energy, but their formation process spans millions of years, classifying them as non-renewable resources. Estimates suggest that global oil reserves could last around 50 years at current consumption rates, while coal supplies may persist for over 100 years. However, these projections depend on factors like new discoveries, technological advancements, and shifts in energy demand. Some sources estimate coal could last 300–400 years, but this varies based on extraction rates and global energy transitions (Data source: Energy Institute—Statistical Review of World Encrgy).

Beyond the issue of depletion, fossil fuels pose significant environmental challenges, primarily through the release of pollutants. The combustion of fossil fuels is responsible for nearly all major air pollution, including acid rain, smog, carbon monoxide, and greenhouse gases. While technologies like smokestack devices and fuel cleansing can mitigate some of these pollutants, there is no cost-effective method to address the massive amounts of carbon dioxide produced. Burning just one gallon (3.785 L) of gasoline produces over 20 pounds (9 kg) of carbon dioxide, with coal being even more carbon intensive. Fossil fuels also cause many other pollution problems, such as seepage from storage tanks into groundwater and ocean spills.

Solar energy offers an abundant, renewable, and clean alternative to fossil fuels. The amount of solar energy reaching Earth each month far exceeds the total energy stored in known fossil fuel reserves, making it a highly promising solution for sustainable energy generation. Unlike fossil fuels, which are finite and contribute to greenhouse gas emissions, solar power is virtually inexhaustible and environmentally friendly. Advancements in solar technology—including improved photovoltaic efficiency, energy storage solutions, and large-scale solar farms—continue to enhance its viability as a primary energy source. Investing in solar infrastructure can help mitigate climate change, reduce dependence on non-renewable resources, and promote long-term sustainability.

4.2.3 Three Key Traits of the Earth System

The Earth system is a dynamic and interconnected network of components that shape our planet's environment. By analyzing openness, integration, and complexity of the system, we can define how its components interact and evolve over time.

4.2.3.1 Openness

Openness refers to whether a system interacts with its environment by exchanging energy, matter, or information. An open system exchanges matter or energy with other systems, a closed system exchanges energy but does not exchange matter with its surroundings, while an isolated system exchanges neither energy nor matter with its surroundings.

Based on the law of entropy, energy cannot be recycled, meaning any system without an external energy source will eventually cease to exist. The Earth is an open system in terms of energy, as it receives a continuous supply of solar energy and radiates some of it back into space (Figure 4.2). In contrast, the current Earth is a closed system in terms of matter, as it retains nearly all the matter it will ever have, with only negligible additions from meteorites and losses of light gases like hydrogen. This matter cycles continuously among the Earth's spheres-lithosphere, hydrosphere, atmosphere, and biosphere-driven by solar and internal energies, shifting between gas, liquid, and solid states, and supporting life processes.

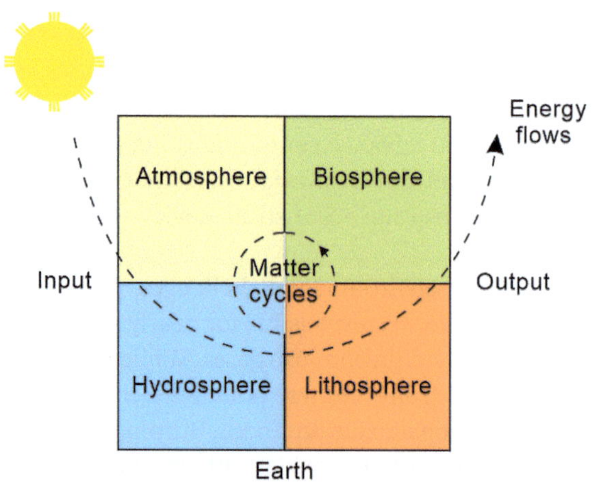

Figure 4.2 The Earth is open to energy, which flows into and out of it. Essentially, however, the Earth is closed to matter, which cycles over and over through the four spheres.

4.2.3.2 Integration

Integration describes the strength of interactions among the parts of a system. Some systems, like the human body, are highly integrated, with cells and organs interdependent but in close communication. The disruption of critical components such as the heart or brain, and the loss of any component can lead to systemic failure. In contrast, weakly integrated systems, such as colonies of single-celled organisms (e.g. green alga *Volvox*), can lose many cells without significantly affecting the rest.

The degree of integration in the global environmental system is debated. Some scientists, such as proponents of the Gaia hypothesis, view Earth as a superorganism, with its spheres intricately connected through complex pathways, resembling the interdependence of cells in a body (Lovelock and Margulis 1974). Others argue that the global system is less integrated than this extreme view suggests. While not as interconnected as a superorganism, the Earth's spheres are far more integrated than a loose collection of independent parts, like a colony of single-celled organisms. The exact level of integration in the global system remains a topic of ongoing debate, given its vast scale and intricate interconnections.

4.2.3.3 Complexity

Complexity refers to the number and variety of components and their interactions in a system. This definition conforms to our intuition: a tiny insect seems more complex than a large rock because it has more components. Insects have more complex molecules, and these are used to construct cells and organs. This example also illustrates that complexity is often hierarchical, with smaller components being used to construct larger ones.

The Earth system is enormously complex, with trillions of components operating across multiple spatial and temporal scales. The four subsystems-lithosphere, hydrosphere, atmosphere, and biosphere-interact in intricate ways, from the dynamics of weather and climate to the cycles of matter and energy. Even within individual spheres, complexity is staggering; for example, the atmosphere's complexity makes long-term weather or climate prediction challenging despite advanced computing. This unpredictability is a characteristic of complex systems and contributes to much of the controversy and inaction surrounding environmental issues.

4.3 Biogeochemical Cycles

4.3.1 Concept

Biogeochemical cycles describe the movement and transformation of chemical elements between living organisms (the biosphere) and non-living reservoirs (the atmosphere, hydrosphere, and lithosphere). These cycles are essential for sustaining life and maintaining Earth's systems. By focusing on the cycles of specific chemical elements, scientists can simplify complex environmental processes and better understand the interconnectedness of Earth's four subsystems: the biosphere (the sum of all living things as well as formerly living things that have not yet decomposed), the hydrosphere (the entirety of Earth's water except for vapor in the atmosphere), and the lithosphere. Elements participate in biogeochemical cycles by fulfilling three primary roles in the metabolism of organisms (Nester et al. 2004).

4.3.1.1 Biomass Production

Elements are incorporated into cellular materials as basis for essential biomolecules. For example, nitrogen is crucial for synthesizing amino acids. Plants and some prokaryotes assimilate ammonia

(NH_3) to produce glutamate, which serves as a precursor for other amino acids. Certain prokaryotes convert nitrate (NO_3^-) to ammonia through nitrate reduction. Some prokaryotes can reduce atmospheric nitrogen to form ammonia, which can then be incorporated into cellular material. Animals, in contrast, must obtain amino acids from their diet.

4.3.1.2 Energy Source

Reduced forms of elements are oxidized to generate adenosine triphosphate (ATP), the energy currency of cells. Chemoorganotrophs, such as most animals and fungi, use reduced carbon compounds (e.g. sugars, lipids, and amino acids) as energy sources. Chemolithotrophs, on the other hand, can use energy from inorganic molecules, such as the oxidation of hydrogen sulfide (H_2S), ammonia (NH_3), and hydrogen gas (H_2) to produce energy.

4.3.1.3 Terminal Electron Acceptors

During respiration, electrons from energy sources are transferred to an oxidized form of an element. Oxygen (O_2) is one of the most common terminal electron acceptors in aerobic respiration. In anaerobic conditions, some prokaryotes use alternatives such as nitrate (NO_3^-), nitrite (NO_2^-), sulfate (SO_4^{2-}), carbon dioxide (CO_2), Mn^{4+}, and Fe^{3+} as terminal electron acceptors.

4.3.2 Element Abundance

There are 92 naturally occurring elements, with uranium (atomic number 92) being the heaviest. However, heavier elements like plutonium (Pu) and neptunium (Np) can be found in trace amounts in nature, but they are primarily produced artificially. There are about two dozen artificially created elements, but they are mainly of interest in certain specialties in chemistry and physics. Only a few naturally occurring elements play a significant role in the interactions between living organisms and the inanimate world: hydrogen, carbon, nitrogen, oxygen, phosphorus, and sulfur. These elements account for over 95% of the mass of all living things. As Table 4.1 shows, oxygen alone accounts for more than 62% of the weight of the human body and more than 77% of the weight of the alfalfa plant. Carbon and oxygen together account for more than 80% of the human body's mass. Although hydrogen is proportionally far less abundant on Earth compared to the universe, its role in forming water makes it indispensable for life. Water (H_2O), essential for all known forms of life, is the primary medium through which many biogeochemical cycles operate. Nitrogen and oxygen form the bulk of the atmosphere, providing essential gases for life.

Table 4.1 Atomic composition by weight of three representative organisms.

Element	Human (%)	Alfalfa (%)	Bacterium (%)
Oxygen	62.81	77.90	73.68
Carbon	19.37	11.34	12.14
Hydrogen	9.31	8.72	9.94
Nitrogen	5.14	0.83	3.04
Phosphorus	0.63	0.71	0.60
Sulfur	0.64	0.10	0.32
Total	97.90	99.60	99.72

Table 4.2 The relative abundance by weight of some chemical elements in the Earth's crust.

Element (chemical symbol)	Relative abundance (%)
Oxygen (O)	42.6
Silicon (Si)	27.7
Aluminum (Al)	8.1
Iron (Fe)	5.0
Calcium (Ca)	3.6
Sodium (Na)	2.8
Potassium (K)	2.6
Magnesium (Mg)	2.1
Phosphorus (P)	0.07
Carbon (C)	0.03
Nitrogen (N)	Trace

While the biosphere is dominated by carbon, oxygen, hydrogen, nitrogen, phosphorus, and sulfur, the Earth's crust exhibits significantly different elemental ratios. As shown in Table 4.2, oxygen is the most abundant element in both the crust and living organisms. However, silicon that is rare in biological systems, is the second most abundant element in the crust. Carbon, which is critical to life, is hundreds of times less abundant in the crust compared to its prevalence in organisms. These discrepancies highlight the chemical distinctness of life from its surrounding environment and underscore the importance of biogeochemical cycles in maintaining the balance required for life.

In the following sections, we will focus on elements that play pivotal roles in the interplay between the biosphere and Earth's surface environment (lithosphere, atmosphere, and hydrosphere). Finally, we will explore the rates at which these cycles operate and their implications for life on Earth.

4.3.3 Carbon Cycle

Each of the many biogeochemical cycles has different pathways of transport and temporary storage reservoirs (Post et al. 1990). The carbon cycle in Figure 4.3 is one of the most critical biogeochemical cycles, involving the transfer of carbon between various environmental reservoirs. Like many cycles, the carbon cycle appears to be quite complex with many pathways (arrows), but we can primarily split it into two fundamental processes: withdrawal from and addition to the atmosphere.

4.3.3.1 Carbon Reservoirs

Carbon reservoirs are categorized into surficial and geological reservoirs. Surficial reservoirs include carbon in terrestrial biomass, carbon in soils, carbon present as carbon dioxide in the atmosphere, carbon present as dissolved inorganic carbon (DIC) in ocean waters, carbon in marine biomass, and carbon in the form of methane. These reservoirs cycle carbon rapidly and are relatively small in size compared to geological reservoirs. Geological reservoirs dominate in terms of carbon mass and are stored in the lithosphere (Table 4.3). They include inorganic carbon in carbonate rocks (e.g. limestone) and organic carbon in kerogen and fossil fuels (petroleum, natural gas, and coal).

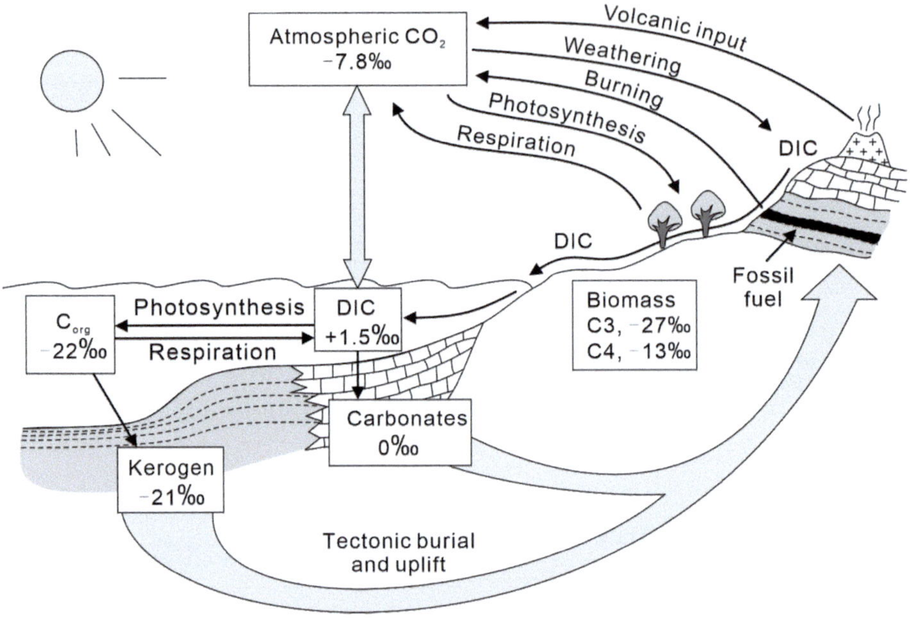

Figure 4.3 A schematic illustration of the pre-industrial global carbon cycle. *Source:* Adapted from Newton and Bottrell (2007).

Table 4.3 Masses of carbon in different reservoirs.

Reservoir		**Mass (Pg C = 10^{15} g C)**
Environmental reservoirs		
Atmosphere		780
Oceans	Inorganic	37,400
	Organic	1000
Terrestrial biomass		500
Soils		2000
Methane hydrates		10,000
Geological reservoirs		
Oxidized	Carbonates	$>60 \times 10^6$
Reduced	Kerogen	15×10^6
Fossil fuels		4130

Source: Newton and Bottrell (2007)/Geological Society of London.

4.3.3.2 Flux: Withdrawal

Carbon withdrawal from the atmosphere is driven by geological and biological processes (Figure 4.3). Weathering of silicate and carbonate rocks releases carbon into the hydrological cycle, acting as a significant sink for atmospheric CO_2. This carbon is transported via the hydrological cycle in the form of DIC (which has minimal carbon storage) to aquatic environments (oceans, lakes, and rivers), where it is precipitated as limestone and marls through biological processes, e.g. clams and other skeleton-bearing organisms withdraw carbon for use in constructing their shells. The reactions are (1) $CaCO_3 + CO_2 + H_2O$ (*weathering*) $\rightarrow Ca^{2+} + 2HCO_3^-$ (*transport*) $\rightarrow$ $CaCO_3 + CO_2 + H_2O$ (*sedimentation*); (2) $CaSiO_3 + 2CO_2 + 2H_2O$ (*weathering*) $\rightarrow Ca^{2+} + 2HCO_3^- + 2H^+ + SiO_3^{2-}$ (*transport*) $\rightarrow CaCO_3 + CO_2 + SiO_2 + H_2O$ (*sedimentation*). In aquatic environments, photosynthetic organisms utilize DIC to produce biomass and induce the precipitation of carbonate, a portion of which is buried in sediments. In terrestrial ecosystems, atmospheric CO_2 is fixed into organic carbon through photosynthesis, accumulating in biomass and soils.

Net increases in terrestrial and marine biomass, along with weathering, drive the withdrawal of CO_2 from the atmosphere. Additionally, carbon is temporarily stored in the lithosphere when organic matter, such as marine plankton or terrestrial plants, is buried and transformed into fossil fuels over millions of years. Recent studies highlight the role of enhanced silicate weathering as a potential mechanism for long-term carbon sequestration. Researchers measured CO_2 fluxes using materials like kimberlite, brucite, forsterite, serpentinite, and wollastonite, finding that mineralogical composition and reactivity significantly impact CO_2 removal rates. Experiments demonstrate that unweathered residues of kimberlites achieved the greatest drawdown rate of -870 g CO_2/m^2/yr at 48% saturation, fine and coarse residues previously exposed to mine process water achieved fluxes of -150 and -160 g CO_2/m^2/yr at 60% saturation, and brucite reached -2940 g CO_2/m^2/yr at 14% saturation compared to forsterite, serpentinite, and wollastonite that achieved fluxes of -500, -260, and -190 g CO_2/m^2/yr, respectively, at higher saturations of 53%–60% (Stubbs et al. 2022).

4.3.3.3 Flux: Addition

Addition of carbon to the atmosphere often occurs from combustion. The burning of the fossil carbon releases CO_2 back to the atmosphere. This process is essentially the reverse of photosynthesis, combining oxygen with organic matter to produce CO_2 and H_2O. Human activities, particularly the combustion of coal, oil, and natural gas, have significantly increased atmospheric CO_2 levels since the Industrial Revolution. Biological combustion occurs through respiration in animals and microorganisms, where organic matter is broken down to release CO_2. This process is a natural part of the carbon cycle, but human activities have amplified its impact through mining, deforestation, and land-use changes.

Volcanic eruptions release CO_2 and other gases into the atmosphere, and CO_2 is the second abundant volcanic gas, with the first abundant being water vapor. While volcanic emissions are relatively small compared to anthropogenic sources, they contribute to the long-term carbon cycle. The research by Kelemen and Manning (2015) challenges traditional views of the carbon cycle by highlighting that subduction zones, rather than solely acting as conduits for carbon recycling into the mantle, play a significant role in storing carbon within the lithosphere and crust. Sediments and carbon-rich materials from the subducting plate may be scraped off and accreted onto the overriding plate, preventing their descent into the mantle. This underplating contributes to crustal thickening and carbon storage. The study suggests that subduction processes are less efficient at transporting carbon deep into the mantle than previously assumed. Factors such as subduction angle, plate composition, and fluid interactions may trap carbon in the lithosphere.

Methane is produced through the anaerobic decomposition of organic matter in soils, wetlands, and marine sediments. Methane hydrates (clathrates) represent a significant but unstable reservoir of carbon in deep-sea sediments (probably ca. 10,000 Pg C). Emerging studies suggest that warming ocean temperatures could trigger large-scale methane release, with profound implications for climate change. As global temperatures rise, methane hydrates collapse, releasing methane—a potent greenhouse gas—into the atmosphere. This, in turn, accelerates warming, leading to even more methane release, creating a vicious cycle. This feedback loop makes climate change more difficult to reverse and highlights the urgency of reducing greenhouse gas emissions (Soukharev 2025).

4.3.3.4 Isotope Fractionation

Carbon occurs naturally as ^{12}C, ^{13}C, and ^{14}C, primarily as ^{12}C, with a small proportion present as ^{13}C. ^{14}C is a radioactive isotope with a half-life of ~5730 years. CO_2 released into the oceans from volcanic sources typically consists of approximately 1.11% ^{13}C and 98.89% ^{12}C. During photosynthesis, enzymes that fix CO_2 preferentially utilize the lighter isotope (^{12}C). As a result, the fixed carbon becomes enriched in ^{12}C and depleted in ^{13}C relative to the initial substrate. The degree of this depletion is quantified as isotopic fractionation using the formula:

$$\delta^{13}C = \left\{\left[\left({}^{13}C/{}^{12}C\right)_{\text{sample}} - \left({}^{13}C/{}^{12}C\right)_{\text{standard}}\right] / \left({}^{13}C/{}^{12}C\right)_{\text{standard}}\right\} \times 1000\ (\text{PDB})$$

Here, the standard ($^{13}C/^{12}C_{standard} = 0.011237$) is derived from a belemnite sample from the Cretaceous PeeDee Formation in Carolina, USA.

The isotopic composition of a sample preserves a record of its past biological activity. Figure 4.4 illustrates the range of $\delta^{13}C$ values for various substances. It is evident that plant material and petroleum (derived from biological sources) exhibit similar isotopic compositions; carbon from both sources is isotopically lighter than the standard because it was fixed through photosynthesis, which

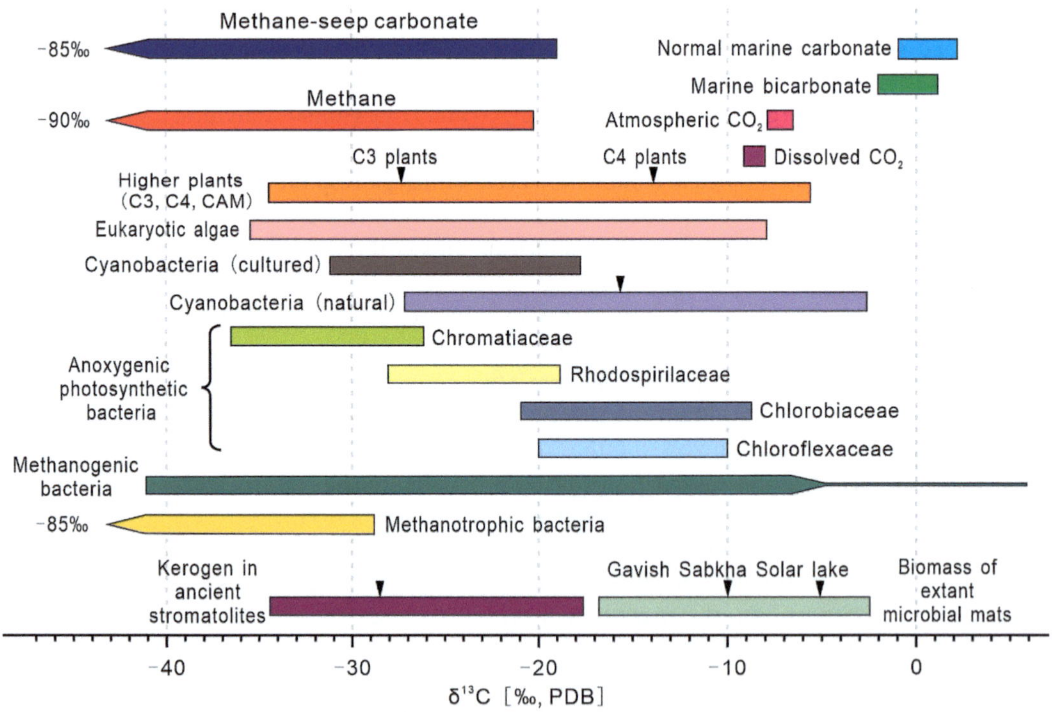

Figure 4.4 $\delta^{13}C$-isotopic variations of organic and inorganic carbon. Black arrows indicate approximate means. *C_3 plants (e.g. rice, wheat, and soybeans) use the Calvin cycle for the initial steps that incorporate CO_2 into organic material, forming a three-carbon compound as the first stable intermediate. *C_4 plants preface the Calvin cycle with reactions that incorporate CO_2 into four-carbon compounds, the end product of which supplies CO_2 for the Calvin cycle. *Source:* Adapted from Schidlowski (2001).

discriminates against $^{13}CO_2$. Methane of biological origin can be extremely light in carbon isotope composition, indicating that CO_2-reducing methanogens significantly discriminate between the isotopic forms of CO_2. In contrast, marine carbonates are isotopically much heavier than biogenic carbon, as this large carbon reservoir has not yet been significantly processed by phototrophic organisms. Due to the differing proportions of ^{12}C and ^{13}C in carbon of biological and geological origin, the $^{13}C/^{12}C$ isotopic ratio in various geological strata has been employed to trace the evolution of life and environmental conditions on Earth. Examples include estimating the burial rate of carbon (Figure 4.5) and understanding the evolution of biological productivity. Carbon isotopic analyses have also been used to distinguish biogenic origins from abiogenic origins. For instance, carbon microgranules in rocks as old as 3.8 Ga show isotopic lightness, suggesting that autotrophy may have evolved very early in the history of life (Rosing 1999).

The isotopic compositions of carbon reservoirs are governed by the isotopic fractionations associated with key processes in the carbon cycle (Newton and Bottrell 2007 and references therein). These processes include the exchange of CO_2 between the oceans and atmosphere and the kinetic fractionation during photosynthesis. Understanding these isotopic distributions provides critical insights into carbon cycling and its feedback mechanisms in Earth's systems.

The exchange of CO_2 between the oceans and atmosphere is a major driver of carbon isotope distribution across these reservoirs. While kinetic effects influence this exchange, it largely approximates isotopic equilibrium between bicarbonate in ocean surface waters ($\delta^{13}C \approx +1.5‰$ V-PDB) and atmospheric CO_2. The ocean's DIC reservoir, being significantly larger, dictates the isotopic composition of atmospheric CO_2, which is depleted in ^{13}C by approximately 8‰ relative to bulk ocean DIC. Bulk ocean DIC typically has a $\delta^{13}C$ value close to 0‰, while atmospheric CO_2 exhibits $\delta^{13}C$ values around $-7.8‰$ V-PDB.

Photosynthesis imposes a kinetic isotope fractionation that depletes ^{13}C in organic carbon relative to the CO_2 source. This occurs because $^{12}CO_2$ molecules are processed slightly faster than $^{13}CO_2$ molecules resulting in an average fractionation of ~20‰. In terrestrial ecosystems, atmospheric

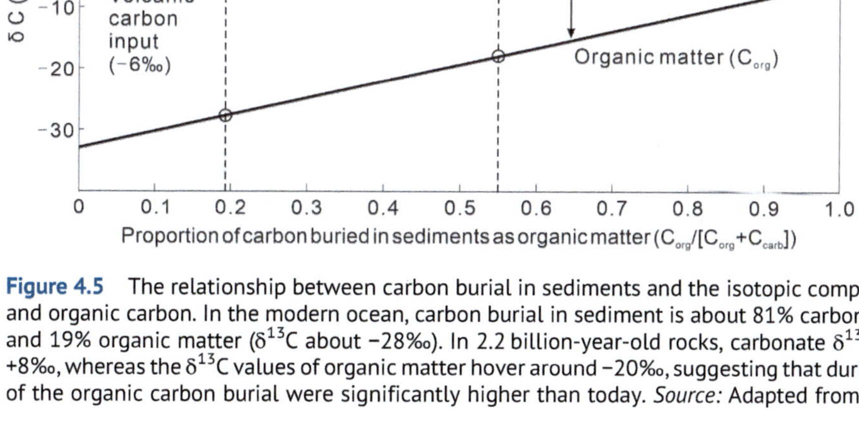

Figure 4.5 The relationship between carbon burial in sediments and the isotopic composition of carbonates and organic carbon. In the modern ocean, carbon burial in sediment is about 81% carbonate ($\delta^{13}C$ about 0‰) and 19% organic matter ($\delta^{13}C$ about −28‰). In 2.2 billion-year-old rocks, carbonate $\delta^{13}C$ values are about +8‰, whereas the $\delta^{13}C$ values of organic matter hover around −20‰, suggesting that during this interval, rates of the organic carbon burial were significantly higher than today. *Source:* Adapted from Hayes et al. (1999).

CO_2 ($\delta^{13}C \approx -7.8‰$) is used in C_3 photosynthesis, producing organic carbon with $\delta^{13}C$ values around −27‰. In marine environments, where ocean DIC ($\delta^{13}C \approx 0‰$) is the carbon source, the fractionation results in marine photosynthetic carbon with $\delta^{13}C$ values around −22‰. ^{13}C-depleted photosynthetic carbon is returned to the atmosphere through respiration, which adds ^{13}C-depleted carbon to this reservoir. Other pathways, such as the reductive TCA (Tricarboxylic Acid) cycle (used by green sulfur bacteria), the acetyl-CoA pathway, and the hydroxypropionate pathway, also exhibit smaller isotopic fractionations than C_3 photosynthesis.

During the Paleogene and Neogene, the evolution of C_4 photosynthesis in dryland grasses introduced a pathway with a smaller isotopic fractionation (~4–6‰), producing biomass that is ^{13}C-enriched compared to C_3 plants. This adaptation is reflected in the $\delta^{13}C$ values of herbivore teeth, providing a record of dietary shifts and ecosystem changes. Other pathways, such as the reductive TCA cycle (used by green sulfur bacteria), the acetyl-CoA pathway, and the hydroxypropionate pathway, also exhibit smaller isotopic fractionations than C_3 photosynthesis.

Precipitation of $CaCO_3$ from marine DIC involves minimal fractionation of carbon isotopes, resulting in limestones inheriting the $\delta^{13}C$ values of contemporaneous marine DIC ($\delta^{13}C \approx 0 \pm 3‰$). In contrast, organic carbon buried in sediments is significantly ^{13}C-depleted ($\delta^{13}C \approx -21‰$) due to the large fractionation during photosynthesis. The isotopic composition of bulk Earth carbon, derived from mantle-derived igneous rocks, ranges from −7‰ to −5‰. The isotopic balance between carbonate ($\delta^{13}C \approx 0‰$) and organic carbon ($\delta^{13}C \approx -21‰$) implies that approximately twice as much carbon is stored in carbonate form as in organic form, maintaining the bulk Earth isotopic signature.

Methane (CH_4) is a significant reservoir in the carbon cycle with a distinct isotopic signature. Biogenic methane is produced in anaerobic sediments and waterlogged soils via two primary mechanisms: acetoclastic methanogenesis and CO_2 reduction (see Section 1.4.6). Acetoclastic methanogenesis dominates freshwater environments, producing methane with $\delta^{13}C$ values ranging from −65‰ to −50‰. In marine environments, CO_2 reduction is more prevalent, yielding methane with $\delta^{13}C$ values between −110‰ and −60‰. Both mechanisms produce methane that is strongly depleted in ^{13}C relative to the organic carbon substrate.

4.3.4 Oxygen Cycle

Oxygen is a versatile element found in numerous forms across Earth's spheres. In the atmosphere, it exists primarily as molecular oxygen (O_2), essential for aerobic respiration, and as O_3, which creates ozone and protects life from harmful ultraviolet (UV) radiation. In the hydrosphere, oxygen is present in water (H_2O) and dissolved oxygen (O_2), which supports aquatic life. In the lithosphere, oxygen is bound in minerals such as silicates (e.g. SiO_2) and oxides (e.g. Fe_2O_3), forming the bulk of Earth's crust. In the biosphere, oxygen is a key component of organic molecules, including carbohydrates (e.g. $C_6H_{12}O_6$), proteins, and nucleic acids, which are vital for life processes. Additionally, oxygen participates in redox reactions, transitioning between oxidized (e.g. CO_2 and SO_4^{2-}) and reduced (e.g. H_2O) forms, driving biogeochemical cycles and maintaining Earth's chemical balance.

One of the defining features of Earth as a planet capable of sustaining active and diverse life is the presence of free molecular oxygen (O_2) in the atmosphere. The concentration and distribution of O_2, both temporally and spatially, are governed by a complex interplay of biological, chemical, and physical processes occurring on and beneath the Earth's surface. These processes drive substantial fluxes of O_2 into and out of the atmosphere through mechanisms of production and consumption. In this section, adapted from Petsch (2005), we will explore the distribution of O_2 among various reservoirs, mechanisms of O_2 production and consumption, and finally, global budgets and dynamic cycle of O_2.

4.3.4.1 O_2 Reservoirs

The distribution of molecular oxygen (O_2) across Earth's surface reservoirs is a dynamic interplay of physical, chemical, and biological processes. Understanding these distributions is essential for comprehending global biogeochemical cycles and the feedback mechanisms that regulate Earth's habitability.

The Earth's atmosphere contains ~0.21 bars of O_2, amounting to a total mass of $\sim 34 \times 10^{18}$ mol. This O_2 is uniformly mixed with other major atmospheric gases (N_2, O_2, Ar) up to ~80 km altitude, encompassing the troposphere, stratosphere, and mesosphere. Turbulent mixing dominates over molecular diffusion in this region, ensuring homogeneity. However, atmospheric pressure decreases exponentially with altitude, concentrating the majority of O_2 within the lower several kilometers of the surface. Above ~120 km, in the thermosphere, molecular O_2 is photodissociated by UV radiation, forming atomic oxygen, which becomes the dominant form of oxygen at these altitudes.

Dissolved O_2 in seawater exhibits significant spatial and temporal variability. Photosynthesis in surface waters can lead to O_2 supersaturation and degassing to the atmosphere, while aerobic respiration in deeper layers can cause O_2 depletion or even anoxia. In surface mixed layer of oceans (<60 m water depth), O_2 concentrations are typically saturated to supersaturated, with higher levels in colder, high-latitude waters compared to equatorial regions. Seasonal variations are pronounced in high-productivity zones. Found in many ocean basins, O_2 concentrations drop below the surface mixed layer to form O_2 minimum zones (OMZs) resulting from biological O_2 consumption outpacing resupply via advection and diffusion (Stramma et al. 2010). Below OMZs, O_2 concentrations gradually increase due to global thermohaline circulation. Anoxia occurs in restricted basins like the Black Sea and Cariaco Basin, as well as in fjords and certain estuarine environments (e.g. Chesapeake Bay, Schelde estuaries, and river mouth of the Mississippi River). Additionally, several regions of the open ocean (the equatorial Pacific along Central and South America and the Arabian Sea) are also associated with strong O_2-depletion. These regions are associated with deep-water upwelling, high rates of surface water primary productivity, and high dissolved oxygen demand in intermediate waters. Recent research highlights the expansion of OMZs in response to ocean warming and deoxygenation, with significant implications for marine ecosystems.

Air-sea O_2 exchange is tightly coupled with surface water O_2 concentrations and seasonal productivity patterns. High rates of marine primary productivity drive net O_2 outgassing in spring and summer, while net ingassing occurs in fall and winter. This exchange exhibits latitudinal and seasonal patterns, with the northern hemisphere outgassing coinciding with the southern hemisphere ingassing during respective high- and low-productivity periods. Low-latitude oceans show minimal net O_2 exchange, reflecting stable conditions year-round.

O_2 concentrations in the pore fluids of marine sediments are governed by sediment deposition, diffusive exchange with the water column, and biological respiration. In coastal and shelf sediments, rapid organic matter burial depletes O_2 within millimeters to centimeters of the sediment–water interface. In abyssal sediments, where organic matter delivery is reduced, O_2 can penetrate several meters before being fully consumed. Whilst bioturbation plays a crucial role in oxygenating sediments by physically mixing sediment layers and enhancing oxygen penetration. Organisms such as worms, crustaceans, and burrowing fish disturb sediments through their movements, creating pathways for oxygen-rich water to infiltrate deeper layers.

In flowing freshwater systems, turbulent mixing ensures O_2 concentrations remain near air-saturated values. Static water bodies, however, can develop O_2 depletion, particularly in ice-covered lakes or high-productivity, turbid environments. Shallow aquatic systems often exhibit sharp O_2 gradients, with supersaturation at the surface transitioning to near anoxia within a few meters to the bottom. The high concentration of labile dissolved and particulate organic matter in many freshwater environments leads to rapid O_2-depletion where advective resupply is limited. The role of eutrophication and climate change has a significant impact in exacerbating freshwater O_2 depletion, with cascading effects on aquatic biodiversity (Kernan et al. 2011). Eutrophication leads to an oxygen crisis in freshwater environments by triggering excessive algal growth, which disrupts the natural balance of aquatic ecosystems. This eutrophication-driven oxygen crises can lead to fish kills, habitat degradation, and harmful algal toxins, posing risks to both wildlife and human health.

Soil water O_2 concentrations depend on gas diffusion, rainwater infiltration, air-gas exchange, and organic matter respiration. Organic-matter-rich temperate soils often exhibit reduced but not depleted O_2 levels, while deeper groundwater, water-saturated soils, and wetlands are generally O_2-poor. High-latitude soils and groundwater retain higher O_2 concentrations due to lower temperatures and reduced biological demand. Tropical soils show distinct patterns: dry soils are oxidized but show lower O_2 concentration compared to the air, while wet tropical forests experience near-surface O_2 depletion due to rapid organic matter oxidation. Soil permeability, particularly in clay-rich soils, further influences O_2 availability. Localized O_2 anomalies could be caused by subsurface oxidation of reduced metal sulfides in the subsurface.

4.3.4.2 O_2 Flux: Production

The transfer of O_2 between the atmosphere and oceans is governed by air-sea gas exchange, influenced by factors such as wind speed, sea surface roughness, and primary productivity. High O_2 concentrations in surface water can exceed equilibrium, leading to supersaturation and degassing. Conversely, calm seas limit gas exchange, creating localized O_2 accumulation. Despite advances in observational and modeling techniques, the fluxes involved in O_2 production and consumption remain poorly constrained. Figure 4.6 and Table 4.4 summarize the dominant processes and their estimated fluxes. One calculation estimated that it would take approximately 3000 years for photosynthesis to generate

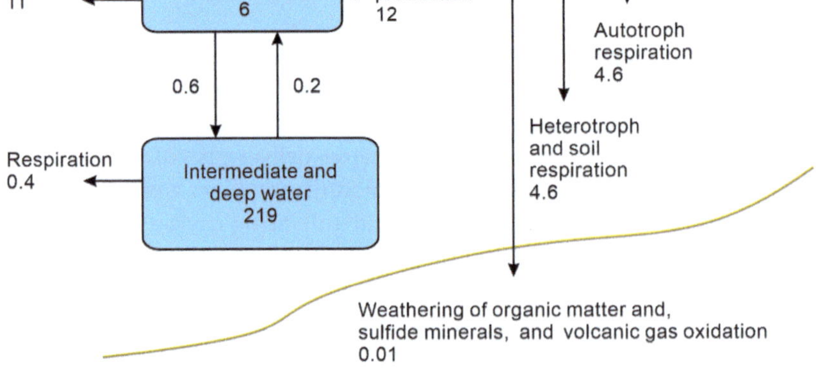

Figure 4.6 Global budget for molecular oxygen, including gas and dissolved O_2 reservoirs. *Source:* Modified from Keeling et al. (1993) and Petsch (2005).

Table 4.4 The modern geological O_2 budget.

O_2 production	Unit: 10^{12} mol O_2/yr
Burial of C^0	10 ± 3.3
Burial of FeS_2	7.0 ± 3.6
Burial of FeO	0.9 ± 0.4
Loss of hydrogen to space	0.024
O_2 Loss	
Oxidation of C^0 during weathering	7.5 ± 2.5
Oxidation of FeS_2 during weathering	7.0 ± 3.6
Oxidation of FeO during weathering	1.0 ± 0.6
Surface outgassing of H_2	0.1
Surface outgassing of SO_2 and H_2S	0.2
Submarine outgassing of H_2S	1.0
Submarine H_2 from serpentinization	0.2
Total	17.0

Source: Adapted from Kasting and Canfield (2012).

the current amount of oxygen in Earth's atmosphere, when assuming no oxygen consumption (Borisov and Björn 2018). The main sources and sinks of O_2 are presented below.

Oxygenic photosynthesis is the dominant mechanism of O_2 production, carried out by plants, eukaryotic algae, cyanobacteria, and prochlorophytes. Global net primary production (derived from atmospheric O_2 abundance and isotopic composition) ranges between 23×10^{15} and 26×10^{15} mol/yr. Terrestrial ecosystems contribute $\sim 14 \times 10^{15}$ mol/yr, while marine ecosystems account for $\sim 12 \times 10^{15}$ mol/yr. Additionally, approximately 50% of global photosynthetic CO_2 fixation occurs in marine surface waters by phytoplankton, highlighting the oceans' significant role in O_2 production. Marine primary production is highly sensitive to nutrient availability, light penetration, and ocean warming. On land, primary production exhibits strong seasonal and regional variability, driven by temperature, precipitation, and land-use changes. These patterns directly influence atmospheric O_2 concentrations, with seasonal fluctuations linked to photosynthesis rates in terrestrial and marine ecosystems.

In the upper atmosphere, a minor source of O_2 is the photolysis of water vapor, a process that also contributes to the trace O_2 detected on icy moons such as Ganymede and Europa. On early Earth, water vapor photolysis may have been the primary source of O_2 before the evolution of oxygenic photosynthesis. Also, the oxidation of water at the surface of quartz produces molecular oxygen, which can possibly be an important O_2 source, particularly during the Earth's history (He et al. 2021). However, the O_2 produced would have reacted with reduced gases like methane and carbon monoxide, preventing their accumulation in the atmosphere.

4.3.4.3 O_2 Flux: Consumption

Aerobic respiration is the primary mechanism of O_2 consumption, occurring in terrestrial and aquatic ecosystems. It involves the oxidation of organic substrates to produce ATP (Adenosine Triphosphate) and NADH, (Nicotinamide Adenine Dinucleotide) driven by aerobic cellular respiration of eukaryotes and prokaryotes. Photorespiration is another process consuming O_2 without producing ATP and is enhanced in hot, dry environments where stomatal closure reduces CO_2 availability. In some plants, as much as 50% of the carbon fixed is respired through photorespiration, but photorespiration does not occur in prokaryotes. Beyond aerobic cellular respiration and photorespiration, there are several other biological processes common in prokaryotes that consume oxygen (Section 1.3.5). For example, methylotrophic organisms can metabolize compounds containing only one carbon, such as methane, methanol, formaldehyde, and formate, which are common in soils and sediments as the products of anaerobic fermentation reactions. In addition, some chemolithotrophic microorganisms are able to oxidize inorganic compounds rather than organic substrates as a source of energy and consume oxygen. For example, hydrogen-oxidizing bacteria oxidize hydrogen, nitrifying bacteria oxidize ammonia, some chemolithoautotrophic bacteria can oxidize nitrite, and non-photosynthetic bacteria can oxidize reduced sulfur compounds. Macro-scale patterns of aerobic respiration are discussed below.

In terrestrial ecosystems, aerobic respiration degrades a significant fraction of O_2 produced from primary production, with partially degraded organic matter incorporated into soils. Degradation of organic matter breaks larger biomolecules into smaller compounds, which, if transported into aerobic zones of soil or sediment, can be degraded by aerobic microorganisms. Research has shown that soil organic matter can be preserved for several millennia, and riverine export of aged terrestrial organic matter may be a significant source of dissolved and particulate organic carbon to the oceans.

In marine ecosystems, aerobic respiration dominates the water column, where O_2 resupply is efficient. Sediment trap and particle flux studies reveal that most marine primary production is degraded before reaching the seafloor, with O_2 consumption nearly balancing photosynthetic O_2 release. In nearshore sediments, O_2 penetration is limited (1 mm or less), except in bioturbated coastal zones where the movement of animals facilitates deeper O_2 diffusion. For example, O_2 penetration is facilitated by the recharging of pore fluids through animals burrowing into sediments, which normally leads to an O_2 penetration of several centimeters below sediment–water interface. Conversely, in pelagic sediments, where the supply of organic matter is minor, temperature is low, and the water column is in fully equilibrium with O_2 in the atmosphere, O_2 may penetrate uniformly 1 m or more into the sediment.

Volcanic emissions (both lava flows and pyroclastic flows) release reduced gases (e.g. CO, H_2, SO_2, H_2S, and CH_4), which are rapidly oxidized in the atmosphere, consuming O_2. This process is predominantly abiotic and highly efficient. Weathering of reduced minerals (e.g. Fe^{2+}-bearing silicates, sulfides) consumes O_2, particularly during soil formation and erosion. Fe^{2+} oxidation is a key component of soil development, while sulfide oxidation generates acidity and poses environmental challenges in mining areas. Iron oxides derived from oxidation of Fe^{2+} in the parent rock accumulate in temperate soils, and extensive laterites consisting of iron and aluminum oxides develop in tropical soils. Ferrous iron oxidation is efficient, leading to the absence of clastic ferrous-iron minerals in soils. Sulfide minerals are extremely susceptible to oxidation in areas where mining has brought underground sulfide minerals to the surface, contacting with the atmosphere O_2 or O_2-charged rainwater. Chemolithoautotrophic microorganisms mediate much of pyrite oxidation, resulting in the oxidation of both ferrous iron and sulfide and generation of acidity (Section 1.3.5). The amount of O_2 consumed annually through oxidation of Fe^{2+} and sulfur-bearing minerals can be estimated based on sulfur isotope mass balance constraints, which is on the order of $(0.1–1) \times 10^{12}$ mol/yr, similar in magnitude to the flux of reduced gases from volcanism (Canfield 2004).

The spreading of lithospheric plates along mid-ocean ridges drives undersea volcanic activity, leading to the release of chemically reduced substances such as metal sulfides. Submarine eruptions generate extrusive lava flows, form pillow basalts composed partly of ferrous-iron silicates, and release dissolved reduced gases to seawater. High-temperature "black smoker" chimneys at the seafloor release metal-rich plumes containing Fe-Cu-Zn sulfides into seawater. O_2 in seawater is consumed through the oxidation of these reduced substances. The oxidation of reduced substances can be both abiotic and biotic, e.g. it is observed that chemolithoautotrophic microorganisms colonize the walls and channels of hydrothermal vents. Isotopic analyses of sulfur and iron in hydrothermal fluids suggest that microbial activity may play a more substantial role in sulfide oxidation than previously recognized (Qiu et al. 2021).

Organic matter can be oxidized in various environments where biological activity is absent or minimal (e.g. photo-oxidation of dissolved and particulate organic matter and fossil fuels, combustion of vegetation and fossil fuels, and atmospheric methane oxidation). Dissolved and particulate organic matter, as well as fossil fuels, can be oxidized by UV light, ozone, gamma radiation, or transition metal catalysts. This process is particularly important in surface waters and the atmosphere. Fires, resulting from the burning of vegetation and fossil fuels, are high-temperature oxidation processes that consume significant amounts of O_2. Most methane on Earth is oxidized during slow, gradual transport upward through soils and sediments. Methane reacts with O_2 in the presence of light or metal catalysts, preventing its accumulation in the atmosphere.

4.3.5 Nitrogen Cycle

Nitrogen is a bio-essential element that is used for the synthesis of essential cellular components. Nitrogen is found in soils, sediments, and aquatic environments as inorganic salts of nitrogen (ammonium or nitrate), or as organic matter. The atmosphere, comprising 78% nitrogen gas (N_2), serves as a major reservoir. However, this form of nitrogen is basically inaccessible to organisms. The interconversion of nitrogenous compounds by microbial communities forms the cornerstone of the nitrogen cycle (Figure 4.7).

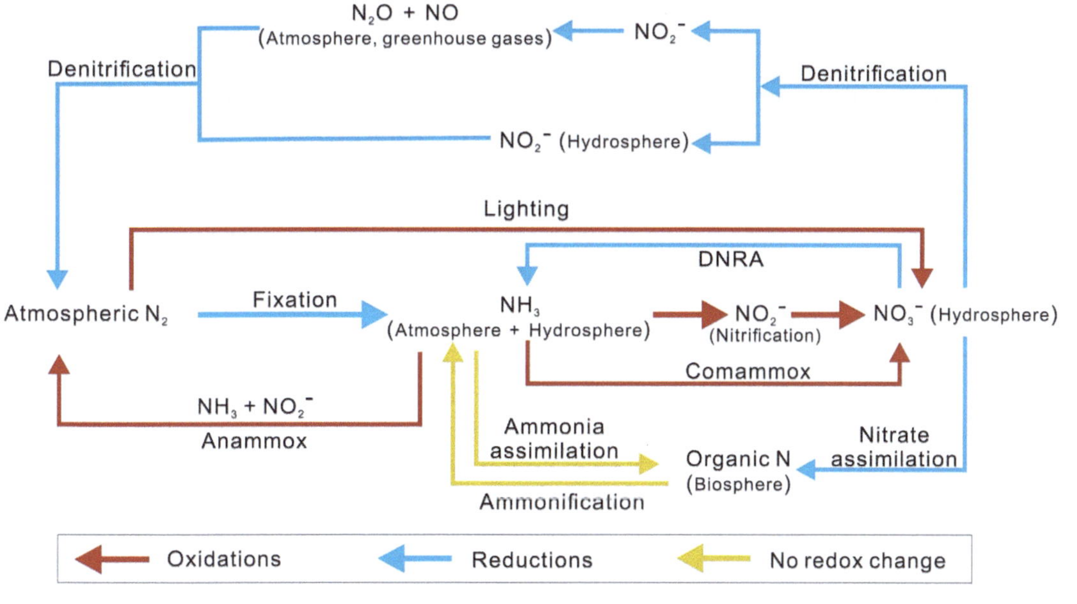

Figure 4.7 The nitrogen cycle. The nitrogen cycle is an important process that takes place in nature and ensures an adequate supply of nitrogenous compounds to living organisms. The cycle comprises a sequence of transformations involving various forms of organic and inorganic nitrogen compounds achieved by the metabolic activities of microorganisms. DNRA, Dissimilatory nitrate reduction to ammonium.

4.3.5.1 Reservoirs

Earth's nitrogen reservoirs contain approximately ~5×10^{21} g N, 80% of which is in the atmosphere. Sedimentary rocks account for most of the remainder, with small amount of N (<1% of total) occurring in oceans and organic matter. In the atmosphere, nitrogen mainly occurs as N_2, except for trace amounts of N_2O, NO_y, NH_x, and organic N. In oceans and soils, it primarily occurs as organic nitrogen, nitrate, and ammonium. Atmospheric and hydrologic transportations are the primary paths for exchanges between reservoirs (Galloway 2005 and references therein). Nitrous oxide is a greenhouse gas that can trap heat near the Earth's surface. It also destroys stratospheric ozone. Nitrous oxide (N_2O) in the stratosphere can be broken down by UV light into nitrogen dioxide (NO_2) and nitric oxide (NO), which can catalytically reduce ozone. Nitrogen oxides are ultimately transformed into N_2, nitrate or nitrite compounds, forming acidic rains which cause ecological problems.

4.3.5.2 Flux

Nitrogen can be transferred from land to the atmosphere through the combustion of biomass/fossil fuel, volatilization of gases from soils and waters, and turbulent injection of particulate matter into the atmosphere. The gases are generated primarily as a result of microbial activity (e.g. nitrification, denitrification, and ammonification). When the partial pressure of gas in seawater is greater than the partial pressure of gas in the atmosphere, gas forms of nitrogen (e.g. NH_3, N_2, and N_2O) can be emitted from the ocean through volatilization. Nitrogen can also be transferred to the atmosphere via aerosols formed by breaking waves or bubbles.

Atmospheric nitrogen wet deposition (e.g. rain, snow, hail) and dry deposition (e.g. gases, aerosols) play a critical role in transferring nitrogen to land and oceans. Gaseous HNO_3 and NH_3 are the two most important dry-deposited species. By comparison, the fluxes of aerosol dry deposition of nitrate and ammonium are generally small relative to gaseous deposition. River and groundwater discharges also transport nitrogen to coastal ecosystems, primarily as nitrate or particulate organic nitrogen.

4.3.5.3 Key Processes

Nitrogen fixation, the conversion of atmospheric N_2 to biologically available ammonia (NH_3), remains a fundamental process mediated by specialized microorganisms. A substantial portion of terrestrial nitrogen fixation is attributed to the symbiotic association between legumes and nitrogen-fixing bacteria. For instance, the leguminous shrub *Sophora davidii*, when forming root nodules in symbiosis with *Mesorhizobium* species (Figure 4.8), contributes significantly to biological nitrogen input in its native habitats. Such associations can fix considerable amounts of nitrogen, potentially exceeding 100 kg of nitrogen/acre/year under favorable conditions. In contrast, *Azotobacter*, a nitrogen-fixing bacterium that is not associated with plants, fixes less than 1 kg of nitrogen/year in the same types of soils. Some actinomycetes fix nitrogen in symbiotic relationships with trees such as alder and bayberry. In aquatic environments, cyanobacteria such as *Anabaena*, *Rivularia*, and *Nostoc* fix as much as 10 kg of nitrogen/acre/year. Nitrogen-fixing microorganisms can also live in extreme environments, such as deep-sea sediments and hydrothermal vents. Once fixed, nitrogen is either oxidized via nitrification or assimilated into biomass via ammonia assimilation (Figure 4.7). A few bacteria use ammonium as a source of energy for their metabolism. For example, the aerobic chemolithotrophic bacteria *Nitrosomonas* and *Nitrosococcus* oxidize ammonium ions, converting them to nitrite. The plants then use it to manufacture amino acids and convert it into proteins.

Ammonification is the decomposition process that converts organic nitrogen (amino acids, nucleotides, and urea) into ammonia (NH_3). In neutral environments, ammonia forms ammonium

Figure 4.8 Nitrogen-fixing bacteria in root nodules. Root nodules formed on *Sophora davidii* in symbiosis with *Mesorhizobium* sp. *Source:* Photo courtesy of Ying Cao.

(NH_4^+), which adheres to negatively charged soil particles. In alkaline environments, gaseous ammonia may volatilize into the atmosphere. A wide variety of organisms, including aerobic and anaerobic bacteria as well as fungi, can degrade protein, one of the most prevalent nitrogen-containing organic compounds. They do this initially through the action of extracellular proteolytic enzymes that break down proteins into short peptides or amino acids. Then, the amino groups are removed, releasing ammonium for organisms to produce biomass.

Nitrification is a critical biogeochemical process that facilitates the oxidation of ammonium (NH_4^+) to nitrate (NO_3^-). This transformation is carried out by a specialized group of chemolithotrophic bacteria, collectively known as nitrifiers. These bacteria utilize ammonium and its intermediate, nitrite (NO_2^-), as energy sources. Nitrification predominantly occurs in well-aerated soils and is pivotal in providing nitrate. The nitrifiers encompass two populations of chemolithotrophic bacteria, the ammonia oxidizers (e.g. *Nitrosomonas* species) and the nitrite oxidizers (e.g. *Nitrobacter* and *Nitrospira* species). Nitrifiers use molecular oxygen as a terminal electron acceptor. Consequently, nitrification does not occur in waterlogged soils or in anaerobic regions of aquatic environments.

Nitrification has some important consequences with respect to agricultural practices and pollution. Farmers often apply ammonium-containing compounds to soils as a source of nitrogen for plants. The ammonium is retained by soils because its positive charge enables it to adhere to negatively charged soil particles. Nitrification converts the ammonium to nitrate, a form of nitrogen that is more readily used by plants but can be rapidly leached from soils by rainwater. Ammonium oxidation can also lead to the accumulation of nitrite in soil when nitrite oxidizers are insufficient. If the nitrite leaches into groundwater, it may contaminate wells used for drinking water. Nitrite is toxic because it combines with hemoglobin of the blood, reducing blood's O_2-carrying capacity.

Denitrification is the reduction of nitrate (NO_3^-) to gaseous nitrogen. Some *Pseudomonas* species can use nitrate as a terminal electron acceptor when molecular oxygen is not available, which is the process of anaerobic respiration. During this process, *Pseudomonas* reduce nitrate ions to nitrite, nitrous oxide (N_2O), and eventually molecular nitrogen. This process depletes nitrate in sediments and releases nitrous oxide (N_2O), which is a potent greenhouse gas. Denitrifying microorganisms may reduce oxidized nitrogen compounds in agricultural fertilizers, resulting in the release of gaseous nitrogen to the atmosphere and leading to economic loss. In some areas, this process may represent 80% of nitrogen lost from fertilized soil. Additionally, denitrification can be applied in wastewater treatment as a means to remove nitrate. *Brocadia anamoxidans* was found to be able to oxidize ammonium under anaerobic conditions, using nitrate as a terminal electron acceptor. This reaction, called anammox (Section 1.3.4), forms N_2 and might provide an economic approach for removing nitrogen compounds for sewage treatment.

4.3.5.4 Isotope Fractionation

Nitrogen has two stable isotopes, ^{14}N and ^{15}N, with atmospheric abundances of approximately 99.63% and 0.37%, respectively. The standard for nitrogen isotopes is atmospheric N_2, with its $\delta^{15}N$ conventionally defined as 0‰.

Nitrogen fixation converts inert N_2 into biologically available nitrogen, typically resulting in organic matter with a slightly negative $\delta^{15}N$ value (lower than atmospheric N_2). Terrestrial nitrogen-fixing bacteria are commonly found in plant root nodules. Breaking the triple bond between N_2 molecules requires substantial energy, making nitrogen fixation an inefficient process with minimal isotopic fractionation (Hoefs 2009).

Nitrification is a multi-step oxidation process mediated by various autotrophic microorganisms, producing several intermediate products. It can be described by two partial oxidation reactions: ammonia oxidation ($NH_4^+ \rightarrow NO_2^-$), carried out by ammonia-oxidizing bacteria (AOB); nitrite oxidation ($NO_2^- \rightarrow NO_3^-$), performed by nitrite-oxidizing bacteria (NOB). Since $NO_2^- \rightarrow NO_3^-$ oxidation occurs rapidly with negligible fractionation, the primary isotopic fractionation arises during the slower $NH_4^+ \rightarrow NO_2^-$ step (Hoefs 2009). Notably, this fractionation assumes abundant reactants. If NH_4^+ availability is limited (e.g. early Earth with sparse biological fixation and reliance on abiotic processes like lightning), little isotopic fractionation would occur.

Denitrification is a multi-step reductive process where microorganisms progressively reduce NO_3^- to N_2 via intermediates (primarily nitrogen oxides). Occurring in anaerobic soils or aquatic environments, it follows the simplified reaction: $5C_6H_{12}O_6 + 24NO_3^- + 24H^+ \rightarrow 30CO_2 + 42H_2O + 12\,N_2$. This process generates a significant fractionation ($\delta^{15}N$ depletion of 10–30‰ in N_2), with slower reduction rates enhancing fractionation. However, at low NO_3^- concentrations, almost all of the nitrate will be reduced, and in this case, there will be little fractionation (Hoefs 2009).

Figure 4.9 shows the $\delta^{15}N$ values among different Earth's reservoirs. Sedimentary organic matter has $\delta^{15}N \approx +7$‰ (Sweeney et al. 1978). During diagenesis, NH_4^+ from organic matter may replace K^+ in clay minerals or micas, inheriting similar isotopic values from organic matter. In contrast, N extracted from mid-ocean ridge basalts and mantle-derived diamonds shows $\delta^{15}N \approx -5$‰, indicating homogeneous mantle N and sharp crust-mantle differentiation. Metamorphic rocks exhibit enriched $\delta^{15}N$ due to volatile loss under high P–T conditions.

4.3.6 Sulfur Cycle

Sulfur cycling is one of the most complicated biogeochemical processes due to its multiple oxidation states. The redox cycle of sulfur and the role of microorganisms in sulfur transformations are given in Figure 4.10. Although a number of oxidation states are possible, the most common sulfur oxidation states are −2 (sulfhydryl, R-SH, and sulfide, HS^-), 0 (elemental sulfur, S^0), and +6 (sulfate, SO_4^{2-}).

Sulfur is an essential component of all living organisms, primarily as part of the amino acids methionine and cysteine, as well as coenzymes like biotin (vitamin H). Like the nitrogen cycle, key steps in the sulfur cycle are mediated by prokaryotes (Figure 4.10), which play pivotal roles

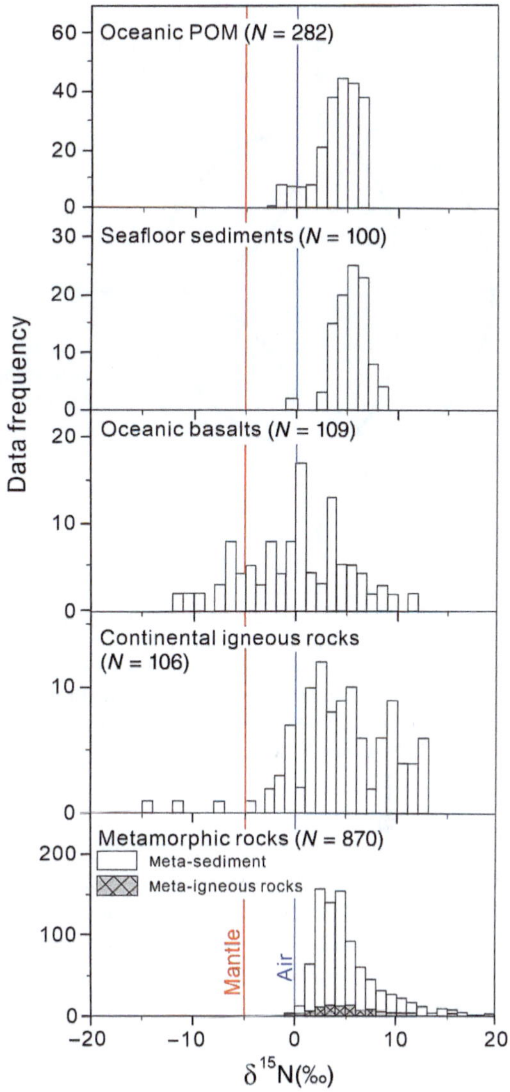

Figure 4.9 The distribution of nitrogen isotopes among different reservoirs. *Source:* Li et al. (2014) / with permission of Elsevier.

in both oxidative and reductive processes. Sulfide-oxidizing bacteria, often chemolithotrophs, convert reduced sulfur compounds (e.g. S^0, HS^-) to sulfate (SO_4^{2-}). This process is critical in environments such as hydrothermal vents, sulfur springs, and anoxic sediments. Sulfate-reducing bacteria (SRB) utilize sulfate as an electron acceptor in anaerobic respiration, producing hydrogen sulfide (H_2S) as a byproduct. Sulfate reduction is a major biogeochemical process in marine sediments, wetlands, and anoxic water bodies, with significant implications for metal cycling and toxicity due to the reactivity of H_2S with metals. Dimethyl sulfide (DMS) is the major organic sulfur compound of ecological significance in nature.

4.3.6.1 Reservoirs

The bulk of the sulfur on Earth is found in sediments and rocks in the form of sulfate minerals (primarily gypsum) and sulfide minerals (primarily pyrite, FeS_2), while sulfur in the oceans (in

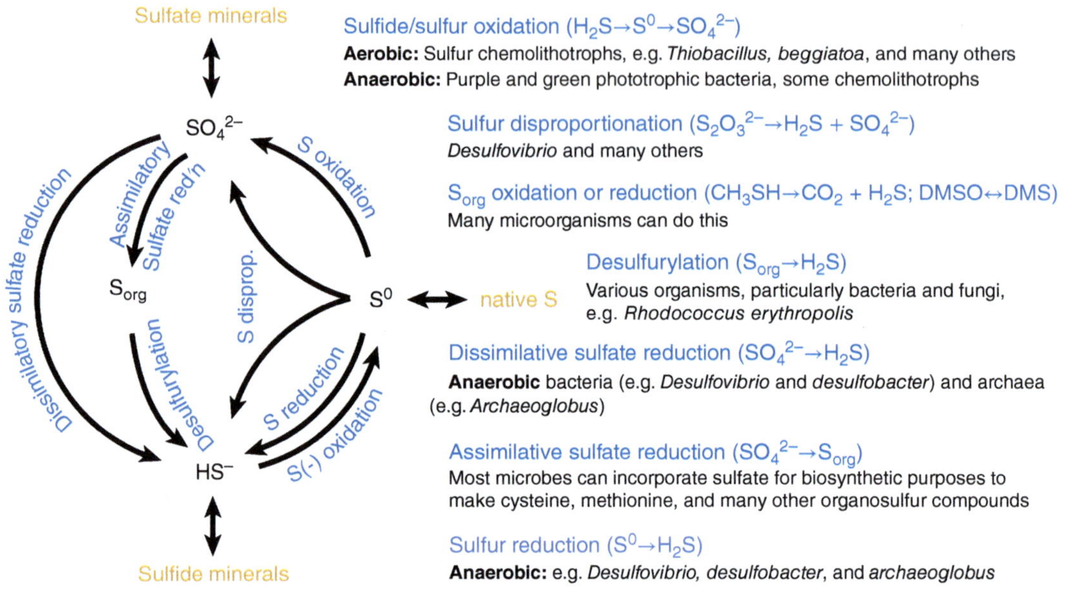

Figure 4.10 Microbial redox reactions in the global sulfur cycle. Microbial reactions are labeled in blue font and solid phases are in orange font. S_{org} represents sulfur in organic compounds, e.g. dimethyl sulfide (DMS, CH_3-S-CH_3) and dimethyl sulfoxide (DMSO, C_2H_6OS). For simplicity, elemental sulfur (S^0) represents one of several sulfur compounds with oxidation state between −2 and +6 that can form during sulfide oxidation. At least some of these intermediate sulfur species can be disproportionated into sulfate and sulfide. *Source:* / Dr. Matthew Kirk / CC BY SA 4.0.

the form of sulfate) is the most significant sulfur reservoir for the biosphere. The global sulfur cycle is mainly controlled by three major reservoirs (Figure 4.11): sulfate dissolved in the world's oceans; sulfate in ancient evaporite deposits; and sulfide (mainly in the form of pyrite, FeS_2) in marine sediments. The mass of sulfate-sulfur in the modern oceans is well constrained at 1.3×10^{21} g, and the mass of pyrite sulfur buried in sediments is around 6×10^{21} g. However, there is little agreement on the size of the reservoir of oxidized sulfate-sulfur stored in buried evaporites, with estimates ranging between 1.3 and 8×10^{21} g. These reservoirs of sulfur are constantly augmented by the release of volcanic sulfur, mainly as gas phase (SO_2 and H_2S) that ultimately rain out to oceans via the hydrosphere. Other reservoirs of sulfur are generally small, including sulfur in atmosphere (1×10^{12} g), soil (3×10^{16} g), terrestrial biomass (1×10^{16} g), and fossil fuels (7×10^{16} g) (see Newton and Bottrell 2007).

4.3.6.2 Flux

Sulfur is cycled among the major geological reservoirs in two ways. Biological reduction of sulfate to sulfide in organic-rich marine sediments leads to the reaction of sulfide and iron in pore water forming pyrite in sediments. Bacteriogenic sulfide can also be fixed in sediments by reaction with sedimentary organic matter to form organosulfur compounds.

Sulfur can also be removed from the oceans as evaporite sulfate minerals, most commonly gypsum ($CaSO_4 \cdot 2H_2O$). There is also a small amount of sulfate precipitation within carbonate from the ocean water column. This is a very small flux (maximum estimate ca. 0.5×10^{13} g S/yr), compared with net pyrite S burial at 6×10^{13} g S/yr.

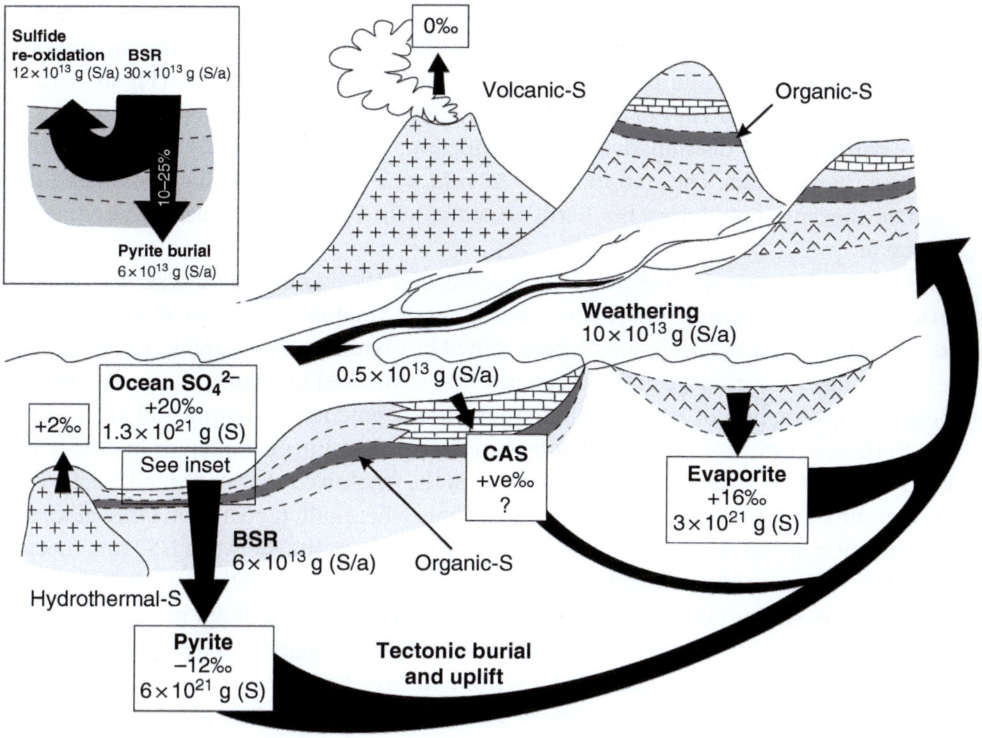

Figure 4.11 A schematic illustration of the global sulfur cycle and its isotopic composition in various reservoirs. BSR, bacterial sulfate reduction; CAS, carbonate-associated sulfate. *Source:* Adapted from Newton and Bottrell (2007).

Sulfur stored as pyrite, evaporite minerals, and organics is essentially stable after burial and is preserved in sedimentary successions on geological time scales. Ultimately, tectonic processes uplift buried sediments, which are then exposed to weathering in the terrestrial environment. During weathering, sulfate in evaporite minerals and in marine carbonate rocks is released directly into solution as their host minerals dissolve. Pyrite reacts in the presence of oxygen, consuming oxygen and generating sulfate ($FeS_2 + 3.5O_2 + H_2O \rightarrow Fe^{2+} + 2SO_4^{2-} + 2H^+$). Sulfate produced during these weathering reactions enters the hydrological cycle and is returned to the oceans.

Recent studies highlight the importance of local organic matter flux in controlling sulfur cycling in marine sediments. For example, during the Ediacaran Shuram excursion, increased organic matter input led to higher microbial sulfate reduction (MSR) rates, as evidenced by isotopic fractionation in pyrite grains (Matsu'ura et al. 2024). This suggests that local redox conditions, rather than global oceanic sulfate concentrations, often dictate sulfur cycling dynamics.

4.3.6.3 Key Processes

A major volatile sulfur gas is hydrogen sulfide (H_2S), formed primarily by the microbial reduction of sulfate. Hydrogen sulfide can be dissolved in aqueous environments and its speciation depends on solution pH. At high pH, the dominant form is S^{2-}. At neutral pH, HS^- predominates, and below pH 6, its speciation is H_2S. Even at neutral pH, some volatilization of H_2S from HS^- can occur because there is an equilibrium between HS^- and H_2S, and as volatilization occurs, the reaction is pulled toward H_2S.

A wide variety of organisms can use sulfate as a sulfur source and carry out assimilative sulfate reduction, converting the newly formed HS^- to organic sulfur (Section 1.3.5). The decomposition of organic sulfur by putrefaction and desulfurization can ultimately generate HS^- (Figure 4.10), and this is a significant source of HS^- in an aqueous environment. Another major source of HS^- in an aqueous environment is dissimilative sulfate reduction, which is normally carried out by SRB. Modern ocean has a sulfate concentration of ~28 mM which is relatively high compared to the oceanic sulfate concentration during the Earth's history, which provides an ideal environment for sulfate reduction.

Due to the necessity of organic electron donors (e.g. molecular H_2, organic matter), sulfate reduction occurs extensively only where significant amounts of organic matter are present. Generally, the rate of sulfate reduction in marine sediments is determined by organic carbon supply. The disposal of sewage, sewage sludge, and garbage to the ocean can lead to marked increases in organic matter in the sediments. The formation of HS^- by sulfate reduction is potentially detrimental because it combines with the iron in cytochromes and other essential iron-containing compounds in cells. The combination of sulfide with iron, leading to the formation of the insoluble FeS, can generally detoxify sulfide in the environment. The black color of many sediments where sulfate reduction takes place is due to the accumulation of FeS. The formation of metal sulfide (e.g. pyrite) between HS^- and metal cations buries sulfur in the sediment which is an important flux to geological S reservoir.

Under oxic conditions, sulfide (HS^-) oxidizes rapidly at neutral pH even without biological catalysis. While sulfur-oxidizing bacteria can accelerate this reaction, their activity is typically confined to interfacial zones where upward-diffusing H_2S from anoxic layers encounters downward-diffusing O_2 from oxic regions. In illuminated anoxic environments (e.g. shallow lake waters where light penetrates sufficiently), phototrophic sulfur bacteria mediate HS^- oxidation, though such settings are spatially limited.

Elemental sulfur (S^0) is chemically stable in most oxic environments but can be readily oxidized by sulfur-oxidizing bacteria. *Thiobacillus* species are the most prevalent sulfur-oxidizing bacteria in nature. Due to the low solubility of elemental sulfur, S^0-oxidizing bacteria are generally hydrophobic and form direct attachments to sulfur particles. The oxidation process generates sulfate and hydrogen ions, resulting in significant environmental acidification. This biochemical property is exploited in agricultural practice, where S^0 amendments are used to lower pH in alkaline soils.

4.3.6.4 Organic Sulfur Compounds

While inorganic sulfur species play key roles in biogeochemical cycling, diverse organic sulfur compounds also participate through biological synthesis and transformation. Most plants and microorganisms assimilate sulfur primarily through sulfate reduction and incorporation into biomass. The dominant organic sulfur species in natural systems is DMS (H_3C-S-CH_3), particularly abundant in marine environments as a degradation product of dimethylsulfonium propionate—an osmoregulatory compound in marine algae. Microbial metabolism converts dimethylsulfonium propionate into DMS and acrylate (a propionate derivative), with the latter serving as a growth substrate.

Microbial production of DMS plays a significant role in global sulfur cycling, with approximately 45 million tons released annually into the environment. In oxic environments, atmospheric DMS undergoes photochemical oxidation, forming methane sulfonic acid ($CH_3SO_3^-$), sulfur dioxide (SO_2), and sulfate (SO_4^{2-}). However, in anoxic habitats, DMS can serve as a substrate for various microbial processes, including: methanogenesis (producing CH_4 and H_2S); photosynthetic CO_2 fixation by phototrophic purple bacteria (yielding dimethyl sulfoxide, DMSO); energy metabolism in chemoorganotrophs and chemolithotrophs (also producing DMSO). DMSO, in turn, can act as an electron acceptor in anaerobic respiration, regenerating DMS and completing a microbial sulfur cycle. While other volatile organic sulfur compounds—such as methanethiol (CH_3SH), dimethyl disulfide (H_3C-S-S-CH_3), and carbon disulfide (CS_2)—also contribute to sulfur biogeochemistry, DMS remains the most quantitatively significant on a global scale.

The sulfur cycle has evolved dramatically over the Earth's history. During the Archean and Proterozoic eons, marine sulfate concentrations were much lower (μM to low mM) compared to modern levels (28 mM). The transition to stable, high-sulfate ocean began in the Phanerozoic, driven by increased oxygen levels and tectonic activity (Section 8.4.3). Oceanic anoxic events (OAEs) temporarily disrupted this trend, leading to feedbacks between sulfur, carbon, and iron cycles that are still not fully understood.

Also, human activities show a significant impact on sulfur cycle. SO_2 from fossil fuel combustion and industrial processes has increased atmospheric sulfur levels, contributing to acid rain. Meanwhile, the use of sulfur-containing fertilizers has disrupted soil sulfur cycling, affecting plant and microbial communities.

4.3.6.5 Isotope Fractionation

Sulfur has four stable isotopes, with ^{32}S being the most abundant in nature. However, small amounts of ^{33}S, ^{34}S, and ^{36}S are also present. Their natural abundances are as follows: ^{32}S, 95.02%; ^{33}S, 0.75%; ^{34}S, 4.21%; and ^{36}S, 0.02%.

Dissimilatory sulfate reduction by SRB preferentially utilizes the lighter isotope, ^{32}S, due to the lower bond energy of ^{32}S-O compared to the ^{34}S-O bond. This effect results in sulfides being depleted in ^{34}S and enriched in ^{32}S relative to the initial sulfate pool. Similarly, biological sulfide oxidation also fractionates sulfur isotopes, showing a preference for the lighter isotope, although the fractionation is not as significant as sulfate reduction (Table 4.5). However, a recent study suggests that *Desulfurivibrio alkaliphilus* can oxidize sulfide and generates an enrichment of ^{34}S in the produced sulfate by +12.5‰ or greater (Pellerin et al. 2019). The degree of fractionation is calculated using the formula:

$$\delta^{34}S = \left\{\left[\left(^{34}S/^{32}S\right)_{sample} - \left(^{34}S/^{32}S\right)_{standard}\right] / \left(^{34}S/^{32}S\right)_{standard}\right\} \times 1000$$

Table 4.5 Cell-specific sulfate reduction rates and fractionation factors of investigated sulfate-reducing prokaryotes. csSRR is cell-specific sulfate reduction rate and the enrichment factor (ε) is the relative fractionation between $\delta^{34}S$ of SO_4^{2-} and $\delta^{34}S$ of S^{2-}.

Microorganisms	Isolated from	Electron donor (mM)	csSRR (fmol cell^{-1}/day)	ε (‰)
Complete oxidizing				
Desulfonema magnum	Marine mud	Benzoate (3)	5.9	42.0
Desulfobacula phenolica	Marine mud	Benzoate (3)	125.0	36.7
Desulfobacterium autotrophicum	Marine mud	Butyrate (20)	64.8	32.7
Desulfobacula toluolica	Marine mud	Benzoate (3)	40.3	28.5
Desulfotomaculum gibsoniae	Freshwater mud	Butyrate (20)	13.0	27.8
Desulfospira joergensenii	Marine mud	Pyruvate (20)	0.9	25.7

(Continued)

Table 4.5 (Continued)

Microorganisms	Isolated from	Electron donor (mM)	csSRR (fmol cell^{-1}/day)	ε (‰)
"Desulfarculus baarsii"	Lake mud	Butyrate (20)	11.5	23.2
Desulfotignum balticum	Marine mud	Butyrate (20)	4.2	23.1
Desulfofrigus oceanense	Arctic sediment	Acetate (20)	7.6	22.0
Desulfobacter sp. ASv20	Arctic sediment	Acetate (20)	6.6	18.8
Desulfobacca acetoxidans	Anaerobic sludge	Acetate (20)	17.0	18.0
Desulfococcus sp.	Marine mud	Pyruvate (20)	41.8	16.1
Desulfosarcina variabilis	Marine mud	Benzoate (3)	11.1	15.0
Incomplete oxidizing				
*Desulfonatronum lacustre**	Alkaline lake mud	Ethanol (20)	16.2	18.7
Archaeoglobus fulgidus strain Z*	Submarine hot spring	Lactate (20)	63.8	17.0
Thermodesulfovibrio yellowstonii	Thermal vent water	Lactate (20)	24.0	17.0
"Desulfobotulus sapovorans"	Freshwater mud	Lactate (20)	26.0	16.5
*Desulfotomaculum thermocisternum**	Oil reservoir	Lactate (20)	310.0	15.0
Desulfomicrobium baculatum	Manganese ore	Lactate (20)	4.8	12.7
*Desulfotomaculum geothermicum**	Aquifer	Lactate (20)	8.1	12.5
Desulfohalobium redbaense	Saline sediment	Lactate (20)	434.0	10.6
Desulfocella halophila	Great Salt Lake	Pyruvate (20)	10.2	8.1
Desulfobulbus "marinus"	Marine mud	Propionate (20)	28.9	6.8
Desulfotalea arctica	Arctic sediment	Lactate (20)	4.2	6.1
Desulfobulbus elongatus	Digester	Propionate (20)	35.3	5.5
Desulfovibrio sp. strain X	Hydrothermal vent	Lactate (20)	36.0	5.4
Thermodesulfobacterium commune	Thermal spring	Lactate (20)	45.4	5.0
"Desulfovibrio oxyclinae"	Hypersaline mat	Lactate (20)	69.1	4.5
Desulfotalea psychrophila	Arctic sediment	Lactate (20)	6.3	4.3
Desulfovibrio profundus	Deep sea sediment	Lactate (20)	17.0	4.1
Desulfovibrio halophilus	Hypersaline microbial mat	Lactate (20)	33.1	2.0
Hydrogen and formate				
Desulfobacterium autotrophicum	Marine mud	H_2 (10^5 Pa)	34.8	14.0
Desulfonatronovibrio hydrogenovorans	Lake mud	Formate (20)	12.7	5.5

Under experimental conditions, strains followed by an asterisk are considered incomplete oxidizers.
Source: Detmers et al. (2001) / The American Society for Microbiology.

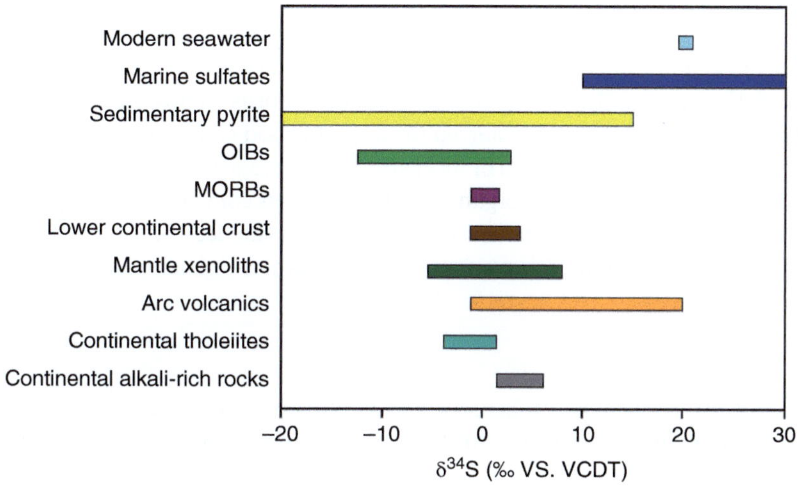

Figure 4.12 $\delta^{34}S$-isotopic variations of some geological material in different terrestrial reservoirs. OIBs, Ocean island basalts; MORBs, mid-ocean ridge basalts; VCDT, Vienna-Canyon diablo troilite. *Source:* Hammerli et al. (2021) / Elsevier / CC BY 4.0.

The standard material used is either troilite (FeS) from the Cañon Diablo meteorite (CDT) or IAEA-S-1 (Ag_2S, which is referenced to the V-CDT). As shown in Figure 4.12, sedimentary sulfide is highly enriched in ^{32}S, compared to the meteoritic sulfate standard. In contrast, sulfur in igneous rocks does not exhibit this bias toward the lighter isotope (Figure 4.12). Therefore, the $\delta^{34}S$ values of sulfides (e.g. pyrite) in sedimentary rocks provide insights into the activity of SRB and the conditions under which the sulfides formed. Current available data demonstrate no simple relationship between the degree of fractionation and the cell-specific sulfate reduction (Table 4.5) or sulfur oxidation (Rickard 2012).

Sulfur isotopes can be used to differentiate between biogenic and abiogenic processes. Understanding the fractionation patterns associated with sulfate reduction and subsequent sulfide oxidation by phototrophic or chemolithotrophic microbes enables the monitoring of key processes in the sulfur cycle. Sulfur isotopic analyses have also provided evidence for the absence of biological activity on the Moon. For example, the sulfide isotopic composition of lunar rocks, $\delta^{34}S$ ranging from 0 to +1‰ closely resembles that of meteoritic standard of sulfide (0‰), rather than the isotopic signature typical of biogenic sulfide (Figure 4.12). Additionally, sulfur isotopic fractionation has been used to infer MSR during the Paleoarchean (Shen et al. 2001).

The isotopic composition of sulfur in the various reservoirs of the global sulfur cycle is controlled principally by the large isotope fractionation that accompanies bacterial reduction of sulfate. The lighter ^{32}S isotope forms slightly weaker bonds than the heavier isotope and therefore has a more rapid rate of reduction. The sulfide produced from sulfate reduction is thus depleted in ^{34}S relative to the sulfate being reduced. The large reservoir of ^{34}S-depleted sulfur fixed as pyrite in sediment is balanced by the enrichment of ^{34}S in the residual ocean sulfate. At present, marine sulfate has a $\delta^{34}S$ of +20.1‰ and a reservoir size of 1.3×10^{21} g S; there is also an estimated 3×10^{21} g S of buried evaporite sulfate with an average $\delta^{34}S$ of +16‰. These pools of ^{34}S-rich sulfate can be offset by a sedimentary sulfide reservoir, estimated at 6×10^{21} g S with an average $\delta^{34}S$ of −12‰ (as bulk Earth sulfur has $\delta^{34}S$ of close to the meteoric standard value of 0‰).

4.3.7 Phosphorus Cycle

Phosphorus, an essential nutrient for all life forms, is a critical component of nucleic acids, phospholipids, and ATP. Plants and microorganisms primarily absorb phosphorus as orthophosphate (PO_4^{3-}), assimilating it into biomass, where it is transferred through the food chain. It is also suggested that polyphosphate granules are the main form of inorganic P for storage in marine bacteria and diatoms (Omelon et al. 2013). Upon the death of organisms, decomposers convert organic phosphorus back to inorganic phosphate. Most microorganisms do not alter the valence of the phosphorus atom. This section presents a brief overview of the various components of the global phosphorus cycle (Figure 4.13).

4.3.7.1 Reservoirs

Phosphorus resides in three major terrestrial pools: bedrock and sediments, soil, and living organisms (biomass). Atmospheric P is relatively minor (9×10^8 mol). The majority of phosphorus in the Earth system is in rocks and sediment, constituting the largest P reservoir, which is estimated to range between 0.27 and 1.3×10^{20} mol. Land soils contain dissolved and particulate P (organic and inorganic), with reservoir sizes ranging from 3.1 to 6.45×10^{15} mol. P stored in terrestrial biomass is about 8.39–9.68×10^{13} mol.

4.3.7.2 Flux

Weathering of phosphorus-bearing minerals, primarily apatite [$Ca_{10}(PO_4)_6(OH,F,Cl)_2$], is driven by microbial activity and natural acids, releasing phosphate into soils. This phosphorus sustains terrestrial vegetation, with a weathering flux of approximately 6.45×10^{11} mol P/yr. Atmospheric P deposition is comparatively insignificant. Phosphate sorption by soil constituents, particularly ferric iron and aluminum oxyhydroxides, critically controls terrestrial P bioavailability.

Phosphorus is transferred from the continent to the oceanic reservoir primarily by rivers. Riverine phosphorus derives from weathered continental rocks and soils in the form of dissolved P (3.2–5.8×10^{10} mol/yr) and particulate P (5.9–6.5×10^{11} mol P/yr). Because phosphorus is particle reactive, most riverine-P is associated with particulate matter. By most estimates, well over 90% of the phosphorus delivered by rivers to the ocean is as particulate-P. Available data suggest that ~20%–40% of phosphorus in suspended particulate matter is organic; inorganic forms are partitioned mainly between ferric oxyhydroxides and apatite. Aluminum oxyhydroxides and clays may also be significant carriers of phosphorus. Dissolved phosphorus in rivers occurs in both inorganic and organic forms. In fresh water, dissolved P mainly exists as $H_2PO_4^-$ and is converted to HPO_4^{2-} in oceanic water because of ionic strength and pH. The scant data on dissolved organic P suggest that it may account for 50% or more of dissolved riverine-P. Deposition of atmospheric aerosols (2–5×10^{10} mol/yr) is a minor flux. Groundwater seepage to the coastal ocean is a potentially important but poorly documented flux.

The concentration of dissolved P in the oceanic water column varies with water depth (Figure 4.13), ranging from only a few ppb in surface waters (depleted by biological uptake) to values of 50–100 ppb at depth greater than about 200–400 m. The surface ocean (0–300 m) contains about 8.74×10^{13} mol of total dissolved P, while the deep ocean (300–3300 m) holds about 2.81×10^{15} mol. The oceanic upwelling is a steady supply of phosphate to the surface from the large deep-ocean reservoir, supporting marine primary productivity. Phosphorus in its simplest form, dissolved orthophosphate, is taken up by photosynthetic organisms at the base of the marine food web. The marine phosphorus cycle is linked to the marine carbon and nitrogen cycles through the photosynthetic fixation of these elements by the microscopic floating marine plants, or

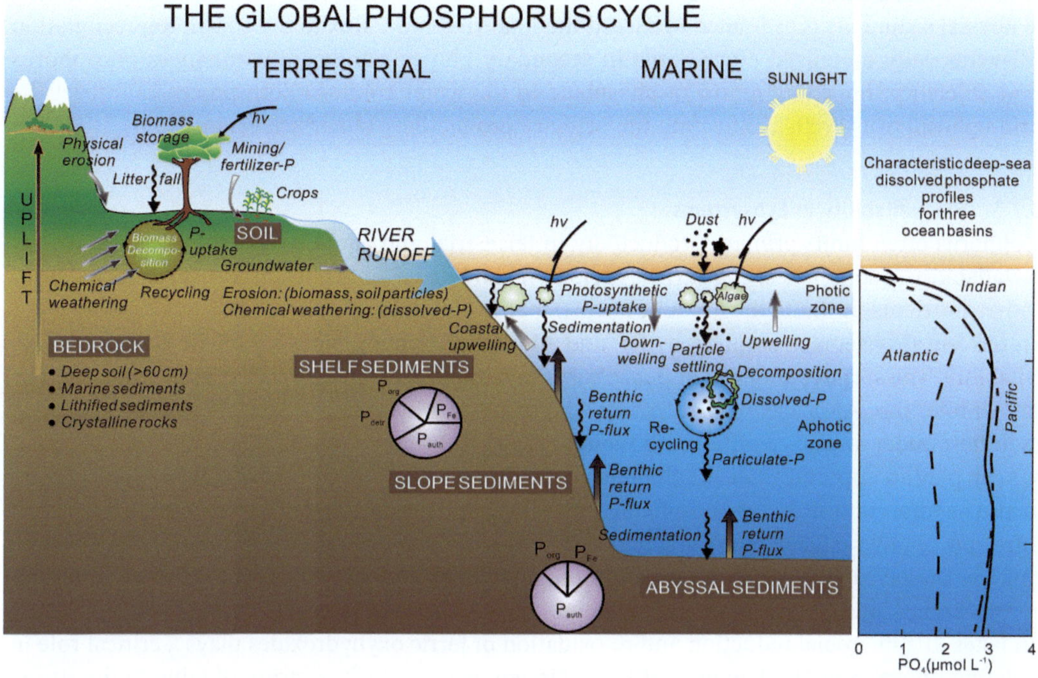

Figure 4.13 The global phosphorus cycle. The oceanic photic zone, idealized in the cartoon, is typically thinner in coastal environments due to turbidity from continental terrigenous input, and deepens as the water column clarifies with distance away from the continental margins. The distribution of phosphorus among different chemical/mineral forms in marine sediments is given in the pie diagrams, where the abbreviations used are: organic phosphorus (P_{org}), iron-bound phosphorus (P_{Fe}), detrital apatite (P_{detr}), and authigenic/biogenic apatite (P_{auth}). The P_{org}, P_{Fe}, and P_{auth} reservoirs represent potentially reactive-P pools, whereas the P_{detr} pool reflects mainly detrital apatite weathered off the continents and passively deposited in marine sediments (note that P_{detr} is not an important sedimentary phosphorus component in abyssal sediments, far from continents). *Source:* Adapted from Ruttenberg (2005).

phytoplankton. It was recognized that marine phytoplanktonic organisms have an average C:N:P molar ratio of 106C:16N:1P. In the open ocean, most phosphorus associated with biogenic particles is recycled within the upper water column.

The critical pathway of phosphorus removal from the oceans is burial with marine sediments. The P-flux to shelf and slope sediments is larger than the P-flux to the deep sea. Terrigenous-dominated shelf and slope (hemi-pelagic) sediments and abyssal (pelagic) sediments have distinct P-distributions. While both are dominated by authigenic P associated with Ca (P_{auth}: mostly authigenic apatite), this reservoir comprises a larger fraction of total phosphorus in pelagic sediments. The remaining phosphorus in hemipelagic sediments is partitioned between ferric iron-bound-P (P_{Fe}: mostly oxyhydroxides), detrital apatite (P_{detr}), and organic-P (P_{org}); in pelagic sediments, detrital apatite is unimportant, most likely due to the limited transport of this heavy mineral beyond the near-shore. Certain coastal environments characterized by extremely high, upwelling-driven biological productivity and low terrigenous input are enriched in authigenic apatite; these are protophosphorite deposits. A unique process contributing to the pelagic sedimentary Fe-P reservoir is the sorptive removal of phosphate onto ferric oxyhydroxides in mid-ocean ridge hydrothermal systems.

Mobilization of sedimentary phosphorus by microbial activity during diagenesis causes dissolved phosphate buildup in sediment pore waters, promoting benthic efflux of phosphate to bottom waters or incorporation in secondary authigenic minerals. The combined benthic flux from coastal and abyssal sediments is estimated to exceed the total riverine-P flux to the ocean. Reprecipitation of diagenetically mobilized phosphorus in secondary phases significantly enhances phosphorus burial efficiency, impeding return of phosphate to the water column. Both processes impact the marine phosphorus cycle by affecting the primary productivity of surface waters.

4.3.7.3 Bioavailability in Ecosystems

Phosphorus often limits primary productivity in terrestrial and aquatic ecosystems due to its low bioavailability in soils, lakes, and surface oceans. When bioavailable phosphorus is exhausted prior to other nutrients, it limits the amount of sustainable biological productivity.

In terrestrial systems, P sorption by Fe- and Al-oxides/oxyhydroxides significantly reduces bioavailability. The sorptive removal of phosphorus can impact the production level of crops and forests. Unlike the Al-oxides/oxyhydroxides, Fe-oxides/oxyhydroxides are subject to reductive dissolution under anoxic conditions, in which Fe-bound P can be released, and thus redox conditions can play an important role in soil P-bioavailability. Iron and Al complexes with organic matter can also be important sinks for phosphate. Mineralogy and morphology of oxyhydroxides also exert control on relative efficiency of P-sorption, e.g. goethite shows higher P sorption capacity than hematite. Dissolved organic phosphorus compounds can also be sorptively removed from soil solution by various solid-phase components, including clays, oxides, and oxyhydroxides.

In lakes, the microbial reduction and re-oxidation of ferric oxyhydroxides plays a critical role in regulating phosphorus (P) dynamics at the sediment–water interface, primarily due to the strong sorption affinity of phosphate to iron mineral surfaces. Under oxic conditions, ferric oxyhydroxides (e.g. ferrihydrite, goethite) retain P via adsorption, limiting its release into the water column. However, when microbes reduce Fe(III) to Fe(II) under anoxic conditions, these minerals dissolve, releasing adsorbed phosphate. It was suggested that orthophosphate concentrations in natural waters may be 2–3 orders of magnitude lower than traditionally assumed (Hudson et al. 2000), approaching levels typical of micronutrients like dissolved iron and zinc.

Due to its extreme particle reactivity, most phosphorus in rivers is associated with particulate matter, primarily through sorption processes. River suspended sediments control dissolved inorganic phosphorus (DIP) levels through sorption equilibrium. This mechanism leads to a constant DIP in rivers, streams, and estuaries through adsorption and desorption of DIP because suspended sediments serve as a substantial reservoir of potentially bioavailable phosphorus.

The role of phosphorus as a limiting nutrient in the oceans remains debated. Geochemical studies traditionally argue that P is the ultimate limiting nutrient on geological timescales because its oceanic supply depends exclusively on continental weathering and riverine inputs. This continental weathering (and its underlying tectonic drivers) creates a fundamental constraint on marine primary productivity over long-term periods.

4.3.7.4 P Supply and Sink

Marine sediments receive phosphorus through multiple sources, including detrital inorganic/organic material, biogenic particles, and atmospheric dust. Refractory P phases (e.g. detrital apatite) are buried, while reactive particulate phases undergo biogeochemical transformations. Key biogeochemical processes include (1) microbial breakdown of particulate P, releasing dissolved inorganic and organic P; (2) phosphate sorption onto iron (oxyhydr)oxides and other mineral surfaces; (3) authigenic mineral formation, where secondary phosphate minerals precipitate *in situ*; and (4) benthic efflux of dissolved P back to overlying waters (Figure 4.14). Iron (oxyhydr)oxides (e.g. ferrihydrite) play a critical role in P sequestration due to their high surface area and affinity for phosphate adsorption/coprecipitation. However, under reducing conditions, these phases may dissolve, potentially releasing P back into the system.

Organic matter undergoes microbial breakdown at the sediment–water interface, transforming it into phosphate, dissolved organic phosphorus, nitrate, and carbon dioxide. These reactions accumulate phosphate in pore waters, with additional phosphate released from Fe-oxyhydroxides surfaces when they are buried into suboxic/anoxic sediment zones. Phosphate in pore water may escape through diffusion, resuspension, or benthic irrigation, or be retained through authigenic mineral formation. The phosphate concentration profile typically shows either: (1) continuous exponential increase with depth which is dominated by organic matter breakdown (km = 0) or (2) constant concentration (C_{eq}) with rapid authigenic phosphate precipitation (km = ∞). In intermediate cases (solid curve profile), the phosphate concentration increases exponentially at the beginning and then decreases, indicating phosphate removal during mineral authigenesis (Figure 4.14).

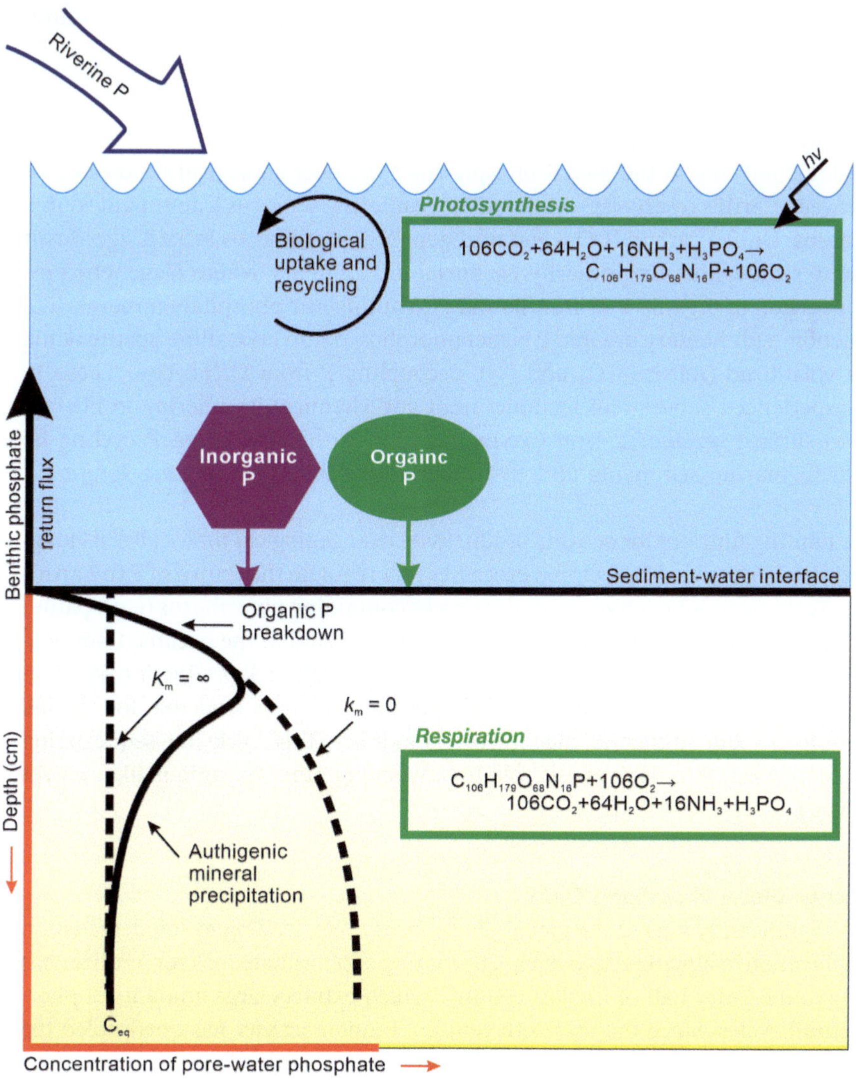

Figure 4.14 Early diagenetic transformations of phosphorus in marine sediments. *Source:* Ruttenberg (2005) / with permission of Elsevier.

Authigenic phosphate minerals in marine sediments are primarily represented by carbonate fluorapatite (CFA) or francolite, commonly found in phosphorite deposits. During diagenesis, these CFA have a tendency to recrystallize to the thermodynamically more stable fluorapatite phosphorites. Vivianite [$Fe_3(PO_4)_2 \cdot 8H_2O$] serves as another significant phosphate sink, particularly in iron-rich muddy marginal sediments. Authigenic aluminophosphate is usually found in marine sandstones and is normally enriched in rare Earth elements and phosphate. Florencite [(REE) $A1_3(PO_4)_2(OH)_6$] is the primary phase alongside crandallite [$CaAl_3(PO_4)_2(OH) \cdot 5H_2O$], gorceixite [$BaAl_3$ $(PO_4)_2(OH) \cdot 5H_2O$], and xenotime (YPO_4).

The retention and release of phosphorus in sediments serve as critical regulators of biological cycles with significant effects on the global carbon cycle. Over short timescales, phosphorus released from sedimentary pore water promotes primary productivity. On geological timescales, sedimentary rocks containing phosphorus can weather upon tectonic uplift, thereby controlling its reintroduction to biogeochemical cycles. This interlinking of early diagenesis (redox-driven P mobilization), marine productivity, and long-term carbon cycling has driven significant research into marine phosphorus biogeochemistry.

4.3.7.5 Tectonic Controls on Global Phosphorus Cycle

Tectonics plays a fundamental role in the global phosphorus cycle over geological timescales by regulating P fluxes between Earth's reservoirs. Continental weathering serves as a dominant source of bioavailable P to oceans. Unlike carbon (CO_2) and nitrogen (N_2), phosphorus lacks a significant gaseous phase. Marine P removal occurs primarily via burial in sediments, where diagenetic processes mediate the conversion of organic-P or iron-bound P to authigenic phosphate minerals like apatite. During subduction, sedimentary organic P is metamorphosed into crystalline apatite while organic C and N are volatilized (release CO_2 and N_2), decoupling P from C/N cycles. These P-containing minerals experienced subsequent tectonic uplift and chemical weathering to liberate bioavailable P back to surface reservoirs. Iron (oxyhydr)oxides further modulate P cycling by scavenging phosphate in marine sediments and hydrothermal systems, influencing long-term P sequestration.

If phosphorus is the limiting nutrient for oceanic productivity over geological timescales, its long-term availability is ultimately governed by tectonic processes that regulate the return of P to Earth's surface environments. Tectonic uplift exposes crystalline rocks containing phosphorus (e.g. apatite) to chemical weathering, releasing bioavailable P that can be transported to the oceans. Over geological timescales, this slow but persistent tectonic cycling of P—through rock production, subduction, uplift, and weathering—controls the total exogenic phosphorus mass available for marine productivity. Iron (oxyhydr)oxide minerals may further modulate this cycle by sequestering P during sedimentary diagenesis (similar to their role in scavenging other oxyanions like molybdate) (Guidry et al. 2000).

4.3.7.6 Human Impact on Global Phosphorus Cycle

Human activities have significantly altered the global phosphorus cycle, leading to soil degradation and widespread eutrophication in aquatic ecosystems. The mining of phosphate rock for fertilizer has increased dramatically in the latter half of the last century, which extracts large amounts of phosphorus, disrupting natural cycles. Since the mid-20th century, human activity has quadrupled the global mobilization of phosphorus. Phosphorous deposits have been extensively mined to produce agricultural fertilizers, establishing a near-irreversible transfer of phosphorus from geological reserves to croplands and ultimately into aquatic ecosystems. This disruption has severely degraded freshwater and coastal marine environments. One of the most visible consequences is the proliferation of oxygen-starved "dead zones" in coastal waters, fueled by nutrient-driven algal blooms. The Gulf of Mexico's dead zone, for instance, has exceeded 17,000 km^2 (Elser and Bennett 2011).

In addition to fertilizer use, deforestation, increased cultivation, urban, and industrial waste disposal all have enhanced phosphorus transport from terrestrial to aquatic systems, leading to elevated phosphorus concentrations in rivers, lakes, and coastal waters. Increased erosion due to deforestation and widespread cultivation has increased riverine suspended matter concentrations and thus has increased the riverine particulate-P flux. These and other human activities eventually cause elevated phosphorus in water bodies. Excessive phosphates stimulate growth in bacteria, algae, and other green plants, which increases organic matter flux to the bottom of aqueous system. Decomposers consume the remains of organisms, leading to the depletion of oxygen in aqueous system. As a result, those species die off, to be replaced by others that are more tolerant of lowered oxygen levels—for example, worms. Phosphorus pollution remains a pressing environmental issue, necessitating sustainable management practices to mitigate its impacts.

4.3.8 Iron Cycle

Iron is the fourth most abundant element in Earth's crust (4.3% by mass), yet it is relatively scarce in modern aquatic systems due to its low solubility under oxic conditions. Despite this, iron is an essential trace element for nearly all living organisms, playing a critical role in proteins, cytochromes, and other Fe-S containing biomolecules. Iron primarily exists in two oxidation states, Fe(II) (ferrous iron) and Fe(III) (ferric iron), with reduction potential of Fe^{3+}/Fe^{2+} couple being +0.77 V and transitions between these states driven by both biotic and abiotic processes. These transformations influence nutrient availability, microbial ecosystems, and environmental redox conditions, making iron one of the key players in global biogeochemical cycles. Iron redox cycling has been a globally significant process throughout Earth's history, as evidenced by the massive deposition of banded iron formations (BIFs) during the Precambrian era. These formations testify to the interplay between iron, oxygen, and microbial life during Earth's early history (Section 6.2.4). Modern iron cycling continues to shape ecosystems and environmental processes, from nutrient dynamics in marine and freshwater systems to the mobilization and immobilization of pollutants.

Microorganisms are central to iron redox cycling, catalyzing both the oxidation of Fe(II) and the reduction of Fe(III), and are of great geological and ecological importance (Sections 1.3.4 and 1.3.5). Iron-oxidizing bacteria, such as *Acidithiobacillus ferrooxidans*, obtain energy by oxidizing ferrous iron into ferric iron and sulfide to sulfate, which contributes to acid mine drainage (AMD) in coal-mining regions. These chemolithotrophs oxidize Fe(II) to Fe(III), generating energy while producing ferric hydroxide deposits. *Gallionella ferruginea*, another iron-oxidizing bacterium, oxidizes ferrous iron to the ferric iron and coats its sheath with insoluble ferric hydroxide. These microbial activities are responsible for the formation of bog-iron deposits in anaerobic swamps, which were historically mined during the early Industrial Revolution. Iron-reducing bacteria, such as *Shewanella* and *Geobacter* species, are active in anoxic environments. They reduce Fe(III) to Fe(II), mobilizing iron from sediments and soils. This process is critical in environments like swamps, wetlands, and iron-rich aquatic habitats, where it influences nutrient cycling and carbon degradation.

4.3.8.1 Flux

In the Earth's crust, iron is primarily found in silicate minerals such as olivine, pyroxenes, and amphiboles. Under oxic conditions and at near neutral pH, iron released during the weathering reprecipitates rapidly as oxides and hydroxides, as expressed here by the oxidative weathering of fayalite (Fe_2SiO_4) to goethite (FeOOH) by reaction (Canfield et al. 2005):

$$Fe_2SiO_4 + 0.5O_2 + 3H_2O \rightarrow 2FeOOH + H_4SiO_4$$

Most of the iron transported to the oceans is associated with river-borne particulates, while glaciers and wind-borne dust represent minor fluxes (Figure 4.15). Approximately 90% of the riverine inputs deposit on the continental shelves, with atmospheric dust providing a critical source of iron for primary production in offshore regions. Atmospheric dust deposition is relatively important in alleviating iron limitation in high-nutrient, low-chlorophyll (HNLC) regions of the ocean (Fennel et al. 2003). This process enhances phytoplankton growth, which in turn drives carbon sequestration and influences global climate dynamics. Hydrothermal fluids also deliver small amounts of soluble iron to ocean basins, where it is oxidized microbially or abiotically, forming precipitates around vents. Iron also interacts with methane cycling in anoxic environments. In methanogenic sediments, iron-reducing bacteria compete with methanogens for electron donors, influencing methane production and greenhouse gas emissions. Recent discoveries of novel electron carriers, such as methanophenazines, have shed light on the competitive dynamics between these microbial processes.

Iron undergoes dynamic redox cycling at the boundary between oxic and anoxic environments, particularly in sediments. Once buried in anoxic sediments, iron can be mobilized through

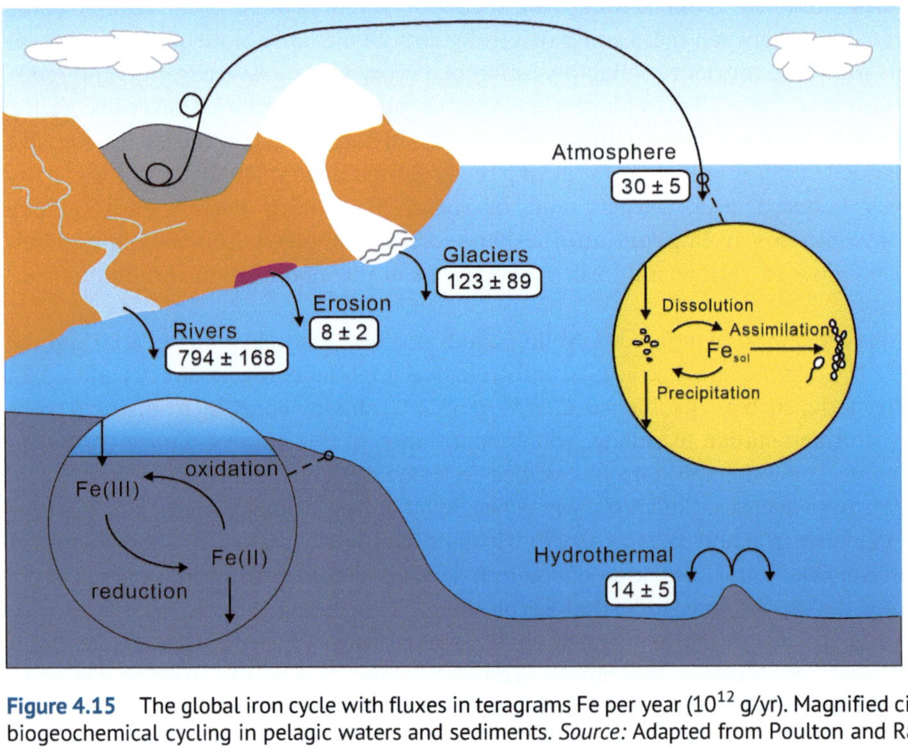

Figure 4.15 The global iron cycle with fluxes in teragrams Fe per year (10^{12} g/yr). Magnified circles illustrate biogeochemical cycling in pelagic waters and sediments. *Source:* Adapted from Poulton and Raiswell (2002).

microbial respiration and abiotic reduction processes. However, soluble iron (Fe^{2+}) is efficiently trapped by oxidation in the thin oxic surface layers of sediments, typically only a few millimeters thick. As a result, only a small fraction of mobilized iron escapes into the overlying bottom waters, leading to intensive redox cycling of iron at the oxic–anoxic interface. This process occurs in a variety of sedimentary environments, from coastal zones to deep-sea sediments.

While benthic iron fluxes are relatively small compared to the rates of iron cycling within sediments, they play a critical role in supplying iron to shelf waters, where it can support primary production. In offshore regions, however, dissolved iron concentrations are extremely low, often below 1 nM. In these iron-limited environments, organisms in surface waters rely on scavenging trace metals from atmospheric particles that settle into the ocean. The solubilization of iron from these particles is facilitated by organic chelators produced by marine plankton, such as siderophores, which bind iron and enhance its bioavailability.

4.3.8.2 Key Processes

The bacterial reduction of Fe(III) to Fe(II) is a key mechanism for iron solubilization in nature. This process is prevalent in waterlogged soils, bogs, and anoxic lake sediments, where anaerobic conditions favor microbial iron reduction. When iron-rich groundwater from these environments flows into oxic regions, ferrous iron is rapidly oxidized to ferric iron, leading to the precipitation of insoluble ferric hydroxides and the formation of characteristic brown deposits. Although ferric hydroxides are highly insoluble, some ferric iron can remain in solution in natural waters through complexation with organic ligands. The oxidation of these organic compounds by microorganisms causes iron precipitation.

At neutral pH, bacteria such as *Gallionella* and *Leptothrix* drive the oxidation of ferrous iron, likely through energy-yielding reactions. These microorganisms are often found in environments where oxygen is present, such as freshwater streams and lake sediments. Additionally, anoxygenic phototrophic bacteria, which thrive in anoxic but light-exposed environments, can oxidize Fe(II) to Fe(III) as part of their photosynthetic metabolism (Sections 1.3.2 and 1.3.4). In acidic environments, the chemical oxidation of ferrous iron is slow, but acidophilic chemolithotrophs like *A. ferrooxidans* catalyze this process efficiently. This bacterium oxidizes ferrous iron as its primary energy source, and because the energy yield per iron atom oxidized is low, *A. ferrooxidans* must oxidize large quantities of iron to sustain growth. Consequently, even a small population of these bacteria can precipitate significant amounts of iron. *A. ferrooxidans* thrives in sulfuric acid-dominated environments, where ferric iron does not precipitate as hydroxide but instead forms complex sulfate minerals such as jarosite [$HFe_3(SO_4)_2(OH)_6$].

Pyrite (FeS_2), a common mineral found in bituminous coals and various geological deposits, forms through the reaction of sulfur with ferrous sulfide (FeS), resulting in a highly insoluble crystalline structure (Section 6.2.4). The oxidation of pyrite is a complex process involving both chemical and microbial catalysis, with molecular oxygen (O_2) and ferric ions (Fe^{3+}) serving as the primary electron acceptors. Ferric ions, however, are only stable in highly acidic conditions ($pH < 2.5$), as they precipitate as ferric hydroxide ($Fe(OH)_3$) at higher pH values. When pyrite is exposed to oxygen, such as during mining operations, a slow chemical reaction occurs:

$$FeS_2 + 3.5O_2 + H_2O \rightarrow Fe^{2+} + 2SO_4^{2-} + 2H^+$$

This reaction generates ferrous iron (Fe^{2+}), sulfate ions (SO_4^{2-}), and protons (H^+), leading to the acidification of the environment. Under these acidic conditions, ferrous iron remains stable in the presence of oxygen. Then, *A. ferrooxidans* plays a critical role in pyrite oxidation by catalyzing the oxidation of ferrous iron to ferric iron:

$$Fe^{2+} + 0.25O_2 + H^+ \rightarrow Fe^{3+} + 0.5H_2O$$

The ferric ions produced under acidic conditions are soluble and can further oxidize pyrite through the following reaction:

$$FeS_2 + 14Fe^{3+} + 8H_2O \rightarrow 15Fe^{2+} + 2SO_4{}^{2-} + 16H^+$$

This creates a self-sustaining cycle, where ferric irons repeatedly oxidize pyrite, generating more ferrous iron, which is then re-oxidized by the bacteria. This cycle leads to a rapidly accelerating rate of pyrite oxidation and the associated acidification and sulfate generation. In natural systems, some of the ferrous iron generated by bacterial activity is transported by groundwater into surrounding streams. In aerobic environments, bacterial oxidation of ferrous iron occurs, leading to the precipitation of insoluble ferric hydroxides. This process contributes to the formation of iron-rich deposits in streams and the acidification of water bodies, a phenomenon commonly known as AMD.

Bacterial oxidation of sulfide minerals, particularly pyrite (FeS_2), is the primary driver of AMD, a pervasive environmental issue in coal-mining regions. This process involves the action of *A. ferrooxidans*, following the steps outlined above. Mixing of acidic mine waters with natural waters in rivers and lakes causes serious degradation in the quality of the natural water because both the acid and the dissolved metals are toxic to aquatic life. In addition, such polluted waters are unsuitable for human consumption and industrial use. Bioremediation strategies can be applied to AMD, such as using iron-reducing bacteria to immobilize metals and neutralize acidity.

The bacterial oxidation of pyrite is a self-sustaining sequence of chemical reactions (propagation cycle) and operates by multiple processes in steps. (1) Initial oxidation by oxygen, in which pyrite reacts with O_2 and H_2O to produce Fe^{2+}, $SO_4{}^{2-}$, and H^+, and contributes to acidity. (2) Fe^{2+} is oxidized by O_2 to Fe^{3+}, which at acid pH is a rate-limiting step but can be accelerated by bacteria. (3) Fe^{3+} as a secondary oxidant, then reacts with more pyrite, oxidizing it much faster than oxygen alone, which produces more Fe^{2+}, more sulfate, and more acidity. (4) Acidophilic bacteria like *A. ferrooxidans* can boost the oxidation rate of Fe^{2+} to Fe^{3+} by up to a million times, which keeps the cycle going even in low-oxygen environments. (5) Fe^{3+} precipitate as $Fe(OH)_3$ or jarosite, releasing additional H^+ ions; these precipitates can form crusts or coatings on rock surfaces.

The breakdown of pyrite leads ultimately to the formation of sulfuric acid and ferrous iron, and pH values can be as low as pH 2. The acid formed attacks other minerals in the rock, causing the breakdown of the whole rock fabric. Aluminum ions are commonly found at low pH, and often several grams of $A1^{3+}$ can pose a severe threat to freshwater ecosystems.

4.4 Major Features of Biogeochemical Cycles

Biogeochemical cycles are fundamental processes that govern the movement and transformation of elements through Earth's spheres-atmosphere, hydrosphere, biosphere, and lithosphere. These cycles can be analyzed based on their pathways, rates of cycling, and the extent to which human activities disrupt them.

4.4.1 Diversity of Pathways

Each biogeochemical cycle involves a multitude of pathways, driven by various chemical, physical, and biological processes. For instance, the carbon cycle transports carbon through all four spheres. From the atmosphere, CO_2 dissolves in water in the hydrosphere, where photoplanktons assimilate it as HCO_3^- to synthesize organic molecules (CH_2O molecule), transferring carbon into the biosphere. When these organisms die and are buried, their organic matter transforms into fossil fuels like oil, storing carbon in the lithosphere. Upon combustion, fossil fuels release CO_2 back into the atmosphere, where it may enter alternative pathways, such as being absorbed by terrestrial vegetation instead of marine organisms (Section 4.3.3). Unlike carbon, elements such as phosphorus and iron rarely cycle through the atmosphere because they do not readily form gaseous compounds. Each element follows distinct pathways based on its chemical properties and interactions with living organisms and the environment. At the same time, the same element can move through multiple interconnected cycles. This interconnectedness is crucial because changes in one cycle can impact others.

4.4.2 Variable Rates of Cycling

Biogeochemical cycles vary significantly in their rates of cycling, ranging from a few 100 years to tens of millions of years. Figure 4.16 illustrates the average cycling times of water, oxygen, and carbon dioxide across Earth's four spheres, though actual rates vary significantly. In the water cycle, rapid processes like evaporation, precipitation, and river flow occur over days to years, whereas slower processes—including deep groundwater movement, glacier dynamics, and ocean circulation—span centuries to millennia. Over geological timescales (millions of years), plate

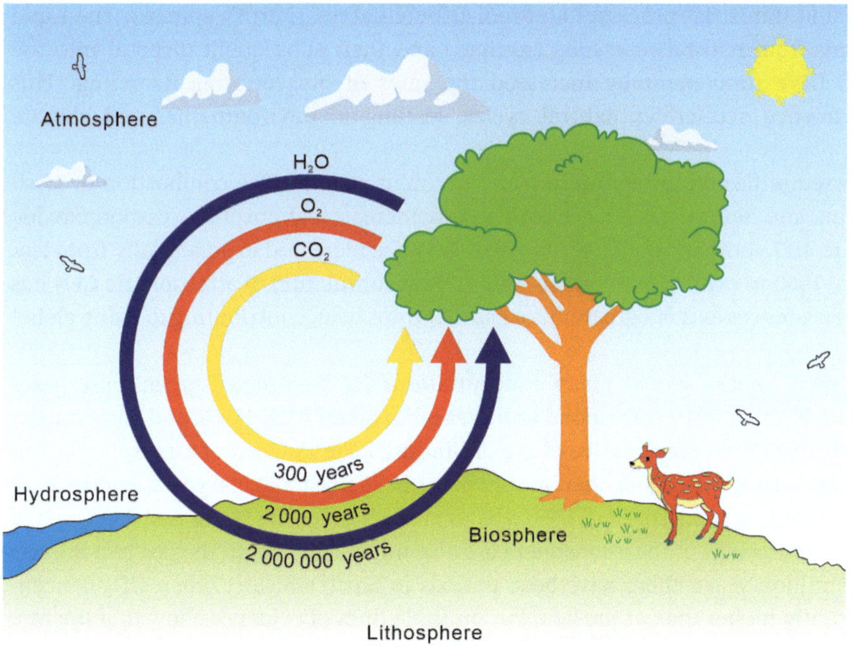

Figure 4.16 Recycling rates of water, oxygen, and carbon dioxide through the four spheres.

tectonics and volcanic outgassing further shape the global water cycle. Similarly, the carbon cycle operates at different speeds depending on the process. Rapid exchanges such as photosynthesis, respiration, ocean surface interactions, and organic matter decomposition occur within hours to centuries, while slower processes—such as sedimentation, volcanic activity, and rock weathering—take place over millions of years.

Two key factors determine the rate of biogeochemical cycling: (1) the chemical reactivity of the substance and (2) whether it has a gas phase (occurs in the atmosphere) in the cycle. Carbon can participate in numerous chemical pathways, leading to faster cycling. Carbon's versatility in forming organic and inorganic compounds accelerates its movement through the biosphere, hydrosphere, and atmosphere. Substances that exist as gases, like carbon dioxide (CO_2), oxygen (O_2), and water vapor, cycle more rapidly because gas molecules diffuse and transport more quickly than those in liquids or solids.

Phosphorus exemplifies a substance with an extremely slow cycling rate due to its absence of a gas phase and low chemical reactivity. Aqueous transportation is the primary pathway for phosphorus, even though it is mostly insoluble. Large quantities of phosphorus become sequestered in ocean sediments and Earth's crust, remaining locked away for tens of millions of years. Upwelling ocean currents or the weathering of phosphorus-rich rocks, are primary pathways to recycle this nutrient back into active biogeochemical cycles. This slow cycling rate severely limits phosphorus availability, making it the limiting nutrient in most ecosystems. Its scarcity has a profound impact on life on Earth, as phosphorus is essential for DNA, RNA, ATP, and other critical biological molecules.

4.4.3 The Effects of Human Activity

Biogeochemical cycles, essential to sustaining life on Earth, are increasingly disrupted by human activities. With the development of civilization, vast quantities of materials are extracted through mining, agriculture, and industrial processes and redistributed across Earth's spheres. The rapid extraction of materials from natural reservoirs (sources) and their subsequent disposal into the environment (sinks) have fundamentally increased the rates of biogeochemical cycling. This anthropogenic interference accelerates natural cycles, leading to environmental and climate challenges.

The carbon cycle exemplifies the profound impact of human activity. The combustion of fossil fuels—coal, petroleum, and natural gas—has exponentially increased atmospheric carbon dioxide concentrations. Figure 4.17 shows how global release of CO_2 has increased exponentially from less than 10 Gt s/yr before 1960 to over 35 Gt/yr since 2000. The accumulation of atmospheric CO_2 has significant global consequences as it is a greenhouse gas that traps heat, contributing to rising global temperatures.

According to the IPCC report, global mean concentrations for well-mixed greenhouse gases (GHGs: CO_2, CH_4, and N_2O) in 2019 correspond to increases of about 47%, 156%, and 23%, respectively, above the levels in 1750 (representative of the pre-industrial era). Current atmospheric concentrations of the three GHGs are higher than at any point in the last 800,000 years, and in 2019 reached 409.9 parts per million (ppm) of CO_2, 1866.3 parts per billion (ppb) of CH_4, and 332.1 ppb of N_2O. Current CO_2 concentrations in the atmosphere are also unprecedented in the last two million years. In the past 60 million years, there have been periods in Earth's history when CO_2 concentrations were significantly higher than at present, but multiple lines of evidence show that the rate at which CO_2 has increased in the atmosphere during 1900–2019 is at least 10 times faster than at any other time during the last 800,000 years, and 4–5 times faster than during the last 56 million years.

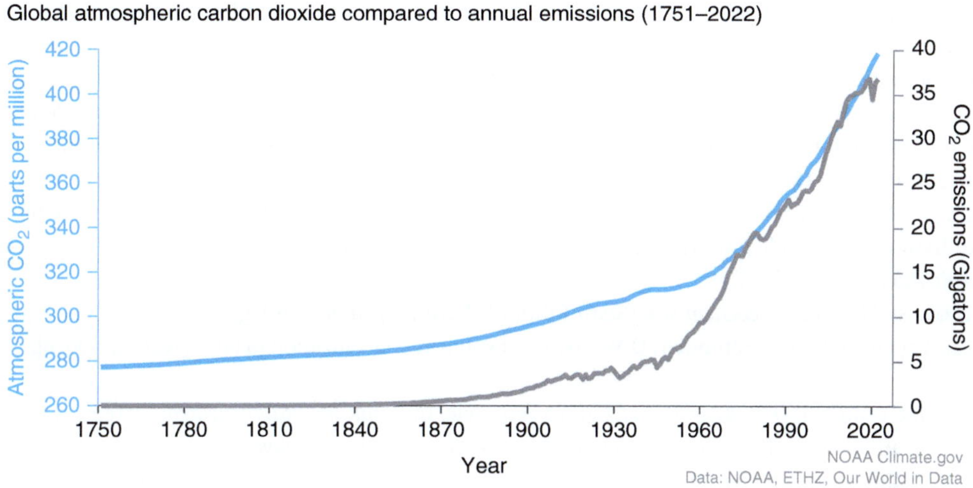

Figure 4.17 The amount of CO_2 in the atmosphere (blue line) has increased along with human emissions (gray line) since the start of the Industrial Revolution in 1750. *Source:* Adapted from NOAA Climate.gov graph.

Human activities are increasingly disrupting multiple biogeochemical cycles beyond the carbon cycle, with severe environmental consequences. Excess phosphorus from agricultural runoff leads to eutrophication, causing harmful algal blooms and oxygen depletion in aquatic ecosystems. Similarly, anthropogenic nitrogen emissions contribute to photochemical smog formation, while sulfur emissions from fossil fuel combustion result in acid rain that damages soils and freshwater systems. These disturbances demonstrate how human-driven perturbations to natural element cycles can have cascading ecological impacts.

References

Borisov, A.Y., and Björn, L.O. 2018. On oxygen production by photosynthesis: a viewpoint. *Photosynthetica*, 56, pp. 44–47.

Canfield, D.E. 2004. The evolution of the Earth surface sulfur reservoir. *American Journal of Science*, 304, pp. 839–861.

Canfield, D.E., Kristensen, E., and Thamdrup, B. 2005. *Aquatic geomicrobiology*, pp. 640. Amsterdam, Elsevier.

Detmers, J., Brüchert, V., Habicht, K.S., and Kuever, J. 2001. Diversity of sulfur isotope fractionations by sulfate-reducing prokaryotes. *Applied and Environmental Microbiology*, 67, pp. 888–894.

Elser, J.J., and Bennett, E. 2011. A broken biogeochemical cycle. *Nature*, 478, pp. 29–31.

Fennel, K., Abbott, M.R., Spitz, Y.H., Richman, J.G., and Nelson, D.M. 2003. Impacts of iron control on phytoplankton production in the modern and glacial Southern Ocean. *Deep Sea Research Part II: Topical Studies in Oceanography*, 50, pp. 833–851.

Galloway, J.N. 2005. The global nitrogen cycle. In: *Biogeochemistry. treatise on geochemistry* (Vol. 8) (ed. W.H. Schlesinger), pp. 557–583. Oxford, Elsevier-Pergamen.

Guidry, M.W., Mackenzie, F.T., and Arvidson, R.S. 2000. Role of tectonics in phosphorus distribution and cycling. In: *Marine authigenesis: from global to microbial* (Vol. 66) (eds. C.R. Glenn, L. Prevot-Lucas, and J. Lucas), pp. 35–51. SEPM Special Publication.

Hammerli, J., Greber, N.D., Martin, L., Bouvier, A.-S., Kemp, A.I.S., Fiorentini, M.L., Spangenberg, J.E., Ueno, Y., and Schaltegger, U. 2021. Tracing sulfur sources in the crust via SIMS measurements of sulfur isotopes in apatite. *Chemical Geology*, 579, pp. 120242.

Hayes, J.M., Strauss, H., and Kaufman, A.J. 1999. The abundance of 13C in marine organic matter and isotopic fractionation in the global biogeochemical cycle of carbon during the past 800 Myr. *Chemical Geology*, 161, pp. 103–125.

He, H., Wu, X., Xian, H., Zhu, J., Yang, Y., Lü, Y., Li, Y., and Konhauser, K. 2021. An abiotic source of Archean hydrogen peroxide and oxygen that pre-dates oxygenic photosynthesis. *Nature Communications*, 12, pp. 6611.

Hoefs, J. 2009. *Stable isotope geochemistry* (Sixth Edition). Berlin, Springer Verlag.

Hudson, J.J., Taylor, W.D., and Schindler, D.W. 2000. Phosphorus concentration in lakes. *Nature*, 406, pp. 54–56.

Kasting, J.F., and Canfield, D.E. 2012. The global oxygen cycle. In: *Fundamentals of geobiology* (eds. A.H. Knoll, D.E. Canfield, and K.O. Konhauser), pp. 94–104. West Sussex, Blackwell.

Keeling, R.F., Najjar, R.P., Bender, M.L., and Tans, P.P. 1993. What atmospheric oxygen measurements can tell us about the global carbon cycle? *Global Biogeochemical Cycles*, 12, pp. 361–372.

Kelemen, P.B., and Manning, C.E. 2015. Reevaluating carbon fluxes in subduction zones, what goes down, mostly comes up. *Proceedings of the National Academy of Sciences of the United States of America*, 112, pp. E3997–E4006.

Kernan, M., Battarbee, R.W., and Moss, B.R. 2011. *Climate change impacts on freshwater ecosystems*. John Wiley & Sons.

Li, L., Zheng, Y.F., Cartigny, P., and Li, J. 2014. Anomalous nitrogen isotopes in ultrahigh-pressure metamorphic rocks from the Sulu orogenic belt: effect of abiotic nitrogen reduction during fluid–rock interaction. *Earth and Planetary Science Letters*, 403, pp. 67–78.

Loeb, N.G., Johnson, G.C., Thorsen, T.J., Lyman, J.M., Rose, F.G., and Kato, S. 2021. Satellite and ocean data reveal marked increase in Earth's heating rate. *Geophysical Research Letters*, 48, pp. e2021GL093047.

Lovelock, J., and Margulis, L. 1974. Atmospheric homeostasis by and for the biosphere: the Gaia hypothesis. *Telus*, 26, pp. 2–10.

Matsu'ura, F., Sawaki, Y., Komiya, T., Han, J., Maruyama, S., Ushikubo, T., Shimizu, K., and Ueno, Y. 2024. Oceanic and sedimentary microbial sulfur cycling controlled by local organic matter flux during the Ediacaran Shuram Excursion in the Three Gorges Area, South China. *Geobiology*, 22, pp. e12617.

Nester, E.W., Anderson, D.G., Roberts, C.E., Pearsall, N.N., and Nester, M.T. 2004. *Microbiology-A human perspective* (Fourth Edition). New York, McGraw-Hill.

Newton, R., and Bottrell, S. 2007. Stable isotopes of carbon and sulphur as indicators of environmental change: past and present. *Journal of the Geological Society of London*, 164, pp. 691–708.

Omelon, S., Ariganello, M., Bonucci, E., Grynpas, M., and Nanci, A. 2013. A review of phosphate mineral nucleation in biology and geobiology. *Calcified Tissue International*, 93, pp. 382–396.

Pellerin, A., Antler, G., Holm, S.A., Findlay, A.J., Crockford, P.W., Turchyn, A.V., Jørgensen, B.B., and Finster, K. 2019. Large sulfur isotope fractionation by bacterial sulfide oxidation. *Science Advances*, 5, pp. eaaw1480.

Petsch, S.T. 2005. The global oxygen cycle. In: *Biogeochemistry: treatise on geochemistry* (Vol. 8) (ed. W.H. Schlesinger), pp. 515–555. Oxford, Elsevier-Pergamen.

Post, W.M., Peng, T.H., Emanuel, W.R., King, A.W., Dale, V.H., and DeAngelis, D.L. 1990. The global carbon cycle. *American Scientist*, 78, pp. 310–326.

Poulton, S.W., and Raiswell, R. 2002. The low-temperature geochemical cycle of iron: from continental fluxes to marine sediment deposition. *American Journal of Science*, 302, pp. 774–805.

Qiu, W.J., Zhou, M.F., Li, X., Huang, F., and Malpas, J. 2021. Constraints of Fe-S-C stable isotopes on hydrothermal and microbial activities during formation of sediment-hosted stratiform sulfide deposits. *Geochimica et Cosmochimica Acta*, 313, pp. 195–213.

Rickard, D. 2012. Sulfidic sediments and sedimentary rocks. *Developments in Sedimentology*, 65, pp. 1–801.

Rosing, M.T. 1999. ^{13}C-depleted carbon microparticles in > 3700 Ma sea-floor sedimentary rocks from West Greenland. *Science*, 283, pp. 674–670.

Ruttenberg, K.C. 2005. The global phosphorus cycle. In: *Biogeochemistry: treatise on geochemistry* (Vol. 8) (ed. W.H. Schlesinger), pp. 585–643. Oxford, Elsevier-Pergamen.

Schidlowski, M. 2001. Carbon isotopes and microbial sediments. In: *Microbial sediments* (eds. R.E. Riding and S.M. Awramik), pp. 84–95. Berlin, Springer.

Shen, Y.A., Buick, R., and Canfield, D.E. 2001. Isotopic evidence for microbial sulfate reduction in the early Archaean era. *Nature*, 410, pp. 77–81.

Skinner, B.J., and Murck, B.W. 2011. *The blue planet: an introduction to Earth system science*, pp. 672. Hoboken, John Wiley & Sons, Inc.

Soukharev, B. 2025. How positive climate feedbacks amplify global warming. In: *Global warming and mass migration* (ed. B. Soukharev), pp. 49–112. Cham, Springer.

Stramma, L., Schmidtko, S., Levin, L.A., and Johnson, G.C. 2010. Ocean oxygen minima expansions and their biological impacts. *Deep Sea Research Part I: Oceanographic Research Papers*, 57, pp. 587–595.

Stubbs, A.R., Paulo, C., Power, I.M., Wang, B., Zeyen, N., and Wilson, S.A. 2022. Direct measurement of CO_2 drawdown in mine wastes and rock powders: implications for enhanced rock weathering. *International Journal of Greenhouse Gas Control*, 113, pp. 103554.

Sweeney, R.E., Liu, K.K., and Kaplan, I.R. 1978. Oceanic nitrogen isotopes and their use in determining the source of sedimentary nitrogen. In: *Stable isotopes in the earth sciences* (Vol. 220) (ed. B.W. Robinson), pp. 9–26. DSIR Bull.

5

Biomineralization and Its Origin

Luoyang Li and Xingliang Zhang

5.1 Introduction

The processes by which organisms form solid inorganic substances are referred to as biomineralization. Almost all living organisms can influence the precipitation of mineral materials, spanning from simple bacteria to complex metazoans, whether alive or dead, by direct or indirect pathways, in crystalline or amorphous forms, and either inside or outside their bodies. Biomineralizers such as magnetotactic bacteria (MTB) deposit iron oxides within their cells, molluscs form protective armor shells, and mammals produce apatite crystals in their bones and teeth (Lowenstam and Weiner 1989). Over billions of years of evolution, Earth's life has adapted to produce and utilize an incredible diversity of minerals through a wide range of pathways, but they follow the general principles of biomineralization. The critical roles of biological activities in mineral formation distinguish biomineralization from pure chemical/geological mineralization. Biomineralization is one of Nature's best designs, and has constantly provided us inspiration for creating novel materials that are tougher, stronger, more sustainable, and environmentally friendly. Biomineralization stands out as a typical interdisciplinary research field. In particular, with the development of biogeochemistry and the urgency of understanding the evolution of the Earth-life system as a whole, biomineralization is moving to the forefront of various fields within Earth sciences, including geobiology and paleontology.

In nature, we observe a vast majority of biomaterials (see Figure 5.1), and the study of these internal (endo-) and external (exo-) skeletons has a long history (Dauphin 2022; Weiner and Dove 2003). Historically, the first book on biomineralization was published in 1924 in German by W.J. Schmidt (1924), and the subject has continued to intrigue a dedicated community of scientists for many years. Until the early 1970s, the field remained largely known as "calcification," reflecting the predominance of biogenic calcium-containing minerals, mainly calcium carbonate and calcium phosphate groups. As more and more biogenic minerals were discovered that contained other cations, the field became known as "biomineralization," a relatively recent term. During the following several decades, great advances in analytical methods and cutting-edge techniques have significantly expanded our understanding of biomineralization. Since the influential work of Lowenstam and Weiner (1989), an invaluable knowledge base has been established through extensive literature and publications, and model species for biomineralization studies are currently being developed, for example, the pearl oyster *Pinctada fucata* (Zhang et al. 2018). Growing interests and extensive

Introduction to Geobiology, First Edition. Xingliang Zhang, Zisheng Guo, Wei Liu, Ruliang He, Luoyang Li, and Weiduo Hao.

Companion website: www.wiley.com/go/Zhang_1e

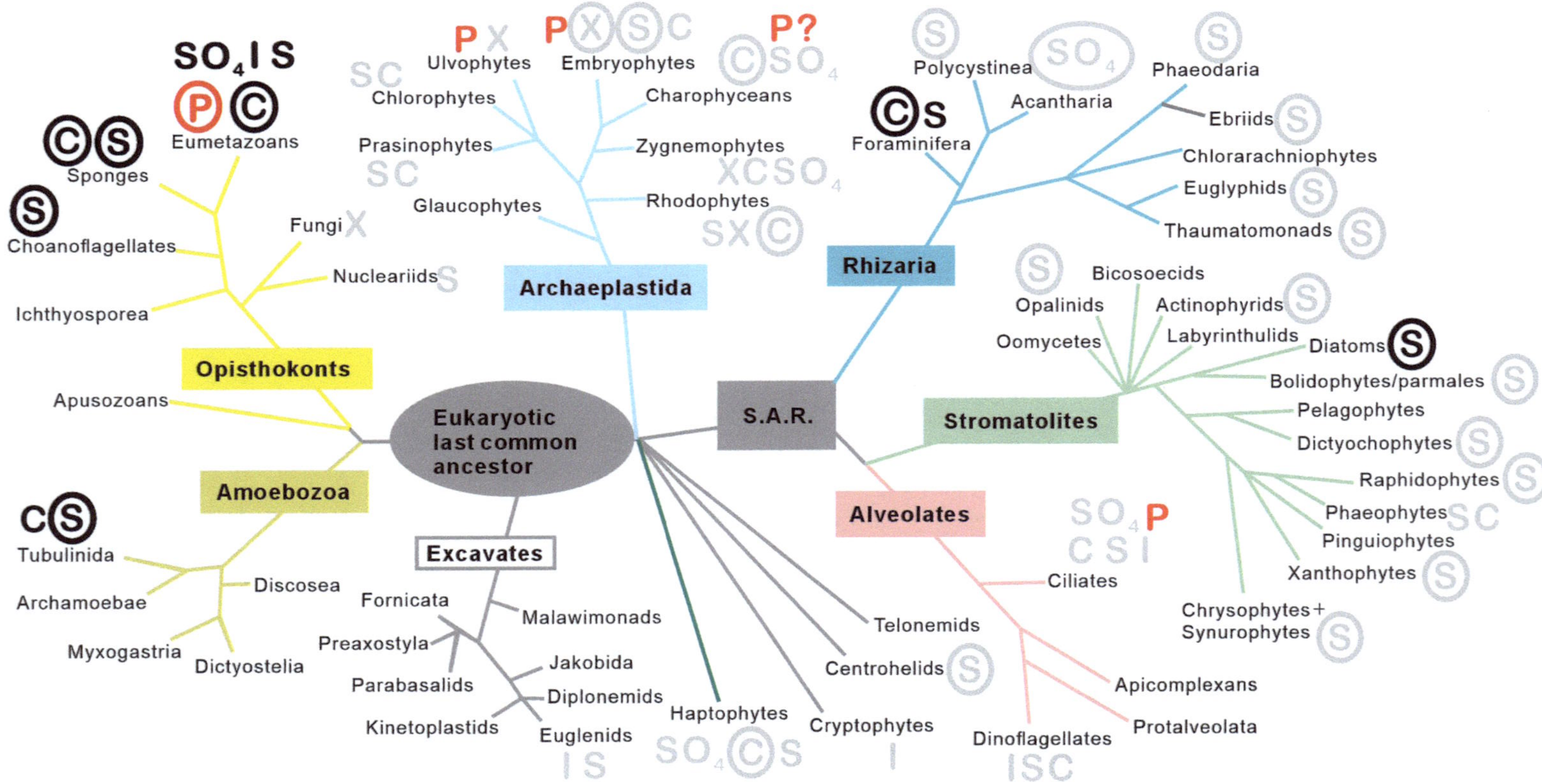

Figure 5.1 The diversity of biomineralization across the tree of life. Letters next to taxon names denote the presence of biomineralization, with circled letters indicating prominent biomineral types. S, biosilica; C, calcium carbonate; P, calcium phosphate; I, iron (magnetite); SO_4, sulfates (calcium/barium/strontium); X, calcium oxalate. *Source:* Derived from Hendry et al. (2018) / Frontiers Media S.A / CC BY 4.0.

surveys on biomaterials cover various aspects of biomineralization, providing outstanding opportunities for specialists and non-specialists of various disciplines to explore the diversity and fundamental principles of biomineralization, evolutionary history, and practical implications.

This chapter, from a geobiological and paleontological perspective, gathers the current state of knowledge on biomineralization with emphasis on the diversity of biominerals/organominerals, classification, and basic principles of biomineralization. A mechanistic understanding of biomineralization is essential for deciphering the underlying evolutionary mechanism with respect to the origin and early evolution of biominerals through Earth's history. This chapter also uses molluscs as a model system to help in-depth understand the rise of animal skeletonization during the Ediacaran-Cambrian geological period, over half a billion years ago. The last section of this chapter focuses on the taphonomy and preservation of primary biomineral structures in ancient fossils and rocks, and their critical role in reconstructing the past of the Earth-life system.

5.2 Biominerals and Organominerals

5.2.1 Concept and Unique Characteristics

Biomineral refers to a type of mineral produced by organisms through their physiological processes. Biominerals meet the criteria for being true minerals, but they also possess other properties that are markedly different from their abiogenic equivalents formed purely through geological/chemical processes. With the involvement of biological activities in the processes of mineralization, biominerals exhibit unique characteristics in terms of morphology, crystallography, isotopic and trace element compositions, micro-nanoscale organizations, etc.

Firstly, biominerals have unusual external morphologies. For example, coccolithophores have an astonishing ability to guide the arrangement of crystal building blocks into a coccolith with incredible hierarchical complexity that is by far still an impossible task for humankind to replicate. A Cambrian extinct epitheliozoan called chancelloriid possessed a rosette-like aragonitic sclerite on their bodies, giving it a cactus-like appearance (see Figure 5.2). The complex molecular and genetic mechanisms of biomineralizers for constructing unusual external morphologies are far from fully understood. Remarkably, genetic regulatory network runs the controlled-mineralization "program" in skeletogenesis with great fidelity again and again, and from generations to generations (Clark 2020; Weiner and Dove 2003).

Secondly, biominerals are organic-mineral biocomposites (Weiner and Dove 2003). Mineral parts often exist as small discrete bodies within a complex framework of macromolecules assembled with collagen or chitin (Addadi et al. 2003). From a material science perspective, these organic macromolecules are strong, flexible, and fracture-resistant, while mineral phases are hard and brittle. Biomineral incorporates the best of these mechanical properties and minimizes the weaknesses significantly. The nacre (mother of pearl) produced by molluscs is about 3000 times more fracture-resistant than that of monolithic calcium carbonate. They are both hard and strong, the best design of Nature!

Thirdly, the distinctiveness is also observable in their crystallography, as the crystalline biominerals may behave as a single crystal with coherent crystallographic orientation regardless of their external morphological constraints. The single-crystal nature of biominerals is evident in echinoderm ossicles, sea urchin spines, molluscan nacreous tablets, etc. But this may not always be the case when observed at higher magnifications on the micro-nanometer scale. Some "single-crystal" biominerals are actually composed of a mosaic of nanogranules (ca. 50–200 nm) enclosed within a

Figure 5.2 A Cambrian chancelloriid sclerite with unique rosette-like construction. *Source:* Photo by Luoyang Li.

thin layer of organic cortex. These nanogranules, considered as the fourth characteristic of biominerals, contain significant amount of organic contents. Assembling of these nanogranules is a key process forming biominerals.

Fifthly, with considerable biological controls, the assembly of crystal building units further forms highly organized and diverse macro-microstructures with complex hierarchical organizations. This is manifested in the crossed-blade microstructure of shell plates of some chitons (Mollusca: Polyplacophora) (see Figure 5.3).

Figure 5.3 Complex hierarchical organization of calcium carbonate biominerals in the shell plate of chitons, molluscs. *Source:* Photos by Luoyang Li.

5.2.2 Major Groups of Biominerals

At the time when life emerged, over ca. 4.0 Ga ago, the simplest life forms (i.e. unicellular prokaryotes) might have acquired the ability to generate various types of biominerals in the reaction with available mineral ions within the early ocean. By the terminal of the Neoproterozoic Era (i.e. Ediacaran Period) to early Cambrian, a disparate group of early metazoans were able to make their skeletons (exoskeletons and endoskeletons) that are certainly composed of biominerals. Among multicellular eukaryotes, biominerals often serve multiple biological functions, including mechanical protection, structural support, movement, grinding, gravity, or magnetic sensing. Among prokaryotes, biominerals are often simply formed as a byproduct of a biochemical activity, during which the organisms oxidize or reduce transition metals or other species in solution for metabolic energy generation.

Many different types of biominerals have been identified. H. A. Lowenstam has periodically compiled and monitored the list of biominerals *sensu stricto* for more than 30 years, which has been expanded from just over 10 in the early 1960s to over 70 today (Weiner and Addadi 2002) (see Table 5.1). The biomineral atacamite ($Cu_2Cl(OH)_3$), a latest addition to this growing list, is a type of copper mineral that is formed under a biologically controlled mechanism by some carnivorous marine polychaete worms to reinforce the sharp tips of their jaws (Bekele et al. 2022).

Table 5.1 The diversity and chemical compositions of biogenic minerals.

Carbonates	
Calcite	$CaCO_3$
Mg-calcite	$(MgxCa1\text{-}x)CO_3$
Aragonite	$CaCO_3$
Vaterite	$CaCO_3$
Monohydrocalcite	$CaCO_3{\cdot}H_2O$
Protodolomite	$CaMg(CO_3)_2$
Hydrocerussite	$Pb_3(CO_3)_2(OH)_2$
Amorphous calcium carbonate	$CaCO_3$
Phosphates	
Octacalcium phosphate	$Ca_8H_2(PO_4)_6\ 5H_2O$
Brushite	$CaHPO_4{\cdot}2H_2O$
Francolite	$Ca_{10}(PO_4)6F_2$
Carbonated-hydroxylapatite, dahllite	$Ca_5(PO_4CO_3)_3(OH)$
Whitlockite	$Ca_{18}H2(Mg,Fe)_2{}^{+2}(PO_4)_{14}$
Struvite	$Mg(NH_4)(PO_4){\cdot}6H_2O$
Vivianite	$Fe_3{}^{+2}(PO_4)_2{\cdot}8H_2O$
Amorphous calcium phosphate	Variable
Amorphous calcium pyrophosphate	$Ca_2P_2O_7{\cdot}2H_2O$
Sulfates	
Gypsum	$CaSO_4{\cdot}2H_2O$
Bassanite	$CaSO_4{\cdot}0.5H_2O$

(*Continued*)

Table 5.1 (Continued)

Barite	$BaSO_4$
Celestite	$SrSO_4$
Jarosite	$KFe_3^{+3}(SO_4)_2(HO)_6$
Sulfides	
Pyrite	FeS_2
Hydrotroilite	$FeS{\cdot}nH_2O$
Sphalerite	ZnS
Wutzite	ZnS
Galena	PbS
Greigite	Fe_3S_4
Mackinawite	$(Fe,Ni)_9S_8$
Amorphous pyrrhotite	$Fe_{1-x}S$ ($x = 0$–0.17)
Acanthite	Ag_2S
Hydrated silica, arsenates, chlorides, fluorides, and sulfur	
Orpiment	As_2S_3
Amorphous silica	$SiO_2\ nH_2O$
Atacamite	$Cu_2Cl(OH)_3$
Fluorite	CaF_2
Hieratite	K_2SiF_6
Sulfur	Element S
Oxides, hydroxides, and hydrous oxides	
Magnetite	Fe_3O_4
Amorphous ilmenite	$Fe^{+2}TiO_3$
Amorphous iron oxides	Fe_2O_3
Amorphous manganese oxides	Mn_3O_4
Oethite	α-FeOOH
Lepidocrocite	γ-FeOOH
Ferrihydrite	$5Fe_2O_3{\cdot}9H_2O$
Todorokite	$(Mn^{+2}CaMg)Mn_3^{+4}O_7{\cdot}H_2O$
Birnessite	$Na_4Mn_{14}O_{27}{\cdot}9H_2O$
Organic compounds (three subgroups) (Nickel-Strunz classification)	
Salts of organic acids (e.g. formates, acetates, oxalates, benzene salts, cyanates) Hydrocarbons (e.g. Carpathite) Miscellaneous organic minerals (e.g. ambers)	

Source: Data from Weiner and Addadi (2002), Weiner and Dove (2003).

5.2.2.1 Carbonate Biominerals

The calcium carbonates ($CaCO_3$) are the most prevalent biominerals in terms of the production quantity and their widespread distribution across a diverse array of organisms (Lowenstam and Weiner 1989). Calcareous biominerals are also found in plants and fungi (Gadd 2021). The widespread adoption of calcium carbonate for biocalcification by organisms is attributed not only to the critical role of Ca^{2+} in numerous physiological processes but also to evolutionary pressures that favored the acquisition of calcareous skeletons within specific clades during their early evolution.

Calcium carbonates possess several polymorphs, but only the calcite, aragonite, and vaterite, as well as the amorphous phase that are utilized for biomineralization. Each specific $CaCO_3$ polymorph, mainly calcite and aragonite, constructs particular biomineralized structures, such as the calcitic prismatic-aragonitic nacre of molluscan shells. Yet multiple polymorph phases may also co-exist in close proximity within a single skeleton or the same structure. In addition, their mineralogical composition may change over time during the ontogeny of organisms. In many metazoan species like molluscs, the initially secreted shell during embryonic and larval stages (i.e. shell secretion is formed by shell gland and shell field) is exclusively aragonitic with a large amount of amorphous phase, while the adult typically produces bimineralic shells with aragonite and calcite. Different mineral phases of biomineralized hard tissues play specific roles in the proper functioning of the organisms (Skinner and Jahren 2003). Calcite and aragonite are the two most common biominerals and have equal importance employed by organisms. Vaterite, most commonly found in the molluscan nacre and repaired shells, is thermodynamically unstable and easily transforms into calcite under ambient conditions over time.

The isostructural calcite group comprises several carbonate minerals where calcium (Ca^{2+}) is replaced by other divalent cations, including magnesium (Mg^{2+}), iron (Fe^{2+}), manganese (Mn^{2+}), cobalt (Co^{2+}), zinc (Zn^{2+}), cadmium (Cd^{2+}), and nickel (Ni^{2+}). Each calcium-substituted mineral corresponds to a distinct type of mineral with specific chemical composition, for example, dolomite ($CaMg(CO_3)_2$) is a notable member of this group. Rhombohedral siderite ($FeCO_3$) minerals with equant microcrystals ca. 20 μm in size were known to be produced in laboratory cultures with bacterium *Geobacter metallireducens* (Konhauser 2007). These siderite microcrystals produced under laboratory conditions are morphologically similar to naturally occurring minerals, highlighting the enormous diversity of carbonate minerals formed through both abiogenic processes and biologically mediated pathways. Similarly, the calcium ion in aragonite lattice is substituted by other elements like strontium (Sr^{2+}), lead (Pb^{2+}), and barium (Ba^{2+}), leading to the formation of distinctive minerals: strontianite ($SrCO_3$), cerussite ($PbCO_3$), and witherite ($BaCO_3$), respectively (Skinner and Jahren 2003). These broad variations reflect the chemical flexibility of carbonate minerals and their capacity to incorporate various trace elements under specific environments, by both biogenic and abiogenic processes.

Looking back at the Earth's history, the vast depositions of ancient sedimentary dolomites in Precambrian remain a 200-year-old unresolved enigma in geology, the so-called "Dolomite Problem." This problem arose from the fact that geologists, so far, have consistently failed to synthesize perfectly ordered dolomite, specifically the protodolomite, under Earth's surface conditions (below ca. 80 °C) in the laboratory. Protodolomite is formed under a more "unnatural" environment that is often associated with Mg-enriched fluids and highly evaporative environments, a process termed dolomitization (Roberts 2024). A recent laboratory experiment demonstrated that frequent cycling of solutions between supersaturation and undersaturation, aided with pH and salinity fluctuations, can significantly accelerate protodolomite crystallization, which may offer a solution to the "dolomite problem" (Kim et al. 2023).

In efforts to synthesize highly ordered protodolomite under laboratory conditions, some microorganisms such as the sulfate-reducing bacteria (SRB) and methanogens are often used to mediate protodolomite precipitation under anoxic and hypersaline conditions (e.g. Roberts et al. 2004; Vasconcelos et al. 1995). The SRB are able to effectively induce the precipitation of the same protodolomite crystals as in the natural systems from which the bacterium was originally isolated. The metabolic process of sulfate reduction overcomes the kinetic barrier to primary dolomite precipitation by increasing pH (i.e. alkalinity) and removing sulfate ions that inhibit mineral nucleation. Extracellular polymeric substances (EPSs) secreted by SRB contribute to the concentration of Ca^{2+} and Mg^{2+} cations around the cell, which serve as favorable adsorption sites for CO_3^{2-} ions, thereby promoting protodolomite precipitation (Konhauser 2007). Protodolomite precipitated via microbial activities is often called the microbial dolomite, tentatively considered to be a type of carbonate biomineral.

5.2.2.2 Phosphate Biominerals

Phosphate minerals comprise principally the apatite group ($Ca_{10}(PO_4)_6(OH,F,Cl)_2$), which includes those found in biomineralized hard tissues known as bioapatite. Bioapatite typically represents an impure carbonate-substituted apatite ($Ca_5(PO_4CO_3)_3(OH)$) resembling hydroxyapatite ($Ca_{10}(PO_4)_6(OH)_2$) that forms at lower temperatures and under biological controls (Elliott 2002; Skinner 2005). Non-carbonate–containing hydroxyapatite cannot be formed through biological process, and thus belongs to an abiogenic mineral. Both carbonate and phosphate biominerals belong to the calcium-bearing biomineralization system.

Bioapatite is a selected biomineral type for vertebrate skeletons including bones and teeth, but it is also found in various other organisms. These include bacteria, invertebrates (e.g. octocorals, arthropods, some bivalve molluscs, linguliform brachiopods), and even plants (Weigend et al. 2018). A typical bone, such as the human femur, is composed of ca. 55-60 wt% mineral, ca. 30 wt% collagen, and ca. 10-15 wt% water, in which nanometer-scaled bioapatite crystallites are intimately interwoven with collagen fibrils (Pasteris 2016). Among invertebrates, linguliform brachiopods represent a key group known for secreting calcium phosphate biominerals within an organo-phosphatic framework (Zhang et al. 2024) (see Figure 5.4). The remaining brachiopods, including craniiforms and rhynchonelliforms, produce calcitic shells under a biomineralizing mechanism that is similar to and deeply conserved with molluscan shell formation (Wernström et al. 2022). Calcium phosphates were recently discovered in the stinging hairs of flowering plant *Loasaceae* from South America, and subsequent surveys confirmed a more widespread distribution that has previously been overlooked in plants (Weigend et al. 2018).

Figure 5.4 Photomicrograph of organophosphate brachiopod *Lingula* from the Chinese Yellow Sea. (a) SEM image; (b) backscattered image. Organic tissue with distinct muscular threads. *Source:* Photos by Luoyang Li.

5.2.2.3 Silica Biominerals

Anhydrous silica, quartz, and its microcrystalline varieties, e.g. chert, are one of the most common rock-forming minerals, comprising ca. 28% of the Earth's crust. Yet, it is the low-temperature, hydrated, highly insoluble variation of silica—opal ($SiO_2 \cdot nH_2O$) that is synthesized by living organisms for structural and functional utilization. Opal is composed of different amounts and arrangements of structural units of amorphous silica and water. The abiogenic opal contains amorphous silica particles of regular size packed in a well-ordered array which may extend over many mm or cm in length. Biogenic opal (Opal-A or Biosilica in general) is an extremely fine-grained, virtually amorphous variety of opal, and has very limited order. Opal-A, either biogenic or abiogenic genesis, is metastable and crystallizes over time to Opal-CT, which is eventually converted into microquartz by a solution-reprecipitation process (Liesegang and Tomaschek 2020). Biosilica consists of minor organic components and is formed biologically.

On contrary to its limited structural diversity compared to those more "popular" calcium-bearing biominerals, biosilicification occurs surprisingly more widespread in unicellular bacteria, eukaryotic protists (choanoflagellates, radiolarians, and diatoms), plants, and basal invertebrates (e.g. sponges) (see Figure 5.5). Biosilica is rare in vertebrates but still exists in fish otoliths and human kidney stones. In contrast, biosilicification is more prevalent in microorganisms, where the cell provides a large reactive surface for silicon adsorption, facilitating nucleation by lowering activation energy barriers. The bacterial mediation of biosilica precipitation is not limited to any specific species but is widespread across diverse functional groups (Konhauser 2007).

Some eukaryotes secrete siliceous skeletons with delicate architectures, as seen in diatoms, radiolarians, and sponges (Demospongiae and Hexactinellida). Unlike unicellular bacteria that passively precipitate amorphous biosilica from supersaturated solution, these eukaryotes exert a certain degree of biological control over the silicification process. Biomineralized silica bundles have also been found in bamboos (e.g. *Bambusa*) and hairs of grasses, in which the high concentration of silicon offers structural advantages to these plants (Collin et al. 2012). Fossil evidence for

Figure 5.5 Photomicrograph of microalgae diatom with delicate biosiliceous skeleton.
Source: Photo by Luoyang Li.

silica biomineralization (i.e. phytoliths) was recently found in Permian lycophyte plants, indicating their early origin and critical role in plant evolution (Feng et al. 2024).

5.2.2.4 Sulfate Biominerals

Sulfate minerals precipitate easily in evaporative environments through purely chemical processes. Sulfate-containing biominerals are also known in some bacteria, algae, protozoans, and even animals. Some modern gypsified stromatolites consist of alternating laminar layers of microbial mats and gypsum ($CaSO_4{\cdot}2H_2O$). Each set of laminae results from (1) growth of cyanobacterial mats during periods of low salinity, and (2) interstitial crystallization of gypsum occurs when trapped brines reach salinity levels too high for growth of most microorganisms.

Laboratory experiment shows that some cyanobacteria such as *Synechococcus* not only induce the formation of gypsum but also precipitates celestite ($SrSO_4$) enveloping the cell wall. Similarly, the heterotrophic bacterium *Myxococcus* has been reported to precipitate barite ($BaSO_4$) by exposing it to a solution enriched with Ba^{2+} (Konhauser 2007). Interestingly, some desert colonizing microorganisms extract water from gypsum rocks in tackling the harsh environment, causing mineral phase transformation from gypsum to anhydrite ($CaSO_4$) (Huang et al. 2020). The two water molecules in gypsum can be removed under laboratory conditions, but this requires heating above 190 °C, the only method to produce anhydrous $CaSO_4$ phase.

Sulfate biominerals occur almost exclusively in marine microbial communities (Bosselmann and Epple 2008). However, there are a few non-bacterial species that utilize sulfate minerals to form functional hard tissues. The jellyfish has varying-sized statoliths (mineralized crystals within statocysts for gravity sensing) of calcium sulfate hemihydrate (bassanite, $CaSO_4{\cdot}0.5H_2O$). In addition, a group of marine protozoans called acantharians is able to precipitate crystalline strontium sulfate (celestite, $SrSO_4$) to construct a remarkably delicate skeleton (test). These tests vary greatly in construction, however, each of them strictly bears 20 radially arranged spines that is a diagnostic feature of this group.

5.2.2.5 Sulfide Biominerals

Metal sulfide is one of the major groups of minerals. For biogenic metal sulfides, the formation of low-temperature metal sulfide biominerals is mostly related to microbial activities; thus, this process is often called microbial biomineralization. Dissimilative SRB play a leading role, as these bacteria utilize sulfate as a terminal electron acceptor during metabolism under anaerobic conditions, producing highly reactive sulfide ions (S^{2-}) (Konhauser 2007). It is the S^{2-}-molecular compound that induces metal sulfide precipitation under appropriate environmental conditions with available iron or other metal cations (Frankel and Bazylinski 2003).

Some sulfate-reducing MTB are known not only to produce iron sulfide particles extracellularly, but also to synthesize intracellular crystals of greigite magnetosome (Sakaguchi et al. 1993). Sulfide ions react with iron forming magnetic particles of greigite (Fe_3S_4) and pyrrhotite (Fe_7S_8), as well as a number of other non-magnetic iron sulfides including mackinawite (tetragonal FeS), pyrite (cubic FeS_2), and marcasite (orthorhombic FeS_2). The most prevalent sulfide metal species in sediments is pyrite. However, the synthesis of low-temperature pyrite in sediments involves a number of intermediate phases: mackinawite (FeS) → greigite (Fe_3S_4) → pyrite (FeS_2), and its direct precipitation is unfavorable due to slow kinetics of mineral phase transformation (Schoonen and Barnes 1991). Microorganisms play a crucial role in the dynamics of iron sulfide mineralization by accelerating the formation of intermediate mineral phases.

Within environments that lack iron, some bacteria tend to precipitate other sulfide mineral types, e.g. millerite (NiS), galena (PbS), and sphalerite (ZnS), showcasing the capacity of microbial biomineralization and the diversity of sulfide biominerals (Park and Faivre 2022). While metal sulfides are commonly formed in microbial communities either extracellularly or intracellularly, they also occur in some animals with specific biological functions. The "scaly-foot" gastropod inhabiting deep-sea hydrothermal vents has a dense cover of imbricating scales on the sides of its foot (Sun et al., 2020; Warén et al. 2003) (see Figure 5.6). These sclerites are mineralized with the iron sulfides that include pyrite and greigite minerals (Suzuki et al. 2006).

5.2.2.6 Oxide and Hydroxide Biominerals

Oxide biominerals primarily consist of metal oxides, such as iron oxides and manganese oxides. Of these, biogenic iron-bearing oxide biominerals are particularly common during microbial biomineralization. The adoption of iron for biomineralization in Fe(II)-oxidizing bacteria is probably due to the essential role of iron in some physiological processes, the toxicity of excessive ferrous iron (Fe^{2+}) products, and the easy oxidization of iron with O_2 into insoluble ferric iron (Fe^{3+}) at neutral pH conditions (Frankel 1991). With respect to the insolubility of Fe^{3+}, the vast amount of insoluble ferric iron formed during the Great Oxidation Event in the Paleoproterozoic Ocean had caused the massive precipitation of Banded Iron Formations in geological record during the period (see Section 6.2.4).

Fe(II)-oxidizing bacteria, the so-called iron bacteria, are responsible for the precipitation of oxides and hydroxides at acidic and neutral pH conditions, which form a diverse group of iron biomaterials, e.g. ferric oxyhydroxide ($Fe(OH)_3$), goethite (α-FeO(OH)), lepidocrocite (γ-FeO (OH)), ferrihydrite ($5Fe_2O_3 \cdot 9H_2O$), magnetite (Fe_3O_4), maghemite (γ-Fe_2O_3), schwertmanite ($Fe_8O_8SO_4(OH)_6$). Fe(II)-oxidizing acidophiles encompass several functional groups, including thermotolerant Gram-positive bacteria, mesophilic archaea, and thermophilic archaea (Frankel and Bazylinski 2003). Under low pH conditions, a wide range of acid-tolerant iron-oxidizing bacteria are particularly active in precipitating Fe(III) oxides. In contrast, only a few bacterial groups are responsible for Fe(II) oxidation under neutral pH conditions. Iron biomineralization plays a fundamental role in the global biogeochemical cycling of iron.

Figure 5.6 Deep-sea hydrothermal vent "scaly-foot" snail, a gastropod of mollusc *Chrysomallon squamiferum*. *Source:* Courtesy of Chong Chen and Jin Sun.

Among diverse iron biominerals, magnetite has attracted particular interests due to its contribution to the magnetism of sediments and its unique biological functions for organisms. Magnetite biomineralization is one of the most ancient biologically controlled mineralizing systems exploited by a broad form of life for magnetoreception (Lin et al. 2020). In prokaryotes, magnetite particles can be formed either extracellularly or intracellularly. By using Fe(III)-reducing bacteria, Fe(III) raw minerals were transformed into magnetic Fe(II)-bearing nanoparticles at ambient temperature (Joshi et al. 2018). Magnetic nanoparticles formed extracellularly by Fe(III)-reducing bacteria are generally irregular-shaped, poorly crystallized, and loosely organized (Kirschvink and Hagadorn 2000).

MTB synthesize either iron-oxide (magnetite) or iron-sulfide (greigite) in the form of magnetosomes (Frankel and Bazylinski 2003). MTB precipitate individual magnetite nanoparticles within an intracellular phospholipid-membraned vacuole, forming elaborate structures called "magnetosomes" that are arranged into bead-like chains (see Figure 5.7). Chains of these magnetosomes act as simple compass needles that align the bacterial cells with the Earth's magnetic field. Magnetite-bearing magnetosomes or akin products also exist in some invertebrates (e.g. desert ants, lobsters), higher animals, including migratory birds, fish, and even the brain of human beings, which perform a magnetic sensing function. Furthermore, the radular teeth of chitons (Mollusca: Polyplacophora) are constructed with magnetite (Mohs hardness is 6) and other iron oxides in complex architectures, making it one of the hardest and most wear-resistant biomaterials known to date (Joester and Brooker 2016).

Manganese oxides (MnOx) are another common type of oxide biomineral formed through microbial metabolic activity, primarily by bacteria and fungi. These oxides are among the strongest natural oxidants and are highly sensitive to Earth's redox conditions, making them valuable geochemical proxies for reconstructing past ocean oxygenation dynamics. Mn(II)-oxidizing microorganisms, through enzymatic catalysis, dramatically accelerate the rate of manganese biomineralization by several orders of magnitude compared to abiotic processes, leading to the accumulation of substantial manganese deposits. A prominent example is the widespread deep-sea manganese

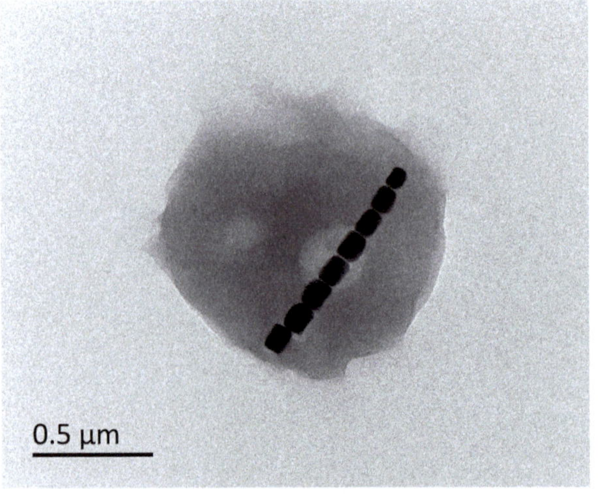

Figure 5.7 A magnetotactic bacterium containing a single chain of magnetosomes.
Source: Courtesy of Kuang He.

nodules, or polymetallic nodules, whose formation is largely driven by complex microbial communities residing within them (Tebo et al. 2004).

5.2.2.7 Organominerals

Abovementioned diverse biominerals are generated by living organisms in either direct/indirect, passive/positive pathways, extracellular/intracellular spaces, yet, dead organisms, even decomposed organic matters can influence mineral precipitation under some circumstances, forming the so-called organominerals. Organominerals are minerals that form through the interaction of organic matter with mineral components. Organominerals are defined as "any minerals precipitated by interaction with organopolymers, bioorganic, and/or non-biological organic compounds, without evidence of direct skeletal, intracellular or extracellular biological control" (Perry et al. 2007).

Unlike biogenic minerals (i.e. biomaterials *sensu stricto*), which are formed via metabolic activities of living organisms, the formation of organominerals is heavily influenced by extrinsic geological processes, hence reflecting interactions between life and its environments (Perry et al. 2007). With this concept, microbialites are formed in direct association with acidic macromolecules in microbial biofilms, thus considered as typical organominerals (Dupraz et al. 2009). Organominerals are often considered as forgotten group of organic components in ancient rocks, due to the low potential for fossilization, but their presence will be highly informative (Jehlička and Edwards 2008). This is manifested with a recent exciting finding of two-million-year-old DNA sequences of mastodons from frozen soils in Greenland (Kjær et al. 2022), and the permineralization of organic matters in more ancient dinosaur bones (Wiersma et al. 2021).

Some organic materials themself also exhibit properties of traditional crystalline minerals. Organic minerals are naturally occurring compounds that contain carbon and are formed through biological processes of living organisms, thus distinguishable from organominerals. According to the composition-based Nickel-Strunz classification of minerals, organic minerals constitute one of the 10 mineral groups and are defined as organic compounds intermixed with other materials that exhibit properties similar to traditional crystalline minerals. Under this scheme, organic minerals are generally categorized into three major subgroups: salts of organic acids, hydrocarbons, and miscellaneous organic minerals. Echigo and Kimata (2010) proposed an alternative classification based on electric charge, dividing organic minerals into two major subgroups: Ionic organic minerals, where organic anions and cations are held together primarily by ionic bonds; Molecular organic minerals, electrically neutral, are bonded through weak intermolecular interactions. One representative of organic minerals is calcium oxalate that is formed in the human urinary system, leading to the formation of kidney stones. All organic minerals contain organic compounds with carbon (C), but graphite is excluded from the group.

5.3 Classification of Biomineralization

Processes of biomineralization are categorized into several schemes based on different criteria: (1) type of biominerals, e.g. iron biomineralization, biocalcification, biosilicification, phosphate biomineralization, and organomineralization; (2) degree of biological controls, commonly named as the biologically induced mineralization (BIM), and the organic mediated mineralization (Lowenstam 1981) that was later termed as the biologically controlled mineralization (BCM) (Mann 1983); Organomineralization, also popularly known as the biologically influenced mineralization (BFM)

(Dupraz et al. 2009), is a relatively new term, referring to precipitation of minerals mediated and/or influenced by organic matters, independently of the living organisms which it derives from; or (3) location of biomineral formation, comprising extracellular, intercellular, and intracellular biomineralizing pathways, referring to site of mineralization occurring on outside environments of cell, space between cells, and inside the cell, respectively (Weiner and Dove 2003) (see Figure 5.8).

5.3.1 Biologically Influenced Mineralization (BFM)

The concept of BFM (=organomineralization) evolved from the pioneer work of Mitterer (1968) and Trichet (1968), who studied the composition and function of internal organic matter in non-skeletal grains such as ooids. BFM refers to the process of mineralization in close association with a broad diversity of organic templates such as bacterial cell wall and EPS (Görgen et al. 2021). That is to say, BFM represents a passive mineralization process, which is not necessarily linked to

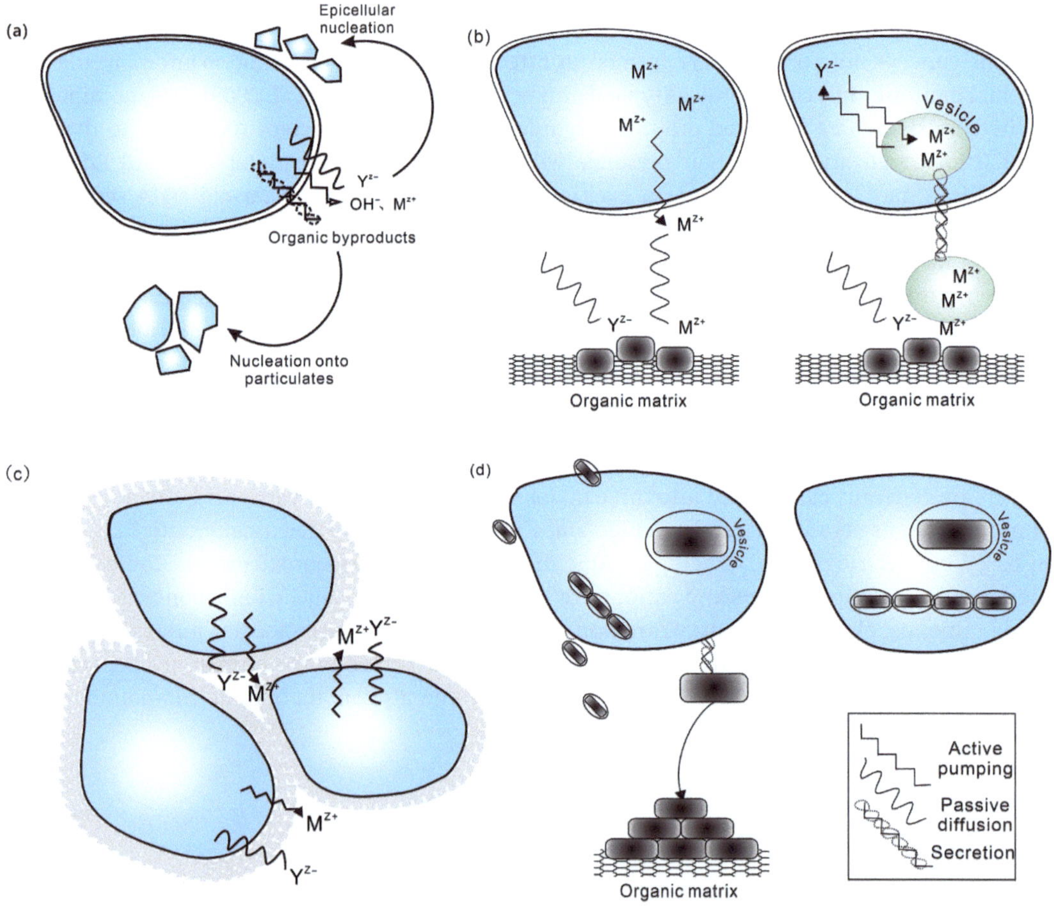

Figure 5.8 Schematic illustration of different biomineralization pathways. (a) BIM, cellular metabolic activities affect environmental pH and chemical condition, inducing mineral precipitation; (b) extracellular BCM, cations are transported outside of the cell through ionic pumping or organic vesicles to the side of biomineralization; (c) intercellular BCM, the intercellular zone is functioned as a diffusion-limited compartment and the surface of cells are used as organic substrate for nucleation and crystallization; (d) intracellular BCM, nucleation and crystallization occur within a specialized organic vesicle, precursory and end-biominerals remain either inside the cells or are expelled extracellularly. *Source:* Adapted from Weiner and Dove (2003).

any living organisms or specific physiological activities, but takes place anywhere in non-living, reorganized macromolecular films or aggregates.

In all BFM cases, external environmental changes such as degassing and evaporation of EPS can promote ion concentration, supersaturation, nucleation, and then the growth of minerals. BFM differs from other biomineralization pathways, especially the BIM process, in the absence of biologically controlled ionic pumping or metabolic functions and enzymatic processes (Trichet and Defarge 1995). The process of BFM is increasingly well understood attributed to the extensive investigations of modern carbonate sediments such as microbial mats and biofilms (see Section 3.2). Moreover, BFM requires organochemical mechanisms rather than advanced biological systems, therefore, it represents a geologically early process of biomineral production, possibly prior to and/or co-existing with life when it first appeared on the Earth.

5.3.2 Biologically Induced Mineralization (BIM)

The formation of biominerals involves a series of physicochemical processes (e.g. crystal nucleation, growth, and crystallographic orientation) and biological processes (i.e. biological control). The BIM refers to the secondary precipitation of biominerals that is produced by interactions between biological activity and surrounding environment (Weiner and Dove 2003). In the BIM pathway, the metabolic processes employed by the living organism within a given environment mediate pH, pCO_2, and the ionic compositions of secreted products. Cell surfaces often act as causative agents for nucleation and subsequent crystal growth, yet the biological system exerts minor control over the type, habit, and location of minerals deposited. The forming biominerals generally exhibit poor crystallinity, broad size distribution, various crystal morphologies, and often have less-ordered specificity and high impurities in the lattice.

The heterogenetic properties of BIM products are sometimes indistinguishable from those of BFM. But differently, BIM involves active metabolic activities by living organisms during the entire process of mineral formation. The sheath calcification of diverse photosynthetic cyanobacteria is typically a BIM model. The photosynthetic uptake of inorganic carbon releases OH^- into the surrounding environments, raising the pH adjacent to the cyanobacterial sheath, which induces $CaCO_3$ precipitation on it (Kah and Riding 2007) (see Figure 5.9). The reaction between Ca^{2+}

Figure 5.9 A confocal laser scanning micrograph showing cyanobacterial filaments (yellow, 5 μm wide) and calcite crystals (green) of a soda lake biofilm. *Source:* Courtesy of Gernot Arp and Christian Böker.

and CO_3^{2-} to form calcium carbonate is more favorable at higher pH levels in seawater. Undoubtedly, cyanobacteria exert some degrees of biological control over sheath calcification during photosynthesis. However, changes in external environments, particularly seawater conditions such as ocean acidification, characterized by decreased pH and altered chemical compositions, may also significantly impact carbonate production.

5.3.3 Biologically Controlled Mineralization (BCM)

BCM emphasizes the extensive biological regulation of biomineralization. Nearly all aspects of mineral precipitation, including nucleation, growth, crystallography, morphology, and the assembly of final skeletons, are governed by intrinsic cellular processes supported by complex genetic regulatory networks. Organisms exhibit varying degrees of biological control, ranging from weak to strict regulations, as exemplified by the differences in skeletal production and properties between serpulid annelids and molluscs. BCM occurs within an isolated compartment, and the resulting biominerals typically serve specific biological functions, in contrasting to the by-products formed through biologically induced and BFM pathways (see Table 5.2).

Due to precise biological regulations, biomineralization through BCM results in highly organized crystal units with uniform sizes and distinct crystal habits. The wide array of biominerals produced by BCM serves numerous biological functions across various organisms, including mechanical protection, structural reinforcement, locomotion, and the detection of gravity or magnetic fields. These skeletal structures exhibit intricate hierarchical architectures extending across multiple microscopic scales, often displaying well-aligned crystallographic orientations. Typically, BCM-derived biominerals are organo-mineral biocomposites, with the organic component making up only a minor fraction of the overall structure. This distinctive combination of properties confers BCM-formed biominerals exceptional mechanical strength and resilience, far exceeding that of inorganically formed equivalents.

5.3.3.1 Weakly vs. Strictly Biological-Controlled Mineralization

Living organisms generally form highly organized and well-ordered biominerals through the BCM pathway, but the extent of biological control over mineralization varies significantly among species and lineages. Consequently, despite their BCM origin, biominerals often display notable differences in crystal structure, habit, chemical composition, and morphology.

Table 5.2 Comparison of biominerals through different biomineralization pathways.

BFM	BIM	BCM
Mineral formation independent of living organisms. Organominerals with properties similar to that of BIM.	Unintended and uncontrolled consequence of metabolism.	Significant biological controls, either weakly or strictly.
	Poor crystallinity, broad size distribution, and the lack of specific crystal morphologies.	Well-ordered, narrow size distribution, taxon or lineage-specific, consistent, crystal habits.
	Poor mineral specificity and impurities in the lattice.	Purity, mostly not sensitive to external environment.
	No specific function, may serving as a solid attachment or a form of protection against certain environment.	As functional hard parts of organisms that cannot maintain survival without them.

Molluscan shells exemplify strictly BCM, where mineralization is tightly regulated by a complex organic matrix within the shell, assembled by the underlying soft tissue, the mantle. This high level of biological control makes the molluscan mineralized shell remarkably resilient to environmental changes, including fluctuations in temperature, pH, and salinity. In contrast, serpulid worms (polychaete annelids) construct their calcareous tubes with the aid of an inner organic sheath lining the tube's inner surface (Vinn 2021). However, serpulid tube calcification occurs in a relatively weakly regulated, semi-enclosed environment. As a result, their calcareous tubes contain high content of impurities, with external sediments frequently incorporated into the tube walls. Despite being classified under BCM, serpulid tubes, like coral reefs are more vulnerable to environmental changes, particularly in the face of global warming and ocean acidification.

The sulfide iron mineralization observed in the "scaly-foot" snails inhabiting the hydrothermal vent indicates environmental influence on the biomineralogical selection of the animal, even though molluscan shell formation is strictly biologically controlled. The advantage with iron sulfides does not mean to be their mechanical or magnetic properties, but that they are physiologically cheap to make in an environment enriched with dissolved sulfide and metal ions (Bengtson and Briggs 2005). Similarly, the preference of many animals for calcium carbonate skeletons over calcium phosphate or other biomineral types is partly attributed to the marine chemical conditions prevalent during the Ediacaran-Cambrian transition, when mineralized skeletons first appeared in the fossil record (see Section 5.5).

5.3.3.2 Biologically Controlled Extracellular Mineralization

In extracellular BCM, the site of mineralization is located within a framework of macromolecules that remains a fair distance outside the cell. This organic framework, in molluscan shells is composed of proteins, polysaccharides, and lipids that are self-assembled into a three-dimensional scaffold, called organic matrix (see Figure 5.10). These organic matrices occluded within mineral phase are genetically programmed to regulate processes essential for the formation of biominerals. Most extracellular mineralized structures develop upon pre-existing organic matrix secreted by specialized epithelial tissues (e.g. the mantle epithelium of molluscs). These tissues serve as templates that facilitate the construction of large skeletal structures, a process previously described as the organic-mediated/templated mineralization.

The transport of mineral precursors from cells to sites of biomineralization invokes two primary pathways (Weiner and Dove 2003): (1) ion pumping and active transport: cells actively transport cations across the membrane into the extracellular space (e.g. the extrapallial space between mineral shells and underlying mantle epithelium in molluscs), where they establish a supersaturated microenvironment; (2) vesicle transport: cations and precursory minerals are concentrated within intracellular vesicles in the form of a highly disordered amorphous phase. These ion/minerals-containing vesicles are transported through the membrane to the extracellular mineralization site, where they destabilize and re-crystallize, and eventually synthesize into end-products (Gilbert et al. 2022). Both mechanisms actively supply and concentrate cations onto the organic matrix, enabling nucleation and growth of biominerals. Organisms that primarily exploit extracellular BCM include but not restricted to molluscan and brachiopod shells, scleractinian corals, serpulid tubes, and vertebrate bones and teeth. Extracellular BCM emphasizes the essential role of genetic, molecular, and cellular activities in the formation of functional, complex, and highly organized skeletal structures.

5.3.3.3 Biologically Controlled Intercellular Mineralization

Intercellular BCM refers to the site of biomineralization that occurs outside the cell wall in the space between adjacent cells. It typically appears in some unicellular/multicellular microorganisms

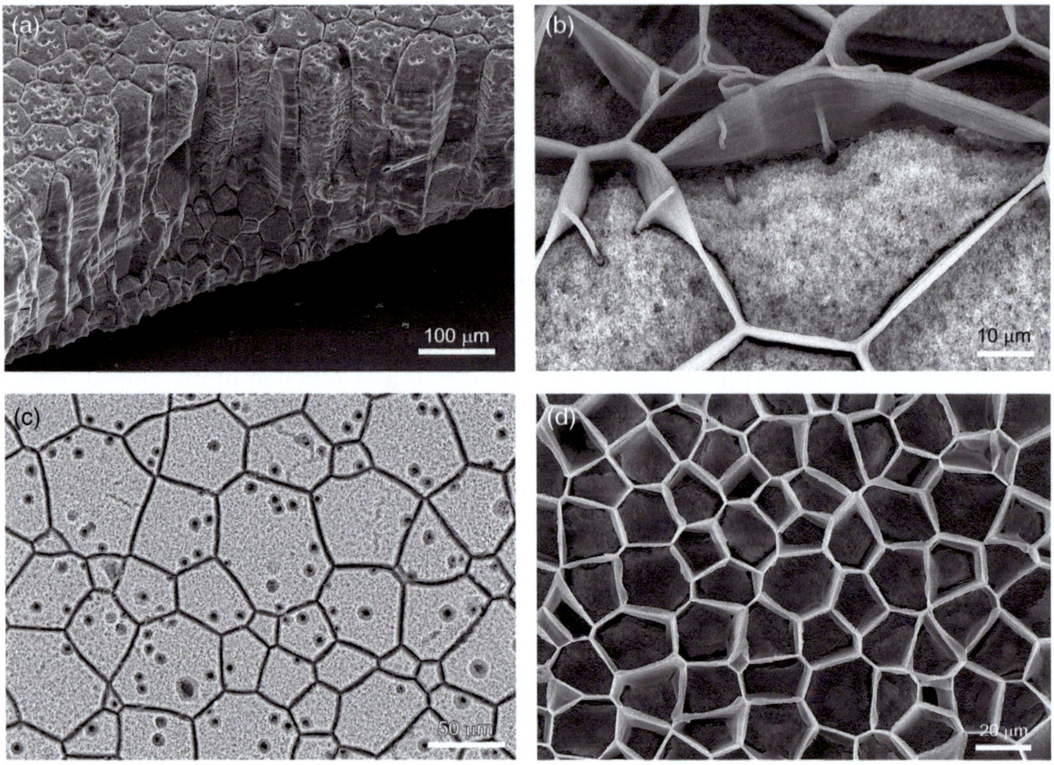

Figure 5.10 The prismatic organic matrix of calcitic prismatic microstructure of bivalve *Pinna*. (a) SEM photomicrographs of fractural surface of prismatic shell; (b), proceed with 5% diluted HCL acid for 10 seconds, showing inter- and intra-prismatic organic matrix surrounding calcitic prism; (c), backscattered image; (d), prismatic organic matrix forming a 3D scaffold that controls the shape of mineral prisms precipitated, proceed with 5% diluted HCl for 1 minute. *Source:* Photos by Luoyang Li.

that form communities, particularly in some calcifying algae (Weiner and Dove 2003). The intercellular space between adjacent cells is filled with mucilage that formes a transitional zone from cells to surrounding environment. Those mucilaginous frameworks are responsible for concentration of ions, nucleation, and growth of biominerals. In some calcareous green algae, precipitation of needle-like aragonites is more or less perpendicular to the cell wall, indicating a certain degree of biological controls over calcification (Granier 2012; Watabe and Kingsley 1989). This is in mark different from the loosely organized by-products of BFM and BIM. Mineralization between cells can become extensive so as to completely fill the intercellular spaces, forming a type of "skeleton."

5.3.3.4 Biologically Controlled Intracellular Mineralization

Intracellular BCM involves a process where cells exert biological control over mineral formation, occurring inside specific compartments within the cell. This is a common strategy allowing cells to regulate critical aspects of mineralization, including nucleation, composition, crystallography, and morphogenesis. The forming biomaterials exhibit species/lineage-specific morphologies (Weiner and Dove 2003).

In many cases, a single crystal is formed within the cell before being transported extracellularly, either as an individual unit or as part of a pre-assembled structure. For example, coccolithophores (haptophyte algae) intracellularly produce and assemble calcitic crystals, which are then secreted to form the protective coccolith surrounding the cell. In other instances, intracellularly mineralized structures remain within the cell throughout the organism's entire lifespan. The latter is less commonly employed, as it often imposes a physiological burden on the cell. However, MTB are an exception, capable of constructing intracellular magnetosome chains that function like a "compass needle" of aligned magnetic crystals. Similarly, some cyanobacteria precipitate amorphous carbonate nanoparticles internally (Couradeau et al. 2012), with nucleation and crystallization potentially occurring through membrane- or vesicle-based processes (Li et al. 2016). Despite this, intracellular calcification in cyanobacteria differs significantly from BCM in organisms such as coccolithophores, diatoms, and foraminifera, particularly in terms of the morphological and crystallographic complexity of the final mineralized structures.

5.4 Principle of Biomineralization

Rome wasn't built in a day, nor by a single way. Biomineralization is a highly intricate process governed by complex genetic, molecular, and cellular activities (i.e. biological processes) that direct the formation of solid minerals (i.e. physicochemical processes). Despite the mechanism underlying biomineralization is far from fully understood, an integrated model of biomineralization is emerging in recent years (Clark 2020; Gilbert et al. 2022). Gaining a mechanistic understanding of biomineralization, from initial nucleation to the formation of skeletal end-products, is essential not only for better understanding the diversity and evolutionary history of biominerals but also for advancing biomimetic applications.

5.4.1 Supersaturation and Nucleation

In the biomineralization system, the skeletal formation occurs within specialized microenvironments, either extracellular or intracellular spaces, that maintain sufficient supersaturation to enable initial nucleation and subsequent crystal growth. Specifically, mineral precipitation takes place in a physicochemically distinct site, isolated from the external environment. Although the exact dimensions of these enclosed zones are sometimes unclear, it is well-known that they are able to restrict ionic diffusion in or out of the system. These spaces often function as membrane-bounded compartments (e.g. vesicles). These enclosed zones are never completely isolated from external environments; the degree of isolation varies across organisms. In molluscs, biomineralization occurs in the extrapallial space, which lies between the mantle epithelium and the mineralized shell with a distance of about tens to hundreds of nanometers. In some bacteria, the isolation is passive and minimal, occurring in diffusion-limited inter- and/or extracellular zones. By contrast, many organisms such as MTB utilize intracellular vesicles to create highly regulated, compartmentalized microenvironments, within which the ionic components are precisely maintained for mineralization.

Crystallization initiates with nucleation at the atomic level from a saturated solution. Traditionally, nucleation was known to occur with a monomer-by-monomer process, namely the classical nucleation model, where a structure-insensitive cluster forms and grows through the attachment of individual ions. Over the past two decades, however, non-classical nucleation pathways have emerged for various biomolecules and biomaterials, challenging the previous classical nucleation model. In biomineral formation, the non-classical nucleation involves the formation of metastable prenucleation ion clusters, amorphous precursors, and particle attachment-based mechanisms, typically occurring at the micro- to nanoscale (Gebauer et al. 2008). These short-range ordered precursors are generally isotropic and more soluble, serving as heterogeneous nucleation sites that transform into a long-range ordered crystalline phase, significantly influencing crystallization kinetics. Both classical and non-classical nucleation processes play a role in skeletal biomineralization.

5.4.2 Amorphous Phase and Solidification

The formation of highly disordered transient precursors is a widely utilized strategy in biological systems for biomineralization (Addadi et al. 2003). The concept of such intermediate mineral phases dates back to the early 20th century, but conclusive evidence emerged in the 1960s when Towe and Lowenstam (1967) discovered that magnetite in chiton teeth precipitates through a disordered iron hydroxide precursor. Subsequent studies confirmed the existence and transformation of amorphous calcium phosphate (ACP) into crystalline dahllite in the radular teeth of chitons (Lowenstam and Weiner 1985). Today, it is widely accepted that nanoparticles of amorphous calcium carbonate (ACC) act as precursors to the more stable crystalline phases of aragonite and calcite, and that ACP serves as a precursor for carbonated apatite in vertebrate enamel and bone (Addadi et al. 2003).

These nanogranules of amorphous precursors are isotropic, soluble, and metastable, which do not tend to crystallize into a long-range ordered crystalline phase directly via solid-state phase transformation. Instead, these amorphous-phased minerals are further destabilized and re-crystalized in the aid with a variety of macromolecules and inorganic additives (e.g. Mg^{2+}), during which a specific crystal phase with distinct crystallographic orientations is determined (see Figure 5.11). Calcite, aragonite, and vaterite are three major polymorphs of calcium carbonate.

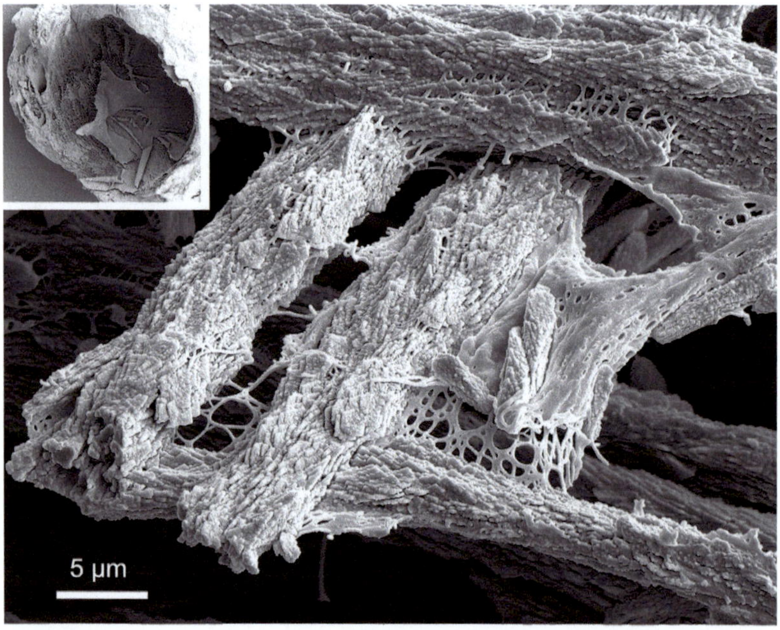

Figure 5.11 Tube microstructure and nanogranular constitutes of serpulid polychaete *Hydroides elegans*. The arrangements and solidification of amorphous nanogranules of calcium carbonate are intimately intertwined with proteinaceous framework of organic sheath lining inside of the tube. *Source:* Photo by Luoyang Li.

The mineral phase transformation of ACC into a particular carbonate polymorph is closely related with pH, organic contents, and magnesium ions (Cartwright et al. 2012). The organism regulates the concentration and speciation of inorganic ions and various proteins, all of which regulate crystal particle formation, phase transitions, growth, and the precipitation of specific mineral polymorph in the final (Gilbert et al. 2022).

Biomolecules play a crucial role in mineralization. At the molecular level, the organic matrix occluded within the mineral phase is negatively charged with carboxyl groups (COO^-) that attract positively charged cations from solution for nucleation. The carboxyl groups are located either along polysaccharide chains or in acidic amino acids. They are nearly always deprotonated, and therefore their carboxyl group terminations are negatively charged. Carboxyl group-rich proteins and/or polysaccharides are the most common and most effective cation-binding macromolecules existed in any organism to bind Ca^{2+} during process of biocalcification. The carboxyl group resembles a molecular "glue" that determines specific polymorph during mineralization (Gilbert et al. 2005) (see Figure 5.12). The negatively charged COO^- groups bind Ca^{2+} from solution, with the spacing between their vertical side chains determining the resulting $CaCO_3$ polymorph: 4.99 Å for calcite, 4.96 Å for aragonite, and 4.13 Å for vaterite (Reitner et al. 2001).

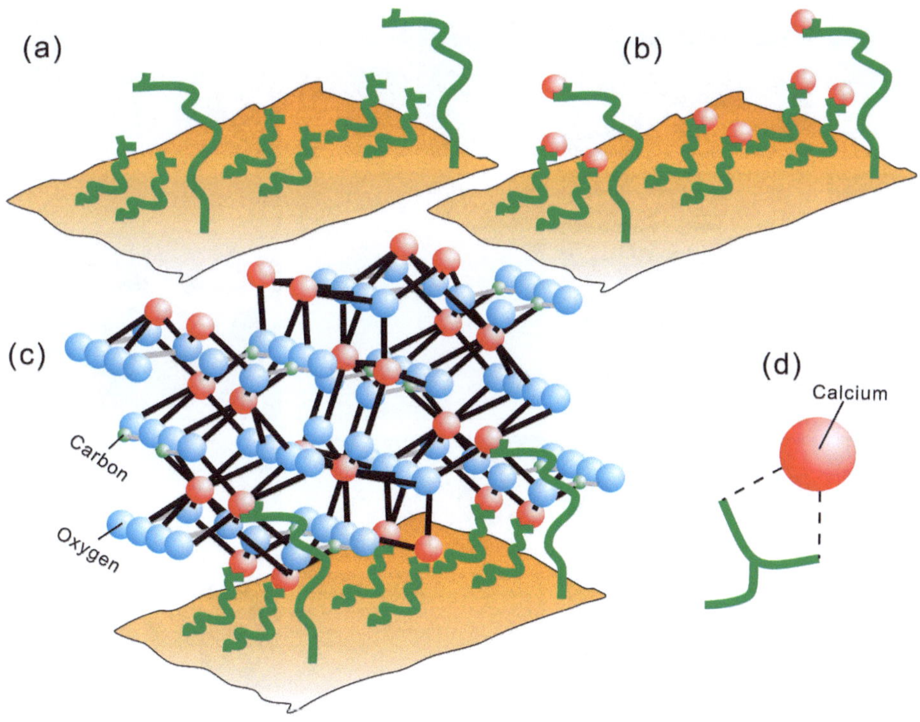

Figure 5.12 Conceptual model for biomineralization. (a) The organic matrix is composed of macromolecules, e.g. polysaccharides and various proteins; (b) the organic components have negatively charged functional groups that attract ions from solution. The steric arrangement of organic macromolecules, their sequence, and folding determines the precise position in three dimensions of the ions; (c) The crystal structure of the mineral aragonite; (d) close-up of the organic-mineral interface, the inter-atomic bonds are indicated by dashed lines. *Source:* Adapted from Gilbert et al. (2005).

5.4.3 Hierarchical Organization

The assemble of crystal building blocks into highly organized 3D structure with complex hierarchical organization fulfills various functions of skeletal end-products in living organisms. Molluscan shells are a classical model system to study the formation-structure-function relationships of biomineralized hard tissues. Molluscs fabricate their protecting armor-shells by secreting polymorphs of calcium carbonate in an extremely diversity of macro- and microstructures with more than thirty types. The most commonly formed microstructural patterns include homogeneous, prismatic, fibrous, foliated, chalk, crossed-foliated, nacre, and crossed-lamellar, of which nacre (mother of pearl) and crossed-lamellar (CL) are the two most evolutionary successful and widespread types (Checa 2018). Gastropods and cephalopods produce columnar nacre in which the aragonitic tablets are stacked on top of each other forming a tower-like pattern, while in bivalves, it has a brick-and-mortar construction (i.e. sheet nacre) (see Figure 5.13).

Shell fabrications in molluscs exhibit an incredible complex hierarchical organization of calcium carbonate minerals with occluded organic matrix. For instance, in the CL microstructure,

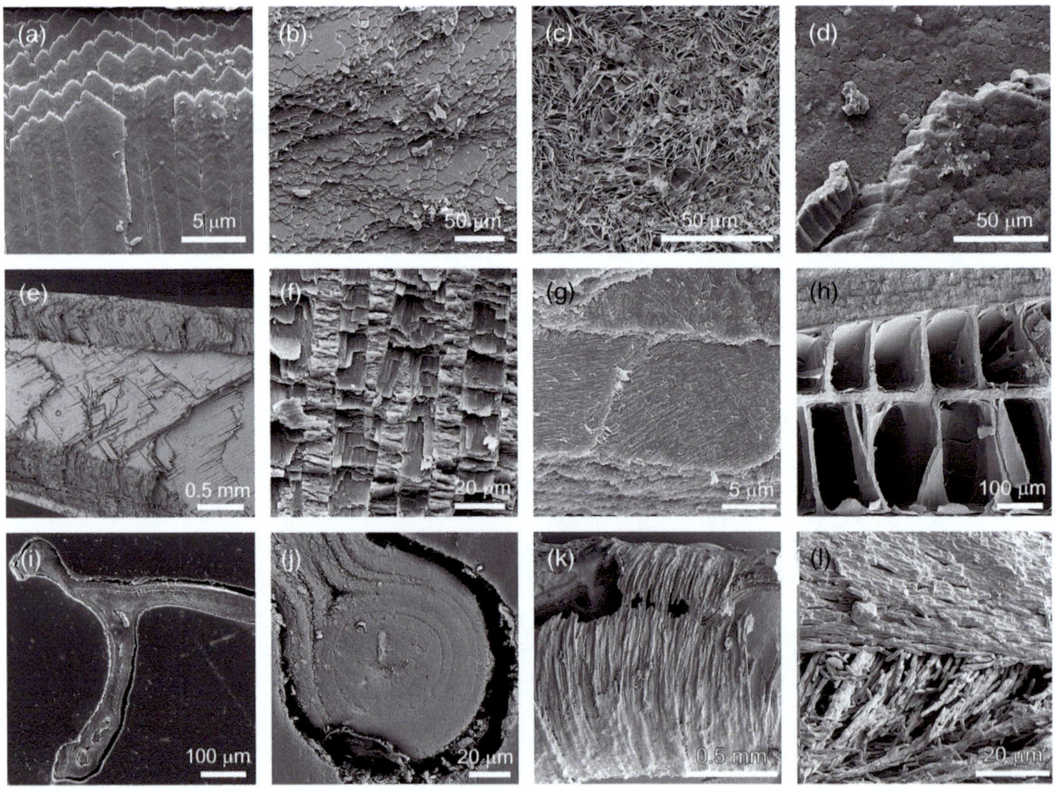

Figure 5.13 Common skeletal microstructures in selected invertebrate phyla. All specimens were collected alive from the Yellow Sea, Qingdao, China. (a–h) Molluscan shell microstructures, (a) foliated calcite in limpet, gastropod; (b) sheet nacre in bivalve shells; (c) calcitic chalk of oyster shells; (d) calcitic prismatic; (e,f) aragonitic crossed-lamellar in transverse section; (f) tangential view; (g,h) aragonitic lamello-fibrillar, also known as lamello-fibrillar nacre in cuttlebone, cephalopod; (i,j) siliceous sponge spicule; (k,l) serpulid tube, polychaete, annelids; (k) external surface of serpulid tube; (l) orientation of single crystal of calcite is aligned with proteinaceous thread. *Source:* Photos by Luoyang Li.

a four-ordered lamellar assembly of aragonitic crystallites has been extensively investigated since its first identification in molluscan shells almost 100 years ago. The commarginally or longitudinally aligned blocks (first-order lamellae), perpendicular to the shell surface, akin to plywood, consist of inclined and stacked "sheets" (second-order lamellae) of parallel micrometer-sized rods or laths (third-order lamellae). The third-order laths, encased within a thin layer of an organic envelope, dip in opposite directions alternating with the first-order lamellae, and are further divided into smaller building blocks of polygonal nanogranular morphologies (fourth-order lamellae). Such complexities of CL microstructure confer outstanding mechanical properties, effectively in limiting propagation of cracks within shells, making it the most widespread microstructural type in molluscan shell biomineralization (Li et al. 2022b). More than 90% modern shells possess a CL microstructure, contributing to the evolutionary success and ecological adaptation of molluscs inhabiting nearly every corner of the Earth.

5.4.4 Genetic and Molecular Systems

Clark (2020) recently provided a comprehensive review of the genetic and molecular regulatory systems underlying molluscan shell biomineralization. Earlier reviews on this topic were also available in Marin et al. (2016) and Zhang et al. (2018). Molluscan shell formation occurs within a three-dimensional organic framework composed of proteins, polysaccharides, and a small proportion of lipids, which collectively account for less than 5% of the shell's weight. Early studies on skeletal organic matrix in the 1950s primarily involved dissolving mineral phases and isolating soluble and insoluble proteins, but confronted significant technical limitations. Since the 1980s, advances in molecular biology have dramatically expanded our knowledge of skeletal organic matrix and associated genes, which have profoundly reshaped the understanding of molecular biomineralization mechanisms, offering critical new insights into their evolutionary history (Marin et al. 2016).

More than 500 biomineralization-associated proteins have been identified from different animal models, mainly molluscan species including pearl oyster *Pinctada*, abalone *Haliotis*, and limpet *Lottia* (Marin et al. 2016). Some specific mineral-associated proteins were identified such as the acidic protein Pif 80 that bound aragonite crystals in the nacre (Suzuki 2020). Large protein sequencing often shows less similarities across different species or lineages, for example, with less than 15% shared shell-forming proteomes, 10% commonality in gene sets between sheet nacre (bivalve) and columnar nacre (gastropod) (Jackson et al. 2010). Even within a single shell, the prismatic and nacreous layers consist of almost entirely distinctive proteins (Marin et al. 2012). This substantial divergence in shell-forming proteomes reflects the high evolvability of the molecular machinery underlying shell formation. However, a set of ancient and highly conserved genes and proteins, often referred to as the molecular biomineralization "toolkit," has been identified, highlighting the evolutionary conservation of molluscan shell biomineralization (Murdock 2020).

5.4.5 Benefits and Costs

Producing biominerals provides obvious advantages to organisms, contributing to their survival, reproduction, and ecological adaptation. These benefits, however, are valuable only on condition that the advantages outweigh the costs, as biomineralization is energetically costly. Biomineralization processes involved in ion transport, organic matrix assemble, and maintaining conditions suitable for mineral growth are metabolically more costly than the mineral itself. Laboratory studies on molluscs estimate that the cellular costs of $CaCO_3$ precipitation range from 1–2 to 17–55 J/mg, while the synthesis of associated organic matrix is even more expensive (~30 J/mg) (Palmer 1992). In corals, the formation of the skeleton, including formation of associated organic matrix, accounts for approximately 30% of overall energetic budget of the organism (Allemand et al. 2011; Gilbert et al. 2022).

Extensive investment of energy and resources in biomineralization inevitably disrupts the energetic balance, reducing resources available for other critical physiological processes such as growth and reproduction (Clark 2020). Therefore, organisms tend to produce mineralized hard tissues using the most physiologically cheap materials available in their environment (Wood and Zhuravlev 2012). Given these trade-offs, it is unsurprising that "scaly-foot" snails inhabiting deep-sea sulfide- and metal-rich hydrothermal vents have adapted to produce unique scaly sclerotomes composed of iron sulfides. What a wise choice! Moreover, biomineralizing organisms adapt to fluctuating environmental conditions by dynamically reallocating energy between the organic and mineral components of their biominerals (Wood 2011). For instance, some cold-water bivalves secrete calcareous shells with a relatively higher organic matrix compared to their warm-water counterparts, suggesting a greater energetic investment in shell formation. This adaptive response to environmental variability has driven the evolution of organic-limited biominerals over hundreds of millions of years. In molluscan shell biomineralization, this trend is evident, as the organic matrix generally comprises less than 5% of total shell weight.

5.4.6 Prokaryotic vs. Eukaryotic Biomineralization

Prokaryotes predominantly engage in BIM. However, MTB stand as a notable exception, alongside coccolithophores, as some of the most extensively studied microscopic organisms that utilize BCM. Biominerals formed through BIM in prokaryotes typically nucleate and grow extracellularly as a byproduct of the organism's metabolic activity. Prokaryotes achieve this by secreting organic macromolecules that interact with environmental ions or compounds, leading to the precipitation of biominerals. These minerals generally form on micro- to nanometer scales, often exhibiting significant variability in particle size and poorly organized. Unlike BCM, where mineralization serves a specific biological function, BIM in prokaryotes is sometimes an unintended consequence of metabolism and even imposes disadvantages to the organism.

Prokaryotes serve as highly effective catalysts of biomineralization. Their high surface/volume ratio enables rapid exchange of nutrients and metabolic byproducts with the surrounding environment, enhancing their efficiency in mineralization processes. Microbial cell walls, sheaths, and filaments carry a strong negative charge, providing multiple binding sites for cations. This facilitates interactions between metal ions and the negatively charged cell surfaces, promoting mineral deposition. Additionally, bacteria can induce mineralization by secreting extracellular polysaccharides and enzymes, which promote ion concentrations in the local environment. These biochemical processes create favorable conditions for nucleation and crystal growth, further driving biomineralization.

Compared to prokaryotic biomineralization, biomineral construction in eukaryotes exhibits greater uniformity in crystal size, habit, and highly ordered micro- to nanostructures, reflecting the BCM. Their formation is tightly regulated by an associated organic matrix, which controls crystal nucleation, growth, and crystallographic orientation. Complex eukaryotic organisms, particularly skeletonized protists and animals, often develop heavily mineralized hard parts. These structures not only provide functional advantages to organisms but also enhance their potential for fossilization, making them more readily identifiable in the fossil record.

5.5 Origin and Evolution of Biomineralization

Answering the origin of biomineralization has never been an easy task. While not all organisms produce biominerals, biomineralization is a fundamental phenomenon to life itself. Biomineralization extends far beyond skeleton formation or serving as an accessory feature of organisms. Biomineralized hard tissues construct the body plan configuration of disparate metazoan biomineralizers. Without a mineralized shell, molluscs can never be recognized as a mollusc as such, even though some cephalopods and land slugs lose their shells through evolution. The long evolutionary journey of biomineralization that created the modern skeletal world reveals a fascinating pattern: microbial biomineralization dominated Earth's oceans for ca. 4.0 billion years during the Precambrian. This was followed by the emergence of protistan skeletons in the Neoproterozoic and subsequent sudden appearance of highly organized BCM skeletons during the Cambrian "explosion" of animal life. Since their first emergence, skeletal animals have left a rich geological record and have profoundly influenced global biogeochemical cycles (Knoll and Kotrc 2015).

In this section, we primarily discuss the origin and rise of animal skeletal biomineralization, explore possible drivers of this critical biological innovation in the animal kingdom during the Precambrian-Cambrian transitional period. In particular, we use molluscs as a model to better understand, but still an incomplete picture with respect to how, why, and under what conditions did non-mineralized molluscan ancestor evolved the ability to produce a mineralized shell.

5.5.1 History of Biomineralization Pathways

The evolution of life and biominerals is deeply intertwined and co-evolved. Biominerals and/or organominerals through the BFM and BIM pathways are often considered as a byproduct of organism's metabolic activity in responding to the surrounding environment. These biomineralization strategies are ubiquitous in microbial world, particular in unicellular prokaryotes and simple multicellular eukaryotes such as fungi and algae. The earliest evidence of life, despite persisted uncertainties, preserved as isotopically light graphite (i.e. chemical fossils), is traced back to the ~4.1 Ga (Bell et al. 2015). Undisputed microbial mats and microbilites are well-known in both carbonate and siliciclastic settings around 3.5 billion years ago (Section 8.2.2). In these ancient deposits, cellular structures of organisms are hardly observable due to limitation of fossil preservation, making the process of mineralization difficult to evaluate. Nevertheless, formation of these deposits was undoubtedly biologically influenced or induced (see Figure 5.14).

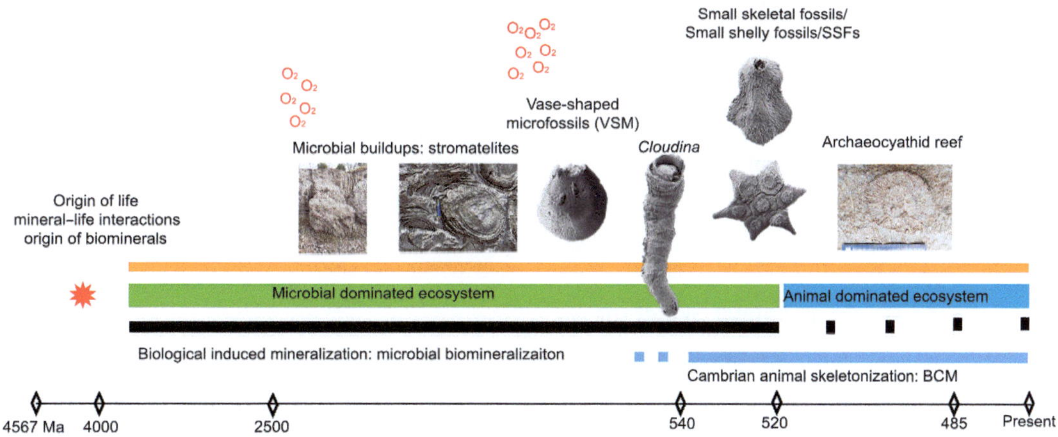

Figure 5.14 Macroevolution of biomineralization throughout the history of Earth's life.

Compelling evidence of organomineralization comes from the 2.72-billion-year-old Tumbiana Formation in the Pilbara Craton, Western Australia. Within the stromatolites of this formation, distinct nanoglobular structures composed of aragonitic crystals are embedded in microcrystalline calcite, likely representing remnants of primary carbonate minerals precipitated in association with microorganisms or organic substances. The earliest known occurrence of cyanobacterial sheath calcification through the BIM pathway is found in *Girvanella*-like filamentous organisms from the Society Cliffs Formation in Arctic Canada, dating to approximately 1.2 billion years ago (Kah and Riding 2007). Calcifying cyanobacteria and algae became widespread on Neoproterozoic carbonate platforms.

Regarding the BCM pathway, the widespread utilization of iron-sulfur and iron oxide biominerals in prokaryotes and basal eukaryotes suggests an ancient origin of iron and sulfur biomineralization. Phylogenetic analyses and molecular clock estimates indicate that magnetosome biomineralization by MTB originated in the Archean or even earlier (Goswami et al. 2022), predating the emergence of biocalcification. Furthermore, some early-branching MTB can form amorphous silica nanoparticles intracellularly, resembling the intracellular carbonatogenesis observed in certain cyanobacteria (Li et al. 2022a). This suggests a deep origin of biosilicification in prokaryotes. The capacity of microorganisms to form biominerals has likely been significantly underestimated.

Generally, biosilica mineralization is prevalent among microscopic eukaryotes and protists, while macroscopic biota tend to utilize Ca-minerals for skeletal construction. The preference of biosilica in microscopic organisms is largely attributed to the material properties of amorphous silica, which can be easily fashioned into strong porous and strutted structures at nano- to microscales, thereby reinforcing fine-scale skeletal architectures of these organisms (Knoll and Kotrc 2015). The earliest fossil evidence for biosilica mineralization comes from the Vase-shaped Microfossils (VSMs) that were found globally in Neoproterozoic deposits. These microfossils were preserved variously as silicified, phosphatized, pyritized, calcified, or cast by siliciclastic infillings through different taphonomic pathways (Tingle et al. 2023). Whether these VSMs were mineralized or not remains an open question, but a number of genera assigned to the testate amoebae seemed to bear siliceous scales (Porter et al. 2003). There is a good record of siliceous sponges and radiolarians-like fossils from the basal Cambrian (Ma et al. 2019). Diatoms first emerged in the fossil record from Jurassic deposits, diversifying throughout the Cretaceous and continuing into the Cenozoic Era (Kooistra et al. 2007).

The earliest known fossil evidence of calcium phosphate biomineralization comes from Neoproterozoic scaly microfossils found in the Fifteenmile Group of Yukon, Canada, approximately 810 Ma (Cohen et al. 2017). These apatitic microfossils exhibit a hierarchical arrangement of fibrous hydroxyapatite crystals with uniform crystallographic orientations, indicating BCM. The accumulation of phosphorus and the precipitation of apatite in marine sediments are closely associated with biological activity or released through the decomposition of organic matter (Omelon et al. 2013). The presence of phosphate minerals in some of Earth's oldest sedimentary rocks such as the Isua Group in Greenland (ca. 3.8 Ga), raises the possibility of an earlier emergence of phosphate biomineralization.

5.5.2 On the Origin of Animal Skeletons

The earliest known skeletonized metazoans occurring in Ediacaran (~550–538.8 Ma) include several representative taxa such as *Cloudina*, *Namacalathus*, *Namapoika*, *Sinotubulites*, and *Corumbella* (Osés et al. 2022; Wood 2011). These skeletal animals with uncertain affinity were likely of diploblastic grade, representing either stem- or crown-group cnidarians or basal stem-group bilaterians, and developed a sessile lifestyle (Wood 2011). Among large and diverse Ediacaran skeletal fauna, *Cloudina* was widely accepted as the first animal to mineralize a skeleton that was composed of calcareous cones nested with one another, giving it a "cone-in-cone" construction (Hua et al. 2005) (see Figure 5.15).

These early calcareous skeletons were generally flexible with somehow soft deformation, characterized by thin skeletal walls and simple, non-hierarchical microstructures made up of either aragonite or high-Mg calcite (Wood et al. 2017). Crystallization during skeletal formation was by the attachment of ACC nanoparticles, the non-classical crystallization pathway (Gilbert et al. 2019). The degree of biological control over biomineralization appears to have been minimal compared to that of Cambrian-type skeletonized metazoans. These early skeletal taxa predominantly inhabited tropical shallow marine carbonate platform where seawater was supersaturated with $CaCO_3$, facilitating calcification at a relatively low physiological cost. In the Ediacaran Nama Group, Namibia, the proliferation of *Cloudina* formed densely aggregating assemblages associated with microbialites, constructing the oldest metazoan-microbial reef complex (Wood and Curtis 2015).

The terminal Ediacaran-Cambrian transition saw the sudden appearance of disparate metazoan phyla with mineralized skeletons, represented in the fossil record by abundant skeletal remains including spicules, tubes, conchs, shells, and a variety of sclerites and ossicles (Bengtson 2004). These skeletal remains are collectively described as the "Small Shelly Fossils" or "Small Skeletal

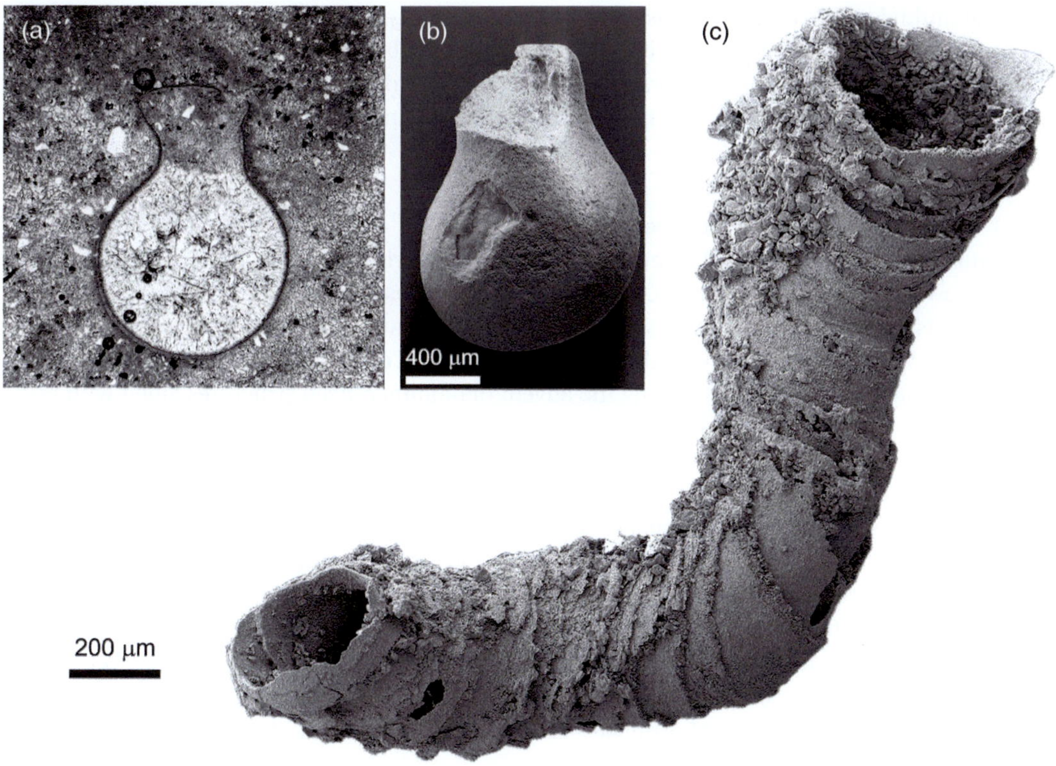

Figure 5.15 Photomicrograph of skeletal fossil *Cloudina* (c) and Vase-shaped microfossil (a,b) from the Ediacaran Gaojiashan biota, Southern Shaanxi, China. *Source:* Courtesy of Hong Hua.

Fossils" (SSFs) that were dominant in lower Cambrian pre-trilobite strata (538.8–521 Ma). The SSFs assemblage primarily consists of fragmented skeletons or disarticulated elements from macroscopic organisms with uncertain affinities, such as hyoliths, chancelloriids, and tommotiids, although some were occasionally preserved as a complete scleritome. Siliceous sponge spicules, calcareous molluscan shells, and organophosphate brachiopods are key members of the Cambrian small shelly fauna.

Unlike the metazoan-microbial buildups of the Ediacaran, some early Cambrian skeletal metazoans, such as archaeocyaths, an extinct group within Porifera, were capable of constructing true reefs on tropical carbonate platforms. Additionally, the gregarious clustering behavior of some lingulate brachiopods occasionally dominated on early Cambrian seafloor, forming biogenic shell beds (Chen et al. 2024). The diverse SSFs, with their cosmopolitan distribution, provide the most direct evidence of the Cambrian explosion. The evolutionary emergence of biomineralization in early animals is commonly known as the Cambrian animal skeletonization event. It was estimated that 89 of the 178 distinct skeletal architectures of marine skeletal animals have appeared within less than 20 million years of the period, and by the middle Cambrian, at least 80% of skeletal design elements recognized have already evolved (Thomas et al. 2000).

5.5.3 Controls on the Onset of Biomineralization

The emergence of biomineralized hard parts represents a pivotal event in the evolution of animal life on Earth. While the evolutionary mechanisms and potential triggers underlying Cambrian animal skeletonization remain debated, it is widely accepted that biomineralization arose independently and multiple times across various animal phyla, primarily through the extensive co-option of a conserved genetic toolkit that preexisted in Ediacaran non-biomineralized ancestors (Murdock 2020). Among possible extrinsic factors, this evolutionary innovation was largely driven by shifts in marine redox conditions, seawater chemistry, and the escalation of predator–prey interactions in marine ecosystems (Wood 2018).

5.5.3.1 Co-option of Ancestral Biomineralization Toolkit

The acquisition of new functions by ancestral characters or derivation of novel characters from old ones is known as co-option (True and Carroll 2002). This process involves the redeployment of preexisting traits, including genes, organs, and other body structures for a new function under new conditions of life or newly acquired habits (Mclennan 2008). Co-option is one of the major drivers of the evolution of novel traits, including the advent of biomineralized structures along with configuration of animal body plans during the Cambrian explosion (Erwin 2020).

Biomineralization has independently evolved multiple times across different animal phyla, both closely and distantly related groups. This includes the acquisition of carbonate skeletons in molluscs, phosphate skeletons in brachiopods, and siliceous skeletons in sponges. With the same biomineral groups, for instance, calcareous skeletons have arisen at least twenty times among metazoans with three times within polychaete worms, specifically in serpulids, cirratulids, and sabellids (Porter 2010; Vinn et al. 2008). Despite the frequent disappearance and recurrence of mineralized skeletons during evolution, skeletal animals share an evolutionarily highly conserved genetic toolkit inherited from a common ancestor. The presence of non-biomineralized, soft-bodied antecedents of Cambrian skeletal animals in Neoproterozoic is supported by molecular clock estimate.

At the time when BCM calcareous skeletons first appeared, their independent acquisition among different animal clades was possibly derived from the co-option of pre-existing genetic toolkits that guide the self-assembly of organic matrix where mineral precipitated (Murdock 2020). Theoretically, biomineralization evolved either through the invention of specific biomineralization genes for organic matrix assembly and mineral ion binding, or by the reuse of ancestral genetic toolkits from forming organic scaffolds to mineralizing biological hard tissues (Ben-Tabou de-Leon 2022). Among various ancient proteins, carbonic anhydrase plays a crucial role in the skeletal matrix, promoting biomineralization by rapidly catalyzing the interconversion of CO_2, water, and bicarbonate ions (Gilbert et al. 2022).

5.5.3.2 Oxygen and Marine Redox

Animals cannot survive without oxygen. The widespread occurrence of diverse skeletal taxa in the Ediacaran is tightly in association with microbially dominated carbonate platform where photosynthetic algae and cyanobacteria provided nutrient-rich, oxygenated, and carbonate-saturated conditions necessary for the costly biomineralization processes of early calcified taxa. Skeletonization is energetically expensive which requires higher O_2 for metabolism. Estimates suggest that early animals required a minimum oxygen concentration of >2.0 mg/L, approximately 25% of the present atmospheric level, to sustain biodiversity in Cambrian oceans (Zhang and Cui 2016).

The stepwise appearance of skeletonized animals through the Ediacaran-Cambrian boundary occurs from basal metazoans (terminal Ediacaran) to lophotrochozoans (Cambrian Terreneuvian), and then ecdysozoans and deuterostomes (Cambrian Age 3), which is in align with the pulse of marine oxygenation during the period. The way to utilize biomineralized hard structures was a remarkable evolutionary innovation for early animals, a process that demand a vast amount of energy and resource investments. Therefore, the rise of oxygen was vital for the biomineralization of skeletal phyla, particularly enabling early skeletal animals to produce large and elaborate mineralized skeletons.

5.5.3.3 Biomineralization as a Detoxification Mechanism

Biomineralization likely initiated as a cellular detoxification mechanism. Under conditions of excessive ion stress, organisms may utilize biomineralization as a strategy to detoxify, effectively excluding or expelling toxic ions from their cells. This phenomenon frequently occurs in microbial communities within environments enriched or contaminated with metal elements. The rise of Ca^{2+} in seawater coincided with the onset of animal skeletonization across the Ediacaran-Cambrian transition, marking a crucial point in the evolution of animal biomineralization. This period saw the extensive use of Ca^{2+} in the formation of calcium carbonate and calcium phosphate skeletons. The critical role of calcium ions in a variety of cellular processes is well established; however, excessive Ca^{2+} is pharmacologically disruptive to normal cell functions, thus ion concentration within cells must be carefully regulated. Extra Ca^{2+} is toxic, and hence early animals are forced to find a way to prevent its unlimited accumulation within cells (Simkiss 1977), otherwise, they cannot maintain survival under such environments.

Geochemical data revealed dramatic shifts in seawater chemistry during the Ediacaran-Cambrian transition due to intensified tectonic activities and increasing continental weathering after the demise of the "Snowball Earth" (Section 8.4.4). Primary fluid inclusions from marine halite indicated that seawater Ca^{2+} concentrations had increased approximately threefold during the early Cambrian, rising from 14 mmol/L in the terminal Ediacaran (ca. 544 Ma) to 37 mmol/L in the early Cambrian (ca. 515 Ma), compared to 11 mmol/L in present-day oceans (Brennan et al. 2004). In response to the calcium crisis in Cambrian seas, soft-bodied ancestors might have adapted either by modifying existing cellular waste-removal systems or by evolving novel genes/proteins that were responsible for binding Ca^{2+} and forming insoluble biominerals/organominerals. These early

Ca-detoxification processes represent the starting point for the more complex biocalcification mechanisms that eventually led to the formation of elaborate calcareous skeletons in metazoans. Analogous to the origin of biocalcification, a significant rise of dissolved phosphate in seawater occurred between 800 and 635 Ma, which is compatible with the advent of calcium phosphate biomineralization during the period. But phosphate is known as a limiting nutrient in marine ecosystems, thus bioapatite is relatively rare among animal skeletal biomineralization, both modern and Cambrian skeletal taxa (Wood 2018).

5.5.3.4 Mineralogical Stability and Changing Seawater Chemistry

The mineralogical composition of biominerals within organisms may vary during their lifespan or over evolutionary timescales. Biominerals and organominerals formed through BFM and BIM are particularly susceptible to environmental changes, such as pH, temperature, and chemical composition, due to relatively weak biological controls (Ries 2010). Among basal animal biomineralizers, mineralized tissues like serpulid tubes and coral reefs are produced within a semi-enclosed extracellular microenvironment, making them particularly sensitive to and vulnerable in the face of environmental perturbations as well. Animal lineages with strict BCM skeletons, such as molluscan and brachiopod shells, rarely switch their mineralogy once skeletons were acquired within the clade (Porter 2007). However, the evolutionary stability of biomineralogy within BCM skeletons does not mean that modern taxa produce the same biomineral types as their ancestors. For instance, an extinct group of brachiopods from the Ordovician–Silurian was known to produce aragonitic shells in contrasting to the characteristic low-Mg calcitic shells of modern brachiopod representatives (Balthasar et al. 2011).

The secular changes of seawater chemistry, especially the magnesium/calcium molar ratio (mMg:mCa) through Phanerozoic have profoundly influenced the precipitation of $CaCO_3$ polymorphs (low- and high-Mg calcite and aragonite) in non-skeletal carbonates (e.g. ooids and marine cements), hyper-calcifying algae and cyanobacteria, and basal animals (e.g. rugous corals). Generally, within the so-called "calcite sea," mMg:mCa < 2, biocalcifying organisms tend to precipitate low-Mg skeletons, while high-Mg calcite and aragonite skeletons are preferred in the so-called "aragonite sea," mMg:mCa > 2 (Stanley 2006). The modern ocean is typically defined as the "aragonite sea" with mMg:mCa of approximately 5.2, in which aragonite and high-Mg calcite minerals are commonly formed in marine sediments and non-strict BCM biomaterials.

The long-term chemical changes of seawater Mg/Ca composition throughout the Phanerozoic are primarily dictated by hydrothermal alteration between oceanic basalts and seawater, which leaches additional Ca^{2+} from basalts while removes Mg^{2+} from seawater. Thus, rapid oceanic floor spreading, associated with active tectonic activity, may lead to a significant decrease in the seawater mMg:mCa ratio, while slow oceanic floor spreading results in an "aragonite sea" environment with high mMg:mCa ratio. Direct evidence of seawater chemical composition comes from primary fluid inclusions occluded within ancient marine evaporite deposits, which documented a sharp transition from an "aragonite sea" environment during the later Ediacaran-Cambrian Terreneuvian to a subsequent "calcite sea" by late Cambrian Age 2 and beyond (Petrychenko et al. 2005; Porter 2010).

The evolution of biomineralogy in early skeletal animals followed a pattern that mirrored the fluctuations in seawater chemistry (see Figure 5.16). During the Cambrian Terreneuvian, a geological time of "aragonite sea," calcareous skeletons were principally composed of high-Mg calcite and aragonite, whereas low-Mg calcitic skeletons emerged slightly later and became prevalent in the subsequent "calcite sea" (Zhuravlev and Wood 2008). This pattern suggests that during the Ediacaran-Cambrian interval, when disparate metazoans began mineralizing hard tissues, biological controls over biomineralization were relatively limited than their modern representatives.

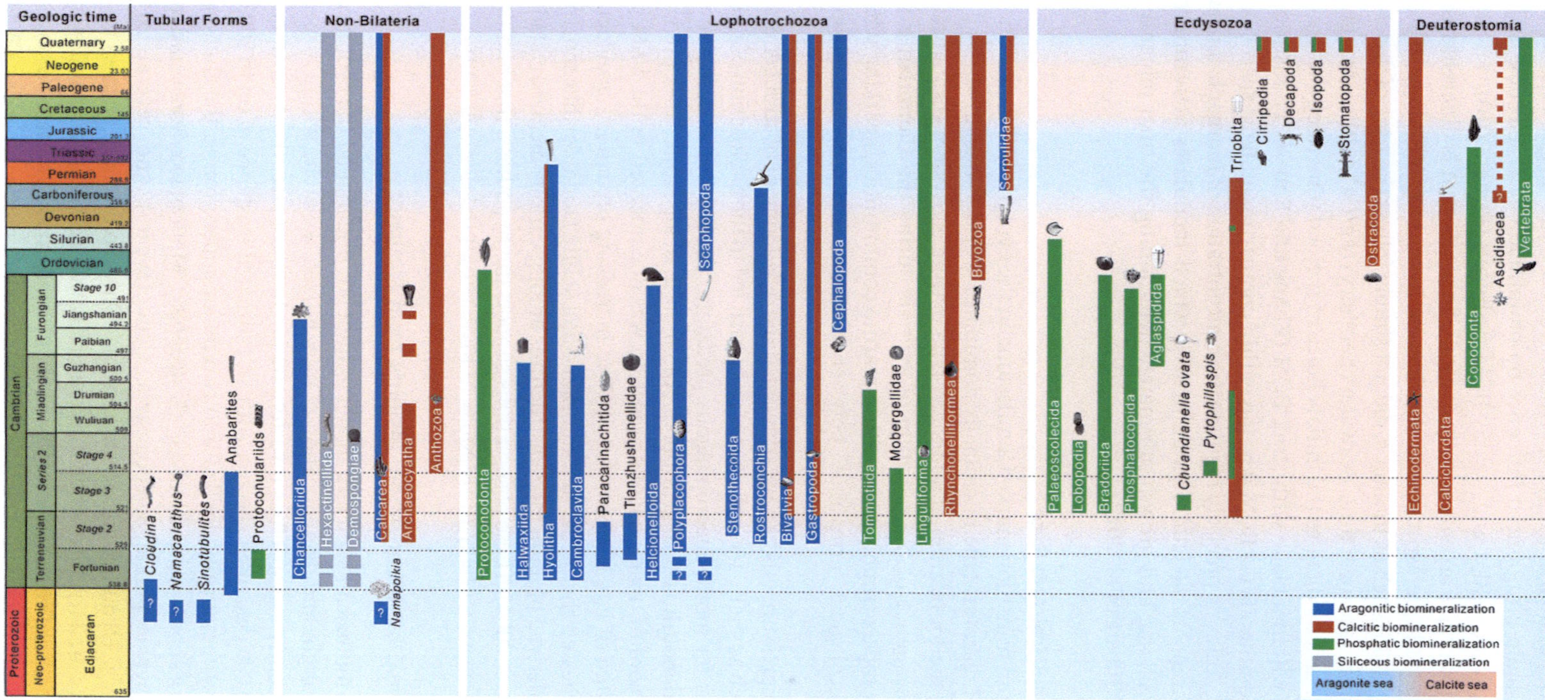

Figure 5.16 Co-evolution of seawater chemistry and biomineralogy. Mineralized skeletons were acquired independently and multiple times across different metazoan clades. Changes of calcium carbonate polymorphs from aragonite to calcite consisting with the switch of seawater chemistry from "aragonite sea" to "calcite sea," with tipping point at about the late Cambrian Age 2. *Source:* Zhang and Shu (2021) / Springer Nature / CC BY 4.0.

Conversely, extrinsic environmental agents, specifically the seawater magnesium/calcium composition, likely exerted a more pronounced influence on biomineralogical selection within a clade, especially those formed via the BIM and weakly BCM pathways (Porter 2010).

However, biomineralization is an enzymically controlled process that is different from the abiogenic mineralization in multiple aspects, and the first choice of biomineralogy, especially those formed through strictly BCM, such as molluscs, did not simply follow the pace of the changing seawater magnesium/calcium ratio over time. Seawater chemistry (Mg/Ca) neither strongly influences skeletal mineralogy nor strictly dictates the initial choice of skeletal mineralogy when mineralized skeletons first evolved within a clade with strict BCM (Zhang and Shu 2021). Instead of forming directly in seawater, biominerals via BCM develop within compartmentalized spaces isolated from external environments. Animals precisely regulate the concentration of inorganic ions within their cells and construct a complex skeletal organic matrix, which modulates activation barriers for nucleation, phase transitions, and growth, ultimately determining the final mineral polymorph in skeletal structures (Gilbert et al. 2022).

5.5.3.5 The Rise of Biological Arm-Race

The Ediacaran-Cambrian transition marked a milestone in the evolution of the marine ecosystem, transforming from an Ediacaran-type microbially dominated world to one driven by a diversity of motile animals, laying the foundation of a modern-type marine ecosystem throughout the Phanerozoic Eon. One consequence of the radiation of animals during the Cambrian explosion was the establishment of complex trophic webs within marine communities, sparking what has often been described as an ecological "arms race" between predators and preys (Bengtson 2002; Vermeij 1987).

The rise of predation pressure might have stimulated the development of stronger defensive strategies among preys, such as shells and spines, which in turn led to the exploration of more powerful phenotypic adaptations for motile macrophagous predators, as seen in the development of giant frontal appendages and mouthparts of arthropods *Anomalocaris*, an extinct apex predator in Cambrian oceans. Fossil record of microphagous predation was documented in the VSM and the first skeletal animal, *Cloudina* of late Neoproterozoic. The increasing predation, as evident with drill holes, bite marks, and healed shell scars in Cambrian small shelly fauna, further spurred the emergence of complex and derived biomineralized structures among various metazoan clades (Bicknell and Paterson 2018; Porter 2004). These observations support the hypothesis that biomineralization evolved partly as a defensive strategy in response to increasing predation pressure.

5.5.4 Cambrian Animal Skeletonization: Insights from Molluscs

In this section, we use molluscan shell as a model to in-depth understand the evolutionary mechanism underlying the origin and diversification of mineralized skeletons within a specific animal clade. Molluscs, the second most specious invertebrate phylum just after the Arthropoda, currently include more than 100,000 described species, with an estimated existence of up to 200,000 species. Modern molluscs are systematically divided into two major super-groups, the shell-bearing Conchifera (Scaphopoda, Cephalopoda, Bivalvia, Gastropoda, Monoplacophora) and the sclerite-bearing Aculifera (Polyplacophora, Solenogastres, Caudofoveata), and several extinct groups including Helcionelloida, Hyolitha, and Rostroconchia. The evolutionary success and ecological adaptability of molluscs are largely attributed to the protective advantages offered by their mineralized shells. The mineralized common ancestor of all molluscs likely possessed a single calcareous shell that was derived from their non-biomineralized antecedent in the Ediacaran (Wanninger and

Wollesen 2019). Unfortunately, no conclusive fossil evidence has yet been discovered to confirm the existence of such Precambrian hypothetical soft-bodied antecedent.

The molluscan shell consists of an outer, organic cuticle-like layer called the periostracum and inner calcareous shells that consist of several superimposed layers, each with distinct microstructures (Checa 2018). Among the most studied microstructures are the nacre (mother of pearl) and the crossed-lamellar, both have obtained much attention for their exceptional mechanical properties and the economic value of pearls. Most microstructures known in modern shells can be traced back to the early Cambrian, except that the earliest undisputed evidence of nacreous shells was found in the Ordovician, a geological period of time known for the Great Ordovician Biodiversity Event. Molluscan shell biomineralization is an ideal model for deciphering the origin and evolution of mineralized skeletons in early metazoans in general, the diversification of shell microstructures, and their interaction with the fluctuating marine environments during the Ediacaran-Cambrian transition.

5.5.4.1 Genetic Co-option Underpinning Shell Diversity

The evolution of molluscan shell biomineralization is primarily driven by the independent co-option of ancient genes and the rapid acquisition of novel biomineralization genes. Molluscan shells are secreted by specialized mantle epithelial cells derived from the ectoderm, with the fabrication of diverse microstructures by the coordinated expression of associated genetic regulatory network. These genes encode proteins and other macromolecules comprising the "mantle secretome" which include a significant proportion of novel genes encoding lineage-specific proteins and combinations of evolutionarily conserved proteins. Both ancient and novel encoded proteins of the mantle secretome contain a suite of simple repetitive low-complexity domains (RLCD).

Different RLCD-containing proteins perform different functions, including binding to chitins and/or calcium ions (when the repetitive motif is acidic). These low-complexity domains are highly evolvable and are susceptible to gain, loss, and swapping of domains (Kocot et al. 2016). The gain of novel genes in the mantle regulatory network and their encoded proteins results in species- or lineage-specific mantle secretome that underpins shell diversity. For instance, despite both abalone and pearl oysters consistently producing nacreous shells, their mantle secretome shares only 15% similarity, highlighting the unique and rapid evolvability of the molecular machinery amongst these lineages (Jackson et al. 2010).

Despite their remarkable evolvability and dissimilarity, molluscan shells also exhibit stability and conservation through their long evolutionary history. The acquisition of novel genes for shell formation in molluscs occurs through either the co-option of pre-existing genes or the incorporation of novel lineage-specific genes into the mantle genetic regulatory network. Ancient and conserved genes, such as those to encode carbonic anhydrase, form the core of the biomineralization toolkit and predate bilaterian and molluscan cladogenesis, but they likely have performed different functions unrelated to biomineralization in soft-bodied antecedents. According to a study of Aguilera et al. (2017), 50 out of 57 new gene families expressed in the mantle of gastropods were resulted from co-option events, and 109 out of 130 bivalve mantle secretome families were from co-options. Therefore, the independent co-option of ancient genes appears to be more prevalent than the recruitment of novel lineage-specific genes in molluscan shell biomineralization (Aguilera et al. 2017).

5.5.4.2 Fossils and Oldest Molluscan Shells

Molluscs originated and diversified very early in the Cambrian period. The first mineralized molluscs appeared alongside other small shelly fossils in the lowermost Cambrian, around 535 million years ago. By the Cambrian Age 2 (ca. 529 Ma), molluscs had already become widespread and constituted a major member of Cambrian marine communities. Among these early molluscs were a variety of shell forms, including simple caps, cones, slipper-shaped, and coiled, often found as millimeter-scale "micromolluscs" in secondarily phosphatized small shelly fauna on carbonate platforms, or as silicified/calcified "macromolluscs" in siliciclastic deposits. These ancient uni-shelled molluscs, tentatively described as helcionellids, differ significantly from post-Cambrian taxa in various aspects such as shell size, shape, and ontogeny, which often leads them to be classified as stem-group members of conchiferans. Additionally, a plethora of scale-shelled molluscs such as sachitids are known from the lower Cambrian and include forms such as the halkieriids and siphogonuchitids, representing the stem-group of anuliferans.

The fossil record provides robust evidence of a rapid and progressive transformation of molluscan shells from scaly shells to fully fused uni-shells in a geologically short period of time (ca. 20 Ma), with several intermediate maikhanellid species illustrating this evolutionary progression. Despite the morphological diversity of early Cambrian molluscs, these ancient shells in the Cambrian Terreneuvian were generally thin-walled, smooth, fragile, and weakly mineralized. It was not until the Cambrian Age 3 when large and heavily calcified shells with prominent radial-ridged sculptures, resembling those of some extant limpets, were first recovered in the 518-million-year-old *Qingjianglepas* from the Qingjiang Biota of South China (Li et al. 2021) (see Figure 5.17). Note that trilobite fossil record exhibited a markedly different evolutionary trajectory, characterized by the absence of gradual evolutionary changes from a hypothesized antecedent in the Terreneuvian, instead, a diverse group of trilobites with skeletonized cuticles emerged synchronously during Cambrian Age 3, the peak of the Cambrian explosion (Paterson et al. 2019; Zhang et al. 2017).

Shell morphologies and microstructural architectures co-evolved in early molluscs. Examination of Cambrian molluscan shell microstructures revealed an evolutionary trend from fibrous and loosely organized prismatic microstructure of the earliest shells to the more highly organized, lath-like lamellar types (e.g. foliated aragonite), indicating increasing biological controls over shell biomineralization. The prismatic microstructure was the most commonly formed type in Terreneuvian molluscs worldwide, whose shells were composed of a single layer of a primitive form of prismatic pattern, in which crystal building blocks were loosely arranged and randomly oriented. Advanced microstructures such as foliated aragonite and crossed-lamellar constituting multiple-layered shells emerged slightly later in more derived molluscan taxa. The crossed-lamellar microstructure with complex hierarchical organization first appeared in the stem-group gastropod *Pelagiella* by the Cambrian Age 3, approximately 515 million years ago (Li et al. 2022b).

The thickening and increased hardness of mineralized molluscan shells from the early Cambrian onward reflect a reinforcement of biomineralization in response to increasing predation pressures. This adaptive evolution in shell microstructure and morphology coincides with the rise of animal predation within a modern-type animal-dominated marine ecosystem established during the Cambrian explosion. Finally, a modern-type and heavily calcified shell, commonly encountered when walking on the beach, appeared in molluscs after about 20 million years of evolution since its first emergence at the dawn of the Cambrian.

5.5.4.3 Mineralogy and Seawater Chemistry

Molluscs are able to utilize a wide range of biominerals. They are exceptional among calcifying metazoans in terms of their exquisite biological controls and the ability to fabricate elaborate hierarchical microstructures by using polymorphs of calcium carbonates. Phosphate mineralization is

Figure 5.17 SEM photomicrograph and artistic reconstruction of early Cambrian limpet-like mollusc *Qingianglepas* from the Qingjiang biota, Hubei, China. *Source:* Photo by Luoyang Li.

notably rare in molluscs, but forming bioapatite occurs sporadically in some bivalve species. Iron mineral goethite (α-FeOOH) and silica in the form of hydrated amorphous opal ($SiO_2{\cdot}nH_2O$) are produced in the radular teeth of limpets and chitons, and iron sulfide pyrite and greigite (Fe_3S_4) in "scaly-foot" snails inhabiting deep-sea hydrothermal vents.

Because of high degrees of biological controls, molluscs hardly switch their shell mineralogy during their lifespan or through evolution. But this is not always the case. Some bivalve species, such as *Mytilus* living in tropical seas with high seawater temperature, generally precipitate more aragonite within their shells than cold-water individuals (i.e. aragonite is more easily precipitated under warm conditions). Living species cultivated in laboratory artificial seawater have been forced to switch from precipitating low-Mg calcite to aragonite by introducing additional magnesium in the solution in higher proportions (Checa et al. 2007). The forced switch of biomineralogy in *Mytilus* under laboratory conditions is a non-genetic phenomenon.

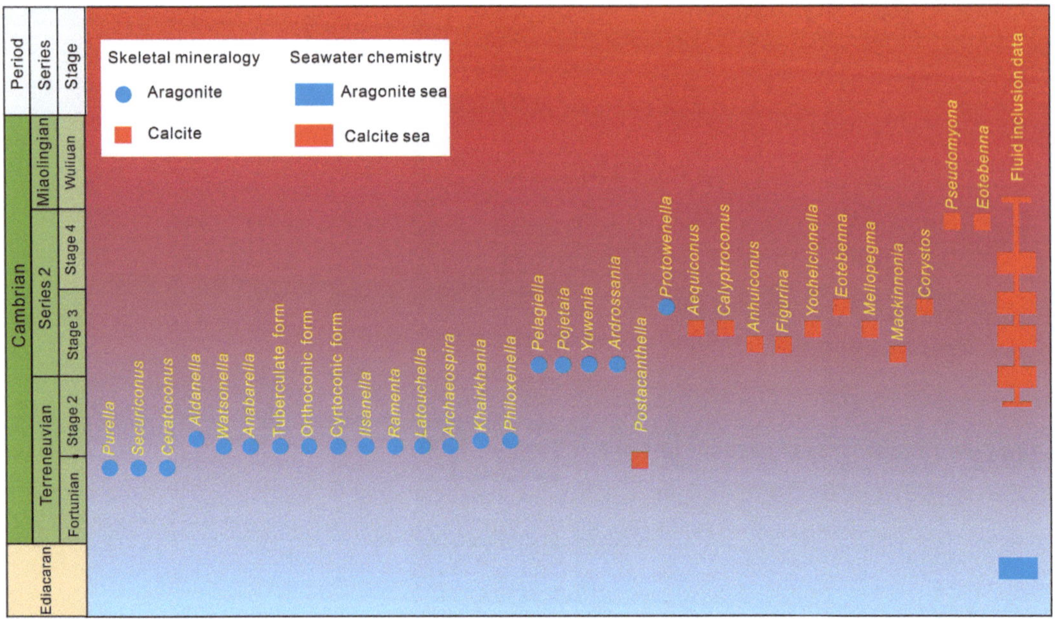

Figure 5.18 Mineralogical evolution of molluscan shells on genus level during the Cambrian. Biominerals are almost exclusively aragonite during the Cambrian Terreneuvian, with one exception that a simple cap-shaped *Postacanthella* shell consisting of low-Mg calcite, while about 70% of them during the Cambrian Series 2 and 3 are composed of calcite. *Source:* Adapted from Li et al. (2017).

When molluscs first acquired mineralized shells, biological control over shell formation was minimal, as evidenced by the simple, loosely organized shell microstructures observed in the Cambrian Terreneuvian. Statistical analysis of molluscan shell biomineralogy show a notable shift from the dominance of aragonitic shells in the Terreneuvian to the rise of calcitic shells by Cambrian Age 3 and beyond (Li et al. 2017) (see Figure 5.18). This evolutionary trend aligns with the broader mineralogical evolution of Small Shelly Fossils, where aragonite and high-Mg calcite skeletons were common during the Ediacaran-early Cambrian "aragonite sea," followed by the appearance and dominance of low-Mg calcite skeletons in the subsequent "calcite sea" (Zhuravlev and Wood 2008). Despite this correlation, any influence of seawater chemistry on the shell mineralogy of early Cambrian molluscs appears to be taxon-specific, as calcitic molluscan shells also existed in the Terreneuvian "aragonite sea," where the shells were formed under strict biological controls (Li et al. 2023). It is clear that seawater chemistry (Mg/Ca) did not significantly influence molluscan shell mineralogy, nor did it strictly dictate the choice of carbonate polymorphs for post-Cambrian heavily calcified mollusc crowns.

5.5.4.4 Microbial Attacks and Defensive Strategy

Over hundreds of millions of years of evolution, molluscs have developed various protective strategies to defend themselves against shell-crushing predators like sea stars, crabs, and seabirds, as well as extensive bioerosion caused by microorganisms, including cyanobacteria, algae, and fungi (Vermeij 1989). Post-Cambrian molluscan shells have become heavily calcified with thick-walled shells, robust external sculptures, and complex microstructures that impede crack propagation and allow the soft tissue retracted deeply inside the shell. The secretion of thick-walled shells and the presence of rigid organic periostracum outside the mineral shell layers are particularly effective in countering drilling predation and microborers on shallow water environments, as those euendolithic microorganisms generally require sunlight for photosynthesis. Moreover, the organic matrix occluded within mineralized shells also plays a role in limiting perforation, as some microborers are unable to penetrate through the intricate organic framework.

During the Ediacaran-Cambrian transition, the origin and diversification of molluscan shell biomineralization took place within a world teemed with microbes, predating the appearance of shell-crushing macroscopic predators. The close interactions between early skeletal animals and microorganisms confer significant evolutionary and ecological benefits, improving resistance to environmental stress and facilitating the process of forming intricate bio-calcified structures. In this Ediacaran-type microbial "Green Ocean," the widespread presence of microbial mats, carpeting the seafloor, provides a favoring habitat not only for Ediacaran soft-bodied organisms but also for shell-secreting molluscs by forming nutrient-rich, oxygenated, and carbonate-saturated conditions. This scenario is similar to the adaptation of "scaly-foot" gastropods found in deep-sea hydrothermal vents, both of which developed highly efficient physiological mechanisms of biomineralization.

The ecological balance between skeletal animals and microorganisms within this unique microbial "Green Ocean" could be easily destroyed with the unconstrained flourishing and rampant growth of microalgae and cyanobacteria. Similar to present-day microbial blooms, which impose significant challenges to the health of ecosystems and animals in the context of early Cambrian greenhouse environments. Evidence of this is seen in the Cambrian small shelly fauna, which were often heavily infested and overwhelmingly penetrated by euendolithic cyanobacteria on the microbial "matgrounds." Skeletal carbonate was invaded soon after the emergence of mineralized skeletons, and they were often more heavily attacked in productive water than elsewhere. The flourishing of euendolithic cyanobacteria and algae, in turn, has exerted substantial influence on the early evolution of molluscan calcification, which is far surpassing the impact of metazoan predation, particularly in the microbe-dominated ecosystems of the Ediacaran-Cambrian transition.

The Ediacaran-type microbial ecosystem has largely persisted throughout the Fortunian (~539–529 Ma) of the Cambrian Period, and was eventually replaced by the modern-type animal-dominated ecosystem during the Cambrian Age 3 (~521–514 Ma). Molluscs certainly played a role, as in the absence of motile predators, molluscs could become notorious for overgrazing onto the microbial mats, significantly altering environment they inhabited. To overcome extensive microbial attacks, the initially skeletonized molluscan ancestors creeping on mat ground developed complex, high-density tubule systems within their shells as a defensive strategy (see Figure 5.19). The tubule system is narrow canals within shells that hosts epithelial cellular extensions, which serve multiple biological functions, including respiration, shell repair, mineralization, protection against boring organisms, and mechanical support. The development of an anti-microboring mechanism and its prevalence in early molluscan shells enabled nascent mineralized molluscs to survive and exploit the seafloor microbial mat environments, which further contributed to the reform of seafloor substrates from microbe- to animal-dominated types, a substrate revolution that profoundly changed the biosphere (Li et al. 2024). This complex interaction between early shell-forming molluscs and boring microorganisms set the biological "arms race" that drove the early evolution of molluscan shell biomineralization.

Figure 5.19 An artist reconstruction of mollusc in the early Cambrian microbial "Green Ocean" ecosystem. The shell recovered from the Cambrian Terreneuvian in southwestern Mongolia has a high density of tubule system that functioned as a defensive strategy against extensive microbial attacks. *Source:* Li et al. (2024)/with permission of Elsevier.

5.6 Releasing Biomineralization Signatures from Fossils

The history of Earth and its life is recorded in fossils/rocks. Well-preserved skeletal remains, such as shells, bones, teeth, carapaces, eggs, and spicules, serve as exceptional archives of Earth's past environments and the evolution of life. Diverse fossil biomaterials provide crucial insights into ancient environments, evolutionary processes, and key geological and biological events from mass extinctions to bioradiation. However, fossil preservation is inevitably biased and incomplete, strongly shaped by geological processes. In particular, complex diagenesis imposes significant challenges in retrieving primary biomineralization signatures from ancient skeletal remains.

Distinguishing primary biomineral structures, either direct or indirect evidence, from diagenetically altered fossil skeletons is crucial for accurately reconstructing Earth's history and understanding the intricate relationship between life and the environment. These primary biomineral features, identifiable in fossils, include: (1) hierarchical organization; (2) mineral-organic biocomposites; (3) unique mineralization mechanisms; (4) biological crystallographic controls; and (5) common nanostructure nature (Pérez-Huerta et al. 2018). Molecular and genetic toolkit for biomineralization is most vulnerable to postmortem degradation, thus hardly fossilized in ancient fossils. Nevertheless, two-million-year-old DNA was recently recovered from frozen soils in Greenland (Kjær et al. 2022). Despite diagenetic overprinting, "relic" primary biogenic structures in ancient skeletal remains still provide valuable insights into the biomineralization processes of early organisms, including those involved in the Cambrian animal skeletonization.

5.6.1 Diagenesis, Permineralization, and Phosphatization

Diagenesis is known as a destructive force that involves a series of physical, chemical, and biological processes that may change primary biogenic properties of skeletal remains in sediments. Diagenesis refers to "all those changes that take place in sediment near the Earth's surface at low temperature and pressure and without crustal movement being directly involved" (Pérez-Huerta et al. 2018).

The diagenetic history of skeletal remains, to a less degree, is independent of their geological age, instead, it is strongly influenced by sedimentary environments and the nature of biomineral themselves (Pérez-Huerta et al. 2018). These controlling factors include: (1) the original biomineralogy; (2) microstructures, morphology, and porosity; (3) organic-mineral constitutes and proportions; (4) chemistry of the aqueous environment; (5) the porosity and nature of surrounding rock/sediment; and (6) physicochemical conditions of the diagenetic environment. Actually, the longer a skeletal fossil was buried, the less its biomineralization information tends to be remained due to lithification and diagenetic overprinting, thus it is considered as another factor (7) (the effect of age) that influences the quality and reliability of primary biogenic structures in ancient fossils. Notably, the metastable phase of ACC, vaterite, and aragonite would eventually transform into thermodynamically more stable low-Mg calcite under Earth's surface conditions through time. The oldest remnants of original aragonite were reported in an extinct group of brachiopod from Ordovician and Silurian rocks (Balthasar et al. 2011); all extant brachiopods produce either low-Mg calcitic or organophosphate shells.

In contrast to the destructive force of diagenesis, permineralization plays a constructive role in preserving skeletal remains in sediments. Permineralization is a common fossilization process in which the primary biological structures, such as shells, bones, or organic tissues, are filled and replaced with minerals like silica, calcium carbonate, or others (Mustoe 2017). Examples of fossils formed through this process include dinosaur bones, petrified wood, and exceptionally preserved pyritized trilobites from the Late Ordovician Beecher's Trilobite Bed. Additionally, early molluscan shells from Cambrian phosphatized SSF assemblages exhibit exquisite preservation, retaining microstructures and associated organic matrix on surface of phosphatic steinkerns. Despite that their primary mineralogical composition has been altered, permineralized fossils maintain the pseudo-morphology of the organisms and skeletons in fine details (see Figure 5.20).

The replacement of original skeletons with calcium phosphate minerals (i.e. phosphatization) plays a critical role in the preservation of Cambrian SSFs. Phosphatization produces highly accurate pseudomorphs, replicating intricate features of the original organisms through processes such as infilling, coating, and encrusting. Cambrian SSFs are predominantly preserved through secondary phosphatization, and their exceptional preservation depends on, or is at least significantly enhanced by global phosphogenesis during the Ediacaran-Cambrian transition. Their decline in post-Terreneuvian strata coincided with an obvious reduction in global phosphorite deposits, marking the close of the phosphatization window.

5.6.2 Recognizing Primary Biomineral Structures

5.6.2.1 Identification of the Original Mineralogy

The three major biomineral groups, namely carbonate, phosphate, and silica, respond very differently to diagenesis within given environments. Siliceous biogenic structures are highly resistant to post-mortem degradation and diagenetic overprinting due to the low solubility of silicon in seawater and the inherent stability of siliceous biominerals. Sponge spicules are ubiquitous in lower Cambrian strata worldwide. Organophosphate linguliform brachiopod shells also occur globally in

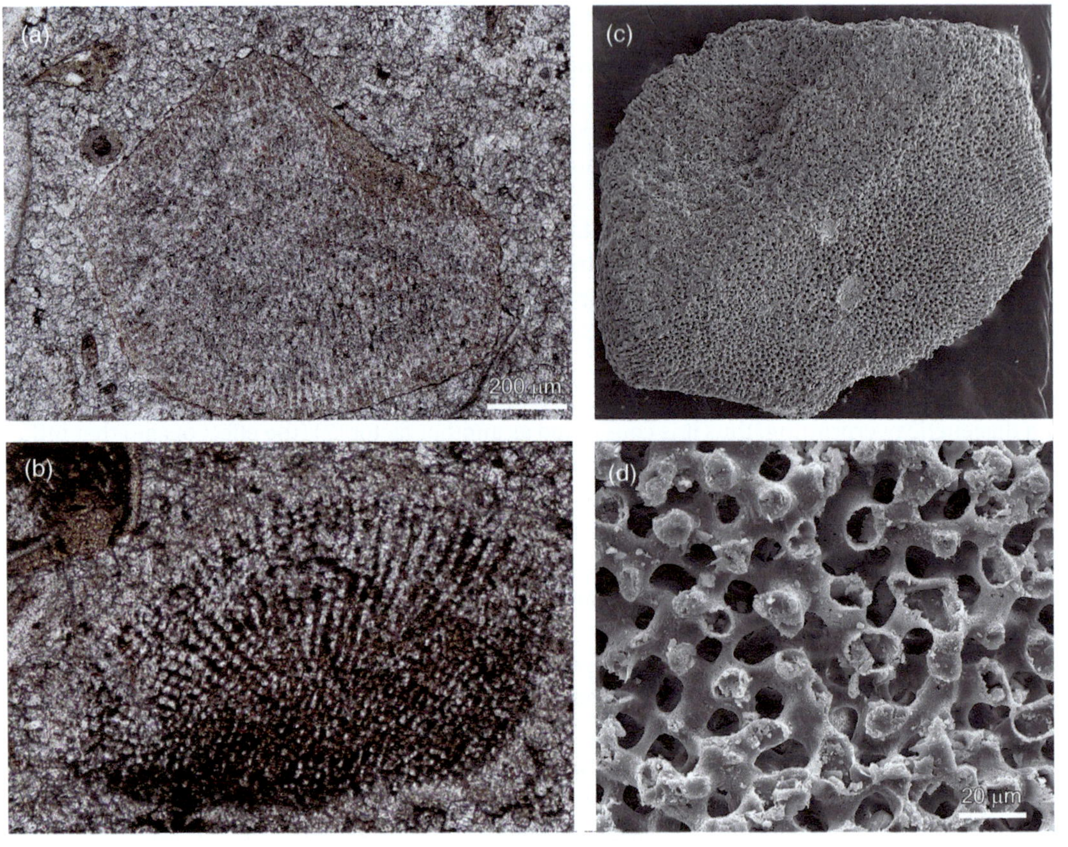

Figure 5.20 Phosphatization and permineralization of echinoderm ossicles in Cambrian rocks. (a,b) Thin sections showing the mesh-like microstructure and behaving as a single crystal of calcite; those dark and reddish parts are enriched with phosphorus; (c) phosphatized fossil extracted from carbonate rocks after laboratory process by using 10% acetic acid solution; (d) characteristic stereom microstructure of echinoderm ossicle. *Source:* Photos by Luoyang Li.

Cambrian carbonate and siliciclastic settings, and their primary mineralogical composition is more or less retained. Organophosphate brachiopods are commonly found in early Cambrian sites of exceptional preservation, such as the Chengjiang Biota in Yunnan Province, South China, which was preserved in siliciclastic sedimentary environments without associated phosphorite deposits. Phosphogenesis typically occurs in low-latitude, shallow-water epicontinental seas at depths of approximately 50–300 m (Trappe 1998).

Carbonate biominerals are highly susceptible to diagenesis, with metastable polymorphs such as aragonite and vaterite gradually transforming into more stable low-Mg calcite over time. Consequently, Cambrian calcareous or calcified skeletons consistently exhibit a low-Mg calcite mineralogy, regardless of their original carbonate polymorphs. These fossils are typically preserved as calcified or silicified shells in siliciclastic settings or undergo secondary phosphatization during periods of global phosphogenesis on carbonate platforms.

Notwithstanding persisting uncertainties and challenges, there are five criteria widely accepted for determining the original carbonate biomineralogy of fossil skeletons (Porter 2010). These criteria are as follow: (1) part or whole of fossil preserved as aragonite; or comparison with co-occurring aragonitic fossils; (2) phylogenetic inference based on biomineralogy of younger members of the taxon, e.g. the crossed-lamellar microstructure of molluscan shells is exclusively composed of aragonite through the evolution of the clade; (3) quality of preservation of original microstructures in calcite; (4) secondarily replicated microstructures and crystal morphologies, e.g. the foliated aragonite shell microstructure is characteristic with straight growth front of lath-like aragonitic crystallites, while an arrow, teeth-like pattern in foliated calcite microstructure, a typical characteristic of calcite mineralogy; and (5) trace element chemistry, particularly, aragonite crystals often incorporate high amount of strontium (Sr) within its lattice, while high Mg content in calcite biominerals (e.g. high-Mg calcite of echinoderm ossicles).

5.6.2.2 Preservation of Skeletal Organic Matrix

Biominerals are organic-mineral biocomposites, with the mineral phase embedded within a complex three-dimensional framework of macromolecules across different length scales. Due to its organic nature, the skeletal organic matrix is generally less resistant to diagenetic degradation and must decay before the enclosed mineral phase exposed to external environments. Laboratory experiments on diagenesis have shown that postmortem diagenetic alterations start immediately after burial, with organic degradation followed by a series of fluid–solid interactions through time (Pederson et al. 2019). Ancient fossil skeletons have undergone longstanding diagenetic alterations, leading to substantial modifications in terms of their chemical and structural properties.

The potential preservation of organic compounds (nucleic acids, proteins, carbohydrates, and lipids) has obtained much attention in biomolecular paleontology (Briggs et al. 2000). Although the idea of cloning a dinosaur from preserved blood in a fossilized mosquito remains firmly in the realm of science fiction, exciting discoveries have been made. Peptide bonds and chitin-protein complexes have been discovered in Jurassic dinosaur bone collagen (Thomas and Taylor 2019), demonstrating an abundance of amino acids that are highly resistant to diagenesis (Myers et al. 2018). Some exceptionally preserved bivalves in the Late Cretaceous retain original aragonite and calcite crystals in their nacre and prismatic shell layers, respectively. Additionally, Delicate preservation of skeletal organic matrix traced back to some early Cambrian molluscs was exquisitely replicated by aggregates of apatite through permineralization (see Figure 5.21). The three-dimensional inter- and intra-prismatic organic framework of early Cambrian molluscs is morphologically comparable to that of modern shells, providing critical insights into the evolution of molluscan biomineralization.

5.6.2.3 Primary Structures and Diagenetic Growths

The recognition of primary skeletal microstructures relies on five key criteria (Pérez-Huerta et al. 2018; Vendrasco et al. 2016). These include: (1) Regularity of microstructure, in particular, skeletal micro- and nanostructures formed through BCM exhibit a regular pattern of crystal organization; (2) Occurrence across taxa and ages, the same microstructure is observed in taxa spanning different geological ages and in multiple specimens; (3) Consistently reoccurring across various taphonomic pathways, identical microstructures are always found in specimens from different sedimentary settings (e.g. siliciclastic, carbonate); (4) Presence in repaired skeletal parts: the same microstructure remains to be identical in repaired parts, underscoring their biogenic genesis and biological controls; and (5) Consistent basic building units, the building blocks of given microstructures exhibit a consistent and narrow size range, reflecting their biologically controlled nature.

The Cambrian small shelly fauna is typically characterized by secondary phosphatization, which often results in diagenetic growths and ghost structures formed by aggregates of apatite. A common form of these diagenetic apatite deposits is spherulitic, often oriented perpendicular to the skeletal

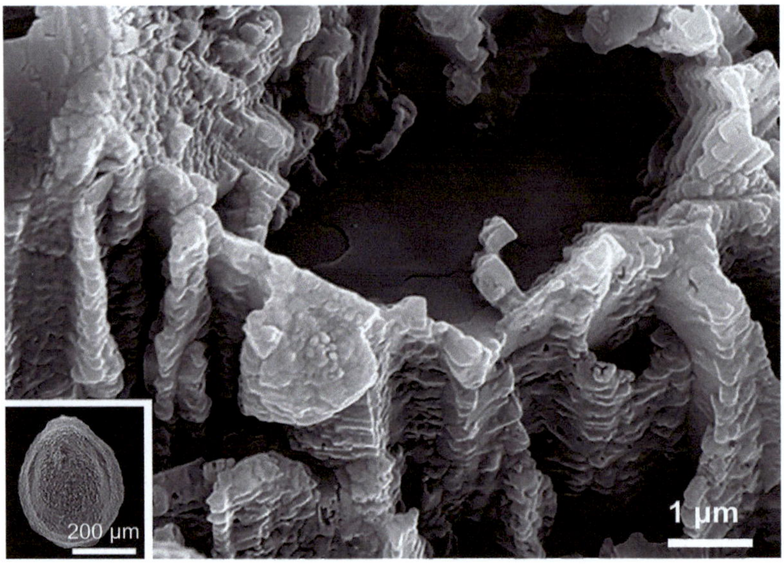

Figure 5.21 Preservation of organic matrix remnants of Cambrian molluscan shell. SEM photomicrograph of prismatic organic matrix of Cambrian Terreneuvian *Postacanthella* that was replicated by aggregates of abiogenic apatite minerals during phosphatization. *Source:* Photos by Luoyang Li.

surface. This orientation can make them morphologically indistinguishable from primary spherulitic prismatic structures of preserved skeletal remains. In such cases, additional criteria are necessary to accurately differentiate primary structures from diagenetic overgrowths.

Diagenetic spherulitic growths are widespread, occurring in phosphatic external and internal coatings as well as in recrystallized skeletal structures. Randomly arranged and irregular polygonal textures are also commonly observable on surface of phosphatic steinkerns, primarily formed during diagenesis. Furthermore, microbial degradation of organic matter releases reactive phosphorus, which, along with phosphorus from the surrounding environment, contributes to the phosphatization of SSFs. As a result, diverse microbial communities, including filamentous and spherical forms, are frequently found associated with the phosphatization of Cambrian SSFs.

5.6.3 Biomineralization in Phylogenetic Systematics

Biomineralized structures, like any other morphological and anatomical traits, provide important insights into phylogeny (tree of life) and taxonomy (classification). The paleontological interest in studying biogenic structures of fossil skeletons initiated largely on the implications of biomineralization for phylogenetic systematics. Phylogenetic systematics focuses on reconstructing the diversification pattern and mutual relationship among organisms, whereby providing a conceptual framework for understanding the evolutionary history of life. The primary biomineral structure is particularly valuable for assessing phylogenetic inter- and intra-group relationships among skeletal animals across various taxonomic levels. Generally, closely related lineages/clades often exhibit skeletal homologies and high microstructural similarities, while distantly related groups tend to share fewer skeletal commonalities.

Modern brachiopods are classified into two major groups based on their biomineralogical and microstructural differences: organophosphate brachiopods (Linguliformea) and calcareous brachiopods (Craniiformea and Rhynchonelliformea). In contrast, molluscs exhibit remarkable diversity in shell microstructures, with some bivalves secreting aragonite in the form of sheet-like nacreous structure in contrasting to the tower-like columnar nacre in gastropod shells. The biomineralization of shells in both molluscs and brachiopods is regulated by the mantle, a highly evolutionarily conserved mechanism, indicating deep homology in their shell-forming processes (Wernström et al. 2022). Despite this shared regulatory framework (deep homology), molluscan and brachiopod shells differ significantly in mineralogy and microstructure; for example, brachiopods possess a calcitic semi-nacre structure, whereas molluscs typically produce true nacre composed of aragonite.

Biomineralization is useful in resolving the puzzle with respect to the uncertain phylogenetic affinities of many Cambrian problematic skeletal animals, such as hyoliths, tommotiids, and halkieriids—three extinct groups of Cambrian communities. Of these, hyoliths produced a calcareous exoskeleton consisting of a conical conch, a lid-like operculum, and, in some species, an additional pair of elongate lateral spines called "helens." From the perspective of biomineralization, hyoliths likely have inherited their calcareous skeletons from a non-mineralized ancestor shared with molluscs, as their skeletal microstructures bear striking similarities to those of contemporaneous molluscan shells, suggesting a close phylogenetic link between the two groups (Li et al. 2022b). Similarly, tommotiids exhibit skeletal similarities to phosphatic brachiopods (Balthasar et al. 2009), and halkieriids share phylogenetic affinities with aculiferan molluscs (Vinther and Nielsen 2005). These skeletal resemblances underscore the implication of biomineralization characters in elucidating evolutionary relationships among early skeletal metazoans.

5.6.4 Skeletal Sediments and Paleoenvironmental Reconstruction

The sudden appearance of biomineralization during the Ediacaran-Cambrian interval has fundamentally changed marine ecosystems and sedimentary environments. This evolutionary milestone contributed significantly to the deposition of marine sediments as seen in abundant coastal shell beds and deep-sea biogenous sediments. With this concept, Cambrian SSFs could be associated with any shell beds observed on the beach, and the rise of biomineralization permanently reshaped the interaction between the biosphere, hydrosphere, and geosphere, altering global biogeochemical cycles forever. Biominerals have left a rich fossil record in sediments, providing an excellent archive of past oceanic conditions and long-term climate change throughout the Phanerozoic Eon.

Biogenous marine sediments are primarily composed of the skeletal remains of once-living organisms, including shells, skeletons, and other biological materials. Biogenous marine sediments specifically refer to offshore deposits where these skeletal remains sink to the seafloor after death and accumulate over time. A sedimentary deposit is classified as a biogenous zone when skeletal remains constitute at least 30% of the sediment. The two primary types of biogenous sediments are calcium carbonate ($CaCO_3$, forming the minerals calcite or aragonite) and biosilica (SiO_2). Biosilica-rich sediments mainly originate from algae such as diatoms and protozoans like radiolarians, while calcium carbonate sediments are primarily derived from foraminifers and coccolithophores. These microscopic and protozoan biomineralizers play a critical role in the "biological pump," facilitating the transfer of carbon and silicon from the surface ocean to the deep sea. The burial of large quantities of skeletal remains within marine sediments serves as a significant "sink" for global carbon and silicon cycles, influencing long-term changes in Earth's climate and ocean chemistry. Beyond their geochemical significance, biogenous sediments also hold considerable biological importance. For instance, corals, which are highly sensitive to ocean salinity and surface temperature, serve as valuable biological proxies for monitoring modern ocean acidification and climate change.

In contrast to biological signatures, geochemical proxy records refer to indirect indicators such as chemical and isotopic compositions that are used to infer unobservable environmental variables. One key parameter is the surface water temperature, which is a critical environmental parameter for reconstructing past ocean conditions. Common temperature proxies include the abundance of surface-dwelling microfossils, the oxygen isotope composition ($\delta^{18}O$) of planktic foraminifera, and elemental ratios like magnesium/calcium (Mg/Ca) and strontium/calcium (Sr/Ca) in calcareous shells (Wefer et al. 1999). However, the accuracy of many geochemical and isotopic proxies is limited by diagenetic overprinting, which affects ancient fossils, making these proxies more reliable for relatively recent specimens. Nonetheless, $\delta^{18}O$ data from phosphatic brachiopods from the Cambrian are resistant to diagenesis and preserve primary $\delta^{18}O$ signatures of seawater (Hearing et al. 2018). Geochemical data suggests that during the Cambrian explosion, ocean surface temperatures at high southern latitudes (~65–70°S) were about 20–25 °C, indicating a greenhouse climate. Suitable temperature and environment are important prerequisites for the rapid diversification of skeletonized animals in the Cambrian.

References

Addadi, L., Raz, S., and Weiner, S. 2003. Taking advantage of disorder: amorphous calcium carbonate and its roles in biomineralization. *Advanced Materials*, 15(12), pp. 959–970.

Aguilera, F., McDougall, C., and Degnan, B.M. 2017. Co-Option and de novo gene evolution underlie molluscan shell diversity. *Molecular Biology and Evolution*, 34(4), pp. 779–792.

Allemand, D., Tambutté, É., Zoccola, D., and Tambutté, S. 2011. Coral calcification, cells to reefs. In: *Coral reefs: an ecosystem in transition* (eds. Z. Dubinsky and N. Stambler), pp. 119–150. Amsterdam, Springer.

Balthasar, U., Skovsted, C.B., Holmer, L.E., and Brock, G.A. 2009. Homologous skeletal secretion in tommotiids and brachiopods. *Geology*, 37(12), pp. 1143–1146.

Balthasar, U., Cusack, M., Faryma, L., Chung, P., Holmer, L.E., Jin, J., Percival, I.G., and Popov, L.E. 2011. Relic aragonite from Ordovician–Silurian brachiopods: implications for the evolution of calcification. *Geology*, 39(10), pp. 967–970.

Bekele, S., Singh, K., Helton, E., Farajollahi, S., Naik, R.R., Dennis, P., Kelley-Loughnane, N., and Berry, R. 2022. Molecular dynamics investigation into pH dependent metal binding of the intrinsically disordered worm jaw protein, Nvjp-1. *The Journal of Physical Chemistry B*, 126(35), pp. 6614–6623.

Bell, E.A., Boehnke, P., Harrison, T.M., and Mao, W.L. 2015. Potentially biogenic carbon preserved in a 4.1 billion-year-old zircon. *Proceedings of the National Academy of Sciences of the United States of America*, 112(47), pp. 14518–14521.

Bengtson, S. 2002. Origins and early evolution of predation. *The Paleontological Society Papers*, 8, pp. 289–318.

Bengtson, S. 2004. Early skeletal fossils. *The Paleontological Society Papers*, 10, pp. 67–78.

Bengtson, S., and Briggs, D.E.G. 2005. Mineralized skeletons and early animal evolution. In: *Symposium on evolving form and function, fossils and development in honor of Adolf Seilacher* (Vol. 101) (ed. D.E.G. Briggs), pp. 124. New Haven, Yale Peabody Museum of Natural History.

Ben-Tabou de-Leon, S. 2022. The evolution of biomineralization through the co-option of organic scaffold forming networks. *Cells*, 11(4), pp. 595.

Bicknell, R.D., and Paterson, J.R. 2018. Reappraising the early evidence of durophagy and drilling predation in the fossil record: implications for escalation and the Cambrian explosion. *Biological Reviews*, 93(2), pp. 754–784.

Bosselmann, F., and Epple, M. 2008. Sulfate-containing biominerals. In: *Biomineralization: from nature to application* (eds. A. Sigel, H. Sigel, and R.K.O. Sigel), pp. 207–217. Wiley.

Brennan, S.T., Lowenstein, T.K., and Horita, J. 2004. Seawater chemistry and the advent of biocalcification. *Geology*, 32(6), pp. 473–476.

Briggs, D.E.G., Evershed, R.P., and Lockheart, M.J. 2000. The biomolecular paleontology of continental fossils. *Paleobiology*, 26(S4), pp. 169–193.

Cartwright, J.H., Checa, A.G., Gale, J.D., Gebauer, D., and Sainz-Díaz, C.I. 2012. Calcium carbonate polyamorphism and its role in biomineralization: how many amorphous calcium carbonates are there? *Angewandte Chemie International Edition*, 51(48), pp. 11960–11970.

Checa, A.G. 2018. Physical and biological determinants of the fabrication of molluscan shell microstructures. *Frontiers in Marine Science*, 5, pp. 353.

Checa, A.G., Jiménez-López, C., Rodríguez-Navarro, A., and Machado, J.P. 2007. Precipitation of aragonite by calcitic bivalves in Mg-enriched marine waters. *Marine Biology*, 150(5), pp. 819–827.

Chen, F., Brock, G.A., Zhang, Z., Topper, T.P., and Zhang, Z. 2024. Biogenic brachiopod shell concentrations from the Hongjingshao Formation (Cambrian series 2, Stage 3) in Malong area of eastern Yunnan, South China. *Gondwana Research*, 134, pp. 209–221.

Clark, M.S. 2020. Molecular mechanisms of biomineralization in marine invertebrates. *Journal of Experimental Biology*, 223(11), pp. jeb206961.

Cohen, P.A., Strauss, J.V., Rooney, A.D., Sharma, M., and Tosca, N. 2017. Controlled hydroxyapatite biomineralization in an ~810 million-year-old unicellular eukaryote. *Science Advances*, 3(6), pp. e1700095.

Collin, B., Doelsch, E., Keller, C., Panfili, F., and Meunier, J.D. 2012. Distribution and variability of silicon, copper and zinc in different bamboo species. *Plant and Soil*, 351(1), pp. 377–387.

Couradeau, E., Benzerara, K., Gérard, E., Moreira, D., Bernard, S., Brown, G.E., Jr., and López-García, P. 2012. An early-branching microbialite cyanobacterium forms intracellular carbonates. *Science*, 336 (6080), pp. 459–462.

Dauphin, Y. 2022. A brief history of biomineralization studies. *ACS Biomaterials Science & Engineering*, 9(4), pp. 1774–1790.

Dupraz, C., Reid, R.P., Braissant, O., Decho, A.W., Norman, R.S., and Visscher, P.T. 2009. Processes of carbonate precipitation in modern microbial mats. *Earth-Science Reviews*, 96(3), pp. 141–162.

Echigo, T., and Kimata, M. 2010. Crystal chemistry and genesis of organic minerals: a review of oxalate and polycyclic aromatic hydrocarbon minerals. *The Canadian Mineralogist*, 48(6), pp. 1329–1357.

Elliott, J.C. 2002. Calcium phosphate biominerals. *Reviews in Mineralogy and Geochemistry*, 48(1), pp. 427–453.

Erwin, D.H. 2020. Evolutionary dynamics of gene regulation. *Current Topics in Developmental Biology*, 139, pp. 407–431.

Feng, Z., Sui, Q., Wei, H., and Chen, J. 2024. Fossil evidence for silica biomineralization in Permian lycophytes. *National Science Review*, 11(12), pp. nwae368.

Frankel, R.B. 1991. Iron biominerals: an overview. In: *Iron biominerals* (eds. R.B. Frankel and R.P. Blakemore), pp. 1–6. Boston, Springer.

Frankel, R.B., and Bazylinski, D.A. 2003. Biologically induced mineralization by bacteria. *Reviews in Mineralogy and Geochemistry*, 54(1), pp. 95–114.

Gadd, G.M. 2021. Fungal biomineralization. *Current Biology*, 31(24), pp. R1557–R1563.

Gebauer, D., Volkel, A., and Colfen, H. 2008. Stable prenucleation calcium carbonate clusters. *Science*, 322(5909), pp. 1819–1822.

Gilbert, P.U.P.A., Abrecht, M., and Frazer, B.H. 2005. The organic-mineral interface in biominerals. *Reviews in Mineralogy and Geochemistry*, 59(1), pp. 157–185.

Gilbert, P.U.P.A., Porter, S.M., Sun, C.-Y., Xiao, S., Gibson, B.M., Shenkar, N., and Knoll, A.H. 2019. Biomineralization by particle attachment in early animals. *Proceedings of the National Academy of Sciences of the United States of America*, 116(36), pp. 17659–17665.

Gilbert, P.U.P.A., Bergmann, K.D., Boekelheide, N., Tambutté, S., Mass, T., Marin, F., Adkins, J.F., Erez, J., Gilbert, B., Knutson, V., and Cantine, M. 2022. Biomineralization: integrating mechanism and evolutionary history. *Science Advances*, 8(10), pp. eabl9653.

Görgen, S., Benzerara, K., Skouri-Panet, F., Gugger, M., Chauvat, F., and Cassier-Chauvat, C. 2021. The diversity of molecular mechanisms of carbonate biomineralization by bacteria. *Discover Materials*, 1, pp. 2.

Goswami, P., He, K., Li, J., Pan, Y., Roberts, A.P., and Lin, W. 2022. Magnetotactic bacteria and magnetofossils: ecology, evolution and environmental implications. *npj Biofilms and Microbiomes*, 8(1), pp. 43.

Granier, B. 2012. The contribution of calcareous green algae to the production of limestones: a review. *Geodiversitas*, 34(1), pp. 35–60.

Hearing, T.W., Harvey, T.H.P., Williams, M., Leng, M.J., Lamb, A.L., Wilby, P.R., Gabbott, S.E., Pohl, A., and Donnadieu, Y. 2018. An early Cambrian greenhouse climate. *Science Advances*, 4(5), pp. eaar5690.

Hendry, K.R., Marron, A.O., Vincent, F., Conley, D.J., Gehlen, M., Ibarbalz, F.M., Quéguiner, B., and Bowler, C. 2018. Competition between silicifiers and non-silicifiers in the past and present ocean and its evolutionary impacts. *Frontiers in Marine Science*, 5, pp. 22.

Hua, H., Chen, Z., Yuan, X., Zhang, L., and Xiao, S. 2005. Skeletogenesis and asexual reproduction in the earliest biomineralizing animal *Cloudina*. *Geology*, 33(4), pp. 277–280.

Huang, W., Ertekin, E., Wang, T., Cruz, L., Dailey, M., DiRuggiero, J., and Kisailus, D. 2020. Mechanism of water extraction from gypsum rock by desert colonizing microorganisms. *Proceedings of the National Academy of Sciences of the United States of America*, 117(20), pp. 10681–10687.

Jackson, D.J., McDougall, C., Woodcroft, B., Moase, P., Rose, R.A., Kube, M., Reinhardt, R., Rokhsar, D. S., Montagnani, C., Joubert, C., and Piquemal, D. 2010. Parallel evolution of nacre building gene sets in molluscs. *Molecular Biology and Evolution*, 27(3), pp. 591–608.

Jehlička, J., and Edwards, H.G. 2008. Raman spectroscopy as a tool for the non-destructive identification of organic minerals in the geological record. *Organic Geochemistry*, 39(4), pp. 371–386.

Joester, D., and Brooker, L.R. 2016. The chiton radula: a model system for versatile use of iron oxides. In: *Iron oxides* (ed. D. Faivre), pp. 177–206. New York, Wiley.

Joshi, N., Jain, N., Pathak, A., Singh, J., Prasad, R., and Upadhyaya, C.P. 2018. Biosynthesis of silver nanoparticles using carissa carandas berries and its potential antibacterial activities. *Journal of Sol-Gel Science and Technology*, 86(3), pp. 682–689.

Kah, L.C., and Riding, R. 2007. Mesoproterozoic carbon dioxide levels inferred from calcified cyanobacteria. *Geology*, 35(9), pp. 799–802.

Kim, J., Kimura, Y., Puchala, B., Yamazaki, T., Becker, U., and Sun, W. 2023. Dissolution enables dolomite crystal growth near ambient conditions. *Science*, 382(6673), pp. 915–920.

Kirschvink, J.L., and Hagadorn, J.W. 2000. A grand unified theory of biomineralization. In: *The biomineralisation of nano- and microstructures* (ed. E. Bäuerlein), pp. 139–150. Berlin, Wiley-VCH Verlag GmbH.

Kjær, K.H., Winther Pedersen, M., De Sanctis, B., De Cahsan, B., Korneliussen, T.S., Michelsen, C.S., Sand, K.K., Jelavić, S., Ruter, A.H., Schmidt, A.M., and Kjeldsen, K.K. 2022. A 2-million-year-old ecosystem in Greenland uncovered by environmental DNA. *Nature*, 612(7939), pp. 283–291.

Knoll, A.H., and Kotrc, B. 2015. Protistan skeletons: a geologic history of evolution and constraint. In: *Evolution of lightweight structures: analyses and technical applications* (ed. C.C. Hamm), pp. 1–16. Dordrecht, Springer.

Kocot, K.M., Aguilera, F., Mcdougall, C., Jackson, D.J., and Degnan, B.M. 2016. Sea shell diversity and rapidly evolving secretomes: insights into the evolution of biomineralization. *Frontiers in Zoology*, 13(1), pp. 23.

Konhauser, K. 2007. *Introduction to geomicrobiology*. Oxford, Blackwell Publishing.

Kooistra, W.H., Gersonde, R., Medlin, L.K., and Mann, D.G. 2007. The origin and evolution of the diatoms: their adaptation to a planktonic existence. In: *Evolution of primary producers in the sea* (eds. P. Falkowski and A.H. Knoll), pp. 207–249. London, Academic Press.

Li, L., Christian, C.B., Dai, T., Yun, H., Fu, D., and Zhang, X. 2021. *Qingjianglepas* from the Qingjiang biota, an evolutionary dead-end of Cambrian helcionelloid mollusks? *Palaeogeography, Palaeoclimatology, Palaeoecology*, 575(1), pp. 110480.

Li, J., Liu, P., Menguy, N., Zhang, X., Wang, J., Benzerara, K., Feng, L., Sun, L., Zheng, Y., Meng, F., and Gu, L. 2022a. Intracellular silicification by early-branching magnetotactic bacteria. *Science Advances*, 8 (19), pp. eabn6045.

Li, J., Margaret Oliver, I., Cam, N., Boudier, T., Blondeau, M., Leroy, E., Cosmidis, J., Skouri-Panet, F., Guigner, J.M., Férard, C., and Poinsot, M. 2016. Biomineralization patterns of intracellular carbonatogenesis in cyanobacteria: molecular hypotheses. *Minerals*, 6(1), pp. 10.

Li, L., Skovsted, C.B., and Topper, T.P. 2022b. Deep origin of the crossed-lamellar microstructure in early Cambrian molluscs. *Palaeontology*, 65(4), pp. e12620.

Li, L., Zhang, X., Yun, H., and Li, G. 2017. Complex hierarchical microstructures of Cambrian mollusk *Pelagiella*: insight into early biomineralization and evolution. *Scientific Reports*, 7(1), pp. 1935.

Li, L., Topper, T.P., Betts, M.J., Dorjnamjaa, D., Altanshagai, G., Enkhbaatar, B., Li, G., and Skovsted, C. B. 2023. Calcitic shells in the aragonite sea of the earliest Cambrian. *Geology*, 51(1), pp. 8–12.

Li, L., Topper, T.P., Betts, M.J., Altanshagai, G., Enkhbaatar, B., Li, G., Li, S., Skovsted, C.B., Cui, L., and Zhang, X. 2024. Tubule system of earliest shells as a defense against increasing microbial attacks. *iScience*, 27(3), pp. 109112.

Liesegang, M., and Tomaschek, F. 2020. Tracing the continental diagenetic loop of the opal-A to opal-CT transformation with X-ray diffraction. *Sedimentary Geology*, 398, pp. 105603.

Lin, W., Zhang, W., Paterson, G.A., Zhu, Q., Zhao, X., Knight, R., Bazylinski, D.A., Roberts, A.P., and Pan, Y. 2020. Expanding magnetic organelle biogenesis in the domain bacteria. *Microbiome*, 8, pp. 152.

Lowenstam, H.A. 1981. Minerals formed by organisms. *Science*, 211(4487), pp. 1126–1131.

Lowenstam, H.A., and Weiner, S. 1985. Transformation of amorphous calcium phosphate to crystalline dahllite in the radular teeth of chitons. *Science*, 227(4682), pp. 51–53.

Lowenstam, H.A., and Weiner, S. 1989. *On biomineralization*. Oxford, Oxford University Press.

Ma, Q., Feng, Q., Cao, W., Zhang, L., Ye, Y., and Gu, S. 2019. Radiolarian fauna from the Chiungchussuan Shuijingtuo Formation (Cambrian series 2) in Western Hubei Province, South China. *Science China Earth Sciences*, 62, pp. 1645–1658.

Mann, S. 1983. Mineralization in biological systems. In: *Inorganic elements in biochemistry* (eds. P.H. Connett, H. Folłmann, M. Lammers, J.D. Odom, and K.E. Wetterhahn), pp. 125–174. Berlin, Springer.

Marin, F., Le Roy, N., and Marie, B. 2012. The formation and mineralization of mollusk shell. *Frontiers in Bioscience*, 4(3), pp. 1099–1125.

Marin, F., Bundeleva, I., Takeuchi, T., Immel, F., and Medakovic, D. 2016. Organic matrices in metazoan calcium carbonate skeletons: composition, functions, evolution. *Journal of Structural Biology*, 196(2), pp. 98–106.

Mclennan, D.A. 2008. The concept of co-option: why evolution often looks miraculous. *Evolution: Education and Outreach*, 1(3), pp. 247–258.

Mitterer, R.M. 1968. Amino acid composition of organic matrix in calcareous oolites. *Science*, 162(3861), pp. 1498–1499.

Murdock, D.J.E. 2020. The 'biomineralization toolkit' and the origin of animal skeletons. *Biological Reviews*, 95(5), pp. 1372–1392.

Mustoe, G.E. 2017. Wood petrifaction: a new view of permineralization and replacement. *Geosciences*, 7(4), pp. 119.

Myers, C.E., Bergmann, K.D., Sun, C.Y., Boekelheide, N., and Gilbert, P.U.P.A. 2018. Exceptional preservation of organic matrix and shell microstructure in a Late Cretaceous *Pinna* fossil revealed by photoemission electron spectromicroscopy. *Geology*, 46(8), pp. 711–714.

Omelon, S., Ariganello, M., Bonucci, E., Grynpas, M., and Nanci, A. 2013. A review of phosphate mineral nucleation in biology and geobiology. *Calcified Tissue International*, 93(4), pp. 382–396.

Osés, G.L., Wood, R., Romero, G.R., Prado, G.M.E.M., Bidola, P., Herzen, J., Pfeiffer, F., Stampar, S.N., and Pacheco, M.L.A.F. 2022. Ediacaran *Corumbella* has a cataphract calcareous skeleton with controlled biomineralization. *iScience*, 25(12), pp. 105676.

Palmer, A.R. 1992. Calcification in marine molluscs: how costly is it? *Proceedings of the National Academy of Sciences*, 89(4), pp. 1379–1382.

Park, Y., and Faivre, D. 2022. Diversity of microbial metal sulfide biomineralization. *ChemPlusChem*, 87(1), pp. e202100457.

Pasteris, J.D. 2016. A mineralogical view of apatitic biomaterials. *American Mineralogist*, 101(12), pp. 2594–2610.

Paterson, J.R., Edgecombe, G.D., and Lee, M.S.Y. 2019. Trilobite evolutionary rates constrain the duration of the Cambrian explosion. *Proceedings of the National Academy of Sciences of the United States of America*, 116(10), pp. 4394–4399.

Pederson, C., Mavromatis, V., Dietzel, M., Rollion-Bard, C., Nehrke, G., Jöns, N., Jochum, K.P., and Immenhauser, A. 2019. Diagenesis of mollusc aragonite and the role of fluid reservoirs. *Earth and Planetary Science Letters*, 514, pp. 130–142.

Pérez-Huerta, A., Coronado, I., and Hegna, T.A. 2018. Understanding biomineralization in the fossil record. *Earth-Science Reviews*, 179, pp. 95–122.

Perry, R.S., Mcloughlin, N., Lynne, B.Y., Sephton, M.A., Oliver, J.D., Perry, C.C., Campbell, K., Engel, M. H., Farmer, J.D., Brasier, M.D., and Staley, J.T. 2007. Defining biominerals and organominerals: direct and indirect indicators of life. *Sedimentary Geology*, 201(1–2), pp. 157–179.

Petrychenko, Y., Peryt, T.M., and Chechel, E.I. 2005. Early Cambrian seawater chemistry from fluid inclusions in halite from Siberian evaporites. *Chemical Geology*, 219(1–4), pp. 149–161.

Porter, S.M. 2004. The fossil record of early eukaryotic diversification. *The Paleontological Society Papers*, 10, pp. 35–50.

Porter, S.M. 2007. Seawater chemistry and early carbonate biomineralization. *Science*, 316(5829), pp. 1302–1302.

Porter, S.M. 2010. Calcite and aragonite seas and the de novo acquisition of carbonate skeletons. *Geobiology*, 8(4), pp. 256–277.

Porter, S.M., Meisterfeld, R., and Knoll, A.H. 2003. Vase-shaped microfossils from the Neoproterozoic Chuar Group, Grand Canyon: a classification guided by modern testate amoebae. *Journal of Paleontology*, 77(3), pp. 409–429.

Ries, J.B. 2010. Geological and experimental evidence for secular variation in seawater Mg/Ca (calcite-aragonite seas) and its effects on marine biological calcification. *Biogeosciences*, 9, pp. 2795–2849.

Roberts, J.A. 2024. The problem with dolomite. *Nature Geoscience*, 17(8), pp. 716–716.

Roberts, J.A., Bennett, P.C., González, L.A., Macpherson, G.L., and Milliken, K.L. 2004. Microbial precipitation of dolomite in methanogenic groundwater. *Geology*, 32(4), pp. 277–280.

Sakaguchi, T., Burgess, J.G., and Matsunaga, T. 1993. Magnetite formation by a sulphate-reducing bacterium. *Nature*, 365(6441), pp. 47–49.

Schmidt, W.J. 1924. *Die bausteine des Tierkörpers in polarisiertem Lichte*, pp. 528. Bonn, F. Cohen.

Schoonen, M.A.A., and Barnes, H.L. 1991. Reactions forming pyrite and marcasite from solution: II. Via FeS precursors below 100 C. *Geochimica et Cosmochimica Acta*, 55(6), pp. 1505–1514.

Simkiss, K. 1977. Biomineralization and detoxification. *Calcified Tissue Research*, 24(1), pp. 199–200.

Skinner, H.C.W. 2005. Biominerals. *Mineralogical Magazine*, 69(5), pp. 621–641.

Skinner, H.C.W., and Jahren, A.H. 2003. Biomineralization. In: *Treatise on geochemistry 8* (eds. W.H. Schlesinger and K.K. Holland Turekian), pp. 117–184. Amsterdam, Elsevier.

Stanley, S.M. 2006. Influence of seawater chemistry on biomineralization throughout Phanerozoic time: paleontological and experimental evidence. *Palaeogeography, Palaeoclimatology, Palaeoecology*, 232 (2–4), pp. 214–236.

Sun, J., Chen, C., Miyamoto, N., Li, R., Sigwart, J.D., Xu, T., Sun, Y., Wong, W.C., Ip, J.C., Zhang, W., and Lan, Y. 2020. The scaly-foot snail genome and implications for the origins of biomineralised armour. *Nature Communications*, 11(1), pp. 1657.

Suzuki, M. 2020. Structural and functional analyses of organic molecules regulating biomineralization. *Bioscience, Biotechnology, and Biochemistry*, 84(8), pp. 1529–1540.

Suzuki, Y., Kopp, R.E., Kogure, T., Suga, A., Takai, K., Tsuchida, S., Ozaki, N., Endo, K., Hashimoto, J., Kato, Y., and Mizota, C. 2006. Sclerite formation in the hydrothermal-vent "scaly-foot" gastropod—possible control of iron sulfide biomineralization by the animal. *Earth and Planetary Science Letters*, 242(1–2), pp. 39–50.

Tebo, B.M., Bargar, J.R., Clement, B.G., Dick, G.J., Murray, K.J., Parker, D., Verity, R., and Webb, S.M. 2004. Biogenic manganese oxides: properties and mechanisms of formation. *Annual Review of Earth and Planetary Sciences*, 32(1), pp. 287–328.

Thomas, B., and Taylor, S. 2019. Proteomes of the past: the pursuit of proteins in paleontology. *Expert Review of Proteomics*, 16(11–12), pp. 881–895.

Thomas, R.D.K., Shearman, R.M., and Stewart, G.W. 2000. Evolutionary exploitation of design options by the first animals with hard skeletons. *Science*, 288(5469), pp. 1239–1242.

Tingle, K.E., Porter, S.M., Raven, M.R., Czaja, A.D., Webb, S.M., and Bloeser, B. 2023. Organic preservation of vase-shaped microfossils from the late Tonian Chuar Group, Grand Canyon, Arizona, USA. *Geobiology*, 21(3), pp. 290–309.

Towe, K.M., and Lowenstam, H.A. 1967. Ultrastructure and development of iron mineralization in the radular teeth of *Cryptochiton stelleri* (Mollusca). *Journal of Ultrastructure Research*, 17(1–2), pp. 1–13.

Trappe, J. 1998. *Phanerozoic phosphorite depositional systems: a dynamic model for a sedimentary resource system*, Lecture Notes in Earth Sciences (Vol. 76), pp. 316. Berlin, Springer.

Trichet, J. 1968. Etude de la composition de la fraction organique des oolithes. Comparaison avec celle des membranes des bactéries et des cyanophycées. *Comptes Rendus Hebdo, Séances Académie des Sciences*, 267, pp. 1392–1494.

Trichet, J., and Defarge, C. 1995. Non-biologically supported organomineralization. *Bulletin de l'Institut Océanographique (Monaco) Numéro Spécial*, 14, pp. 203–236.

True, J.R., and Carroll, S.B. 2002. Gene co-option in physiological and morphological evolution. *Annual Review of Cell and Developmental Biology*, 18(1), pp. 53–80.

Vasconcelos, C., McKenzie, J.A., Bernasconi, S., Grujic, D., and Tiens, A.J. 1995. Microbial mediation as a possible mechanism for natural dolomite formation at low temperatures. *Nature*, 377(6546), pp. 220–222.

Vendrasco, M.J., Rodríguez-Navarro, A.B., Checa, A.G., Devaere, L., and Porter, S.M. 2016. To infer the early evolution of mollusc shell microstructures. *Key Engineering Materials*, 672, pp. 113–133.

Vermeij, G.J. 1987. *Evolution and escalation: an ecological history of life*. Princeton, Princeton University Press.

Vermeij, G.J. 1989. The origin of skeletons. *Palaios*, 4(6), pp. 585–589.

Vinn, O. 2021. Biomineralization in polychaete annelids: a review. *Minerals*, 11(10), pp. 1151.

Vinn, O., Ten Hove, H.A., and Mutvei, H. 2008. On the tube ultrastructure and origin of calcification in sabellids (Annelida, Polychaeta). *Palaeontology*, 51(2), pp. 295–301.

Vinther, J., and Nielsen, C. 2005. The early Cambrian halkieria is a mollusc. *Zoologica Scripta*, 34(1), pp. 81–89.

Wanninger, A., and Wollesen, T. 2019. The evolution of molluscs. *Biological Reviews*, 94(1), pp. 102–115.

Warén, A., Bengtson, S., Goffredi, S.K., and Van Dover, C.L. 2003. A hot-vent gastropod with iron sulfide dermal sclerites. *Science*, 302(5647), pp. 1007–1007.

Watabe, N., and Kingsley, R.J. 1989. Extra-, inter-, and intracellular mineralization. In: *Origin, evolution, and modern aspects of biomineralization in plants and animals* (ed. R.E. Rex), pp. 209e223. New York, Plenum Press.

Wefer, G., Berger, W.H., Bijma, J., and Fischer, G. 1999. Clues to ocean history: a brief overview of proxies. In: *Uses of proxies in paleoceanography: examples from the South Atlantic* (eds. G. Fischer and G. Wefer), pp. 1–68. Berlin, Springer.

Weigend, M., Mustafa, A., and Ensikat, H.J. 2018. Calcium phosphate in plant trichomes: the overlooked biomineral. *Planta*, 247, pp. 277–285.

Weiner, S., and Addadi, L. 2002. At the cutting edge. *Science*, 298(5592), pp. 375–376.

Weiner, S., and Dove, P.M. 2003. An overview of biomineralization processes and the problem of the vital effect. *Reviews in Mineralogy and Geochemistry*, 54(1), pp. 1–29.

Wernström, J.V., Gąsiorowski, L., and Hejnol, A. 2022. Brachiopod and mollusc biomineralisation is a conserved process that was lost in the phoronid–bryozoan stem lineage. *EvoDevo*, 13(1), pp. 17.

Wiersma, K., Läbe, S., and Sander, P.M. 2021. Organic phase preservation in fossil dinosaur and other tetrapod bone from deep time. In: *Fossilization: understanding the material nature of ancient plants and animals* (eds. C.T. Gee, V.E. McCoy, and P.M. Sander), pp. 16–54. Baltimore, Johns Hopkins University Press.

Wood, R.A. 2011. Paleoecology of the earliest skeletal metazoan communities: implications for early biomineralization. *Earth-Science Reviews*, 106(1–2), pp. 184–190.

Wood, R.A. 2018. Exploring the drivers of early biomineralization. *Emerging Topics in Life Sciences*, 2(2), pp. 201–212.

Wood, R.A., and Curtis, A. 2015. Extensive metazoan reefs from the Ediacaran Nama Group, Namibia: the rise of benthic suspension feeding. *Geobiology*, 13(2), pp. 112–122.

Wood, R.A., and Zhuravlev, A.Y. 2012. Escalation and ecological selectively of mineralogy in the Cambrian radiation of skeletons. *Earth-Science Reviews*, 115(4), pp. 249–261.

Wood, R.A., Curtis, A., Penny, A., Zhuravlev, A.Y., Curtis-Walcott, S., Iipinge, S., and Bowyer, F. 2017. Flexible and responsive growth strategy of the ediacaran skeletal *Cloudina* from the Nama Group, Namibia. *Geology*, 45(3), pp. 259–262.

Zhang, X., and Cui, L. 2016. Oxygen requirements for the Cambrian explosion. *Journal of Earth Science*, 27, pp. 187–195.

Zhang, X., and Shu, D. 2021. Current understanding on the Cambrian explosion: questions and answers. *PalZ*, 95(4), pp. 641–660.

Zhang, X., Ahlberg, P., Babcock, L.E., Choi, D.K., Geyer, G., Gozalo, R., Hollingsworth, J.S., Li, G., Naimark, E.B., Pegel, T., and Steiner, M. 2017. Challenges in defining the base of Cambrian series 2 and stage 3. *Earth-Science Reviews*, 172, pp. 124–139.

Zhang, R., Xie, L., and Yan, Z. 2018. *Biomineralization mechanism of the pearl oyster, Pinctada fucata*. Singapore, Springer.

Zhang, Z., Zhang, Z., Holmer, L.E., Topper, T.P., Pan, B., and Li, G. 2024. Evolution and diversity of biomineralized columnar architecture in early Cambrian phosphatic-shelled brachiopods. *eLife*, 12, pp. RP88855.

Zhuravlev, A.Y., and Wood, R.A. 2008. Eve of biomineralization: controls on skeletal mineralogy. *Geology*, 36(12), pp. 923–926.

6

Biosedimentation

Wei Liu and Xingliang Zhang

6.1 Introduction

Biosedimentation refers to the process in which biological activities contribute to the accumulation and/or settling of organic and inorganic material in depositional environment and thereby alter the sedimentary pattern. The term is often used in the context of natural systems, where living organisms play multiple roles in the deposition and consolidation of sedimentary rocks.

Biological activity is involved in sedimentation in many ways. Organisms, such as animals and plants, can directly impact sedimentation by producing organic matter, altering the physical properties of sediments, or influencing the chemical composition of water bodies. For example, shells and coral reefs have been known to leave their biomineralized skeletons as bioclastic sediments of various sizes accumulating to form large-scale sedimentary structures. Similarly, microbial life can also play a pivotal role in mineral precipitation. They are responsible for the breakdown of buried organic carbon, mediate the pore water chemistry, and thus induce the mineralization, leading to the formation of features like microbialites in carbonate environments (Figure 6.1) or microbially induced sedimentary structures in siliciclastic sedimentary environments (Figure 6.2).

Microbial activity is the key aspect of biosedimentation. On the one hand, microorganisms may precipitate minerals (e.g. calcium carbonate) as part of their metabolic processes. These minerals become incorporated into sediments, influencing their structure and stability. On the other hand, the metabolism of microorganisms, such as bacteria, fungi, and algae, influences the sedimentation process through the production of extracellular polymeric substances (EPS), which bind particles together. These microorganisms contribute to the aggregation of particles, making them more likely to settle. Furthermore, the decay of organic matter by microbes in sediments produces gases (e.g. methane or carbon dioxide) and other byproducts that can alter the density and cohesion of sediment particles. This decomposition process also contributes to the overall nutrient cycle in ecosystems.

In this chapter, we go into the insight of these processes and products, typically associated with microorganisms, which are discussed separately in biogenic sediments, biological diagenesis, microbially induced sedimentary structures, and their preservation and recognition in sedimentary rocks. Bioturbation, the process by which organisms disturb and mix sediments through activities like burrowing, feeding, and defecation, plays a crucial role in sedimentation by influencing physical and chemical properties of sediments. Bioturbators such as worms, shrimp, and burrowing fish

Introduction to Geobiology, First Edition. Xingliang Zhang, Zisheng Guo, Wei Liu, Ruliang He, Luoyang Li, and Weiduo Hao.

Companion website: www.wiley.com/go/Zhang_1e

Figure 6.1 Stromatolite in limestone from Kalkowsky's type locality, Lower Triassic, Germany. *Source:* Photo by Xingliang Zhang.

Figure 6.2 Polished slide showing the biolaminite structure in sandstone, a type of microbially induced sedimentary structure (MISS) formed by microbial growth, Lower Cambrian, China. *Source:* Photo by Wei Liu.

disturb sediments, redistribute particles, and prevent stratification. Bioturbation, typically forming trace fossils of burrows, tracks, trails (Figure 6.3), and root penetration structures, is a classic topic of paleoecology and hence not discussed herein.

6.2 Biogenic Sediments

Biogenic sediments, which originate from biological processes, include mineral precipitates formed through biologically influenced, induced, and controlled processes (Section 5.3), organic remains of organisms, and particles resulting from bioerosion (Section 7.2). These sediments can be categorized into two main types: pelagic biogenic sediments (pelagites), which are produced by plankton and represented by siliceous and calcareous oozes found on the ocean floor, and benthic sediments,

Figure 6.3 Bioturbation records. (a) *Treptichnus pedum*, a worm burrow from the Fortunian of Cambrian System of South China; (b) *Gyrolithes scintillus*, a type of helminthoid or spiral burrow attributed to the activity of organisms that created a spiral or coiled structure in sediment. Fortune Head, Newfoundland, Canada. *Source:* Photos by Wei Liu.

which consist of the skeletons of benthic organisms or precipitates formed by benthic microbial mats. Here we will examine biogenic sediments based on their chemical composition, focusing on carbonate, phosphatic, siliceous, and iron-rich sediments.

6.2.1 Carbonates

Carbonate rocks (Figure 6.4), primarily limestones and dolostones, account for 20–25% of all sedimentary rocks. They have long been established that "carbonates are born, not made" (Flügel 2004; James 1979). These carbonate sediments can either form as skeletal grains or precipitate directly

Figure 6.4 Platform carbonates featuring various giant domal bioherms and lithoherms, along with diverse columnar and domical stromatolites from the "Ghaap Plateau Facies" of the Campbellrand Subgroup (Transvaal Supergroup, 2.55–2.50 Ga), South Africa. *Source:* Photo by Xingliang Zhang.

within the depositional environment. More than 90% of the limestones in modern marine settings are of biological origin, precipitating through either biologically influenced, induced, or controlled processes. Even some abiotic carbonate precipitation, such as cement formation, is influenced by organic material or the activities of organisms. The significant role of organisms in limestone formation was first recognized by English geologist and microscopist Henry Clifton Sorby (1826–1908) in 1879, who proposed the Sorby Principle: "Limestones are predominantly biogenic sediments" (see Flügel 2004). Additionally, recent studies have focused extensively on microbial contributions to dolomite formation (Kim et al. 2023).

6.2.1.1 Limestones are Biological Sediments

Available data indicate that near-surface waters in modern oceans are oversaturated with respect to calcite by more than six times and with respect to aragonite more than four times (Morse and Mackenzie 1990). However, the precipitation of carbonate minerals in natural aquatic environments is influenced not only by the saturation state but also by pH, which is controlled by the partial pressure of dissolved CO_2 in the water and the concentration of other ions (e.g. SO_4^{2-}, Mg^{2+}). In general, a loss of CO_2 and a decrease in Mg^{2+} concentration promote the precipitation of calcite or aragonite. The limited precipitation of calcite in oversaturated waters is due to the small loss of CO_2 and the presence of Mg^{2+}, while the slower precipitation of aragonite may be attributed to the effect of organic compounds, which can form thin organophosphatic coatings on aragonite seed nuclei, inhibiting their growth and preventing or significantly delaying aragonite formation (Boggs 2001). Nevertheless, organisms play a significant role in carbonate sedimentation in various ways. They can overcome kinetic barriers through biologically induced (BIM) and biologically controlled (BCM) mineralization processes, thereby generating new sediments. Additionally, they contribute to the reduction of carbonate grains into micrite through intensive bioerosion, a process known as micritization. Flügel (2004) provided a comprehensive review of the biological contributions to the formation of limestone matrix and grains, which is summarized below.

Micrite is a fine-grained sediment, typically composed of very small particles (less than 4 μm in size) of calcium carbonate ($CaCO_3$), that is often found in limestone (Figure 6.5). The term

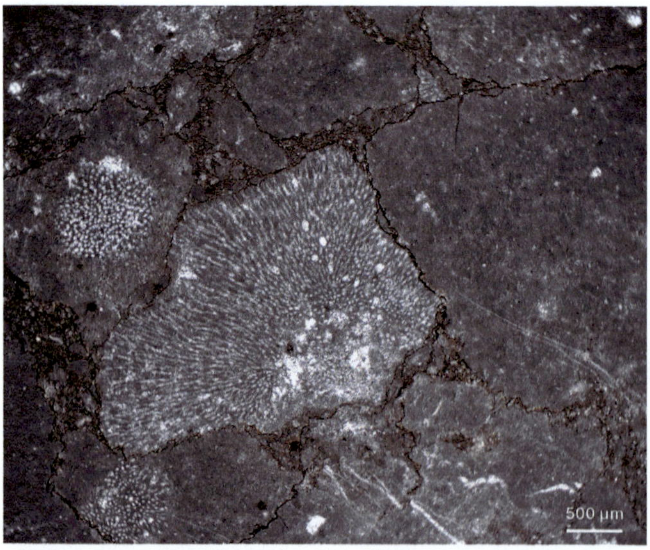

Figure 6.5 Micrite limestone from lagoonal facies, containing fossilized calcified cyanobacteria, found in the Lianglitag Formation, Ordovician, China. *Source:* Courtesy of Lijing Liu.

"micrite" refers to the texture of the material rather than its composition, and it is distinguished by its microscopic grain size. It is an abbreviation for microcrystalline calcite or aragonite and is often used interchangeably with "lime mud." Micrite can occur as a matrix between larger grains or may constitute the majority, or even the entirety, of a limestone.

Micrite is genetically classified into two categories: (1) automicrite refers to autochthonous micrite, which is the result of in-place precipitation of fine-grained calcite or aragonite on the seafloor or within sediments. As an authigenic product, it has a consistent mineralogical and chemical composition. Take diagenetic micrites (or pseudomicrites), for example, which are micro- and cryptocrystalline cements that form in place, corresponding to automicrite. (2) Allomicrite, on the other hand, is allochthonous micrite, derived from the breakdown of various carbonate grains, and may contain mineralogically and chemically diverse constituents.

While both types of micrite can form through abiotic processes, such as physicochemical precipitation triggered by fluctuations in salinity and temperature or the mechanical erosion of limestone, biological processes play a significant role in the formation of both auto- and allomicrites. These biological roles are well documented in modern marine and non-marine depositional environments, as well as in ancient carbonates. These include: (1) non-living organic matrices, which can fix and bind divalent cations (e.g. Ca^{2+} and Mg^{2+}), lead to organomineralization and the formation of organomicrite. (2) Metabolic processes of heterotrophs, chemolithotrophs, and other microbes (e.g. bacterial decay of organic matter) that induce physicochemical changes in microenvironments, resulting in $CaCO_3$ precipitation. (3) CO_2 uptake by phototrophic cyanobacteria and algae, which increases local pH and induces micrite (Mg-calcite and aragonite) precipitation, evidenced by the calcification of filamentous cyanobacteria (Figure 6.6). (4) The disintegration of calcified biological material, such as calcareous algae and invertebrate skeletons, through biological and mechanical processes, produces large quantities of allomicrites. (5) Bioerosion by boring and

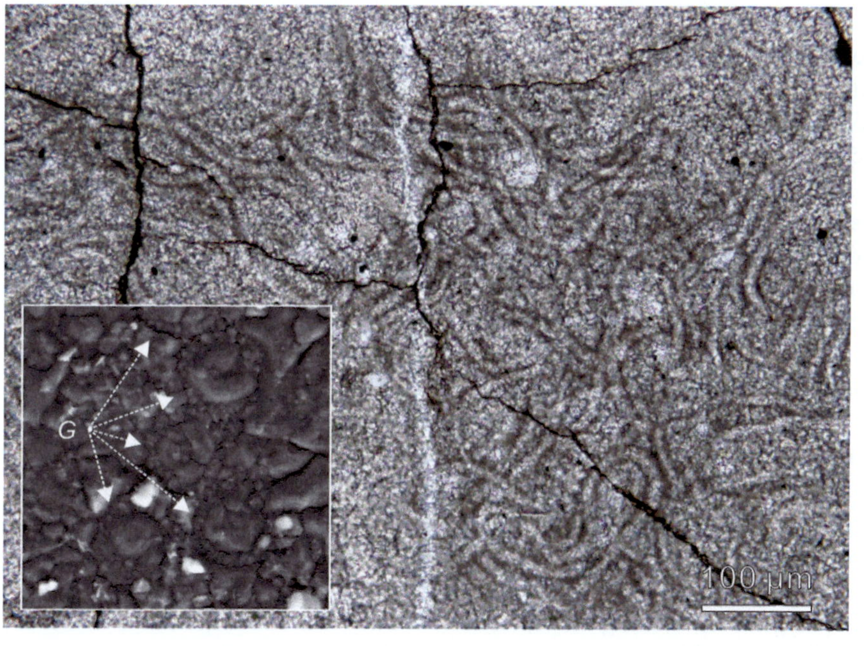

Figure 6.6 *Girvanella* fossils and their microphotographic images under SEM, which appear as flexuous, unsegmented, and unbranched tubular filaments (about 20 μm in diameter) delimited by the tube wall (about 2 μm in thickness), from Cambrian oolitic limestone, China. *G*, *Girvanella*. *Source:* Photos by Wei Liu.

Figure 6.7 Foraminiferal shells from the tropical eastern Indian Ocean. *Source:* Courtesy of Peng Zhang.

grazing organisms generates clay- and silt-sized carbonate particles, significantly contributing to the formation of micrite and calcisiltite. (6) Carbonate ooze, or pelagic micrite (lime mud), forms from the accumulation of calcareous plankton (e.g. planktonic foraminifera and algae) on the deep-sea floor (Figure 6.7).

Calcisilt is defined as detrital, silt-sized lime sediment formed through the breakdown of calcareous skeletons via abrasion and bioerosion processes. The limestone composed of calcisilts is called calcisiltite. Bioeroders, such as organisms that destroy calcareous skeletons, grains, and carbonate substrates, create cavities and produce silt-sized sediments. For example, chemical–mechanical boring by sponges in skeletons and limestone generates large amounts of fine-grained sediment, with most of the boring chips being silt-sized (see Section 7.2).

Bioclasts (skeletal grains) are calcareous fossils that are larger than silt in limestones (Figure 6.8), regardless of whether the fossils are fragmented or fully preserved. Skeletal grains are by far the most common type of grains in carbonate rocks and, in some limestones, are so abundant that they make up most of the rock. Their biological origin is easy to see with the naked eye.

Peloid is a non-genetic term for sand-sized micritic grains that lack distinctive internal structures (Figure 6.9a). Peloids form through various processes, such as micritization or recrystallization of former grains, reworking of semi-consolidated lime mud (small, well-rounded intraclasts), and biological processes. Laboratory studies have shown that peloids can also be created through purely abiotic mechanisms (Bosak et al. 2004). Flügel (2004) identified nine genetic types of peloids, four of which are of biological origin: (1) Fecal pellets, the most obvious type of biological peloids, are produced by organisms that ingest lime mud and then extrude the undigested lime material as pellets. (2) Algal peloids form from the disintegration of calcified algae and microbes, often retaining relicts of algal structures and being irregularly shaped. (3) Bioerosional peloids are created by organisms that bore (e.g. sponges) or rasp (e.g. gastropods and sea urchins) carbonate substrates, releasing chips or clasts that are similar in size and shape to peloids. (4) Microbial peloids are round to irregular micritic grains that precipitate in situ within microbial mats.

Coated grains are polygenetic, meaning they include both abiogenic and biogenic types. **Ooids** are rounded, sand-sized grains with a nucleus (such as a shell fragment, pellet, or quartz grain)

Figure 6.8 Bioclastic limestone from the Sandaokan Formation, Ordovician, China. *Source:* Photo by Wei Liu.

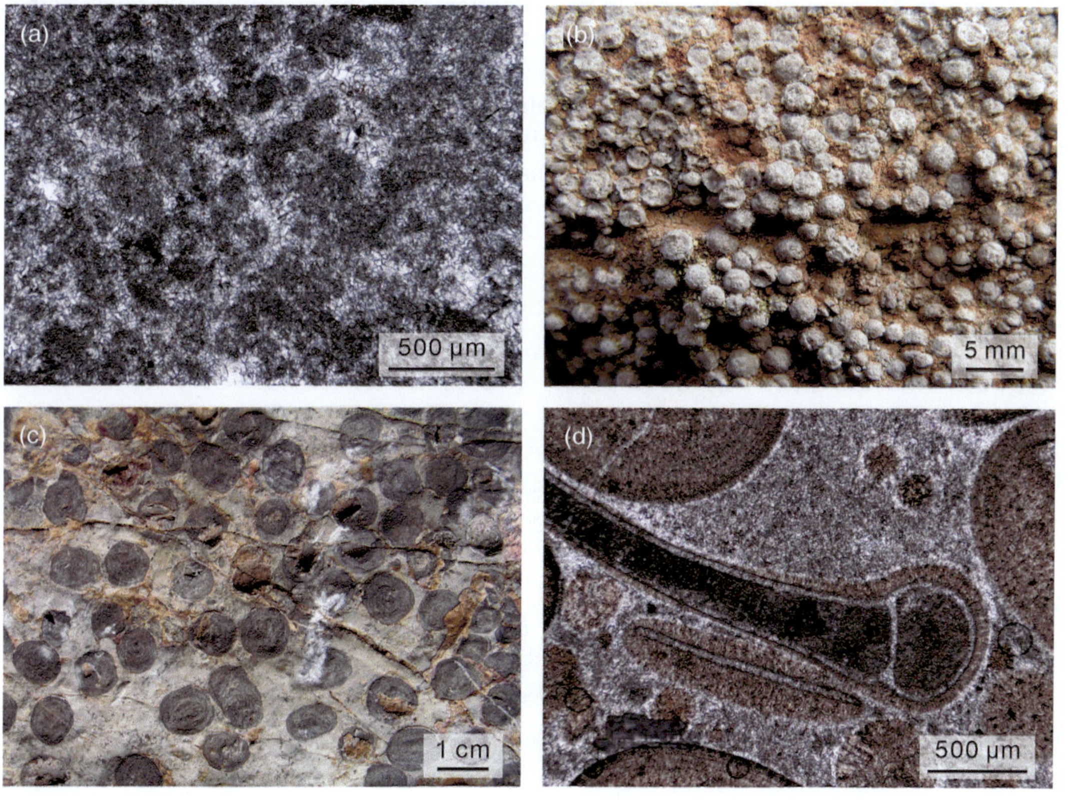

Figure 6.9 Peloids, ooids, and other coated grains. (a) Peloids from the cap carbonate of the Doushantuo Formation, Ediacaran, China; (b) ooids from Kalkowsky's type locality, Lower Triassic, Germany; (c) oncoids from the Mufushan Formation, Lower Cambrian, China; (d) micritic envelope on skeletal grains, Lower Cambrian, China. *Source:* Photos by Xingliang Zhang.

surrounded by an external cortex (Figure 6.9b). For a long time, ooids were thought to be primarily inorganic, forming in shallow, warm, agitated waters that are supersaturated with $CaCO_3$. However, recent studies have introduced the concept of microbial ooids (Flügel 2004; Liu and Zhang 2012; Woods 2013; Zhang et al. 2023) and those formed through organomineralization (Reitner et al. 1997). A case study on modern ooids from the Great Salt Lake shows that, from the nucleus to the cortical layers and even to the outer crust, the organic matter content increases while mineralization gradually decreases (Pei et al. 2024). Similar patterns are also observed in aragonite ooids forming statically within microbial mats from hypersaline ponds of Kiritimati (Suarez-Gonzalez and Reitner 2021) and in the lower Triassic ooids from the Germanic Basin (Figure 6.10). These findings strongly suggest that well-rounded ooids can be generated within microbial mats and organic matter may play a key role in carbonate precipitation of ooids (Pei et al. 2024). When coated grains are larger than 2 mm in diameter but have similar internal structures to ooids, they are called **pisoids**. Some pisoids may be of biological origin, forming through the trapping and binding activities of cyanobacteria, much like stromatolites. Spheroidal coated grains larger than 1 cm are known as oncoids (Figure 6.9c). These are primarily formed by calcimicrobes and algae, but may also involve foraminifera, bryozoans, serpulids, and other organisms. Alternatively, rhodoids are formed by free-living calcareous red algae.

The **micrite envelope** refers to a thin, non-laminated micritic layer surrounding a carbonate grain, most notably visible on skeletal grains (Figure 6.9d). Grains with a micrite envelope are called cortoids. Microbes play a key role in the formation of the micrite envelope through both destructive and constructive micritization: (1) Destructive micritization occurs through repeated boring and filling actions by microborers, such as filamentous cyanobacteria, green and red algae, and fungi. This process destroys the peripheral zone of the grains, ultimately forming a micritic marginal rim with an irregular boundary. (2) Constructive micritization involves the overgrowth of micrite around the surface of grains, often produced by the calcification of a biofilm coating the grains.

Figure 6.10 Ooids and the overlying microbial crusts from the Lower Triassic of Germany. The ooids are generally large, with diameters exceeding 2 mm, and are covered by microbial crusts that exhibit lamination similar to that of the ooids. These crusts are further overlain by another oolitic layer. *Source:* Courtesy of Yu Pei.

6.2.1.2 Microbialites

Microbialite is a general term for microbial carbonates, which are carbonate deposits formed or localized by microbial communities in marine, marginal marine, freshwater, and terrestrial environments. The primary organisms involved include autotrophic bacteria (especially photosynthetic prokaryotes like cyanobacteria), chemoautotrophs, chemoheterotrophs, eukaryotic microalgae, and fungi. These organisms typically form microbial mats or biofilms, and play a key role in the formation of microbialites through two main processes: grain trapping and mineral precipitation (Riding 2000).

Microbial overgrowth stabilizes previously deposited sediments, while trapping processes localize sediments that pass over the mats. Trapping may occur through simple physical blockage (baffling) of grain movement or through adhesion within EPS. Additionally, the metabolic activities of microbial mats or biofilms, such as photosynthetic CO_2 uptake, ammonification, fermentation, denitrification, and sulfate reduction, can increase local alkalinity, promoting carbonate precipitation (Dupraz and Visscher 2005; Riding 2000). Microbially induced precipitation, leading to the calcification of microbes and EPS, is a crucial process in the genesis of primary sediments in microbialites. Based on meso- and macroscale fabric, microbialites can be categorized into four common types including stromatolites, thrombolites, dendrolites, and leiolites (Figure 6.11) (Dupraz et al. 2011; Riding 2000; Shapiro 2004). Oncolites (composed of oncoids) are also considered a type of microbialite (Shapiro and Wilmeth 2020). While, for the purposes of this text, we classify them under the category of coated grains, and therefore, they will not be discussed further here.

Stromatolite is defined as laminated benthic microbial buildups (Riding 2000) (Figure 6.12). While stromatolites are often associated with carbonates, they can also be made of siliceous or evaporite materials. For the purposes of this text, we define **stromatolites** as microbialites with a laminated structure, which can be influenced by biological and environmental factors such as seasonal changes, phototactic responses of organisms, variations in sediment supply, and shifts in species composition during sedimentation pulses. Typical stromatolites in carbonate facies occur as domes, columns, and crusts, and show biogenic laminations. The growth of microbial mats is controlled by such physical and chemical factors as temperature, light, salinity, and moisture. Seasonal and secular rhythmic variations of microbial activity and/or frequent environmental changes contribute to the layered structure of the sediments. Laminations in relatively slowly accreting structures may record seasonal changes, or the frequency of storms that periodically deposit sediments over mats, while minor or more frequent turbulence may incorporate sediment layers in intertidal mats. Fast accreting structures may form laminae daily, following either solar or lunar cycling (Kim et al. 2025; Seong-Joo et al. 2000). The resulting texture consists of biolamination where dark laminae, less than 1 mm in alternation with light laminae, mostly built up by detrital particles.

Stromatolites are classified into five main types based on their origin, components, and lamination characteristics (Riding 2000): (1) Skeletal stromatolites: Formed by the calcification of *in-situ* organisms or biofilms, often containing tubular microfossils. (2) Agglutinated stromatolites: result from the trapping and binding of particulate sediments. (3) Fine-grained stromatolites: known for their distinct lamination originated from either precipitation or agglutination. (4) Freshwater tufa stromatolites: formed by the encrustation of cyanobacteria and algae, with precipitation occurring on the surface of organic tissues rather than within them, setting them apart from skeletal stromatolites. (5) Terrestrial stromatolites: laminar calcretes formed by a combination of evaporation and microbial activity.

Thrombolite is a type of microbialite characterized by a macroscopically unlayered, clotted fabric. The clots differ from the surrounding material in color, form, and texture, creating a blotchy

Figure 6.11 Classification and descriptive terminology for microbialites from macrostructures to microfabrics. Notably, microfabrics are specific to modern microbialites (stromatolites, thrombolites, and leiolites). *Source:* Adapted from Dupraz et al. (2011).

fabric (Riding 2000). Thrombolites formed through microbial calcification are known as calcified microbial thrombolites, which exhibit well-defined clots. Coarse agglutinated thrombolites incorporate sand- and gravel-sized sediments, such as ooids and bioclasts, often displaying complex internal variations. Their blotchy, agglutinated fabric may result from the coarseness of the sediment, uneven accretion, or bioturbation. Arborescent thrombolites are typically associated with a

Figure 6.12 Stromatolite from the Jiudingshan Formation (Neoproterozoic). *Source:* Courtesy of Zhifei Zhang.

decimetric dendritic fabric, which is larger than that of dendrolites. Tufa thrombolites, on the other hand, develop in freshwater lakes and streams.

Dendrolites have a macroscopic, centimetric, bush-like fabric, typically formed by calcified microbes. In hand specimens, the dendritic structure may show areas with variations in color and texture, but it maintains a distinctive bushy or tree-like appearance, with a vertically erect or radial orientation. This sets it apart from the round or irregular clotted fabric found in thrombolites. Like stromatolites and thrombolites, dendrolites can form large domes, columns, and expansive reef-like masses, sometimes organized into large, ovoid, doughnut-shaped calyptrae. However, unlike stromatolites and thrombolites, dendrolites are exclusively formed through microbial calcification and not by the agglutination of particles (Riding 2000). Notably, dendrolite is only observed in ancient sedimentary rocks and has not yet been found in modern environments.

Leiolite has a structureless, aphanitic macrofabric, without clear lamination, clots, or dendritic fabrics (Riding 2000). It forms large domes, often alongside stromatolites and dendrolites, with its development favored by a consistent, uniform supply of well-sorted sediments on surfaces colonized by an even layer of microbial mat.

Microbialites have a long history on Earth, with a significant record throughout various geological periods (Figure 6.13). Marine abundance data indicate that microbialites experienced episodic declines during the Phanerozoic, with a peak around 500 million years ago (Riding 2006). This trend of abundance is generally the opposite of the taxonomic diversity of marine metazoans (Riding and Liang 2004). Stromatolites first appeared around 3.45 billion years ago and were most diverse and abundant between 2.8 and 1.0 billion years ago (Riding 2000). Thrombolites have been found in strata as old as 1.9 billion years and were locally significant during the Proterozoic. They became more prominent near the Precambrian-Cambrian boundary and remained important throughout much of the Cambrian and Early Ordovician periods. Dendrolites first emerged alongside widespread calcimicrobes in the Neoproterozoic (Shapiro 2004), and are most notably found in strata from the Cambrian to Early Ordovician and Late Devonian periods (Riding 2000). Records of leiolites are quite rare, with examples found in the late Jurassic and Miocene (see Flügel 2004).

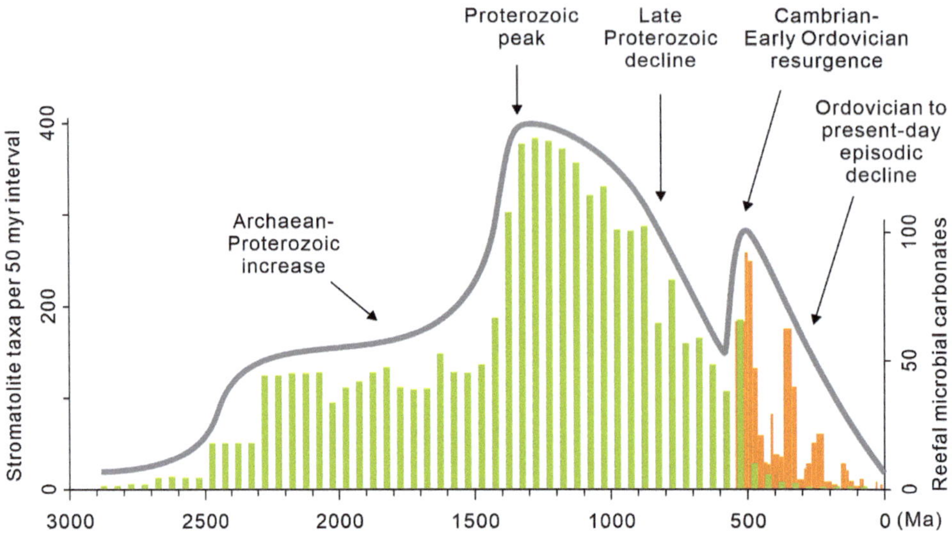

Figure 6.13 Diversity, abundance, and temporal changes of microbialites (green columns) and reefal microbial carbonate (orange columns) throughout geological periods. The smooth curve shows the overall trend of microbialites with the major elements. *Source:* Modified from Riding (2006).

6.2.1.3 Methane-Seep Carbonates

Methane-seep (cold seep) carbonates are carbonates that form in marine environments where methane is being released from the seafloor, typically from deep-sea seeps, mud volcanoes, or hydrocarbon reservoirs. These carbonates form as a result of the interaction between methane, sulfate, and microbial activity. Most methane migrating upward through the marine sedimentary column is intercepted biologically via a process called anaerobic oxidation of methane (AOM), mediated by consortia of methane-oxidizing, sulfate-reducing archaea and bacteria (Section 1.4). This process follows the reaction:

$$CH_4 + SO_4^{2-} \rightarrow HCO_3^- + HS^- + H_2O$$

Methane sources can be abiogenic, thermogenic, or microbial (Krüger et al. 2005). Abiogenic methane forms at hot vents through water-rock reactions of serpentinization, a geological process that occurs when ultramafic rocks, typically rich in minerals like olivine and pyroxene, interact with water at low to moderate temperatures. During this process, these minerals are converted into serpentine minerals, such as chrysotile, antigorite, and lizardite, while also producing hydrogen gas (H_2). This reaction can also lead to the formation of abiogenic methane when hydrogen reacts with carbon dioxide in the environment. Thermogenic methane arises from the geothermal transformation of organic materials buried in marine sediments. Microbial methane production typically occurs in organic-rich, anoxic sediments, especially below the sulfate penetration depth. There are several pathways of microbial methanogenesis that were described in Section 1.3.5 of this book.

A typical zonation of the sediment is observed. In the lower, sulfate-depleted layers, methane is produced; while in the upper, sulfate-containing layers, methane is oxidized by anaerobic methanotrophs. The boundary between these two zones, where methane oxidation and sulfate reduction occur, is known as the methane-sulfate transition zone.

One consequence of high rates of AOM at seeps is an increase in alkalinity, which leads to the precipitation of ^{13}C-depleted carbonate minerals, such as aragonite, calcite, and dolomite. This

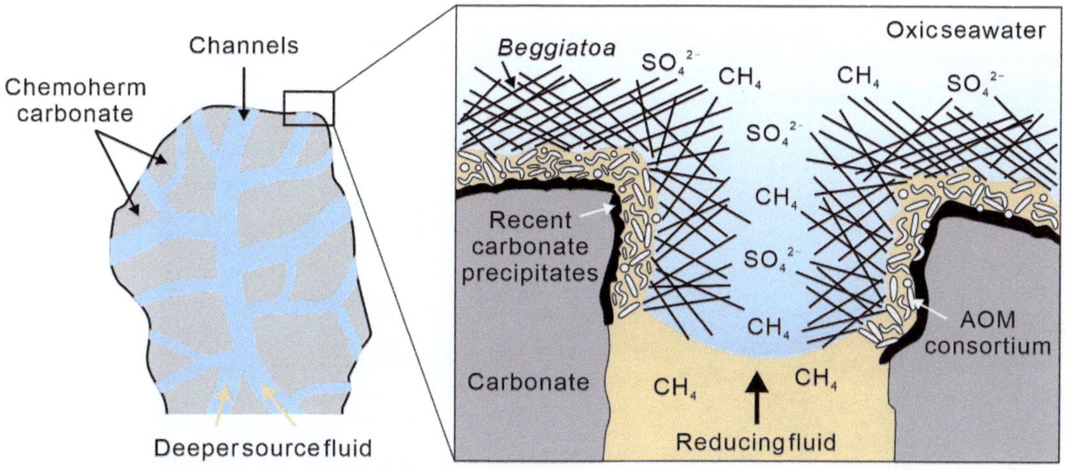

Figure 6.14 Conceptual model of chemoherm carbonate precipitation. The sketch on the left shows a typical chemoherm carbonate with interconnected fluid channels fed from a deeper source. The box is the close-up of a fluid channel, which illustrates the model of chemoherm carbonate precipitation. *Source:* Teichert et al. (2005)/with permission of Elsevier.

occurs even at water depths below the carbonate compensation depth (CCD), with $\delta^{13}C$ values as low as −70‰ PDB, often below −20‰ PDB. These microbially mediated carbonate build-ups can grow into the oceanic water column, rising several tens of meters above the seafloor. They often display a pinnacle-shaped morphology with steep flanks (Figure 6.14), known as chemoherms (Teichert et al. 2005).

Evidence from their characteristic carbonate fabrics, stable isotope values, and lipid biomarkers shows that the AOM-mediating prokaryotes play a key role in forming both modern and ancient methane-seep limestones. As such, methane-seep carbonates are considered microbialites (Peckmann and Goedert 2005). Another consequence of high AOM rates at seeps is the local production of hydrogen sulfide. In some cases, the rates of sulfate reduction exceed the rates of AOM, indicating that sulfate reduction may also be fueled by the oxidation of other organic matter and hydrocarbons, besides methane. As a result, authigenic sulfide minerals, such as ^{34}S-depleted pyrite, are commonly found within seep limestones.

Long-term seepage of methane accompanied by the precipitation of methane-derived authigenic carbonates, the methane-seep carbonates (Figure 6.15), leading to extensive outcrops like carbonate chimneys in the vicinity of cold seeps. From a diagenetic perspective, particularly in terms of microfacies, methane-seep carbonates exhibit the following key criteria and characteristics.

Occurrence, geometry and structural features: (1) the anomalous occurrence of autochthonous limestones in a variety of settings, including deep-sea environments, shallow ocean shelves, and high-latitude regions, even extending to polar areas (Åström et al. 2016); (2) localized limestone formations associated with structural features such as faults, fractures, and diapirs, which facilitate the migration of methane or fluids; (3) mound-shaped geometries ranging from a few meters to tens of meters in size; (4) isolated limestone bodies within non-carbonate sequences like siltstones, shales or marls; and (5) a common association with seafloor features like craters, carbonate mounds, underwater pingos (Zeppilli et al. 2012), or pockmarks—circular depressions formed by the upward seepage of gas, causing sediment collapse, and typical of areas with gas pockets in near-surface sediments (Cathles et al. 2010). These features combine to create

Figure 6.15 Methane leakage in the cold seep area, South China Sea. Long-term seepage of methane-rich fluids accompanied by the precipitation of the methane-seep carbonates. *Source:* Courtesy of Zengfeng Du.

a seafloor habitat that is markedly different from the typical monotypic soft-bottom environments of the deep sea.

Fabric and carbonate cements: (1) layered micritic microbial fabrics; (2) nodular fabrics associated with cylindrical to encircling concretions; (3) laminated, crustose, or void-filling carbonates, occasionally with extensive and thick brecciated carbonate crusts; (4) brecciation on a centimeter scale, resulting in autobreccia; (5) microcrystalline, radial-fibrous, splayed fibrous, and botryoidal cements, along with irregular yellow calcite, Mg-calcite, and aragonite cements; (6) hydrocarbon inclusions within carbonate cements; and (7) cements forming the dominant volume, often overshadowing fossils and resulting in rocks like cementstone (Flügel 2004).

Microfacies: (1) Common microfacies types are bivalve-dominated wackestones, floatstones, and packstones, as well as intraclastic packstones, worm tube-dominated boundstones, rudstones, cementstones, and peloidal grainstones. It is important to note that many textures observed in seep and vent carbonates are diagenetic rather than depositional. Based on microbial contributions, rocks dominated by calcite cement, micrite crusts, and shells can be classified as cementstones or microbial bindstones. (2) Micritic coatings and crusts are prevalent, often alternating with calcite cement layers. (3) Micrite, typically non-ferroan calcite, occurs in several forms: as internal precipitates in burrows and shells; as microcrystalline cement between carbonate or siliciclastic grains and as pelmicrite. It often contains clay- and sand-sized grains, pyrite, and organic matter, and may occasionally recrystallize into microspar. (4) Micritic peloids, including millimeter- to centimeter-sized fecal pellets, are common. (5) Micrite and pelmicrite lithoclasts found in autobreccia are often the result of bioturbation or fluid circulation. (6) Laminated fabrics of irregularly arranged micrite and/or calcite cement layers are frequently observed. (7) Burrows filled and lined with fine-grained carbonates, containing grains similar to those of the surrounding strata, are common. (8) The volumetric contribution of cements (carbonate, sulfate, and sulfide) ranges from medium to high. Common carbonate cements include botryoidal calcite (characterized by rolling extinction under crossed Nicols), fibrous calcite, and irregular, non-isopachous cement layers. (9) Organic matter occurs as dispersed debris, organic inclusions in calcite, or in black bands within the rock. (10) Pyrite is commonly present, occurring as linings, coatings, isolated crystals, or in framboidal form (Flügel 2004).

Biota: (1) Modern cold seeps generally host a wide variety of seep fauna, including chemo-obligate species (Levin et al. 2016). These ecosystems typically support macrofaunal communities with high abundance and biomass, but relatively low diversity compared to surrounding non-seep habitats (Levin 2005). (2) Seep-associated organisms may rely directly on chemoautotrophic symbionts or indirectly on trophic relationships involving AOM and sulfate reduction as alternative energy sources, in place of or alongside photoautotrophy (Thurber et al. 2013). (3) The habitat heterogeneity and chemosymbiotic foundation species in cold seep environments—such as siboglinid tubeworms, clams, and mussels—can attract epifauna and mobile organisms from surrounding habitats, thereby increasing the local diversity of these deep-sea ecosystems (Åström et al. 2018; Levin et al. 2016). (4) Fossils often show strong microbial micritic encrustations and may be enclosed within sedimentary precipitates derived from fluid seepage, such as sulfide/sulfate minerals and isotopically unique authigenic carbonates. (5) *In situ* fossil assemblages typically consist of dissolved, corroded, and broken shells, occasionally coated with pyrite. The frequency of fragmentation and articulation is a key distinguishing factor that differentiates cold seep-related autochthonous shell beds from those formed in non-seep environments (Flügel 2004).

Geochemistry and lipid biomarkers: (1) Authigenic carbonate cements and carbonate shells exhibit extreme $\delta^{13}C$ depletion, ranging from below −20‰ PDB to as low as −70‰ PDB, suggesting that the carbonates formed through microbial oxidation of methane. (2) Specific biomarkers, including the ^{13}C-depleted isoprenoid hydrocarbons crocetane, pentamethylicosane (PMI), and squalane, are present, indicating anaerobic methane oxidation by bacteria (Feng et al. 2018).

6.2.1.4 "Dolomite Problem" and Microbial Dolomite Models

Dolomite (mineral), despite its thermodynamic stability and abundance in the ancient rock record, is rarely observed precipitating in modern environments. It is evident from the presence of relict limestone textures and structures that many dolostones (typically composed of coarse-crystalline dolomites) are secondary, formed by the diagenetic replacement of older limestones (a process known as dolomitization). However, many fine-crystalline dolostones (rock) with primary sedimentary structures lack such textural evidence of replacement, and therefore, cannot be definitively shown to have originated from the diagenetic alteration of limestones. The uneven distribution of dolostones, with their greater prevalence in ancient rocks than in modern carbonate settings, along with the fact that perfectly ordered dolomite has never been seen precipitating under normal surface temperature (~25 °C) and pressure (~1 atm), has made the origin of dolostones a long-standing enigma in sedimentary geology, known as the "Dolomite Problem."

Recent advancements have highlighted the role of microbes in dolomite precipitation, both in natural environments and in laboratory culture experiments conducted under normal temperature and pressure. Vasconcelos and McKenzie (1997) proposed a microbial dolomite model under anoxic, hypersaline conditions to explain their observations of sulfate-reducing bacteria (SRB) mediating dolomite precipitation (Figure 6.16). According to this model, sulfate-reducing microbes overcome the kinetic barriers to forming dolomite, a mineral with a highly ordered crystal structure. The microbes achieve this by altering their microenvironment and providing a surface for dolomite nanocrystals to nucleate. Sulfate-reducing microbes require sulfate (SO_4^{2-}) for their metabolism, while dolomite precipitation necessitates excess magnesium (Mg^{2+}) and increased

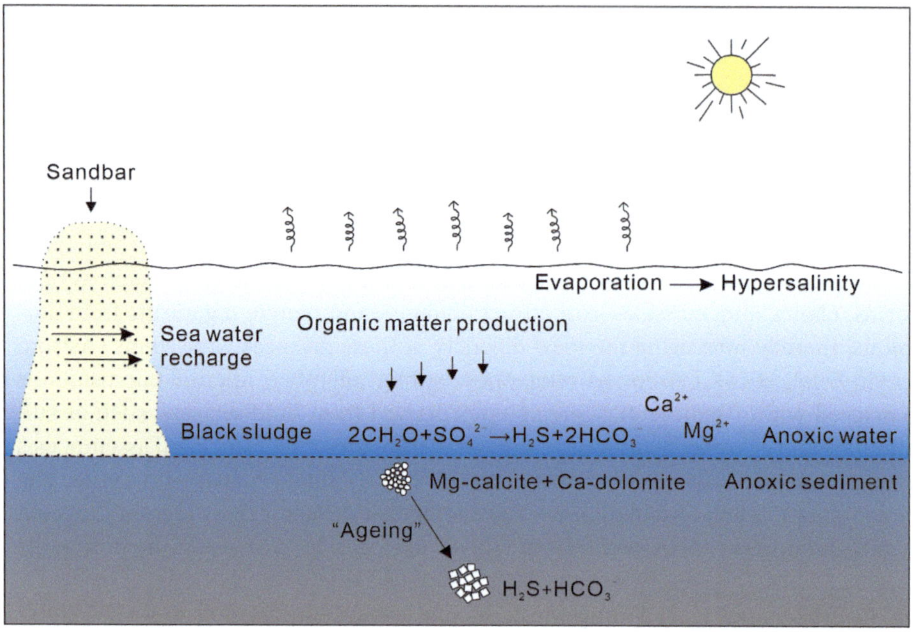

Figure 6.16 Schematic illustration of the microbial dolomite model based on the observation in Lagoa Vermelha, Brazil. Intense evaporation during the dry season lowers the water level, resulting in hypersalinity in the restricted sea and permitting inflow of seawater through the barrier dune. High productivity in this environment leads to an anoxic condition of bottom waters and the formation of a black organic-rich sludge. The hypersaline waters provide high concentrations of SO_4^{2-} for bacterial sulfate reduction and Mg^{2+} for the Mg-calcite and Ca-dolomite precipitates. Microbial activity mediates the precipitation of the carbonate minerals within the sludge layer, which then settle to become part of the anoxic sediments. With deposition, the dolomite undergoes an "aging" process, becoming more stoichiometric. When the sediments buried, the microbial sulfate reduction ceases, but the diagenesis continues as an inorganic process with continued dolomite precipitation occurring on the microbially nucleated crystals. *Source:* Adapted from Vasconcelos and McKenzie (1997).

alkalinity conditions that can be met by hypersaline waters. Once dolomite nucleation begins under sulfate-reducing conditions, the mineral undergoes an "aging" process, where the crystal structure becomes more ordered after deposition. Both the initial precipitation and subsequent early diagenesis are strongly influenced by microbial activity. This model likely accounts for the formation of certain dolostones in the geological record. During Earth's early history, anaerobic microbes may have been especially important for dolomite production, as anoxic conditions were far more widespread than today.

Dolomite formed by microbial/organic catalysis is disordered, with magnesium and calcium ions mixed in the crystal structure, which differs from the ordered dolomite observed in sedimentary rocks. This introduces a new challenge to the "dolomite problem": Is disordered dolomite able to transform into ordered dolomite under ambient temperature conditions? A more recent study on the genesis of ordered dolomite identified that under unsaturated aqueous conditions, the disordered layers of dolomite would dissolve, while under saturated conditions, ordered layers would re-precipitate (Kim et al. 2023). Thus, it was proposed that frequent fluctuations between oversaturation and unsaturation in the solution could accelerate dolomite growth and the turbulent processes of water chemistry in sedimentary systems (such as saturation–unsaturation and fluctuations in salinity or pH) are favorable conditions for the formation of ordered dolomite near the Earth's surface (Kim et al. 2023).

6.2.2 Siliceous Sediments

Siliceous sediments form the foundation of siliceous sedimentary rocks, known as cherts, which are often found as either bedded or nodular deposits in outcrops (Figure 6.17). Chert is typically a fine-grained, dense, and hard rock mainly composed of silica minerals like quartz, chalcedony, and opal, with minor impurities such as siliciclastic grains and diagenetic minerals (e.g. pyrite and calcite). In thin sections, chert displays three primary textures: (1) granular microquartz, made up of nearly equidimensional grains about 8–10 μm in size; (2) chalcedony, forming fibrous silica structures that aggregate into sheaf-like bundles of radiating, extremely fine crystals; and (3) megaquartz, which consists of equant to elongated grains larger than 20 μm.

6.2.2.1 Primary Sediments

Seawater is generally highly undersaturated with silica, making it unlikely for chert to precipitate through purely inorganic processes. Instead, the extraction of silica from ocean water by silica-secreting organisms, such as those that produce opaline skeletal structures, appears to be the primary mechanism for large-scale silica removal from undersaturated seawater (Boggs 2001). The primary siliceous sediments of biological origin are exclusively composed of opal-A, and include the frustules of diatoms, the skeletons of radiolarians, and the spicules of sponges. Many bedded cherts are derived from the remains of these siliceous organisms, which can be classified based on the type and abundance of skeletal material into four main categories: diatomaceous deposits

Figure 6.17 Bedded and nodular cherts in outcrops. (a) Well-bedded chert in the Yurtus Formation, Lower Cambrian, China; (b) nodular chert in the black shale of Doushantuo Formation, Ediacaran, China; (c) lenticular chert in the carbonate of the Shaylik Formation, Middle Cambrian, China. *Source:* Photos by Xingliang Zhang.

(diatomite and diatomaceous cherts), radiolarian deposits (radiolarite and radiolarian chert), siliceous spicule deposits (spicularite and spicular chert), and non-fossiliferous cherts.

Siliceous ooze is a type of pelagic biogenic sediment that blankets the deep ocean floor, typically composed of at least 30% siliceous skeletal remains of marine plankton, often exceeding 60%. This ooze is commonly categorized based on its dominant fossil types: diatomaceous ooze and radiolarian ooze. In modern oceans, radiolarian ooze is widespread in tropical areas of the Pacific, whereas diatomaceous ooze is more common near the Polar Regions. Siliceous ooze is a key source of primary sediment for many banded cherts in the Phanerozoic.

In environments where waters are saturated with silica, such as the hot springs of Yellowstone National Park, microbes are often found encrusted with amorphous silica as a result of the BIM process. A study by Phoenix et al. (2000) demonstrated that, in some cases, the mineral crusts on cyanobacteria can be quite thick, up to 5 μm, significantly enlarging the trichome diameter. Interestingly, these thick silica crusts do not appear to harm the organisms; rather, the cyanobacteria shed their extracellular sheaths once they become heavily encrusted. These mineral crusts can be considered primary siliceous sediments, although their contribution to chert formation remains poorly understood.

Li et al. (2022) isolated a new bacterium, strain WYHC-5 (Figure 6.18), from a freshwater lake in China. This bacterium belongs to a deep-branching group of magnetotactic bacteria (MTB). Strain WYHC-5 is capable of accumulating significant amounts of silicon within its cells, forming amorphous silica globules, as well as magnetite magnetosomes and sulfur inclusions. Metagenomic

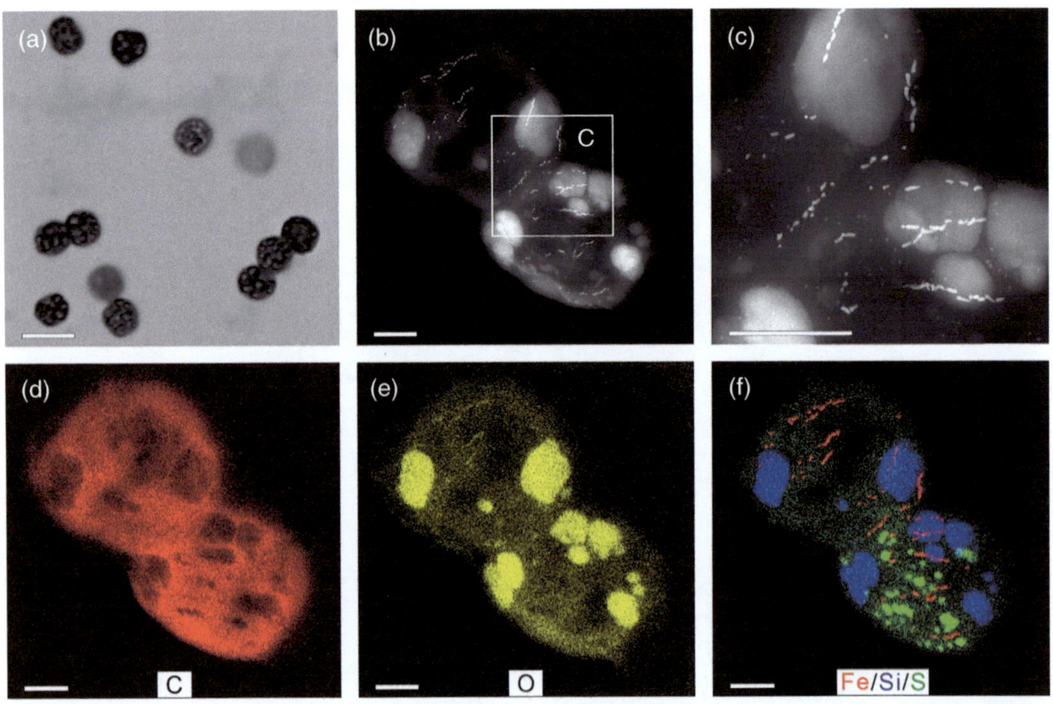

Figure 6.18 Characterization of the magnetotactic strain WYHC-5 producing intracellular amorphous silica globules. (a) Optical micrograph of WYHC-5 cells magnetically collected from sediments in laboratory; (b) STEM-HAADF image of WYHC-5 cells; (c) high-magnification STEM-HAADF image from the cells in (b), showing amorphous silica globules and multiple chains of magnetosome magnetite; (d–f) STEM-EDXS elemental maps of WYHC-5 cells shown in (b). C, Carbon; O, oxygen; S, sulfur; Si, silicon; Fe, iron. Scale bars are 5 μm in (a) and 1 μm in (b–f). *Source:* Reprinted with permission from Li et al. (2022), © 2022 Li et al.

analysis of its genome revealed a key gene likely involved in silica deposition, providing new structural and molecular evidence for intracellular biosilicification within prokaryotes. This discovery has a significant implication for understanding the genesis of banded cherts before the rise of siliceous skeletons. Strain WYHC-5 can accumulate up to 10.9% of its cell weight in silicon, a quantity comparable to some eukaryotic silicifying organisms. Fossil and molecular evidence indicate that MTB have existed since at least 1.9 billion years ago, with molecular clocks suggesting their origin could date back to 3.38 to 3.2 billion years ago. This supports the idea that MTB, even before eukaryotes, may have played a significant role in the Si cycle since the middle Archean (Li et al. 2022).

6.2.2.2 Formation of Chert

While organisms are alive, their siliceous skeletons are protected by physicochemical mechanisms, preventing significant dissolution. After death, this protective system breaks down, and dissolution begins. In areas of the ocean where organisms are particularly abundant, the rate of skeletal production may exceed the rate of dissolution, allowing a sufficient number of siliceous skeletons to accumulate on the seafloor as siliceous sediments before they are entirely dissolved. Once buried, these skeletal materials continue to dissolve, with much of the dissolved silica being trapped in the pore space. This leads to an enrichment of silica in the pore waters, which, in turn, facilitates the diagenetic precipitation of chert.

Biogenic opal-A is metastable and, over time, transforms into opal-C or opal-CT (mainly composed of low-temperature cristobalite and interlayered tridymite), and eventually quartz (Figure 6.19). As a result, opal-A is rarely preserved in ancient rocks, though a notable exception

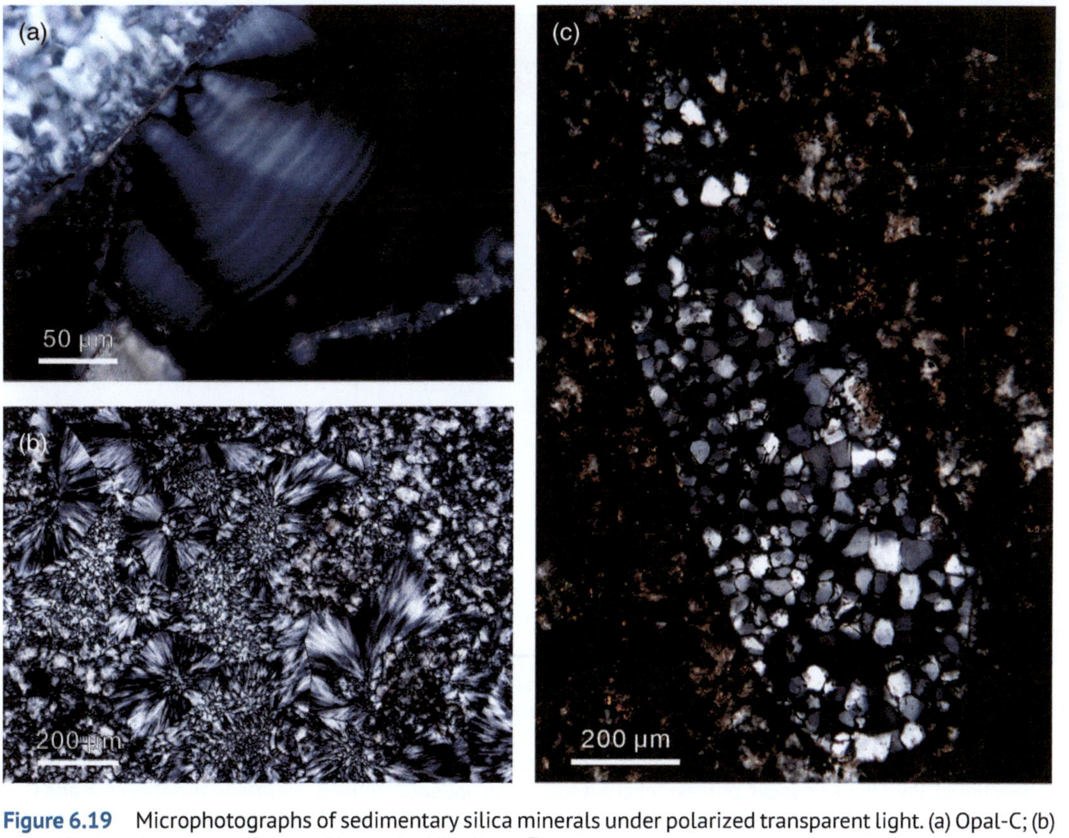

Figure 6.19 Microphotographs of sedimentary silica minerals under polarized transparent light. (a) Opal-C; (b) opal-CT; (c) microquartz. *Source:* Photos by Xingliang Zhang.

exists in the earliest Cretaceous radiolarian claystone with minimal diagenetic alteration (Ogawa and Kawata 1998). The transformation of biogenic opal into chert is controlled by several physico-chemical factors and can occur via a multi-stage process with three possible pathways (Figure 6.20). The conversion of opal-A to opal-CT is a solution–reprecipitation process that typically occurs at burial depths corresponding to temperatures of around 45 °C. In this stage, opal-A dissolves, generating silica-rich pore waters, from which opal-CT precipitates as microcrystalline aggregates, blade-shaped crystals, rim cements, overgrowths, and massive cements (Boggs 2001). Opal-CT is also metastable and further converts to microquartz through solution–reprecipitation at temperatures around 80 °C. Finally, as temperatures approach those found in metamorphic environments, microquartz transforms into megaquartz, although small amounts of megaquartz can also form at lower temperatures as cavity and fracture fillings.

It is widely accepted that most fossiliferous bedded cherts evolve from biogenic siliceous sediments through pathways A or B. The origin of non-fossiliferous bedded cherts, however, is less understood. These may also have originated from biogenic sediments that underwent intense

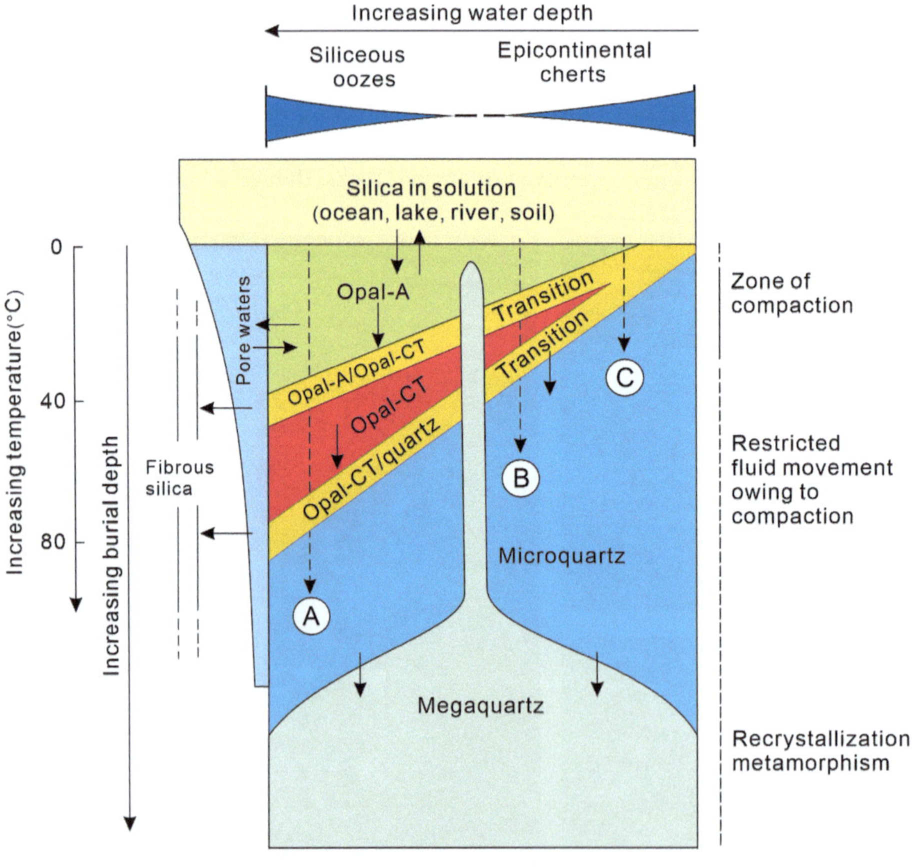

Figure 6.20 Schematic illustration of major silica phases and possible pathways of their diagenetic transformation. The increase of burial depth (vertical) is associated with elevation of temperature and, decrease of porosity and permeability. Horizontal dimension represents relative water depth of the initial environment. In path A, silica initially deposited as opal-A that subsequently transforms to opal-CT and then microquartz via solution–reprecipitation steps. Path C represents early diagenetic chert in which microquartz forms during shallow burial. Path B is a possible process for chert formation under conditions intermediate between A and C. Megaquartz forms by metamorphic recrystallization of microquartz or by direct growth into voids at any stage of burial. *Source:* Modified from Knauth (1994).

diagenetic alteration, which destroyed nearly all the biological structures. No evidence of silica-secreting organisms has been found in Precambrian bedded cherts, leading some researchers to suggest that these cherts formed through inorganic processes, likely following pathway C, which would require higher silica concentrations in the Precambrian Ocean. Nodular cherts, commonly found in carbonates and shales, are thought to form through diagenetic replacement, as indicated by relict textures. Silica is supplied by the dissolution of biogenic opal-A, which causes pore waters to become supersaturated with respect to opal-CT or quartz. The transformation from opal-A to chert may bypass the opal-CT stage altogether, as shown in pathway C (Figure 6.20).

6.2.3 Phosphatic Sediments

Phosphorites are sedimentary rocks that are notably enriched in phosphorus, containing more than 15% P_2O_5 (or 6.5% phosphorus). Rocks with phosphorus concentrations below 6.5%, but higher than typical sedimentary rocks, are classified as phosphatic rocks (Boggs 2001). Examples include phosphatic shales. Phosphorites are primarily composed of various forms of apatite, such as fluorapatite, chlorapatite, and hydroxyapatite, with carbonate hydroxyl fluorapatite ($Ca_{10}(PO_4,CO_3)_6F_{2-3}$) being the most common. Other constituents often found in phosphatic rocks include detrital quartz, authigenic chert, opal-CT, calcite, and dolomite. Phosphatic deposits can be found in various forms, including bedded, nodular, or intraclastic (pebble-bed) phosphorites, and their textures often resemble those of limestones. These deposits may consist of peloids, ooids, bioclasts, and other apatite-rich grains, with the micrite-like sediments referred to as collophane (Figure 6.21).

The activities of organisms, particularly microbes, play a crucial role in promoting calcium phosphate precipitation, and in many cases, the imprint of these biological processes is reflected in the sedimentary patterns of phosphatic rocks. This section provides a brief overview of the sources of phosphorus in phosphatic deposits within marine systems, the microbial processes involved in phosphate formation, and outlines some of the pathways through which these processes occur.

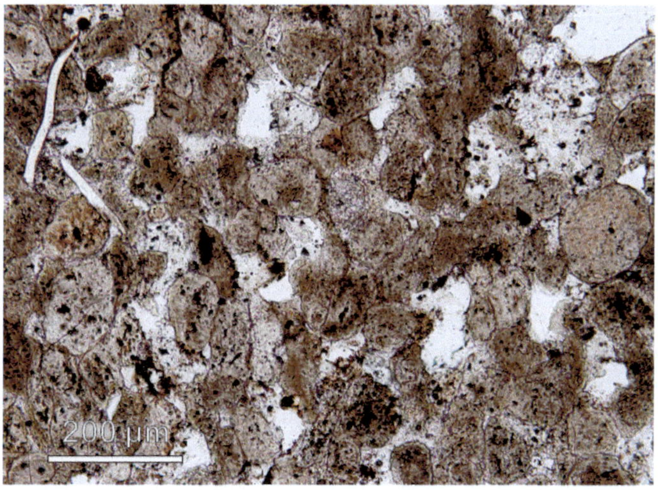

Figure 6.21 Collophane in a phosphorite layer showing peloidal fabric, the Fortunian Stage (~539–529 Ma) of the Cambrian, China. *Source:* Photo by Xingliang Zhang.

6.2.3.1 Phosphorus Source

The global phosphorus cycle consists of four main components: (1) tectonic uplift, which exposes phosphorus-containing rocks; (2) physical, chemical, and biological weathering and erosion, which release dissolved and particulate phosphorus to soils and rivers; (3) phosphorus flux to lakes and the ocean; and (4) the sedimentation of phosphorus, associated with organic and mineral matter, and its burial in sediments (Ruttenberg 2005).

Biological uptake of dissolved phosphorus occurs both on land and in the shallow waters of the ocean. Phosphorus is present only in trace amounts in seawater, with concentrations typically ranging from a few parts per billion (ppb) in surface waters to 50–100 ppb at depths greater than approximately 200–400 m, with an average concentration around 70 ppb. Surface ocean waters are strongly depleted in phosphorus due to biological uptake, while deep waters are enriched, primarily because the sinking of biogenic particles transfers large amounts of phosphorus to the deep ocean.

This raises an intriguing question: assuming that the average phosphorus content in ancient oceans was similar to that of modern oceans, how did the trace phosphorus in ancient seas become concentrated enough to form widespread phosphatic deposits containing up to 40% P_2O_5—an enrichment of up to two million times? One potential mechanism is ocean upwelling, which brings colder, deeper, phosphorus-rich waters to the shallow shelves. This process can supply substantial amounts of phosphorus to form phosphatic deposits. Studies show that most ancient phosphatic deposits are found in lower latitudes, within the trade wind belts along one side of a marine basin, where upwelled deeper water may have reached the continent. Similarly, most nodular phosphatic deposits on the modern seafloor occur in upwelling zones.

However, any phosphorus from deep ocean waters may be rapidly utilized by organisms, preventing the concentration of phosphate in sedimentary environments from reaching saturation. This, combined with the inhibition of phosphate precipitation by Mg^{2+}, suggests that purely inorganic precipitation of apatite from open ocean waters is unlikely (Boggs 2001).

6.2.3.2 Phosphorus Uptake

Phosphorus that is locked up in rocks, soils, and sediments is not directly available to organisms. Only dissolved orthophosphate can be directly assimilated by living organisms (Ruttenberg 2005), a process that requires energy in the form of adenosine triphosphate (ATP). On land, plants and lichens employ various physiological strategies to obtain phosphorus despite its low concentration in the soil solution: (1) increasing root volume and surface area to optimize uptake potential, (2) producing chelating compounds through plant roots and lichens (or associated fungi) that solubilize phosphorus bound to Fe, Al, and Ca, and (3) secreting enzymes or acids near the roots to solubilize phosphate.

In aquatic environments, phosphorus is primarily taken up as inorganic phosphate by almost all microbes for the synthesis of nucleic acids, nucleotides, phospholipids, polyphosphates, some proteins, and other cellular components. Microbes have developed active membrane transport systems to regulate phosphate acquisition from the environment. These primary producers incorporate P into organic tissues, which then serve as the primary source of P for consumers. All organic material of biological origin, including the remains of phytoplankton and zooplankton, coprolites, and fish bones and scales, contains P. For example, phytoplankton have approximately 0.4% P by dry weight.

6.2.3.3 Phosphorus Concentration

Oceanic upwelling provides a continuous supply of dissolved phosphorus from the vast deep ocean reservoir, supporting the growth of large numbers of organisms. In this process, phosphorus is removed from surface waters and concentrated in organic tissues. When organisms die, the phosphorus bound in organic matter is returned to the ocean as the material breaks down. However, catastrophic events or self-burial of microbial biomass can prevent phosphorus from being released back into the surface waters. Under reducing conditions on the seafloor, organic debris is not fully decomposed, allowing it to be preserved long enough to be buried and incorporated into accumulating sediments. As organic particles decay slowly after burial, phosphorus is released into the interstitial water of sediments.

Studies of modern phosphatic deposits under reducing conditions show that pore water P concentrations can reach 1400–7500 ppb, supersaturated with respect to calcium phosphate (Boggs 2001). Organic-bound phosphorus is sequestered in the oxidized microzone at the sediment surface but can be released, up to 25% of the stored P, into the burial zone following the onset of anoxic conditions (Soudry 2000, and references therein). For example, Gächer et al. (1988) demonstrated that under aerobic conditions, bacteria store P in the form of polyphosphate as the final step of oxidative phosphorylation. However, under anaerobic conditions, when oxygen is no longer available, the stored P is used as an energy source to synthesize an organic electron acceptor, leading to the release of P into the aqueous medium. Various organisms, including sulfur-oxidizing bacteria, fungi, and cyanobacteria, are capable of absorbing P when it is in excess. Notably, coccoid cyanobacteria have been shown to surpass eukaryotes in their P uptake capabilities.

6.2.3.4 Apatite Precipitation

Large quantities of phosphorus can be supplied to pore waters, enabling the precipitation of apatite on the surfaces of organisms, sediment grains, organic particles, and bioclasts. Phosphates may also replace calcareous skeletal material (Figure 6.22) and carbonate grains through a process known as phosphatization. Phosphatic deposits thus form within sediments through diagenetic reactions between organic-rich sediments and phosphate-enriched pore waters. In anoxic conditions, magnesium (Mg^{2+}) may be removed from pore waters due to magnesium-iron replacement in clay minerals, facilitating the precipitation of apatite. Additionally, phosphate minerals can precipitate from sediments when Mg^{2+} concentrations are similar to those found in seawater.

The role of microbes in apatite precipitation is somewhat ambiguous, but it is likely that microbes play an indirect role by releasing dissolved P from decaying organic matter. This process accelerates geochemical reactions, shifts chemical gradients in pore waters, and provides microenvironments that enhance the supersaturation of phosphate solutions. Microbial surfaces could also serve as ideal substrates for the nucleation and growth of phosphate minerals. In this context, vacated sheaths and envelopes of filamentous and coccoid microbes may act as biological microniches, where high concentrations of dissolved P could develop locally, promoting rapid mineralization (Soudry 2000).

6.2.3.5 Microbial Structures in Phosphatic Rocks

Fossil microorganisms are abundant in phosphatic rocks (Cui et al. 2020; Zhang et al. 2021) and are preserved in a variety of phosphatized shapes that resemble microbial structures (Figure 6.23), constituting a significant portion of these rocks. Under scanning electron microscopy (SEM), apatite-encrusted filaments, rod-like bodies, capsule-like cells, and solitary or colonial coccus-like forms (Figure 6.24) are commonly observed in phosphorites across different geological ages and environments. In areas where phosphatic deposits are currently forming, microorganisms typically appear

Figure 6.22 Phosphatized Small Shelly Fossils (SSFs) and a cross-section of tubular fossil *Anabarites trisulcatus* preserved on a phosphoritic bedding surface, the Fortunian Stage (~539–529 Ma) of the Cambrian, China. Scale bar is 50 μm. *Source:* Photos by Xingliang Zhang.

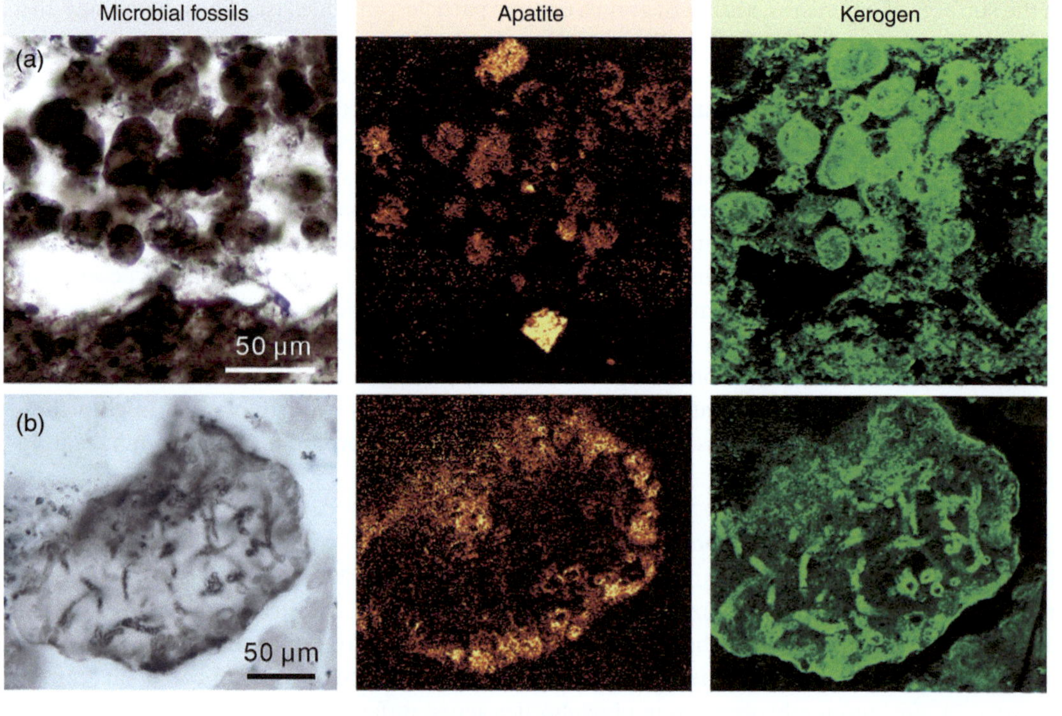

Figure 6.23 Optical photomicrographs and Raman images of *Myxococcoides* sp. (a) *Siphonophycus typicum* (b) from the cherts of the Kuanchuanpu Formation (~535 Ma) in South China. Raman images of apatite (orange) = 954–974/cm and Kerogen (green) = 1240–1680/cm. *Source:* Photos by Xingliang Zhang.

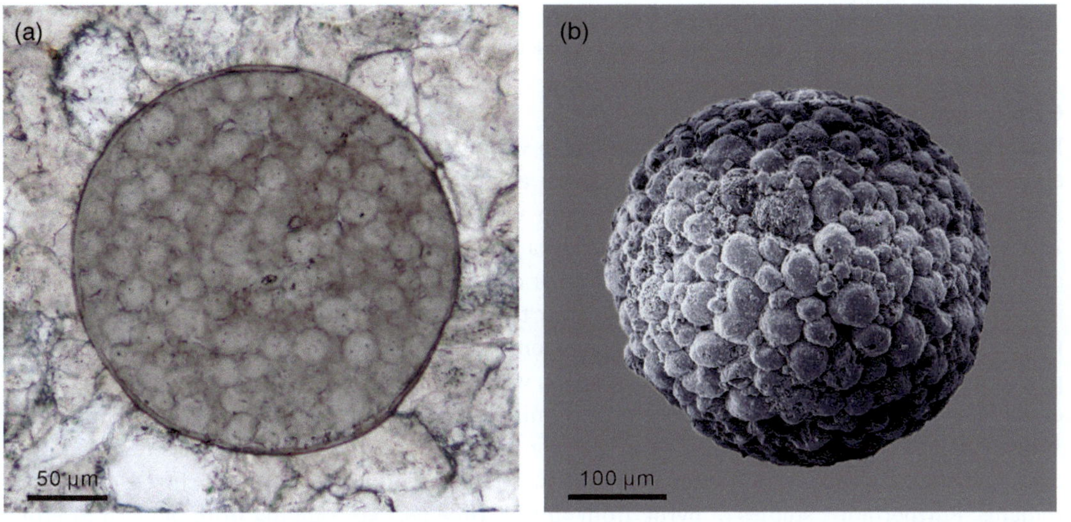

Figure 6.24 Phosphatized embryo-like multicellular microfossil *Megasphaera* from the Ediacaran Dengying Formation, China. (a) Transparent optical micrograph of *Megasphaera*; (b) SEM image of *Megasphaera*. *Source:* Photos by Xingliang Zhang.

as filamentous mats, capping the zone of apatite precipitation. These living mats are primarily composed of sulfur-oxidizing bacteria from the family Beggiatoaceae. Cyanobacterial and fungal mats are also associated with phosphorite formation and are preserved as apatite-encrusted tubules, forming a large portion of the phosphate fraction in both modern and ancient phosphorites. Other structural elements contributing to the phosphate framework of phosphorites include unicellular envelopes of coccoid-dominated microbial mats. Additionally, microbial mats are known to contribute to the formation of phosphate stromatolites (Soudry 2000, and references therein).

6.2.3.6 Grains and Groundmasses

Common phosphatic grains include phosphatized soft parts of organisms, phosphatic skeletons (e.g. bones and scales of fish), peloids (e.g. fecal pellets and end products of bioerosion) (Figure 6.21), micro-oncolites formed through microbial accretion, and stromatolite-like clasts (fragments of early phosphatized microbial mats). A case study showed that matrix phosphorites with high phosphate content are also apatitic in composition, appearing as faint, vermiform phosphate groundmasses under light microscopy (Soudry 2000). SEM observations reveal a meshwork of interlaced apatitic tubules enveloping and binding the grains. Phosphate overgrowths that coat filamentous and globular structures may obscure or completely obliterate the original microbial structures. Over time, the accumulation of heavy minerals gradually masks or eventually transforms the microbial structures into micritic phosphate groundmasses.

6.2.4 Iron Sediments

Iron-rich sedimentary rocks are generally classified into two primary types: ironstones and iron formations. Ironstones typically refer to non-banded, non-cherty, commonly oolitic, iron-rich deposits. In contrast, iron formations are characterized by distinct banded and cherty iron-rich deposits (Boggs 2001). The generation of iron formations is commonly linked to microbial processes, particularly those involving the oxidation of dissolved iron by microbial communities in anoxic environments. Iron sulfides, such as pyrite, are often found as early diagenetic minerals in many sedimentary rocks, which are generally considered to be byproducts of microbial sulfate reduction, and thus merit an inclusion and a discussion here.

6.2.4.1 Sedimentary Pyrite

Pyrite is a pretty common authigenic mineral in various types of sedimentary rocks, particularly abundant in black shales. Its formation is often associated with dissimilatory sulfate reduction, a microbial process, although it can also form under abiotic conditions (Ohfuji and Rickard 2005). Microbial sulfate reduction was traced back to the Paleoarchean Era (Shen et al. 2001) and contributes to the formation of pore-filling or concretionary pyrite crystals. These microbially induced pyrite crystals form early, typically near the sediment surface, where microbial communities are abundant in pore spaces. The involvement of microbes in pyrite formation is further supported by the identification of microbial remains within sedimentary pyrite aggregates (e.g. Gong et al. 2008; Schieber 2002). Schieber (2002) examined pyrite aggregates from the Archean to the Jurassic using scanning and transmission electron microscopy. Most pyrite samples revealed coccoid, rod-shaped, and even filamentous forms, which are interpreted as pyritized microbial remains. Furthermore, studies of pyrite from various time periods, ranging from modern to Proterozoic, have shown that their surfaces are often covered with spheroids approximately 50 nm in diameter (Figure 6.25a). These nano-scale spheroids, present in all the samples studied, were interpreted as the pyritized remains of nano-bacterial cells (Folk 2005). It is further proposed that iron sulfide precipitation was mediated by these small bacteria, which are commonly associated with decaying organic matter, while this interpretation remains confirmed by laboratory pyritization of modern nano-scale sulfate-reducing microbes (Folk 2005).

Pyrite framboids (Figure 6.25b) are microscopic, spheroidal to sub-spheroidal aggregates of equidimensional, equimorphic pyrite microcrystals, often exhibiting remarkable degrees of self-ordering (Liu et al. 2021; Ohfuji and Rickard 2005). Pyrite framboids found in nature range from <1 to 250 μm in diameter and display a mean size of around 6 μm across. Within any single framboid, the individual microcrystals all tend to have the same size, rarely more than 500 nm across. Framboids may be the most abundant mineral texture on Earth. They are abundant in rocks less than 600 Ma old, and decrease through the Proterozoic (>0.6 Ga) until they are rare in Archean rocks (>2.5 Ga). The formation of framboids involves two distinct processes. First, pyrite

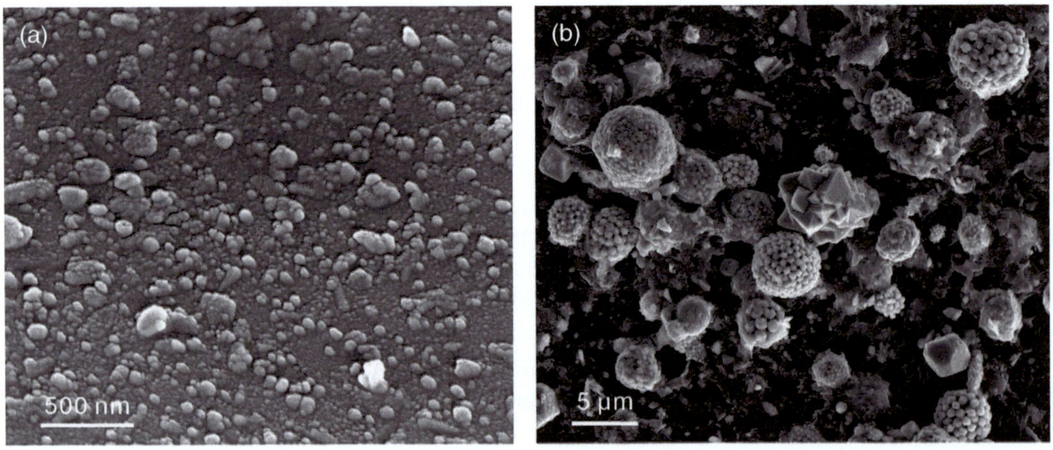

Figure 6.25 SEM images of sedimentary pyrite. (a) Nanoscale spheroidal structures on the surface of diagenetic pyrite from Cambrian limestone, China; (b) pyrite framboids from Cambrian black shales, China. *Source:* Photos by Xingliang Zhang.

microcrystals aggregate into sub-spherical groups through free energy minimization. Second, the microcrystals rearrange themselves into ordered domains through entropy maximization. The initial stage of pyrite microcrystal growth occurs in solutions with large supersaturations, and the less stable octahedral faces develop first. As the pyrite crystals grow, the solution becomes depleted in nutrients and the supersaturation begins to approach saturation and the most stable—or least soluble—cube faces develop. Truncated cubes and octahedra are thus common microcrystal habits. The exact nature of this organic material is unknown. However, it appears that microbial biofilm may be an important contributor. Likewise, the organic residues from framboids often appear similar to sulfur-rich organic geopolymers such as protokerogen. Most of the organic matter in framboids appears to be syngenetic with the framboids, and framboids seem to have grown in organic substrates (Rickard 2021).

Konhauser (2007) distinguished between syngenetic framboids, which form in the water column, and diagenetic framboids. Syngenetic framboids are generally smaller and more uniform in size, reflecting rapid nucleation with limited growth time in the water column. Some microbial colonies form spherical structures that become pyritized, creating aggregates resembling framboids. These pyritized microbial colonies are known as pseudoframboids (Gong et al. 2008). Schieber (2002) explored the differences between pseudoframboids and typical pyrite framboids based on internal texture. True framboids are composed of discrete, euhedral crystallites with no evidence of overgrowth; whereas, pseudoframboids (the mineralized bacterial colonies) usually consist of rounded pyrite grains. Beyond framboids, pyrite can also form loose, irregular aggregates of equant, equidimensional, and equimorphic microcrystals that exhibit an organized structure similar to framboids. However, these aggregates lack the sub-spheroidal morphology typical of framboids and are therefore not classified as such (Ohfuji and Rickard 2005).

6.2.4.2 Banded Iron Formations

Banded iron formations (BIFs) are iron-rich sedimentary rocks that are characterized by alternating layers (or bands) of iron minerals (~20–40%), typically hematite and magnetite, and iron-poor, siliceous (~40–50% SiO_2) mineral layers (Figure 6.26) that primarily precipitated throughout much of

Figure 6.26 BIF from the ~2.11 billion years old Negaunee Iron Formation (Paleoproterozoic), which consists of alternating bands of red chert (SiO_2), red hematite (Fe_2O_3), and silvery-gray specular hematite (Fe_2O_3). *Source:* Photo by Xingliang Zhang.

the Archean (3.8–2.5 Ga) and Paleoproterozoic (2.5–1.8 Ga), but then remerged in the Neoproterozoic (~0.8 Ga) (Posth et al. 2011).

BIFs are globally distributed, with some of the larger formations being those of the Hamersley range in Australia and the Transvaal Supergroup in South Africa (Figure 6.27), and contribute to more than 60% of the global iron resources (Aftabi et al. 2021; Posth et al. 2011). The mineralogy of the least metamorphosed BIFs consists of chert, magnetite, hematite, carbonates (siderite and dolomite–ankerite), greenalite, stilpnomelane, and riebeckite (Klein 2005). It is generally agreed that none of the minerals in BIF are primary, but that instead, the minerals reflect both diagenetic and metamorphic overprinting.

BIFs have been categorized as Algoma or Superior type based on their depositional environment. Algoma-type BIFs are generally small in lateral extent and display a volcanic association, while Superior-type BIFs are found in shelf environments, are vast, and are without a dominant volcanic influence.

The banding within BIFs can be observed on various scales, from coarse macrobandings (several meters thick) to mesobandings (centimeter-scale units) and microbands (millimeter- to submillimeter-thick varve-like laminae). BIFs exhibit a wide range of textures that can resemble those found in carbonates. Recognized textures include micritic, pelleted, intraclastic, peloidal, oolitic, pisolitic, and stromatolitic forms. Common sedimentary structures in BIFs include cross-bedding, load casts, ripple marks, erosional channels, shrinkage cracks, and slump structures, all of which indicate mechanical transportation and deposition (Boggs 2001).

There is no modern environment that mirrors the ancient conditions thought to favor widespread deposition of iron-rich sediments, and various models have been proposed to explain the formation of BIFs. While mineralogy dictates that some form of Fe(II) oxidation was necessary for their formation, which mechanism(s) dominated is still highly uncertain. Thus, an alternative, O_2-independent mechanism is therefore necessary to explain the deposition of the earliest BIF in anoxic conditions. And, it is becoming increasingly accepted that microorganisms were in some way involved in the BIF deposition, including the primary oxidation of Fe(II) to Fe(III), although intact and convincing microbial fossils have not yet been identified in BIFs (Konhauser et al. 2005). Firstly, phototrophs, like cyanobacteria, flourished when nutrients were abundant, producing large amounts of O_2 that oxidized Fe(II) to Fe(III), leading to the passive precipitation of ferric hydroxide. Secondly, some chemolithoautotrophic species, such as *Gallionella ferruginea*, which are known to oxidize Fe(II) as an energy source in environments with limited free photosynthetic O_2. This process may have contributed to ferric hydroxide precipitation across vast ocean basins (Eq. (6.1)). Alternatively, some purple and green bacteria can couple Fe(II) oxidation with CO_2 reduction during anoxygenic photosynthesis (Konhauser et al. 2002). Finally, many other microbes may have acted as nucleation and growth sites for iron mineralization.

$$6Fe^{2+} + 0.5O_2 + CO_2 + 16H_2O \rightarrow [CH_2O] + 6Fe(OH)_3 + 12H^+ \tag{6.1}$$

Konhauser et al. (2002) developed a biological model for BIF deposition and diagenesis (Figure 6.28), calculating the number of Fe^{2+}-oxidizing microbes required to precipitate ferric hydroxide and Fe-reducing bacteria needed for magnetite formation (Konhauser et al. 2005). In this model, it was proposed that direct microbial Fe^{2+} oxidation could have generated the bulk, if not all, of the ferric iron in BIFs, initially in the form of ferric hydroxide. Fe(II) oxidation and Fe(III) hydrolysis likely took place within the euphotic zone, with ferric hydroxide and biomass then settling through the water column to the seafloor. Once buried, ferric hydroxide could have transformed into: (1) siderite or magnetite via microbial Fe(III) reduction, (2) clays, when coprecipitated silica (entrained by iron during deposition) reacted with other cationic species in high pH pore waters, or

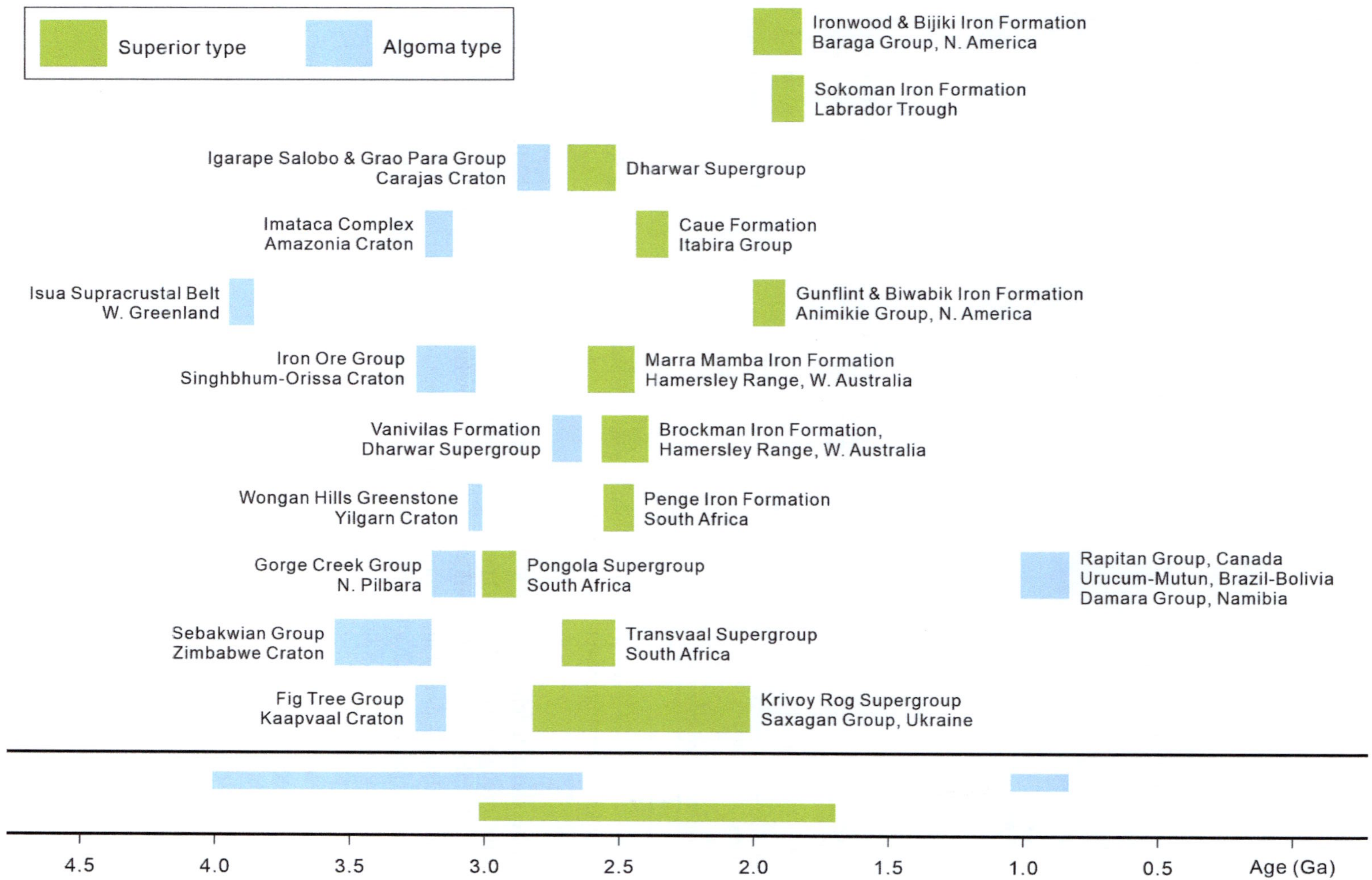

Figure 6.27 Appearance of BIFs in geological time and marked as Algoma and Superior types. *Source:* Posth et al. (2011)/Springer Nature.

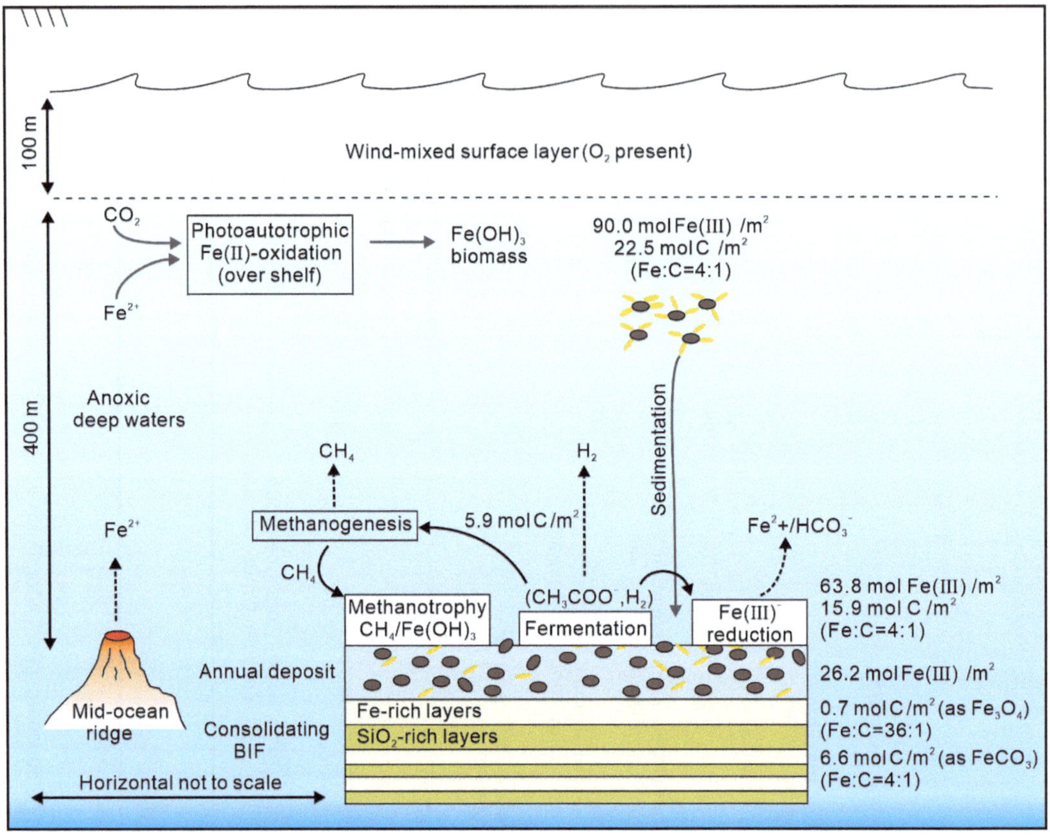

Figure 6.28 A hypothetical model for the microbial role in the formation of late Archean-Palaeoproterozoic BIFs. *Source:* Adapted from Konhauser et al. (2005).

(3) hematite, through dehydration of any residual $Fe(OH)_3$. The majority of Fe(II) in BIFs formed secondarily through biological Fe(III) reduction, meaning several ferrous iron minerals, including magnetite, rhombic siderite, ferrosilicates (e.g. stilpnomelane, chlorite), ankerite, and pyrite, formed during diagenesis and metamorphism. Some of these minerals, however, could also be primary sediments formed within an anoxic water column (e.g. spheroidal siderite). In anoxic bottom and pore waters, ferric hydroxide would have acted as a favorable electron acceptor for the oxidation of cellular remains.

Konhauser et al. (2005) estimated that, under ideal growth conditions, up to 70% of the biologically formed Fe(III) could have been recycled back into the water column through fermentation and organic carbon oxidation coupled with microbial Fe reduction by Fe(III)-reducing bacteria. By comparing the amount of biomass generated by phototrophs with the reducing equivalents needed for Fe(III) reduction and magnetite formation, they also suggested that another anaerobic metabolic pathway might have been used in the surface sediments to oxidize fermentation byproducts (Eq. (6.2)). Assuming the deep ocean was anoxic, this role could have been fulfilled by methanogens, and possibly methanotrophs, utilizing Fe(III) reduction (Eq. (6.3)).

$$CH_3COO^- + 8Fe(OH)_3 \rightarrow 8Fe^{2+} + 2HCO_3^- + 15OH^- + 5H_2O \quad (6.2)$$

$$CH_4 + 8Fe(OH)_3 \rightarrow 8Fe^{2+} + CO_2 + 16OH^- + 6H_2O \quad (6.3)$$

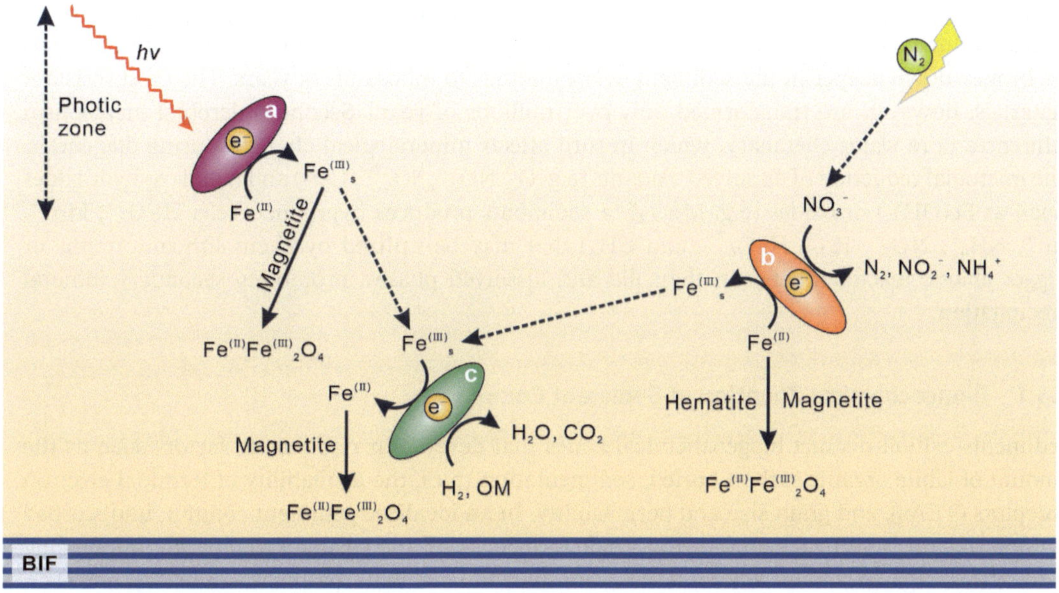

Figure 6.29 A proposed model for the microbial mediation of anoxic deposition of BIF, including phototrophic Fe(II) oxidation (a), nitrate-dependent Fe(II) oxidation (b), and Fe(III) reduction (c). *Source:* Adapted from Weber et al. (2006).

Konhauser's model posits that oxygenic photosynthesis generated sufficient oxygen to drive the precipitation of Fe(III)-rich minerals (Konhauser et al. 2002). However, given the ongoing debate regarding the evolution of oxygen during the early Proterozoic and its potential role in BIF deposition, alternative mechanisms are required to explain the precipitation of Fe(II) and Fe(III) minerals in anoxic environments. Figure 6.29 presents models for the microbial mediation of BIF deposition under anoxic conditions, which suggest that the formation of Fe-rich laminae, hematite, and magnetite may result from microbial metabolic activity in anoxic environments. Specifically, the direct oxidation of Fe(II) by photoautotrophic bacteria and nitrate-dependent Fe(II)-oxidizing bacteria can lead to the production of magnetite, hematite, and solid-phase Fe(III) oxides. Subsequently, these biogenic Fe(III) oxides are reduced by Ferrous Reducing Microorganisms, forming magnetite. This process links the deposition of BIFs to microbial Fe(III) reduction, which is directly associated with the activity of FRMs. Additionally, the bio-oxidation of Fe(II) not only results in the precipitation of Fe(III) oxides, but also directly precipitates magnetite and hematite, potentially linking Fe-oxidizing microorganisms to the formation of BIFs (Weber et al. 2006).

6.3 Biological Diagenesis

Once sediment is deposited, a combination of physical, chemical, and biological processes transforms the unconsolidated material into sedimentary rock. These processes, collectively known as diagenesis, include compaction, mineral transformation, organic matter degradation, hydrocarbon generation, changes in pore water composition, and the cementation of sediments. Throughout the sediment column, microbial activity plays a key role in diagenesis, extending up to the temperature threshold of life over 100 °C. First, microbes contribute to the decomposition of organic matter through various heterotrophic pathways. Labile materials are typically degraded in near-surface sediments over timescales ranging from days to years, while more refractory materials are broken down deeper in the sediment over hundreds to thousands of years. The most resistant materials, however, are transformed only over millions of years. Second, microbial metabolism influences pore water chemistry, which in turn affects mineralogical changes during diagenesis. The microbial reduction of dissolved oxidants (e.g. O_2, NO_3^-, SO_4^{2-}, CO_2) and metal oxyhydroxides (such as $Fe(OH)_3$) or oxides (e.g. MnO_2) in sediments produces byproducts (e.g. HCO_3^-, Mn^{2+}, Fe^{2+}, NH_4^+, NO_2^-, HS^-, HPO_4^{2-}, and CH_4) that may be utilized by chemolithoautotrophs or trigger abiotic reactions between the solid and dissolved phases, promoting secondary mineral precipitation.

6.3.1 Biogeochemical Zonation of Sediment Column

Sediments exhibit distinct biogeochemical zones that develop in response to factors such as the amount of labile organic carbon buried, sedimentation rates, the availability of terminal electron acceptors (TEAs), and grain size and permeability. In an idealized sediment column, undisturbed by bioturbation, a vertical geobiological gradient forms with increasing burial depth, influencing redox states, organic content, microbial communities, and metabolic pathways. In reality, however, changes in organic and inorganic chemistry are complex, dynamic, and involve numerous recycling reactions, making it challenging to fully understand which microbes are metabolically active at any given depth. Froelich et al. (1979) presented a formalized depth-related model that outlines the geobiological features and primary diagenetic reactions occurring in each zone, along with potential diagenetic minerals. This model is applicable to all aquatic environments, though the depth of each zone can vary significantly, from just millimeters or centimeters in coastal sediments to several meters in the deep ocean. Canfield and Thamdrup (2009) and Roberts (2015) developed a schematic redox zonation in sediments, which subdivides the sediment column into oxic (aerobic respiration), nitrogenous (nitrate reduction), manganous (Manganese reduction), ferruginous (Iron reduction), sulfidic (sulfate reduction), and methanic (methanogenesis) zones based on the levels of dissolved O_2 and HS^- in pore waters (Figure 6.30).

Generally, organic matter decreases with burial depth, accompanied by a progressive change in the composition of organic materials. This shift is most noticeable in the C:N ratio, which increases with depth in sediments but decreases with depth in the organic matter itself. The decomposition of freshly deposited organic material in sediments follows a continuous sequence of redox reactions. The most electropositive oxidants are consumed near the surface, while progressively weaker oxidants are consumed at greater depths (Figure 6.30). This process continues until the labile organic fraction is exhausted, leaving deeper sediments with a composition that differs significantly from the original deposition (Konhauser 2007).

In oxic sediments, typically at the sediment-water interface, aerobic respiration is the primary metabolic pathway for organic matter degradation. This pathway is the most efficient form of catabolism, generating the highest amount of free energy. On a global scale, it is estimated that the majority of particulate organic carbon is aerobically respired. Close to the sediment-water interface, in the oxic/suboxic zones, nitrogen cycles through ammonification, nitrification, and denitrification. Denitrifiers resemble aerobic respirers in that they can completely degrade complex organic matter to CO_2. However, neither aerobic respiration nor denitrification significantly affects diagenetic mineralogy (Konhauser 2007).

Mn and Fe^{3+} oxides and hydroxides are stable under oxic conditions but become unstable once buried beneath the oxic layers of sediment. Following denitrification, the reduction of MnO_2 to

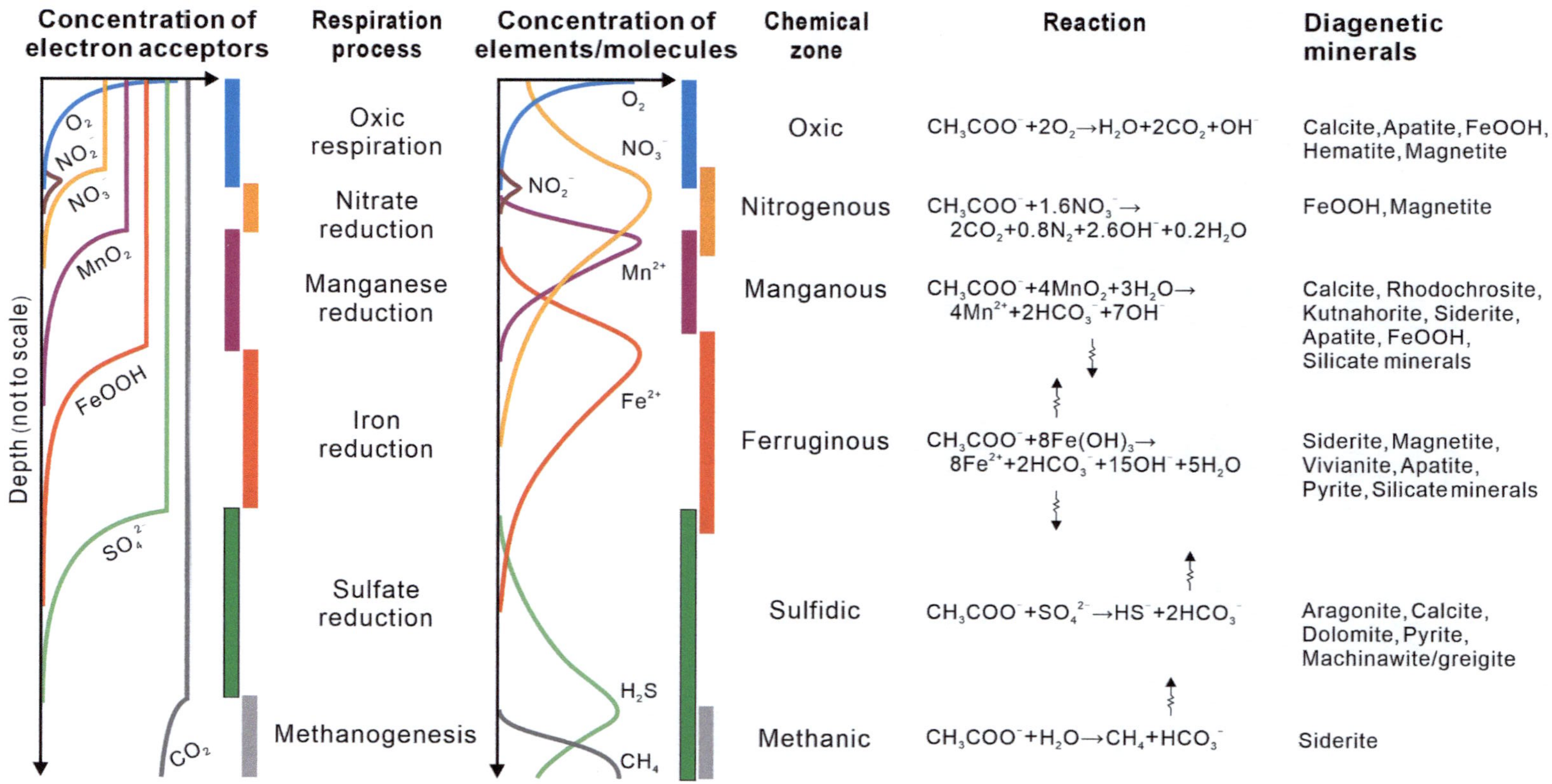

Figure 6.30 Schematic illustration of diagenetically driven redox zonation in an idealized vertical profile of sediment column, showing concentrations of solid phase common electron acceptors, concentration of associated elements/molecules, respiration process, redox zonation, biogeochemical reactions, and possible diagenetic minerals in each zone. *Source:* Modified from Froelich et al. (1979), Canfield and Thamdrup (2009), and Roberts (2015).

dissolved Mn^{2+} becomes the most energy-efficient microbial respiratory process. As Mn^{2+} diffuses downward and reacts with HCO_3^-, it typically forms manganous carbonates, such as rhodochrosite ($MnCO_3$), or mixed Mn and Ca carbonates, like kutnohorite. Fe^{3+} reduction occurs beneath the Mn^{2+} reduction zone and at the depth where nitrate has been completely removed from pore waters. $Fe(OH)_3$ is the most reactive and easily reducible iron mineral, but the rate of its reduction declines rapidly with depth as Fe^{3+} becomes depleted. Fe^{3+} reduction results in an increase in Fe^{2+} concentration in the pore waters, with a peak at the boundary between the Fe^{3+} and sulfate reduction zones (Figure 6.30). Fe^{2+} can diffuse upward, where it may be re-oxidized by reaction with Mn(IV) or NO_3^-, forming $Fe(OH)_3$ coatings on available surfaces, or it can react with HS^- produced by SRB, leading to pyrite formation. When excess Fe^{2+} is present, other mineral phases, including mixed valence oxides, carbonates, phosphates, and silicates, may form, depending on the specific chemical conditions of the sediment. Diagenetic minerals such as magnetite, siderite, and vivianite have been observed. Siderite formation has been linked to the activity of Fe^{3+}-reducing bacteria, which produce both Fe^{2+} and HCO_3^-. When sulfate reduction rates exceed Fe^{3+} reduction rates, monosulfide minerals (such as pyrite precursors) rather than siderite are more likely to precipitate.

Below the Fe^{3+}-reducing layer lies the highly reducing, oxygen-free sulfate reduction zone. Sulfate reduction typically occurs when all other available TEAs have been exhausted, with the exception of CO_2, and its rate decreases with depth. This process leads to the formation of insoluble monosulfides, which eventually transform into pyrite. Pyritization occurs on various scales, ranging from microscopic aggregates and framboids (Figure 6.25) to mesoscale fecal pellets, and even to macroscale features such as burrows, shell infillings, fossils, and concretionary layers (Figure 6.31). The final step in anaerobic degradation of organic material is methanogenesis, which is usually restricted to sediments in rapidly depositing coastal environments, where the accumulation of labile organic material exceeds sulfate availability. The CH_4 produced by methanogens diffuses upward into the sulfate reduction zone, where it is oxidized by methanotrophs under anoxic conditions. Under high pressure (>5 MPa) and low temperature (<10 °C), excess methane can condense to form methane hydrates. The anoxic oxidation of methane (AOM) and other hydrocarbon gases leads to an increase in pore water alkalinity, which in turn promotes the precipitation

Figure 6.31 Pyritized ammonite fossil from the Longtan Formation, Permian, China. *Source:* Courtesy of Guanghui Tong.

of aragonite and Mg-calcite. These carbonates can form as (1) isopachous cements that line cavities created by bedding expansion or relict hydrates, and (2) carbonate nodules that often display repeated zonation, with framboidal pyrite.

6.3.2 Diagenetic Mineralization

From the previous section, we can see that various diagenetic minerals are induced by metabolic activities across different biogeochemical zones. Among these, carbonate, phosphate, and silica minerals are particularly important and can precipitate in any of the biogeochemical zones. Given their significance, it is worthwhile to describe these minerals separately.

6.3.2.1 Diagenetic Carbonate Minerals

Aerobic respiration typically leads to the complete oxidation of organic matter to CO_2, promoting the dissolution of carbonate minerals. In contrast, anaerobic respirations below the oxic zones (Figure 6.30) release HCO_3^- into the pore waters, driving the precipitation of early diagenetic carbonate minerals. These processes become more pronounced at greater depths, particularly within the sulfate reduction zone.

Different oxidation processes, driven by various TEAs in the sediment zones, determine the mineralogy and stable isotopic composition of diagenetic minerals (Figure 6.32): (1) The chemolithoautotrophic oxidation of NH_4^+ and H_2S at the base of the oxic zone may increase the pH, leading to the precipitation of non-ferroan calcite with a $\delta^{13}C$ of 0‰; (2) Mn(IV) and Fe(III) reductions produce rhodochrosite and siderite, respectively, which have more negative $\delta^{13}C$ values (typically −2‰ and −10‰, respectively) due to the incorporation of biogenic carbon ($\delta^{13}C$ typically −20‰ to −30‰); (3) sulfate reduction produces non-ferroan calcite with a $\delta^{13}C$ signature of approximately −15‰, as any available Fe^{2+} preferentially reacts with HS^-; (4) methanogenesis leads to the precipitation of ferroan carbonates, which have a characteristic positive $\delta^{13}C$ signature (> +10‰) due to the preferential removal of ^{12}C from pore waters to methane; (5) AOM-induced carbonate minerals have a strongly ^{13}C-depleted isotopic composition, with $\delta^{13}C$ values as low as −69‰ (Coleman 1985).

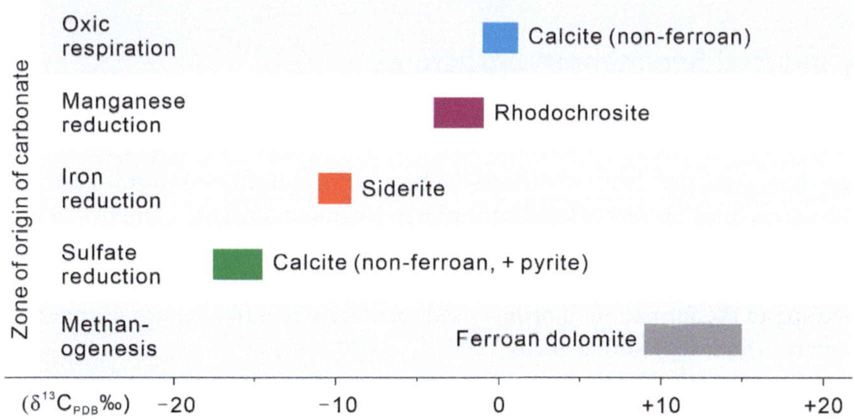

Figure 6.32 Carbon isotopic composition of authigenic carbonates related to the zone of formation. *Source:* Adapted from Coleman (1985).

6.3.2.2 Diagenetic Phosphate Minerals

Respiration processes in sediment zones can release organic-bound phosphorus into the pore waters, typically in the form of dissolved phosphate (HPO_4^{2-}). When the concentration of dissolved phosphorus in the pore waters reaches a certain level, diagenetic phosphate minerals begin to precipitate, reacting with metal cations under both acidic ($H_2PO_4^-$) and alkaline (HPO_4^{2-}) conditions. In most cases, dissolved phosphorus reacts with calcium ions (Ca^{2+}), leading to the formation of calcium fluorapatite or its metastable precursors. Calcium fluorapatite can also form when $CaCO_3$ is replaced through reactions with dissolved phosphorus.

Dissolved phosphorus can interact with ferric (and aluminum) oxyhydroxide minerals, which adsorb the phosphorus and act as temporary traps for upward-diffusing phosphorus. Once these ferric oxyhydroxides are buried within the Fe^{3+} reduction zone, the phosphate is released back into the pore waters. Below the oxycline, polyphosphate carried P would be released and provide additional P source for pore waters.

6.3.2.3 Amorphous Silica

It is clear that metabolic processes within sediments have minimal influence on the formation of diagenetic silica minerals. The primary biological contribution is the supply of silica to the pore waters. Diagenetic silica minerals form through dissolution–reprecipitation reactions. More than 95% of silica shells are recycled, with the remaining shells being incorporated into the sediment where they undergo diagenetic modification.

If the dissolution rate of biogenic opal-A within sediments exceeds the rate of quartz nucleation, the silica concentration in the interstitial waters can rise above the saturation point for opal-CT (Figure 6.20). In this process, as the silica shells are buried, dissolved, and exposed to gradually increasing temperatures (over 30 °C), supersaturation with respect to opal-CT triggers its nucleation and crystallization. Once these transformations begin, almost all biogenic opal-A is converted into opal-CT, often retaining the outline of the original shells. Under near-surface conditions, opal-CT can persist for millions of years. At greater depths, with increased temperature and pressure, the dissolution–reprecipitation process can continue, leading to the formation of quartz, often in the form of banded or nodular cherts. However, the mineral transformation is also influenced by other factors, such as the host sediments and the composition of the pore waters. For example, opal-CT is more likely to be retained in clayey deposits, while quartz is more common in carbonates.

6.4 Microbially Induced Sedimentary Structures

6.4.1 Definition

Modern sedimentology recognizes that benthic microbiota have always been an integral part of sediments, with microbial activities playing a significant role in sediment formation and lithification. Most of these microbes organize into aggregates named as biofilms or larger microbial mats. Biofilms and microbial mats actively influence the erosion, deposition, and deformation of sediments, thereby contributing to the formation of primary sedimentary structures known as microbially induced sedimentary structures (Noffke 2010).

In a more scholarly definition, microbially induced sedimentary structures are "primary sedimentary features that form syndepositionally through the interaction between biofilms, microbial mats, and the physical dynamics of sediment in siliciclastic aquatic environments," commonly simply abbreviated to "MISS" (Noffke 2010; Noffke et al. 2022). Like physical sedimentary structures,

MISS are highly indicative of specific facies and can provide valuable insights for paleoenvironmental reconstruction. In the past decade, there has been a growing interest in modern microbial mat features found in siliciclastic sediments, which has been paralleled by an increasing number of MISS discovered in Precambrian clastic rocks. These findings highlight the significant role that microbial communities have played in Earth's early sedimentation record.

6.4.2 Classification

Gerdes et al. (2000) studied modern textural signatures of microbial activity in two selected areas of siliciclastic sediments along the peritidal coastlines of the North Sea and Tunisia. They identified six types of biologically mediated clastic sedimentary fabrics, which result from: (1) intrinsic biofactors, such as the morphology and behavior of biofilms and microbial mats; (2) biological responses to physical disturbances; (3) trapping and binding of clastic sediments; (4) mechanical deformation of biologically stabilized clastic sediment surfaces; (5) post-burial processes; and (6) bioturbation.

Schieber's (2004) classification scheme covers a range of features that span the entire process continuum, from active mat growth, through mat metabolism, to mat destruction, and ultimately mat decay and diagenesis. Gerdes (2007) distinguished MISS in sandstones (Figure 6.33) and mudstones (Figure 6.34) separately.

Noffke et al. (2001) proposed a formal classification system that groups fourteen main types of MISS into two subclasses: six types of structures found atop bedding planes (leveled bedding surfaces, wrinkle structures, microbial mat chips, erosional remnants and pockets, multidirectional ripple marks, and mat curls) and eight types within the beds (sponge pore fabrics, gas domes, fenestrae structures, sinoidal laminae, oriented grains, benthic ooids, biolaminites, mat-layer-bound grain sizes). Later, a diamond-shaped classification scheme, comprising seventeen main types, was introduced based on microbial activities (Noffke 2010; Noffke et al. 2022), which includes five categories related to the genesis of the structures: (1) structures arising from growth, (2) structures arising from biostabilization, (3) structures arising from baffling and trapping, (4) structures arising from binding, and (5) structures arising from the interference of all microbial activities (Figure 6.35).

6.4.3 Biological Processes in MISS Formation

The MISS produced by microbiotic-physical interactions differ texturally and morphologically from the primary sedimentary structures generated by purely physical dynamics. Microbial activities involved in the sedimentation are different from each other as well. Microbial growth is not the same as biostabilization, and both are distinct from baffling and trapping. The following section takes a closer look at microbial growth, biostabilization, baffling-trapping, binding, and the interference of these biological processes based on Noffke's diamond-shaped classification scheme.

6.4.3.1 Growth

Unconsolidated sediments with moisture that is unaffected by erosion provide the ideal substrate for the growth of a biofilm or microbial mat. As the biofilm or microbial mat continues to grow, its vertical thickness increases. When a microbial mat covers a rough sedimentary surface, it gradually smooths out the uneven topography, eventually creating a level or planar surface, provided the growth remains undisturbed (Noffke et al. 2022). Noffke et al. (2001) referred to these organically smoothed surfaces as "*leveled depositional surfaces.*" In more advanced stages of microbial mat development, the mat surface becomes completely planar, obscuring the topography of the

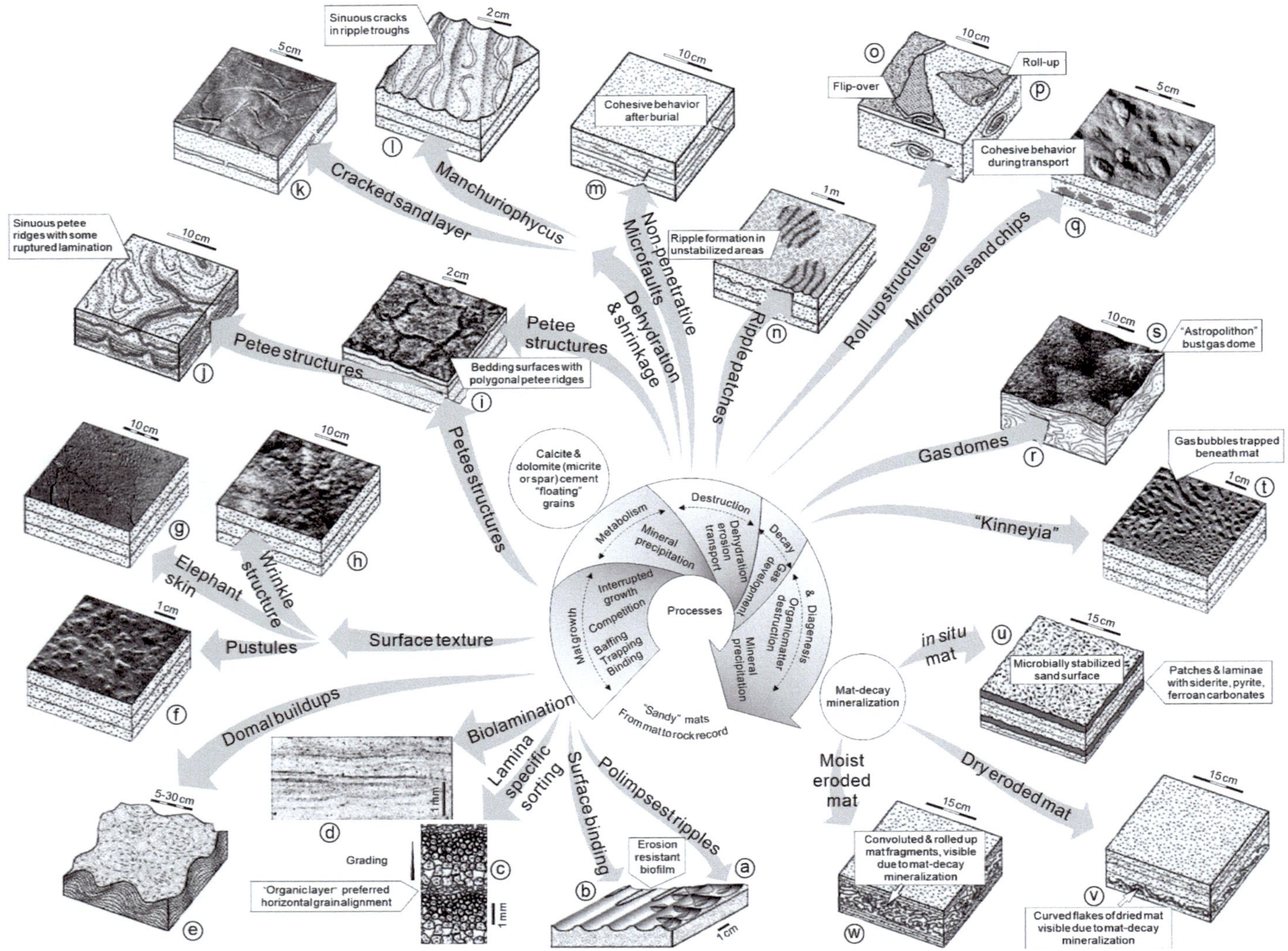

Figure 6.33 MISS found in sandstones where microbial communities flourished in the past. Genetic processes are arranged clockwise along a continuum from mat growth, through metabolism within mats, mat destruction, to organic decay and diagenetic mineral formation. *Source:* Modified from Eriksson et al. (2007).

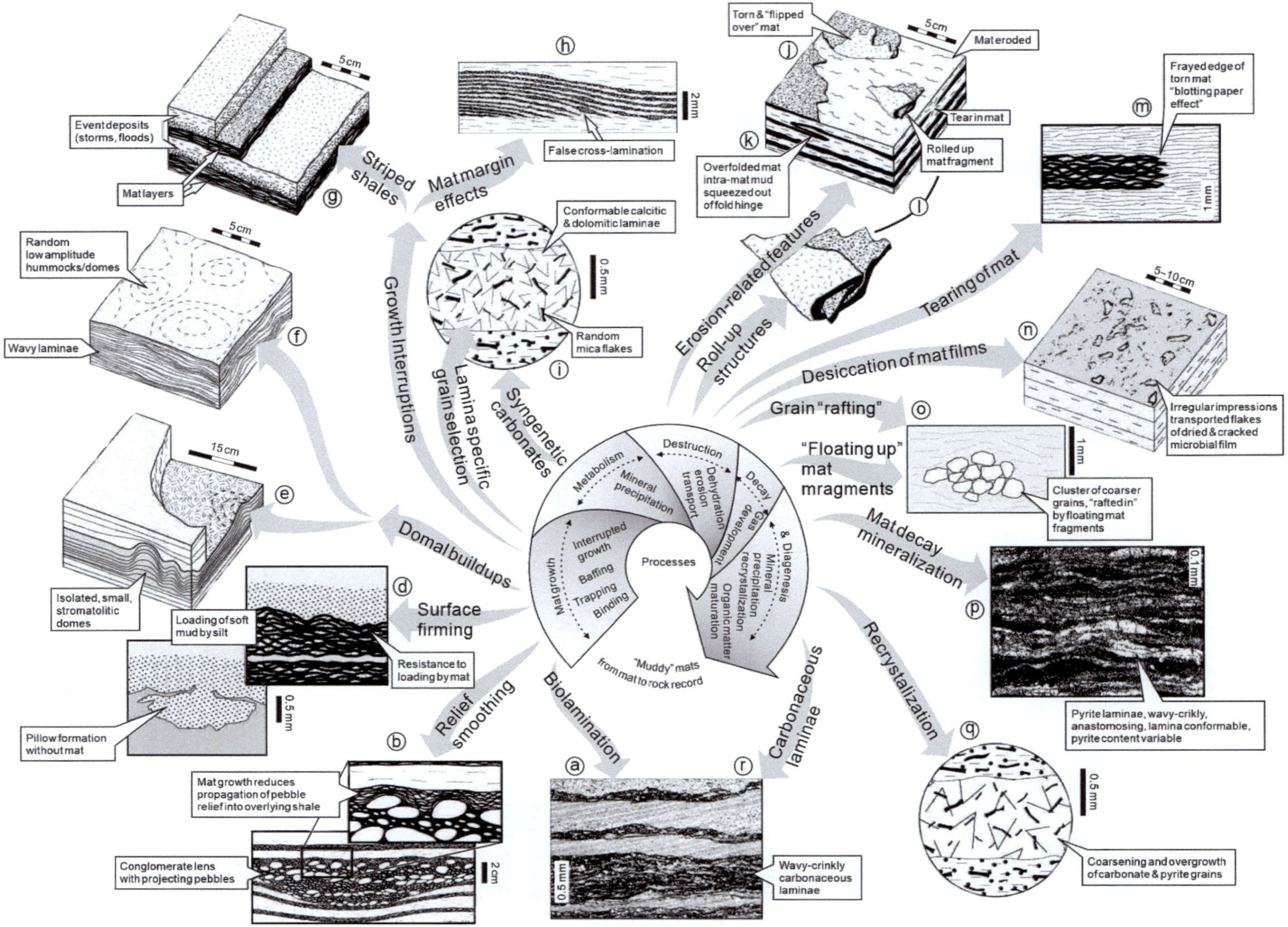

Figure 6.34 MISS found in mudstones where microbial communities flourished in the past. Genetic processes are arranged clockwise along a continuum from mat growth, through metabolism within mats, mat destruction, to organic decay and diagenetic mineral formation. *Source:* Modified from Eriksson et al. (2007).

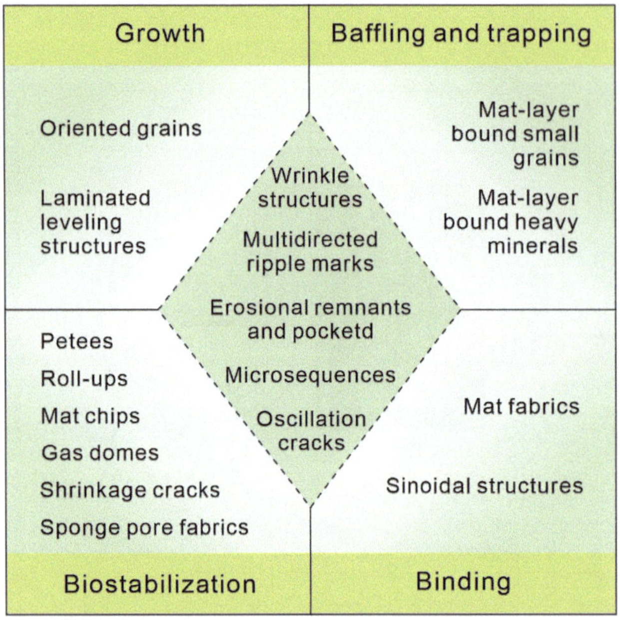

Figure 6.35 Classification of microbially induced sedimentary structures. The diagram shows the formational relationship of all MISS. The structures are induced by growth, biostabilization, baffling, trapping, or binding. The center of the diagram includes MISS caused by the interference of all biotic-physical interactions. *Source:* Modified from Noffke (2010).

underlying substrate. Low-lying areas of the sedimentary surface, such as ripple troughs, promote greater microbial growth due to the longer moisture retention and the better protection from erosion. The organic covering of ripple marks is especially common in the lower supratidal zone, where biomass production is high. As organic material fills the ripple troughs, the amplitude of the ripples decreases, and the ripple index transitions to a modified version. In some areas, the ripples are entirely obliterated by the microbial mats, making it impossible to determine a ripple index. These leveled depressions by growth microbial mats are also present in ancient deposits of different ages and facies. For example, Liu and Zhang (2017) described the microbial laminae structures developed on the undulating sedimentary interface from the Lower Cambrian strata of China, which resulted from the development of microbial mats that were phosphatized in preservation (Figure 6.36).

Oriented grains are another characteristic structure formed during the growth of microbial mats. Initially, thin biofilms adhere to the surfaces of mineral grains. Under favorable ecological conditions, the biofilms grow uniformly around the grain, causing the grains to become elevated and separated by the growth of biomass. Eventually, the grains "float," no longer in contact with each other, within the newly formed microbial mat layer. In thin sections of tidal sediments, it is often observed that the grains in the mat layers are aligned with their long axes parallel to the depositional surface, while the orientation of grains in the underlying substrate is random. This predominant alignment in the mats may suggest that the grains are positioned in an energetically favorable orientation relative to gravity. The soft organic matrix of the mat allows for the rotation of grains, leading to the characteristic texture known as "oriented grains" (Figure 6.37). In carbonate environments, grain rotation may also contribute to the formation of mat-related benthic ooids or sedimentary augen structures (Noffke et al. 2022).

Figure 6.36 Microbial laminae structures from the Zhongyicun Member of the Fortunian Stage, Cambrian, China. The rough sedimentary surface (indicated by the dotted line) is covered and smoothed by microbial mats. The filamentous structures within the mats are highlighted. *Source:* Photos by Wei Liu.

Figure 6.37 Optical microphotographs and Raman mappings of microbial mats from the Guojiaba Formation, Cambrian Stage 3 (~521–514.5 Ma), China. Dark-colored biolaminite preserves the fossil microbial mats interbedded with the light-colored sandstone laminae (left). The middle image shows a close-up view of the fossil microbial mats, while the right displays corresponding Raman component maps for mineral phases and kerogenous material. Note the "floating grains" (indicated by arrows) in the upper part of the Raman map, wrapped in organic matter—these are the oriented grains. *Q*, Quartz; *Fs*, feldspar; *An*, anatase; and *Ker*, kerogen. *Source:* Photos by Wei Liu.

6.4.3.2 Biostabilization

Biostabilization refers to the process by which microorganisms, such as bacteria, fungi, or algae, contribute to the stabilization of sediments by binding or cementing particles together. This occurs through the production of EPS, minerals, or other organic compounds that help bind the sediment grains. In the context of sedimentary environments, biostabilization can enhance the resistance of sediments to erosion, prevent sediments from destabilization by physical forces like water or wind, and promote the development of more complex sedimentary structures.

Biostabilization includes three types of processes, and may be a response to (1) erosion by horizontally directed water currents, but also to (2) intra-sedimentary gas pressure, or to (3) mechanical stress leading to ductile deformation. Noffke et al. (2022) systematically discussed and summarized these three biostabilization responses, and the following description provides an overview.

Biostabilization type-1 refers to the response of benthic microbial mats to erosive forces caused by a horizontally flowing water current passing over the mat surface. Laminar flow, in contrast to turbulent flow, typically exerts much less erosion due to the absence of vertical motion. Endobenthic microbial mats form within the upper millimeters of a sedimentary surface, and under close examination, individual mineral grains appear to project slightly above the surface. This type of biostabilization can also lead to the formation of microsequences—vertical sequences of graded sediment layers, each capped by a microbial mat (Figure 6.38). Once calm conditions set in, the mat can begin to develop, preserving each sediment layer through the mat's stabilizing effect, which prevents sediments from further erosion.

Biostabilization type-2 involves the sealing of sediment by EPS, which restricts gas exchange between the deposits and the surrounding water or atmosphere. As a result, gases such as O_2, CO_2, CH_4, H_2S, and others accumulate in the pore spaces of clastic deposits beneath microbial mats and are unable to escape. This buildup of gas pressure can create millimeter-scale pores, visible in vertical cross-sections of the mat-sealed sediment. These sedimentary features are known as sponge pore fabrics. Sponge pore fabrics are commonly found together with gas domes, which are local, centimeter-scale upheavals, form due to gas accumulation directly beneath the microbial mat.

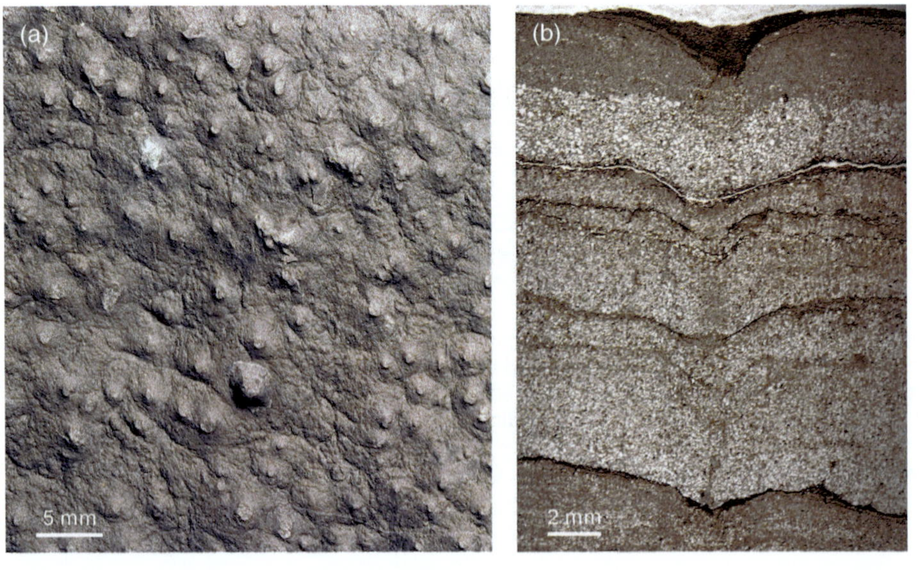

Figure 6.38 Fossil microbial mats from the sandstone of the Nanguanling Formation, Neoproterozoic, China. (a) Polygonal surface texture; (b) photomicrograph of thin-section showing carbonaceous laminae atop graded beddings. *Source:* Photos by Xingliang Zhang.

Biostabilization type-3 refers to the response of mat-stabilized sediment under ductile deformation. This type is common in areas with vertical water movement, such as where oscillating groundwater affects the sediment surface. In such conditions, a desiccating microbial-mat-bound sand layer may contract, curl up, and lose contact with the underlying sediment. Without the influence of biology, loose, unconsolidated sand would typically disperse into individual grains upon desiccation. However, when a microbial mat holds the grains together, the sediment remains intact and deforms in a manner similar to clay (ductile deformation). Deformation can also result from mechanical forces, such as the lateral shear or transport of the mat. Features like roll-ups or over-flips, which are types of MISS, serve as clear examples of this process.

6.4.3.3 Baffling and Trapping

Baffling and trapping both contribute to sediment accumulation through microbial activity and the incorporation of grains into mat fabrics via microbial growth, but they occur through two different processes (Noffke et al. 2022). Baffling refers to the response of the microbenthos to sedimentation. It causes the settling of small grains that, under the same hydraulic conditions, would remain suspended in the absence of microbes. Essentially, microbial baffling increases the deposition rate of finer-grained material compared to ambient conditions. This process can help clear the water column of fine particles that would otherwise cloud the water, reducing light penetration and impairing photosynthesis.

Whereas trapping involves the adhesive effect of EPS from microbial mats, which bind ambient particles. Mineral particles (often silt-sized) and other lithic fragments become both baffled and trapped, adhering to the mat surfaces. The effectiveness of baffling and trapping depends on factors such as the length of filament protrusion above the mat or sediment surface, the size and availability of grains, their weight, the frequency and constancy of current transport, and the angle of the mat's incline. The stickiness or adhesiveness of EPS, which varies between microbial groups, also plays a role in trapping grains. This adhesiveness can be influenced by electrolyte concentration or salinity in the surrounding environment (Noffke et al. 2022).

6.4.3.4 Binding

The process by which a consortium of microbes forms a biofilm or microbial mat is known as binding (Noffke et al. 2022). Lab experiments and field observations have shown active movement of cyanobacteria, both in controlled settings and in nature. Unlike biomass accumulation, which is largely influenced by nutrient availability, nutrient diffusion through the biofilm, and light conditions, binding is controlled solely by sedimentary factors. No biomass buildup is involved in this process. Binding results in structures like reticulate patterns made up of centimeter- to millimeter-scale ridges and tufts, which can cover large areas of microbial mats. Field observations of modern mats show that these patterns are capable of withstanding high-energy events. Sinoidal structures are another feature created by biofilms covering ripple mark troughs, as seen in cross-sections of buried sediment.

6.4.3.5 Interference of Microbial Activities Interacting with Physical Environments

Wrinkle structures, onc of the most common types of MISS, result from the interaction of microbial activities with physical environments and are easily recognized on bedding planes. They are often classified alongside other sedimentary structures, including runzel marken (wrinkle marks), *Kinneyia* ripples, micro-ripples, and related features like "elephant skin." The term "*Kinneyia*" refers to a class of microbially mediated sedimentary structures which is small, ovate grooves or pits found on fine-grained sandy bedding planes, typically on the flat tops of truncated ripples (Figure 6.39).

Figure 6.39 *Kinneyia* ripples preserved on upper surfaces of siliciclastic deposits, characterized by millimeter-scale, winding, flat-topped crests separated by equally sized round-bottomed troughs. *Source:* Photo by Wei Liu.

Thomas et al. (2013) proposed that the key mechanism behind the formation of *Kinneyia* ripples is a Kelvin-Helmholtz type instability, which occurs in a viscoelastic film under flowing water. This instability spontaneously induces ripple corrugations in the film, and these ripples grow in amplitude over time. Experiments with viscoelastic films confirm that the wavelength of the ripple instability is proportional to the thickness of the film. The resulting ripple pattern has a wavelength roughly three times the thickness of the film. Notably, this behavior is independent of the film's viscosity and the flow conditions. Laboratory-analog *Kinneyia* ripples were formed by the sedimentation of glass beads, which preferentially deposit in the troughs of the ripples. Well-ordered patterns emerge, with honeycomb-like and parallel ridges observed depending on the flow speed. These patterns closely resemble those found in *Kinneyia*, with similar morphologies, wavelengths, and amplitudes.

In general, wrinkle structures are characterized by contorted, wrinkled, irregularly pustulose, and sometimes quasi-polygonal surface morphologies, often with steepened ridges. These structures can occur on both bed tops and bottoms. Studies of ancient wrinkle structures have examined their morphology, as well as the microbial filaments that were part of the original mat. Hagadorn and Bottjer (1999) proposed useful parameters for describing these structures (Figure 6.40). Wrinkle surface relief, measured from crest to trough, ranges from 0.03 to 3 mm, with intercrest distances varying between 1 and 5 mm. Crest shapes can be pointed, rounded, or flat-topped, and variations in crest length and ridge steepness result in different surface patterns, including mosaic, polygonal, bifurcating, and elongate subparallel forms.

6.4.4 Distribution and Preservation

6.4.4.1 Temporal and Spatial Distribution

A series of studies has shown that fossil MISS are present in tidal flat and continental shelf sandstones spanning Archean, Proterozoic, and Phanerozoic ages. Notably, these structures appear remarkably consistent over the past 3.5 billion years.

Specimens of MISS have been found in some of the oldest non-metamorphosed sedimentary rock formations, such as the 3.48-billion-year-old Dresser Formation in Western Australia. Other

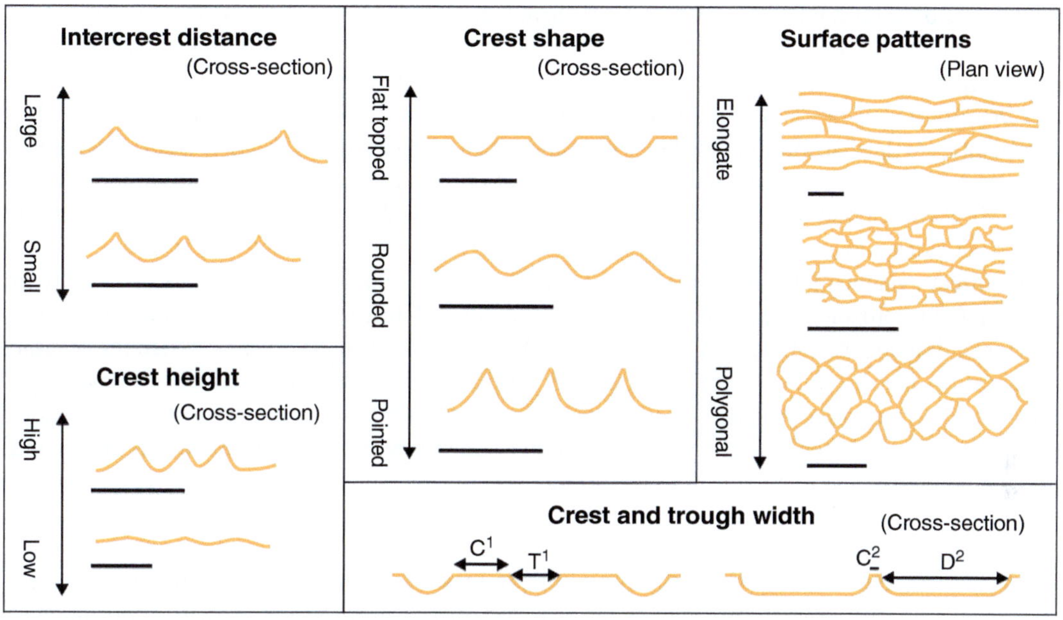

Figure 6.40 Schematic representation of morphological characteristics used to describe microbial wrinkle structures. All scale bars are 0.5 cm. *Source:* Adapted from Hagadorn and Bottjer (1999).

examples include the 2.9-billion-year-old Pongola Supergroup and the Witwatersrand Supergroup, marine stratigraphic successions from the Archean that were once formed by photoautotrophic mats. During the Late Neoproterozoic, microbial mats were widespread across seafloors and textured organic surfaces from this period, which documented interactions between the Ediacara biota (the earliest macroscopic multicellular organisms) and the microbial mats of the time. The exceptional preservation of these soft-bodied organisms underscores the significant role of microbial mats during this period. Terrestrial microbial mats, influenced by aeolian processes, are also found in the Neoproterozoic Venkatpur Sandstone. Phanerozoic examples of MISS have been identified across various periods, including the Cambrian, Ordovician, Silurian, Devonian, Carboniferous, Permian, Triassic, Jurassic, Cretaceous, Neogene, and Quaternary (Noffke et al. 2022).

In both modern and ancient shallow-marine environments, MISS demonstrate that biofilms and microbial mats are not randomly distributed. Microorganisms tend to grow preferentially in habitats where hydraulic dynamics and sediment composition are conducive to the establishment of a biofilm or microbial mat. In particular, continental shelves, lagoons, and tidal flat environments, where average hydraulic conditions and sediment composition allow for favorable exposure to light, support the growth of photoautotrophic microbes. In these environments, fine sand substrates composed of clear (translucent) quartz grains are preferentially colonized by photoautotrophic microorganisms (Noffke 2010).

6.4.4.2 Preservation and Recognition

In older rocks, many original fossils and sedimentary structures of MISS, which represent a largely unexplored archive, offering a significant resource for understanding Earth's early life, are often lost due to processes that alter the deposits from the moment they were laid down to the moment they were studied. To confidently classify certain structures or textures as being of biological origin, a catalogue of criteria for MISS has been established as well: (1) fossil MISS occur in sedimentary rocks that experienced less than low grades of metamorphosis (greenschist facies); (2) in stratigraphic sections, MISS correlate with transgressive phases; (3) MISS occur in a defined "microbial mat depositional facies" that facilitates the development and preservation of photoautotrophic microbiota; (4) the distribution of MISS is not random, but reflects the average hydraulic and meteorological pattern in a defined area; (5) in statistical analyses, the geometries and dimensions of fossil MISS correspond to those of modern MISS; (6) MISS include textures that either represent, have been caused by, or are related to biofilms or microbial mats (Noffke 2010).

The direct evidence for a microbial mat is not always easily detectable in a rock, especially in Archean to Proterozoic samples. The inorganic sedimentary fabrics that arose from microbial activities (granulometric selections and oriented grains) are more likely to be preserved than the organic matter of the mat itself. The organic matter may only become a fossil under conditions favoring a corresponding taphonomic path. In short, the preservation of MISS occurs in two steps, the syndepositional (primary) step, and the post-depositional (secondary) step (Noffke 2010). In the second step, components of a microbial mat, including coccoid cells, filaments, and EPS, are preserved. During early diagenetic biomineralization of organic matter, larger organic molecules are divided into smaller ones through the action of heterotrophic microorganisms. Small organic molecules react with compounds of the surrounding medium and resulting mineral precipitates grow. During progressive diagenesis, such precipitates later recrystallize to form minerals in the consolidated rock. The primary textural and secondary mineralogical evidence is significant for the evaluation of the biogenicity of a possible MISS structure. The more obscure a sedimentary structure is due to weathering, the more important such sediment-internal textures are for the search for ancient life (Callefo et al. 2021).

6.5 Astrobiological Implications

The search for life beyond Earth involves understanding both the nature of life itself and the environments capable of supporting it. However, since the form of life on other planets is still unknown, the challenge lies in detecting life without knowing precisely what to look for. Earthly analogs, such as water and biosignatures, offer valuable clues. These analogues represent life forms and environments on Earth that may resemble those expected on other planets. For example, microbial mats in Antarctic lakes serve as analogs for ecosystems that might exist in the polar regions of Mars.

Fossil biofilms and microbial mats from the 3.3 to 3.5 billion-year-old Onverwacht Group in South Africa (Westall et al. 2003) and MISS from the 3.48 billion-year-old Dresser Formation in Western Australia (Noffke et al. 2013) provide compelling evidence for the formation of microbial mats in shallow water to intermittently subaerial environments. These records, indicating evaporitic conditions, suggest that microbial life was already capable of surviving in exposed environments during the Paleoarchean, a time when the atmosphere lacked significant O_2, was subjected to intense UV radiation, and faced frequent impact events. This resilience of life raises an interesting question: How does this compare to the potential for life on early terrestrial planets like Mars?

While Mars and Earth had distinct early geological evolutions, Mars likely harbored more shallow water habitats conducive to life, particularly before the loss of surface volatiles and the planet's freezing between 3.8 and 3.5 billion years ago. Numerous impact craters and standing bodies of water could have provided suitable environments for potential Martian life. Much like Earth, thick, robust biofilms or mats could have offered protection for microbial colonies against dehydration and UV radiation (Westall et al. 2003).

In the field of astrobiology, biosediments, formed through the interaction of microbes, minerals, and the surrounding ecosystem, provide a unique window into both Earth's ancient past and the ongoing relationship between life and its environment. These biosediments are crucial for understanding how life interacts with the geosphere and the environmental feedback loops that shape planetary habitability. Microbial communities involved in biosedimentation have regulated key global cycles for billions of years. For example, stromatolites serve as an ideal model for investigating fundamental astrobiological questions, such as how microbial life interacts with its environment and influences planetary conditions (Foster et al. 2020).

Stromatolites also offer a valuable environment for studying the feedback between life and its ecosystem. The widespread presence of stromatolites in the geological record suggests that microbial communities responsible for their formation possess a selective advantage, maintaining stable community structures despite environmental disturbances. As stromatolites grow, they create geomorphological features that influence biological processes and amplify physical ones, establishing negative feedback loops. These loops generate self-organizing patterns, which could serve as key indicators of life processes and biosignatures (Foster et al. 2020). The stromatolites of Shark Bay, Western Australia, provide an ideal ecosystem for exploring these biogenic spatial patterns (Suosaari et al. 2019).

In astrobiology, even if biosediments as we know them do not exist beyond Earth, the processes that drive their formation are likely to occur wherever life is present. While there are no perfect analogs for ancient Earth or extraterrestrial environments like Mars or Europa, it's reasonable to expect that microbial life elsewhere in the universe would exist within structured ecosystems, much like on Earth. Many minerals deposited in biosediments, such as carbonates, halite, and gypsum, are found on other terrestrial worlds like Mars. However, these minerals alone do not inherently suggest the presence of life. Chemical alterations, such as isotope fractionation, can help determine whether minerals originated biologically, particularly when considered alongside the depositional context (Foster et al. 2020).

References

Aftabi, A., Atapour, H., Mohseni, S., and Babaki, A. 2021. Geochemical discrimination among different types of banded iron formations (BIFs): a comparative review. *Ore Geology Reviews*, 136, pp. 104244.

Åström, E.K.L., Carroll, M.L., Ambrose, W.G., Jr., Sen, A., Silyakova, A., and Carroll, J. 2016. Arctic cold seeps in marine methane hydrate environments: impacts on shelf macrobenthic community structure offshore Svalbard. *Marine Ecology Progress Series*, 552, pp. 1–18.

Åström, E.K.L., Carroll, M.L., Ambrose, W.G., Sen, A., Silyakova, A., and Carroll, J. 2018. Methane cold seeps as biological oases in the high-Arctic deep sea. *Limnology and Oceanography*, 63, pp. S209–S231.

Boggs, S.J. 2001. *Principles of sedimentology and stratigraphy* (Third Edition), pp. 726. Hoboken, Prentice-Hall.

Bosak, T., Souza-Egipsy, V., and Newman, D.K. 2004. A laboratory model of abiotic peloid formation. *Geobiology*, 2, pp. 189–198.

Callefo, F., Ricardi-Branco, F., Cataldo, R.A., and Noffke, N. 2021. Microbially induced sedimentary structures (MISS). In: *Encyclopedia of geology* (Second Edition) (eds. D. Alderton and S.A. Elias), pp. 545–554. London, Elsevier.

Canfield, D.E., and Thamdrup, B. 2009. Towards a consistent classification scheme for geochemical environments, or, why we wish the term 'suboxic' would go away. *Geobiology*, 7, pp. 385–392.

Cathles, L., Zheng, S., and Chen, D. 2010. The physics of gas chimney and pockmark formation, with implications for assessment of seafloor hazards and gas sequestration. *Marine and Petroleum Geology*, 27, pp. 82–91.

Coleman, M.L. 1985. Geochemistry of diagenetic non-silicate minerals: kinetic considerations. *Philosophical Transactions of the Royal Society A*, 315, pp. 39–56.

Cui, L., Liu, W., and Zhang, X.L. 2020. Phosphatized microbial fossils from the lowest Cambrian of South China and their ecological and environmental implications for the Kuanchuanpu biota. *Precambrian Research*, 338, pp. 105560.

Dupraz, C., and Visscher, P.T. 2005. Microbial lithification in marine stromatolites and hypersaline mats. *Trends in Microbiology*, 13, pp. 429–438.

Dupraz, C., Reid, R.P., and Visscher, P.T. 2011. Microbialites, modern. In: *Encyclopedia of geobiology* (eds. J. Reitner and V. Thiel), pp. 617–635. AA Dordrecht, Springer.

Eriksson, P.G., Shieber, J., Bouougri, E., Gerdes, G., Porada, H., Banerjer, S., Bose, P.K., and Sarkar, S. 2007. Classification of structures left by microbial mats in their host sediments. In: *Atlas of microbial mat features preserved within the clastic rock record* (eds. J. Schieber, P.K. Bose, P.G. Eriksson, S. Banerjee, S. Sarkar, W. Altermann, and O. Catuneanu), pp. 39–52. Amsterdam, Elsevier.

Feng, D., Qiu, J.W., Hu, Y., Peckmann, J., Guan, H.X., Tong, H.P., Chen, C., Chen, J.X., Gong, S.G., Li, N., and Chen, D.F. 2018. Cold seep systems in the South China Sea: an overview. *Journal of Asian Earth Sciences*, 168, pp. 3–16.

Flügel, E. 2004. *Microfacies of carbonate rocks*, pp. 976. Berlin, Springer.

Folk, R.L. 2005. Nannobacteria and the formation of framboidal pyrite: textural evidence. *Journal of Earth System Science*, 114, pp. 369–374.

Foster, J.S., Babilonia, J., Parke-Suosaari, E., and Pamela, R.R. 2020. Chapter 4: stromatolites, biosignatures, and astrobiological implications. In: *Astrobiology and Cuatro Ciénegas basin as an analog of early earth* (eds. V. Souza, A. Segura, and J.S. Foster), pp. 89–106. Cham, Springer.

Froelich, P.N., Klinkhammer, G.P., Bender, M.L., Luedtke, N.A., Heath, G.R., Cullen, D., Dauphin, P., Hammond, D., Hartman, B., and Maynard, V. 1979. Early oxidation of organic matter in pelagic sediments of eastern equatorial Atlantic: suboxic diagenesis. *Geochimica et Cosmochimica*, 43, pp. 1075–1090.

Gächer, R., Meyer, J.S., and Mares, A. 1988. Contribution of bacteria to release and fixation of phosphorus in lake deposits. *Limnology and Oceanography*, 33, pp. 1542–1558.

Gerdes, G. 2007. Structures left by modern microbial mats in their host sediments. In: *Atlas of microbial mat features preserved within the clastic rock record* (eds. J. Schieber, P.K. Bose, P.G. Eriksson et al.), pp. 5–38. Amsterdam, Elsevier.

Gerdes, G., Klenke, T., and Noffke, N. 2000. Microbial signatures in peritidal siliciclastic sediments: a catalogue. *Sedimentology*, 47, pp. 279–308.

Gong, Y.M., Shi, G.R., Weldon, E.A., Du, Y.S., and Xu, R. 2008. Pyrite framboids interpreted as microbial colonies within the Permian *Zoophycos* spreiten from southeastern Australia. *Geological Magazine*, 145, pp. 95–103.

Hagadorn, J.W., and Bottjer, D.J. 1999. Restriction of a late neoproterozoic biotope: suspect microbial structures and trace fossils at the Vendian-Cambrian transition. *Palaios*, 14, pp. 58–72.

James, N.P. 1977. Facies models 7. Introduction to carbonate facies models. *Geoscience Canada*, 4, pp. 123–125. 10.12789/gs.v4i3.3024

Kim, J., Kimura, Y., Puchala, B., Yamazaki, T., Becker, U., and Sun, W.H. 2023. Dissolution enables dolomite crystal growth near ambient conditions. *Science*, 382, pp. 915–920.

Kim, D., Choh, S.J., Liu, W., Zhang, X.L., and Hong, J. 2025. Cambrian Series 2 calcimicrobial crust-cement boundstone in the Yangtze Block, China: a distinctive bioconstruction as a legacy of Precambrian reef evolution. *Sedimentary Geology*, 477, pp. 106804.

Klein, C. 2005. Some Precambrian banded iron formations (BIFs) from around the world: their age, geologic setting, mineralogy, metamorphism, geochemistry, and origin. *The American Mineralogist*, 90, pp. 1473–1499.

Knauth, L.P. 1994. Petrogenesis of chert. *Reviews in Mineralogy*, 29, pp. 233–258.

Konhauser, K. 2007. *Introduction to geomicrobiology*, pp. 448. Oxford, Blackwell Science.

Konhauser, K.O., Hamade, T., Raiswell, R., Morris, R.C., Grant Ferris, F., Southam, G., and Canfield, D. E. 2002. Did bacteria form Precambrian banded iron formations? *Geology*, 30, pp. 1079–1082.

Konhauser, K.O., Newman, D.K., and Kappler, A. 2005. The potential significance of microbial Fe(III) reduction during deposition of Precambrian banded iron formations. *Geobiology*, 3, pp. 167–177.

Krüger, M., Treude, T., Wolters, H., Nauhaus, K., and Boetius, A. 2005. Microbial methane turnover in different marine habitats. *Palaeogeography, Palaeoclimatology, Palaeoecology*, 227(special issue), pp. 6–17.

Levin, L.A. 2005. Ecology of cold seep sediments: interactions of fauna with flow, chemistry, and microbes. *Oceanography and Marine Biology An Annual Review*, 43, pp. 1–46.

Levin, L.A., Baco, A.R., Bowden, D.A., Colaco, A., Cordes, E.E., Cunha, M.R., Demopoulos, A.W.J., Gobin, J., Grupe, B.M., Le, J., Metaxas, A., Netburn, A.N., Rouse, G.W., Thurber, A.R., Tunnicliffe, V., Van Dover, C.L., Vanreusel, A., and Watling, L. 2016. Hydrothermal vents and methane seeps: rethinking the sphere of influence. *Frontiers in Marine Science*, 3, pp. 72.

Li, J.H., Liu, P.Y., Menguy, N., Zhang, X.L., Wang, J., Benzerara, K., Feng, L.J., Sun, L., Zheng, Y., Meng, F.Q., Gu, L., Leroy, E., Hao, J.L., Chu, X.L., and Pan, Y.X. 2022. Intracellular silicification by early-branching magnetotactic bacteria. *Science Advances*, 8, pp. eabn6045.

Liu, W., and Zhang, X.L. 2012. Girvanella coated grains from Cambrian oolitic limestone. *Facies*, 58, pp. 779–787.

Liu, W., and Zhang, X.L. 2017. Possible biogenic structures from the lower Cambrian strata in Yunnan Province, South China. *Geological Magazine*, 154, pp. 1285–1293.

Liu, W., Chang, C., Yun, H., Cui, L.H., and Zhang, X.L. 2021. A laminated microbial ecosystem at the summit of the Cambrian explosion. *Global and Planetary Change*, 205, pp. 103619.

Morse, J.W., and Mackenzie, F.T. 1990. *Geochemistry of sedimentary carbonates*, pp. 707. Amsterdam, Elsevier.

Noffke, N. 2010. *Geobiology: microbial mats in sandy deposits from the archean era to today*, pp. 194. Berlin, Springer.

Noffke, N., Gerdes, G., Klenke, T., and Krumbein, W.E. 2001. Microbially induced sedimentary structures-a new category within the classification of primary sedimentary structures. *Journal of Sedimentary Research*, 71, pp. 649–656.

Noffke, N., Christian, D., Wacey, D., and Hazen, R.M. 2013. Microbially induced sedimentary structures recording an ancient ecosystem in the ca. 3.48 Billion-year-old Dresser Formation, Pilbara, Western Australia. *Astrobiology*, 13, pp. 1103–1124.

Noffke, N., Beraldi-Campesi, H., Callefo, F., Carmona, N., Cuadrado, D.G., Hickman-Lewis, K., Homann, M., Mitchell, R., Sheldon, N., Westall, F., and Xiao, S.H. 2022. Part B, Volume 2, Chapter 5: Microbially induced sedimentary structures (MISS). *Treatise Online*, 162, pp. 1–29.

Ogawa, Y., and Kawata, T. 1998. Preservation of biogenic opal-A in earliest cretaceous radiolarian claystone from the Western Pacific. *Journal of Sedimentary Research*, 68, pp. 435–439.

Ohfuji, H., and Rickard, D. 2005. Experimental syntheses of framboids: a review. *Earth-Science Reviews*, 71, pp. 147–170.

Peckmann, J., and Goedert, J.L. 2005. Geobiology of ancient and modern methane-seeps. *Palaeogeography, Palaeoclimatology, Palaeoecology*, 227(special issue), pp. 1–257.

Pei, Y., Suarze-Gonzalez, P., Duda, J.P., and Reitner, J. 2024. Organic matter influence on ooid formation: new insights into classic examples (Great Salt Lake, USA; Triassic Germanic Basin, Germany). *Sedimentology*, 71, pp. 1419–1435.

Phoenix, V.R., Adams, D.G., and Konhauser, K.O. 2000. Cyanobacterial viability during hydrothermal biomineralization. *Chemical Geology*, 169, pp. 329–338.

Posth, N.R., Konhauser, K.O., and Kappler, A. 2011. Banded iron formation. In: *Encyclopedia of geobiology* (eds. J. Reitner and V. Thiel), pp. 92–102. AA Dordrecht, Springer.

Reitner, J., Arp, G., Thiel, V., Gautret, P., Galling, U., and Michaelis, W. 1997. Organic matter in Great Salt Lake ooids (Utah, USA)-First approach to a formation via organic matrices. *Facies*, 36, pp. 210–219.

Rickard, D. 2021. *Framboids*, pp. 334. New York, Oxford University Press.

Riding, R. 2000. Microbial carbonates: the geological record of calcified bacterial-algal mats and biofilms. *Sedimentology*, 47(Suppl. 1), pp. 179–214.

Riding, R. 2006. Microbial carbonate abundance compared with fluctuations in metazoan diversity over geological time. *Sedimentary Geology*, 185, pp. 229–238.

Riding, R., and Liang, L.Y. 2004. Geobiology of microbial carbonates: metazoan and seawater saturation state influences on secular trends during the Phanerozoic. *Palaeogeography, Palaeoclimatology, Palaeoecology*, 219, pp. 101–115.

Roberts, A.P. 2015. Magnetic mineral diagenesis. *Earth-Science Reviews*, 151, pp. 1–47.

Ruttenberg, K.C. 2005. The global phosphorus cycle. In: *Treatise on geochemistry, volume 8, biogeochemistry* (ed. W.H. Schlesinger), pp. 585–643. Oxford, Elsevier.

Schieber, J. 2002. Sedimentary pyrite: a window into the microbial past. *Geology*, 30, pp. 531–534.

Schieber, J. 2004. Microbial mats in siliciclastic rock record: a summary of diagnostic features. In: *The Precambrian earth: tempos and events, development in Precambrian geology* (eds. D.G. Eriksson, W. Altermann, D. Nelson et al.), pp. 663–672. Amsterdam, Elsevier.

Seong-Joo, L., Browne, K.M., and Golubic, S. 2000. On stromatolite lamination. In: *Microbial sediments* (eds. R.E. Riding and S.M. Awramik), pp. 16–24. Berlin, Springer-Verlag.

Shapiro, R.S. 2004. Neoproterozoic-Cambrian microbialite record. In: *The paleontological society papers, volume 10, Neoproterozoic-Cambrian biological revolutions* (eds. J.H. Lipps and B.M. Waggoner), pp. 5–15. Cambridge, Cambridge University Press.

Shapiro, R.S., and Wilmeth, D.T. 2020. Part B, Volume 1, Chapter 8: Microbialites. *Treatise Online*, 134, pp. 1–24.

Shen, Y.A., Buick, R., and Canfield, D.E. 2001. Isotopic evidence for microbial sulfate reduction in the early Archaean era. *Nature*, 410, pp. 77–81.

Soudry, D. 2000. Microbial phosphate sediment. In: *Microbial sediments* (eds. R.E. Riding and S.M. Awramik), pp. 127–135. Berlin, Springer-Verlag.

Suarez-Gonzalez, P., and Reitner, J. 2021. Ooids forming in situ within microbial mats (Kiritimati atoll, central Pacific). *PalZ*, 95, pp. 809–821.

Suosaari, E.P., Reid, R.P., Oehlert, A.M., Playford, P.E., Steffensen, C.K., Andres, M.S., Suosaari, G.V., Milano, G.R., and Eberli, G.P. 2019. Stromatolite provinces of Hamelin Pool: physiographical controls on stromatolites and associated lithofacies. *Journal of Sedimentary Research*, 89, pp. 207–226.

Teichert, B.M.A., Bohrmann, G., and Suess, E. 2005. Chemoherms on Hydrate Ridge-Unique microbially-mediated carbonate build-ups growing into the water column. *Palaeogeography, Palaeoclimatology, Palaeoecology*, 227(special issue), pp. 67–85.

Thomas, K., Herminghaus, S., Porada, H., and Goehring, L. 2013. Formation of Kinneyia via shear-induced instabilities in microbial mats. *Philosophical Transactions of the Royal Society A*, 371, pp. 20120362.

Thurber, A.R., Levin, L.A., Rowden, A.A., Sommer, S., Linke, P., and Kröger, K. 2013. Microbes, macrofauna, and methane: a novel seep community fueled by aerobic methanotrophy. *Limnology and Oceanography*, 58, pp. 1640–1656.

Vasconcelos, C., and McKenzie, J.A. 1997. Microbial mediation of modern dolomite precipitation and diagenesis under anoxic conditions (Lagoa Vermelha, Rio de Janeiro, Brazil). *Journal of Sedimentary Research*, 67, pp. 378–390.

Weber, K.A., Achenbach, L.A., and Coates, J.D. 2006. Microorganisms pumping iron: anaerobic microbial iron oxidation and reduction. *Nature Reviews Microbiology*, 4, pp. 752–764.

Westall, F., Walsh, M.M., Toporski, J., and Strrle, A. 2003. Fossil biofilms and the search for life on Mars. In: *Fossil and recent biofilms* (eds. W.E. Krumbein, D.M. Paterson, and G.A. Zavarzin), pp. 447–465. Dordrecht, Springer.

Woods, A.D. 2013. Microbial ooids and cortoids from the Lower Triassic (Spathian) Virgin Limestone, Nevada, USA: Evidence for an Early Triassic microbial bloom in shallow depositional environments. *Global and Planetary Change*, 105, pp. 91–101.

Zeppilli, D., Canals, M., and Danovaro, R. 2012. Pockmarks enhance deep-sea benthic biodiversity: a case study in the western Mediterranean Sea. *Diversity and Distributions*, 18, pp. 832–846.

Zhang, X.L., Chang, C., Cui, L.H., and Qiao, Y.H. 2021. Ecosystem reconstruction during the Cambrian explosion. *Paleontological Research*, 25, pp. 305–314.

Zhang, Z., Wang, M., Qi, Y.N., and Liu, B.C. 2023. Microbial origin of micrites in ooids: evidence from the Cambrian in North China. *Carbonates and Evaporites*, 38, pp. 73.

7

Bioerosion and Biological Weathering

Ruliang He and Xingliang Zhang

7.1 Introduction

All geologists are quite familiar with destructive processes of rocks, which form sediments and soil, i.e. weathering and erosion in the rock cycle. Any solid Earth material at or near the surface suffers from such destructive fate when exposed to the sun, atmosphere, water, and activities of organisms. Weathering is defined as physical (or mechanical) breakdown and chemical alteration of Earth material at or near the surface. Weathering involves two processes, i.e. mechanical and chemical, that often work together to break down rocks. Both processes occur in place. No movement is involved in weathering. Erosion is defined as the removal and movement of Earth materials by natural agents. Erosion involves the wearing away of dissolved material, rock particles, and soil material (weathered products); sediments are moved from one place to another. However, erosion is different from transport. Erosion involves all processes that wear away rock components and soils, whereas transport refers only to the phenomenon of carrying eroded material to sedimentary environments. For instance, a stream erodes a canyon but transports eroded material as solids or dissolved substances (Wicander and Monroe 2002).

Weathering and erosion of rocks can be carried out by organisms that span diverse branches of life tree, including animals, plants, and microbial life. For animals, macroborers such as worms, mollusks, sponges, and rodents, consistently erode marine reefs, mix soil sediment particles, and bring material from depth to the surface where further weathering occurs; even rock materials ingested by worms are further reduced in size and new authigenic minerals grow in the animal body (Needham et al. 2004). As for plants, roots of large bushes and trees could wedge themselves into and further widen cracks of rocks through mechanical processes (Figure 7.1) and modify pH and redox conditions in the rhizosphere through chemical processes. Microbe-scale biological activities are perhaps the most important agents of biological rock destruction. In both terrestrial and marine environments, these activities can be produced by bacteria, lichens, algae, fungi, and plant roots (see description later in this chapter).

In this chapter, we highlight the significance of rock destruction caused by living organisms, which are considered as bioerosion and biological weathering in relation to other forms of abiotic weathering and erosion, and draw on well-studied examples for deeper insights. Biological weathering and bioerosion really are not types of their own. Living organisms involve weathering and erosion that can be either chemical or physical in nature. Therefore, bioerosion and biological weathering can be considered as special types of mechanical or chemical processes.

Introduction to Geobiology, First Edition. Xingliang Zhang, Zisheng Guo, Wei Liu, Ruliang He, Luoyang Li, and Weiduo Hao.

Companion website: www.wiley.com/go/Zhang_1e

Figure 7.1 Plant roots wedge into sheet joints of granite and thus increase the breakup of rocks in Mount Hua, China. *Source:* Photo by Xingliang Zhang.

7.2 Bioerosion

Bioerosion is defined as removal of consolidated minerals or lithic substrate by the direct action of organisms (Neumann 1966). Bioerosional processes affect a wide range of geological substrates (e.g. limestones and volcanic glasses) and biological substrates (e.g. corals, shells, and bones). Marine bioerosion includes biotic boring, drilling, rasping, and scraping and it results in a reduction of hard substrates in the oceans or at their margins. There are two primary means of bioerosion: mechanical abrasion and chemical dissolution. Mechanical bioeroders use a series of physical means carving into or excavating a hard substrate, while chemical bioeroders etch away hard substrates (primarily carbonates) with acids and/or chelating agents.

Bioerosion is a crucial process in sculpting the carbonate shoreline. It produces significant amounts of fine-grained sediments and plays a role in carbon and calcium geochemical cycling. The principal controls on bioerosion are productivity, sedimentation, substrate durability, and water depth. Bioerosion typically leaves trace fossils that have important applications in paleoecology (Table 7.1), such as determining relative sea level changes over time along rocky shores. Throughout geological history, borings show distinct ecological and evolutionary changes in endolithic communities. During the Paleozoic and early Mesozoic, bioeroders made mostly small, narrow holes with low density, and had a minimal rate of bioerosion. In contrast, from the later Mesozoic to the Cenozoic, they made larger cavities with high density, resulting in significantly increased rates of bioerosion.

There are three major types of marine bioeroding organisms: microborers, macroborers, and external raspers or scrapers (Table 7.1). Microborers are primarily filamentous endolithic cyanobacteria, algae, and fungi. Macroborers do not eat their host substrates, but excavate dwellings within them, where they are protected from predation. External grazers and raspers erode hard substrate

Table 7.1 Major groups of bioeroding organisms, with names of their trace fossil, description of boring or scraping morphology, age range, and depth range.

Organisms	Trace fossils	Description	Age range	Depth range
Microborers				
Filamentous green algae	Many names	Filaments, 10–50 μm diameter	Pre€-Recent	Photic zone
Cyanobacteria		Filaments, 2–10 μm diameter	Pre€-Recent	Photic zone
Fungi		Filaments, <2 μm diameter	Pre€-Recent	All depths
Macroborers				
Sponges	*Entobia*	Multiple small chambers, connecting filaments	O-Recent	All depths
	Uniglobites	Large rounded chambers, broad papillae	Tr-Recent	Primarily shallow
Bivalves	*Petroxestes*	Facultative borer	O-S	Shallow
Lithophaga, *Gastrochaena*, Pholadidae	*Gastrochaenolites*	Clavate borings in rock, coral, shell, may be lined	O-Recent	Primarily shallow
Teredo, Xylophaginae	*Teredolites*	Elongate carbonate-lined tubes in wood	K-Recent	Primarily shallow
Worms in general	*Trypanites*, *Lapispecus*	Straight cylindrical holes	€-Recent	Primarily shallow
Polydorids	*Caulostrepis*	Small U-shaped tubes	O-Recent	All depths
	Maeandropolydora	Large anastomosing tubes	O-Recent	All depths
Phoronids	*Talpina*	Small branched tubes	O-Recent	All depths
Ctenostome bryozoans	*Ropalonaria*, *Voigtella*, *Spathipora*, *Foraripora*, *Penetrantia*	Small elongate regular chambers, usually in branched networks	O-Recent	All depths
Cheilostome bryozoans	*Leptichnus*	Shallow surface etchings, small ovate units arranged in fan-shaped or branching clusters	K-Recent	Primarily shallow
Brachiopods	*Podichnus*	Pedicle attachment scars	O-Recent	Mostly deep
Acrothoracican barnacles	*Rogerella*, *Zapfella*, others	Large, pocket-shaped, with slit-shaped apertures	S-Recent	Primarily at surface
Upogebia	Similar to *Spongeliomorpha*	Broad U-shaped galleries with 2–3 narrow apertures	J-Recent	Primarily at surface
Beetles and other insects	*Stipitichnus*	Elongate, anastomosing tubes with larval galleries	C-Recent	Terrestrial
Grazers				
Limpets and chitons	*Radulichnus*	Small parallel scrape marks on rocks	€-Recent	Primarily at surface
Regular echinoids	*Gnathichnus*	Larger pentamerous scrape marks	J-Recent	Photic zone
Acanthurid and scarid fish	No name	Gouges, deep scrape marks	Miocene-Recent	Photic zone

Source: Adapted from Edinger (2001).

as they graze on epilithic or endolithic algae. The relative importance of microboring, macroboring, and grazing varies with depth, sedimentation rate, and productivity (Edinger 2001).

7.2.1 Macroborers

Macroborings are excavations made by macroborers more than a millimeter across, in which the occupants dwell for their entire lives. They are different from burrows in definition though the distinction is not always easy to discern. **Borings** are biogenic structures excavated in indurated rock or shell material, whereas **burrows** are biogenic structures emplaced in unconsolidated sediments. Macroboring taxa are quite diverse, which include bivalves, sponges, echinoids, and various worms, and their substrates vary considerably between limestone, mudstone, shells, bone wood, and silicate rocks (Bolotov et al. 2018). However, macroboring trace fossils appear to show an evolutionary conservation; morphologies of the most common macroborings have not changed appreciably since the Early Paleozoic (Ekdale 2006). They have been frequently used as facies indicators. In paleontology, the presence of rocks with boreholes and fossil macroboring assemblage members is one of the primary diagnostic features of shallow marine paleo-environments (Bolotov et al. 2018).

The most prominent macroborings in marine environments can be traced back hundreds of millions of years (Table 7.1). *Trypanites,* cylindrical, finger-like excavations made by sipunculans, polychaetes, and other worms, is the oldest known macroboring in the world, which appeared in localized archaeocyathid reefs in the Lower Cambrian of Labrador (James et al. 1977). *Gastrochaenolites*, flask-shaped borings commonly attributed to bivalves, is the next oldest macroboring trace fossil genus that has been known from the Lower Ordovician of southern Sweden, Russia, Estonia, and western Utah (Benner et al. 2004; Ekdale and Bromley 2001). By the middle and late Ordovician time, bioerosion of carbonate hardgrounds became more common, and macroborings were widespread. The other important macroboring ichnogenus *Entobia* (multi-chambered domichnia of sponges) has been known around since at least the Middle Devonian (Tapanila 2006).

Boring into rock is not an easy task for invertebrates and thus is a remarkable feat for an animal. The ability to excavate lithified material is of evolutionary significance because penetration and removal of indurated material require specialized anatomy, derived behavioral skills, and a considerable expenditure of energy. It is also ecologically important because the novel habitat possibilities have been opened up by an endolithic lifestyle for both the producer and secondary inhabitants (Figure 7.2). It is much easier to burrow in soft sediments than to bore into hard rocks, so why and how did animals begin to exploit the macroboring niche are open questions. A variety of chemical and mechanical means of boring have evolved in numerous animal and plant taxa over the past one-half billion years.

7.2.1.1 Macroboring Groups

Sponge boring is accomplished by amoebocytes that etch and chip minute calcareous fragments from the limestone substrate and results in complex branched multi-apertured networks of large chambers, multi-apertured irregular networks without chambers, and stellate structures with a single aperture (Flügel 2004).

Polychaete worms (the Annelida) that bore into rock are enormously abundant in certain environments. Various species in families Cirratulidae, Eunicidae, Sabellidae, and Spionidae typically form cylindrical borings 0.5–2 mm in diameter that penetrate up to 10 cm deep into the interiors of hard substrates. The boring mechanism varies in taxa. Some eunicids employ their mandibles to excavate. Spionids bore mainly by chemical strategy, with some removal probably due to mechanical abrasion by setae.

Figure 7.2 A novel endolithic habitat opened up by limpets is also occupied by barnacles in Dorset Coast, England. *Source:* Photo by Xingliang Zhang.

The four major groups of boring bivalves vary in their distribution and modes of excavation. They typically make flask-shaped boring with a single aperture (Figure 7.3). In tropical waters, lithophagid and gastrochaenid bivalves corrode chemically into carbonate rocks, corals, mollusk shells, and similar materials. Pholad bivalves, with strongly ridged shells, bore mechanically into a variety of rock types, primarily in temperate regions. Wood-boring bivalves are common in floating and

Figure 7.3 Bivalve borings in a piece of limestone. Note that (a) flask-shaped borings are common on the sample and (b) bivalves dwell within some borings. The specimen was collected by Xingliang Zhang from the Jurassic Coast in England. *Source:* Photo by Ruliang He.

sunken wood in a wide variety of marine, brackish, and some freshwater environments. For example, shipworms—a type of marine mollusk with extraordinarily long calcite-lined siphonal canals—have historically played a role in weakening wooden ships. Some researchers suggest that shipworms may have contributed to the deterioration of Spanish ships while they were stationed in harbors before the battle. These creatures bore into wood, causing structural damage over time, which could have made the Spanish Armada's ships more vulnerable (Edinger 2001).

Several genera of sipunculans (peanut worms) penetrate coral skeletons, while there is no general agreement on the overall importance of these bioeroders. Sipunculan borings are cylindrical and pencil sized or slightly smaller, ranging from straight to sinuous and from near surface to several centimeters deep in skeleton, depending on the species. While feeding, sipunculans extend their introverts outside of their cavities and appear to ingest debris, sand, and algae. The exact manner of boring is not known, but may involve both chemical dissolution and mechanical abrasion.

Crustacean barnacles (Figure 7.4) and shrimp (Arthropoda) are endolithic borers, producing cylindrical chambers. Two groups of barnacles contain species that reside in skeletons of dead corals, namely thoracican *Lithotrya* (erodes 2–10 cm long oval-shaped cavities on the undersides of reef rock and beach rock in shallow, agitated waters) and acrothoracicans (e.g. *Thrypetesa*). Acrothoracican barnacles are the major groups of boring crustaceans, occupying small, millimeter-sized vacancies. Acrothoracican barnacle borings were common in floating substrates like belemnite rostra, and modern acrothoracicans are most abundant at sea level. The basal plate of barnacles is attached at the innermost end of the cavity and the body hangs downward toward the opening with cirri exposed to food-bearing currents. The cavities are formed apparently by mechanical abrasion effected by calcified plates that cover the barnacle's body. Adjacent tubes of boring *Lithotrya* are commonly interconnected, and heavily infested limestones are thoroughly honeycombed and subject to frequent breakage. Pistol shrimp *Alpheus simus* has been known to chemically bore into coral rock on Caribbean reefs and causes considerable erosion on some Costa Rican reefs. The hammer dactylus at the first pereiopod enables the animal via a pressurized hemolymphic chamber to

Figure 7.4 Barnacles from the east coast of North America, showing cirri in the inset. *Source:* Photos by Xingliang Zhang.

mechanically disintegrate the substrate. Male/female pairs excavate 10–15 cm diameter chambers that penetrate as deep as 15 cm into dead coral rock.

7.2.1.2 Insight into Boring Sponges

Among the large number of limestone-eroding organisms, excavating sponges are of special interest because of their efficient means of substrate penetration by cellular etching and because they release characteristically shaped calcium carbonate chips which can be detected in the mud-and silt-size fraction of many sediments. Identifiable trace fossils and sediments are of great ecological and paleoecological significance (Rützler 1975). Their excavations are known in fossil mollusk shells at least as far back as the Devonian Period.

Sponges with capacity for excavating limestone, shells, and other calcium carbonate substrates are called boring sponges which are known mainly from clionaids and taxa within the class Demospongiae. All eight genera of the family Clionaidae (the order Hadromerida) have the habit of boring various types of calcium carbonate substrate. Species of *Aka* and *Siphonodictyon*, which belong to the family Clionaidae, are known for their significant role in bioerosion, particularly in coral reef ecosystems. Other sponge groups that are clearly not related to the family Clionaidae, for instance, the Poecilosclerida, *Paracornulum*, and *Zyzzya*, are also able to efficiently excavate limestones (Calcinai et al. 2005; Rützler 2002).

Boring sponges appear to be eurythermal cosmopolitan animals, which are distributed among a wide variety of ecological habitats, ranging from mean low water to 200 m in depth of Atlantic, Pacific, and Indian coasts. They penetrate calcium carbonate substrate by means of filopodial extensions and etching secretions (Figure 7.5). Boring into the calcareous substrate results in a three-dimensional network of interconnected chambers or galleries below the substrate surface. Erosion chambers are irregular ovoid, sometimes sub-spherical or polygonal in shape, and small, varying in taxa, usually a few millimeters to centimeters in diameter. The chambers are filled with sponge tissues and spicules without orientation.

Limestone excavation is accomplished by chemical etching compounds that are secreted by pseudopodial processes of special archeocytes called etching cells. Each etching cell chemically cuts away a minute chip of the calcareous substrate. The free chip is expelled with the current water into the aquiferous system and eventually discharged at the surface through an osculum (Figure 7.5). In this process, the majority of the calcium carbonate is removed in chips and only 2%–3% is removed in dissolved form. The remaining cavities or series of chambers (centimeters in diameter) have characteristically pitted walls and are occupied by the sponges. Chambers may be connected with each other by papillar canals. The released chips are in micrometer scale, up to about 60 μm, and their shape may vary in both taxa and substrate. Boring into shell or rock may enhance the survival of juveniles and adult sponges because it affords the sponge physical protection from grazing predation.

Studies show that high initial penetration rates result from the mechanical stimulation and the lack of competition, but that the boring rate slows down and remains constant after 6 months, and the optimal long-term erosion of $CaCO_3$ does not exceed 700 $mg/m^2/yr$ (*Cliona lampa* and *C. aprica*). Substrate limitation and competition will further reduce this rate. By monitoring the production of $CaCO_3$ chips produced by *Cliona lampa*, it was possible to link activity patterns to certain environmental factors. Mechanical stimuli, high light intensity, strong currents, and, possibly, low temperature seem to accelerate the boring process. Sponge-generated chips can make up over 40% of coral mud and silt when deposited in the current shadow of the reef framework. Using transect counts and sponge area-biomass conversion factors, the mean abundance of boring sponges on the Bermuda platform was calculated. On suitable hard bottom substrates, it averages 16 g dry weight

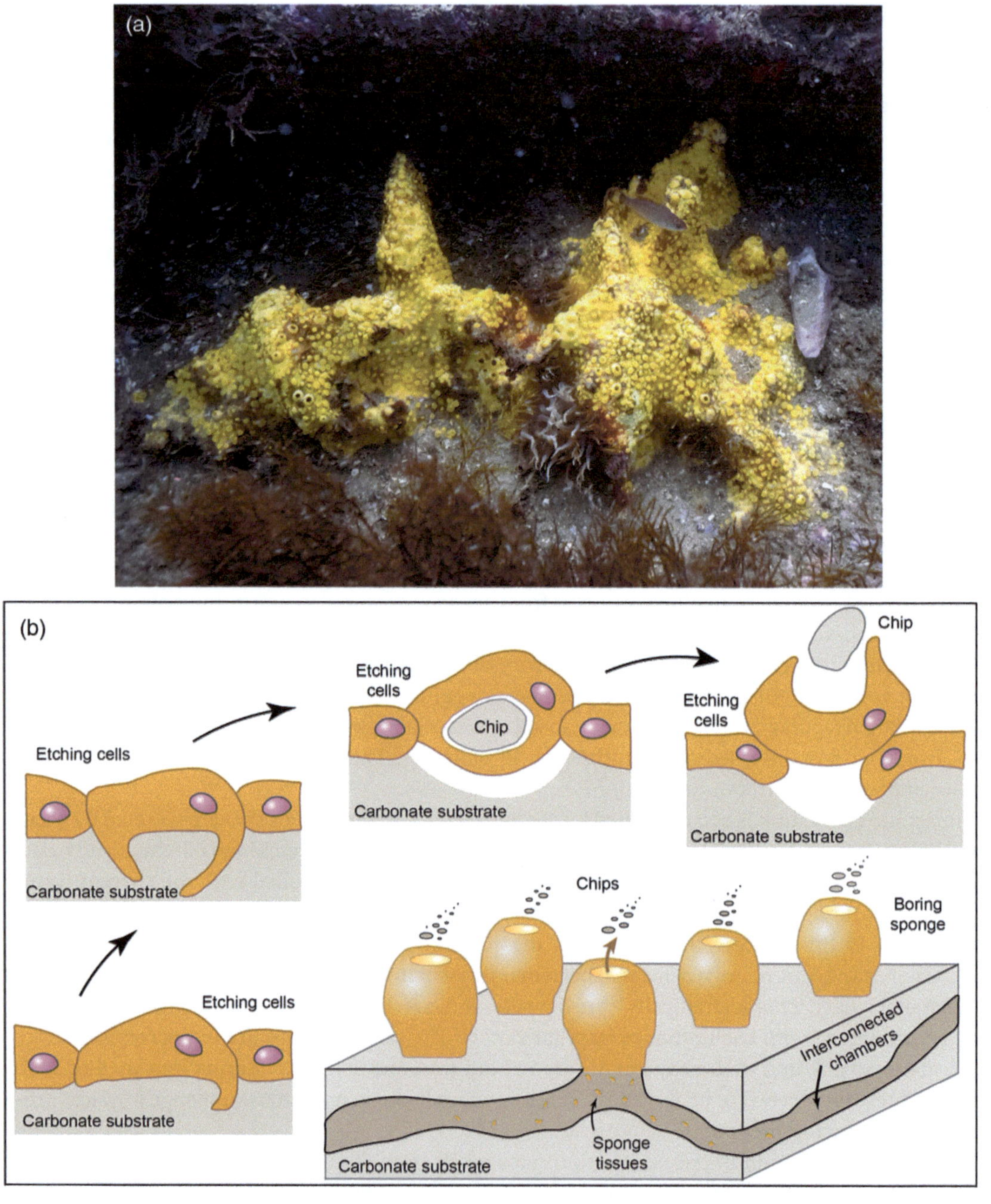

Figure 7.5 Bioerosion by boring sponges. a) *Cliona celata*, a species of demosponge, in the modern ocean. b) The sequence of sponge boring: cells penetrate hard substrates by chemical etching; etched chip is removed from the pitted substrate; chip is released into the aqueous system; chip is discarded in exhalant water. Source for panel a: Diego Delso, delso.photo (http://delso.photo), License CC BY-SA (https://creativecommons.org/licenses/by-sa/4.0/legalcode).

per m^2. From this value, the boring potential of sponges can be estimated as 256 g $CaCO_3$ per m^2 substrate per year. Since 97%–98% of the eroded limestone remains in particulate form, the contribution of fine sediments can amount to 250 g/m^2/yr (Rützler 1975). In coral reef environments, up to 30%–40% of fine sediment particles may be derived from the activities of boring sponges in low-energy environments; lesser proportions (5%–22%) characterize sediments in high-energy environments (Rützler and Rieger 1973).

Calcinai et al. (2003) compared the micro-erosion patterns between the boring genera *Cliona* and *Aka*. The scanning electron microscope analysis of the excavations of the species in the two genera sharing the same substrate clearly indicates that the erosion scars produced on the walls of the boring chambers are different. The pits of *Cliona* are regular, almost circular, and smooth; those of *Aka* are more irregular and show a multi-layered concentric pattern. These differences constitute a successful means of distinguishing the bioerosional agents both in recent and in fossil substrates.

The erosion efficiency of sponges is influenced by substrate characteristics, which, in turn, impact their growth patterns. A field study on *Cliona orientalis* suggests it is not selective about the substrate it inhabits, successfully invading over 90% of the tested blocks. However, its penetration depth is generally limited to slightly over 1 cm, with lateral growth far exceeding downward expansion. This preference benefits the sponge's symbiotic zooxanthellae, which thrive in surface layers exposed to light. Structural irregularities in substrates, such as coral dissepiments, can momentarily alter the sponge's growth trajectory, producing distinctive tissue patterns. Additionally, sponges tend to grow more extensively in denser materials with lower pore volume, likely due to increased protection from predators. Conversely, in porous substrates frequently inhabited by grazers and corallivores, the sponge's tissue is found in deeper layers rather than near the surface (Schönberg 2002).

7.2.1.3 Insight into Boring Bivalves

An important number of bivalved mollusks from five lamellibranch families have the ability to penetrate and live beneath the surface of firm, consolidated substrate, such as peat, mudstone, sandstone, limestone, coral, shell, and wood (Banister and Campbell 1998). These belong to two distinct taxonomic and evolutionary groups but are collectively known as boring bivalves. One group, composed of some members of the Myoida and the Veneroida, arose from soft-bottom infaunal burrowers that gradually evolved the ability to bore into firm substrate. They usually attach to the boring wall by the foot, which has developed a sucker-like ventral surface for this purpose. The Pholadidae, including piddocks, martesias, and angel wings, is the best-known example of this group. The second group, represented by boring mussels (Mytilidae), probably evolved from byssally attached epifaunal bivalves. They attach to the side of their borings by byssal threads (Ruppert et al. 2004). A more recent study revealed multiple origins and high disparity of rock-boring bivalves. Rock-boring bivalves have the greatest range of forms among their relatives, ranging from spheres to cigar and spoon shapes (Figure 7.6). Endolithy in bivalves has originated at least eight times and is evolutionarily accessible from many different morphological starting points (Collins et al. 2023).

Each boring bivalve starts excavating soon after the larva settles on substrate and slowly enlarges and deepens the boring as it grows. Once the boring is constructed, the animal is forever enclosed in it and communicates with the outside via the siphons extended from a small opening at the surface (Figure 7.3). A boring bivalve removed from its chamber cannot excavate a new boring and doom to quick consumption by a predator.

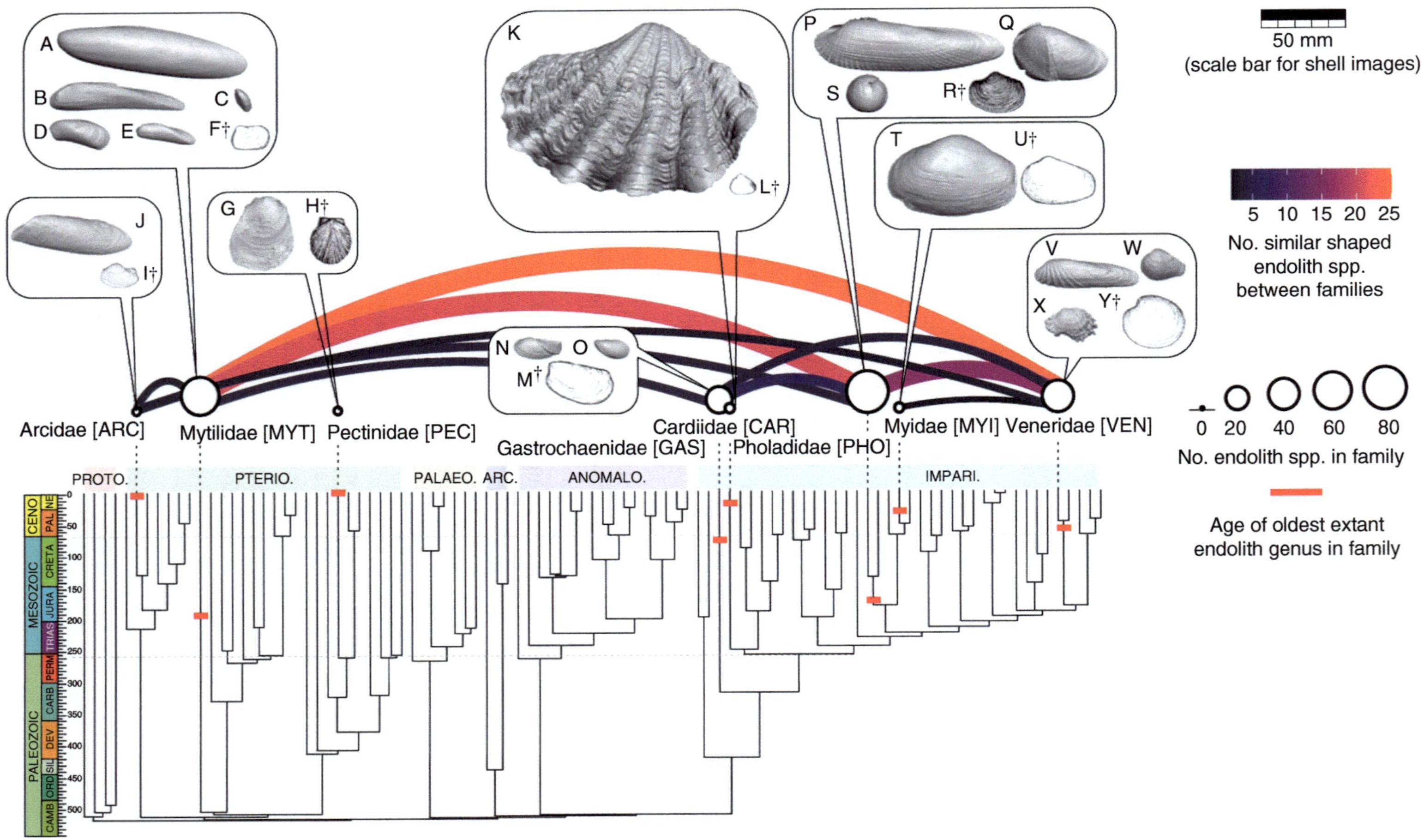

Figure 7.6 Phylogenetic distribution of species richness, shell shape convergence, and exemplar shell morphologies of boring bivalves. *Source:* Collins et al. (2023) / with permission of The Royal Society.

Most rock borers make their borings mechanically by rotation of the shell, often aided by spikes on the shell surface, which erode away materials at the rock. They employ the valves as drills to abrade the substrate and produce a boring. The anterior surfaces of the valves are frequently roughened or toothed to increase their abrasive ability (Figure 7.7). The foot may attach the animal to the end of the boring, as in paddocks. Closure of siphons helps to keep up fluid pressure. Rock raspings are passed out from the mantle as pseudofeces. While some borers produce a boring of even width, flask shells are surrounded by a jacket of cemented shell fragments and live in a rock boring that is narrower at the entrance (Figure 7.3). The contracted siphons dilate outside the shell to form an anchorage during boring. The date-pit mussels of warm seas and other bivalves have an elongated smooth shell. They excavate chemically and make round boring in limestone. A mucoprotein-chelating agent (acid), secreted by glands in the middle fold of the mantle margin, softens the calcareous substrate, which is then scraped away by the valves. The thick, shiny brown periostracum protects the shell from the mollusk's own acid.

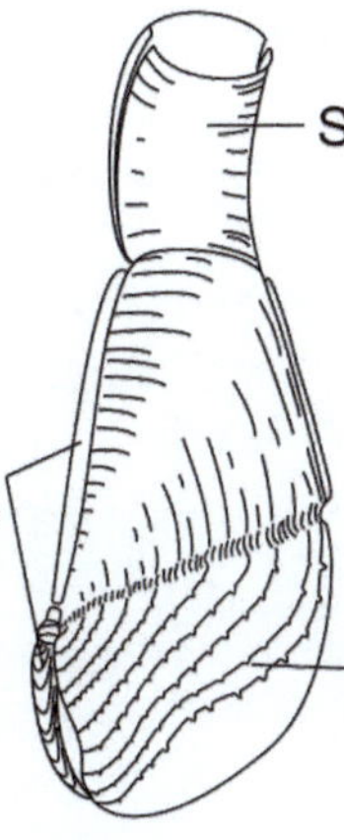

Figure 7.7 A diagram showing structures of a boring bivalve.

Drilling rates vary in species. The piddocks *Penitella penita* and *Chaceia ovoidea* excavate soft shale at rates of 2.6 and 11.4 mm per month, respectively. Some borers rotate on their longitudinal axis within the boring and make round borings. Others remain attached to one wall of the boring and maintain a constant orientation with respect to their longitudinal axis, so the boring tube reflects the cross-sectional shape of the shell. Some of these species, e.g. *Gastrochaena*, create separate openings for inhalant and exhalant siphons at the surface of the rock and thus produce calcareous material between the two siphons. Much of the particulate waste (silt and sand-grain-sized sediments) produced by drilling is transferred to the mantle cavity and then ejected with the pseudofeces through the inhalant siphon (Ruppert et al. 2004).

7.2.2 External Grazers and Scrapers (Raspers)

External or epilithic bioeroders break down the surface layers of carbonates and other sedimentary rocks in a variety of ways. The majority of epilithic bioeroders are herbivorous grazers that scrape and erode substrates while feeding on algae and associated microorganisms residing on and within limestone substrate and their activities are necessarily limited to the photic zone (Figure 7.8).

Grazers are organisms feeding mainly on algal and microbial mats. Most grazers search for algae in living or dead substrates. The most important grazers are limpets, chitons, some regular echinoids, acanthuroids, and scarids. When grazing, they crop very close to a substrate, ingesting substantial portions of living organic tissues, associated with small invertebrates and underlying skeleton, as well as algae. Many grazers are also capable of removing and ingesting calcareous material such as coralline algal skeleton and underlying substrate. A study shows that carbonate removal by grazers is responsible for a high percentage of the total bioerosion (Chazottes et al. 1995). Intensive microbial mat grazing was proposed to start in the late Ediacaran, e.g. *Kimberella* from the White Sea biota (Seilacher 1999). Grazing by conodontophorids and fishes may have been more intensive in the Middle and Late Cambrian (Flügel 2004).

Figure 7.8 Mollusks erode rocks as external grazers and scrapers on the Dorset Coast, England. *Source:* Photo by Xingliang Zhang.

Most bioeroding mollusks are external grazers that abrade and thus weaken the substrate while feeding. Molluscan bioeroders are generally most abundant in the intertidal zone, with some species extending their ranges into supratidal and subtidal habitats. Species abundance also changes horizontally with chitons, which are often most plentiful in areas protected from strong wave assault, and with limpets, certain snails, and echinoids, which are more common in wave-swept habitats. Under quiet to rough water conditions, grazing mollusks are largely responsible for producing the notches and nips on tropical limestone shores (Figure 7.9). Under extremely rough conditions, however, many bioeroders either disappear or their activities are greatly reduced. Limpets and snails (class Gastropoda) often occur with chitons on intertidal carbonate substrate. *Acmaea*, *Cellana*, and *Patelloida* are common limpet genera, and *Cittarium*, *Littorina*, *Nerita*, and

Figure 7.9 Limpets colonize a rock shore and leave round pits known as "home scar" on the Dorset Coast, England. *Source:* Photo by Xingliang Zhang.

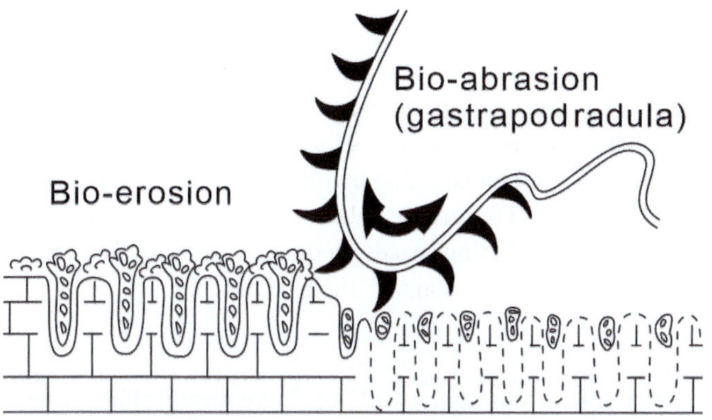

Figure 7.10 Sketch showing that some mollusks graze on epi- and endolithic microflora and rasp away substrate surfaces. Endoliths therefore receive new stimuli to bore to the phototactically controlled light compensation depth (LCD). *Source:* Torunski (1979).

Nodilittorina are some common snail genera. Chitons, limpets, and snails utilize a radula to scrape rock surfaces (Figure 7.10). The radula of patellacean limpets is an especially effective excavating organ with opal ($SiO_2{\cdot}nH_2O$) or goethite (FeO-OH)- sheathed radular teeth. The radula of snails contains proteinaceous teeth, yet still these grazers are capable of erosion because of the frequently weakened condition of the rock substrate upon which they feed.

Sea urchins are the only echinoderms capable of significant bioerosion. Several species in the following genera abrade large amounts of reef rock and defecate carbonate sand while feeding and excavating borings: *Diadema, Echinometra, Echinostrephus*, and *Eucidaris*. Sea urchins possess a highly evolved jaw apparatus, a flexible and protrudable masticatory organ consisting of five radially arranged, calcified teeth. The teeth are mineralized and must be harder than the corroded surfaces they scrape. Sea urchin spines also assist in bioerosion when they are employed in the enlargement of borings.

Grazing fishes are the dominant herbivores in most tropical environments, which are important bioeroders on tropical shallow reefs, up to 7 $kg/m^2/yr$ (Flügel 2004). Numerous fish species erode substrate while grazing on algae and also fragment coral colonies while feeding on live coral tissues or then extracting invertebrates from coral colonies. Surgeonfishes (Acanthuridae) and parrotfishes (Scaridae) are the principal grazing groups. Triggerfishes (Balistidae), filefishes (Monacanthidae), and puffers (Tetraodontidae, Canthigasteridae) are largely carnivorous in feeding habits and are responsible for fragmenting live coral colonies. The jaw muscle and tooth armature are well developed in these fish families. Parrotfishes also have a pharyngeal mill, a gizzard-like organ that further reduces the size of the ingested sediment. Fish teeth are composed of dahllite [$Ca_5(PO_4CO_3)_3(OH)$] or francolite (the fluorinated form, $Ca_5(PO_4CO_3)_3F$), which are apatite minerals that are harder than $CaCO_3$. Parrotfish grazers can produce large amounts of sediment on reefs, especially when their population densities are high. Scarids modify reef carbonate in three ways: (1) by direct erosion, (2) through decrease of particle size due to both erosion and sediment reworking, and (3) by net transport of sediment away from its site of removal.

Scrapers use beaks or claws to scratch substrate in order to remove microbial nutrients or algal/microbial cover. These can have either a denuding or non-denuding effect. The most important denuders are some limpets (mollusks), some echinoids, acanthuroids, rabbitfish (siganids), and damselfish (pomacentrids).

7.2.3 Microborers

Some microbes (bacteria, algae, and fungi) or as microbial biofilms could dissolve the substrate and penetrate its interior (Golubic and Schneider 2003). Depending on their location on substrate, microborers can be classified as: **microbial endoliths** are present within substrate, which etch the hard substrate either by chemical or mechanical process, or a process that involves both mechanisms, and **microbial chasmoliths** grow in pre-existing cracks and fissures in substrates, while cryptoendoliths colonize structural cavities within porous substrates, including spaces produced and vacated by endoliths. Some endoliths can show an epilithic or chasmolithic stage in their life cycle.

7.2.3.1 Microbial Endolithic Groups

Microborers of bacteria, algae, and fungi penetrate substrate by etching. Some foraminiferans and bryozoans penetrate the substrate, leaving borings less than 100 μm in diameter, and thus are attributed to microborers (Flügel 2004).

Both non-photosynthetic and photosynthetic bacteria have groups that live in endolithic habitats, known since the Middle Proterozoic (Zhang and Golubic 1987). Modern endolithic cyanobacteria are represented by more than 20 genera, capable of chemically eroding reef rock from the splash zone to depths of at least 75 m. Species of *Hyella*, *Plectonema*, *Mastigocoleus*, and *Entophysalis*, for example, have been found on limestone surfaces, inside cavities, and penetrating rocks. Cyanobacteria are also associated with the erosion of lagoon-floor sediments on the Great Barrier Reef, accounting for 18%–30% of sediment dissolution.

Some green algae (chlorophyta) have been known to live within carbonate substrates or have endolithic stages (Golubic et al. 1975). Freshly fractured corals often have layers of green banding a few centimeters beneath the live coral surface, which is due to the presence of chlorophyll pigments that intercept light passing through the coral tissues and skeleton (*Ostreobium* band). Although red algae (rhodophyta) are predominantly in marine habitats, endolithic life habits are known from only a few genera.

Fungi are capable of penetrating deeply into coral skeletons by chemical dissolution. The hyphae produce narrow borings and penetrate the deepest recesses of any calcareous skeleton. Fungi have also been implicated in the etching of calcareous surfaces, the weakening and dissolution of calcareous sediments, as well as the calcareous tube linings of various endoliths. Boring fungi occur in shallow and deep marine environments down to 5000 m (Flügel 2004).

Boring foraminiferans occur in groups characterized by agglutinated and calcareous perforate tests. They are known from both cold-water and warm-water environments, and reported from Jurassic to Quaternary.

Microborings of bryozoans closely resemble the producer's morphology and reflect the shape of the body fossils. The trace corresponds to a network that connects elongated chambers and, therefore, can be easily recognized in scanning electron microscope (SEM) images. Most boring bryozoans known from Jurassic to Quaternary limestones are attributed to the Ctenostomata. The Paleozoic record of boring bryozoans is scanty.

7.2.3.2 Diverse Endolithic Habitats

Microbial endoliths not only thrive in clement environments in tropical reefs, but also adapt to extreme ecological conditions. Such extreme conditions include cold regions (such as Antarctic, Alps, high Arctic, and Tibet Plateau), arid habitats (such as Atacama Desert and various salt marshes). Microbial endoliths have to overcome challenges from oligotrophy, desiccation, lethal UV radiation, and limited light penetration (Table 7.2; Sajjad et al. 2022). In such extreme environments, microbes obtain nutrients from the host rock (e.g. via dissolution by organic acids) or

Table 7.2 Environmental stresses and adaptive strategies of endolithic bacteria.

Environmental stresses	Adaptive strategies
Nutrient deficiency	1. Use PolyP to store P.
	2. Nonribosomal peptides and polyketides pathways.
	3. Fix (C) by oxygenic photosynthesis and NH_3 oxidation.
Insufficient colonization space	1. Use rock space, cracks, and fissures.
	2. Surface linked to H_2O and light.
Desiccation	1. Switching metabolic activities during desiccation.
	2. EPS layer for water retention and freeze-thaw tolerance.
	3. H_2O from dews and fog absorber by minerals.
Limited light penetration	1. Rock minerals for light filtration.
	2. Chlorophyll-d for little penetration of light.
	3. Pigments and biliproteins.
Lethal UV-radiation	1. Carotenoids and scytonemin as a shield.
	2. Rock minerals (gypsum, silica, hematite) as a protective layer.

Source: Adapted from Sajjad et al. (2022) / with permission of Elsevier.

from dust deposition and atmospheric precipitation. To avoid UV radiation, endoliths can produce self-protective pigments, such as carotenoids and scytonemin, preventing photo-oxidative damage, or use minerals, such as gypsum and hematite, as protective shields. While endolithic habitats could retain a small amount of water in pores and fractures, microbes could also enhance water retention using EPS and use dew, fog, and even mineral deliquescence as additional water sources (Sajjad et al. 2022). During severe desiccation, microorganisms could even deactivate metabolisms. For photosynthetic microbes, to use light more efficiently in such dark habitats, they could increase concentrations of pigments and biliproteins or take advantage of specific minerals, such as quartz that is transparent to visible wavelengths but reduces UV radiation (Sajjad et al. 2022).

7.2.3.3 Geological Evolution of Microborings

Microbial endoliths have deep geological records. The oldest fossil endolith *(in situ* preserved), *Eohyella campbellii,* was described from the silicified stromatolites of the ~1625 Ma Dahongyu Formation of North China (Zhang and Golubic 1987) (Figure 7.11). The distribution pattern of cell aggregates along the upper edge of silicified horizons and the downward orientation of pseudo-filaments reveals the endolithic habit of the biofilm, which is comparable with the modern endolithic cyanobacterium *Hyella*. Several microbial endoliths were also found in the 800 Ma old ooid assemblages of East Greenland, and compared with modern ones in the ooids of the Bahama Bank and Arabian Gulf (Knoll et al. 1986). Organically preserved body fossils of endolithic algae (Conchocelis-stages of bangiacean rhodophytes) were reported from the Upper Silurian rock of Poland (Kazmierczak and Golubic 1976). Microborings are common throughout the Phanerozoic (Glaub et al. 1999). They can be preserved as natural casts when filled with insoluble materials. However, a significant proportion of microborings are preserved intact, which can be cast in polymerizing resin, studied by SEM, and compared with modern endolithic traces.

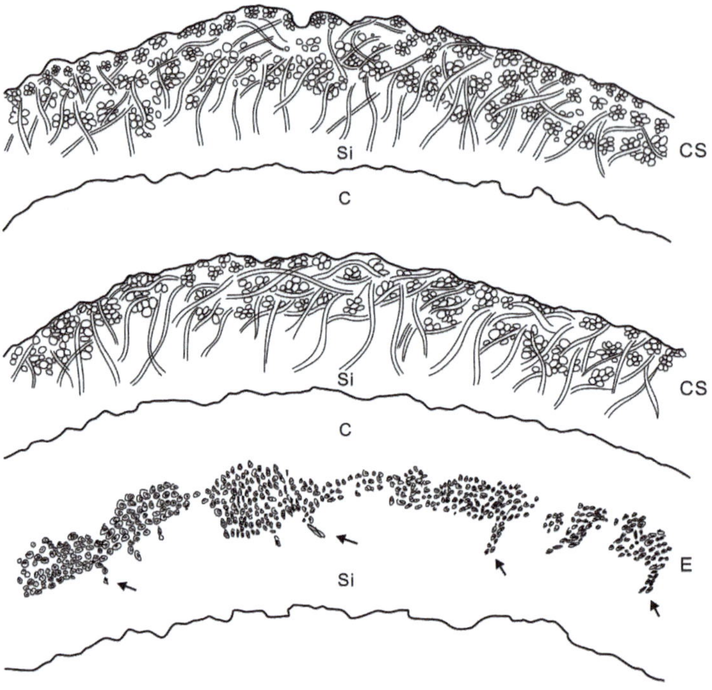

Figure 7.11 A schematic sketch showing distribution of microbial populations in partially silicified stromatolites from the Dahongyu Formation (ca. 1700 Ma), North China. CS, the stromatolite-building, *Coniunctiophycus-Siphonophycus* microbial mat community; E, endolithic *Eophyella campbellii* population, the downward orientation of pseudofilaments (arrows) indicating the endolithic habit of this taxon; C, carbonate layers; Si, silica layers. *Source:* Zhang and Golubic (1987) / Oriprobe Information Services, Inc.

7.2.3.4 Geological and Astrobiological Implications

Boring behavior of microbial endoliths is often specific, and the shape of borings may conform closely to the shape of organisms. Thus, activities of microbial endoliths leave recognizable and fossilizable traces that can be applied to explore past microborers and paleo-environments they once occupied (Golubic and Schneider 2003), on Earth and even other planets.

From geological perspective, studying and analyzing microborings of ancient and modern endoliths are significant in understanding (1) destructive processes resulting in degradation of skeletal elements and the production and accumulation of fine-grained carbonate sediment, (2) processes responsible for formation of micrite envelops, (3) distributional patterns that can be used to differentiate modern and ancient environments, (4) the role of microborers as (paleo-) bathymetric and ecological indicators, and (5) the role of increased nutrient input to reefs that might result in reef demise (Flügel 2004).

From an astrobiological perspective, endolithic habitats have been found in diverse extreme environments on Earth (such as polar and alpine deserts), which broadens our knowledge of the boundary of biosphere. Therefore, understanding the life modes of endoliths would help humans to better constrain habitable environments beyond Earth. In addition, endolithic habitats are great analogs for environments on extraterrestrial bodies; for example, although the surface of Mars is currently considered as inhabitable, it is reasonable to assume Martian endolithic habitats would be potential refugia for the past Martian life if it did exist.

7.3 Biological Weathering

Biological weathering is the breakdown of rocks caused by the action of living organisms, including microbes, plants, animals, and lichens. They involve processes that can be either chemical or physical in nature. Therefore, biological weathering can be considered as a special type of mechanical or chemical weathering. All types of igneous, sedimentary, and metamorphic rocks, either siliceous or calcareous in composition, are susceptible to biological weathering. Generally, the biological weathering of rocks is driven by the nutrient requirements of the biological communities. For example, bacterial communities dissolve the primary rock-forming minerals to obtain essential nutrients and also act as nucleation sites for the precipitation of secondary minerals.

7.3.1 Biological Mechanisms That Enhance Rock Weathering

7.3.1.1 Plant Roots

The effect of plant roots on mineral weathering may result from physical and/or biochemical processes: (1) plant roots induce the formation of macropores, which play a major role in the preferential flow phenomenon and thus the weathering process; (2) plant roots fragment rocks and thus increase the area of reactive surface; (3) plants modify the nutrient concentrations in the rhizosphere and generate zones of depletion or accumulation of nutrients through nutrient uptake; (4) the exudation of organic ligands of high and low molecular weight from plant roots, such as sugars and organic acids, is able to enhance mineral dissolution rate though various processes; (5) plant roots modify the redox condition in the rhizosphere, via oxidation or reduction of metals; (6) plants influence the availability and movement of water in the soil through evapotranspiration intensity; (7) plants could extend their mycorrhiza and promote weathering via symbiosis with fungi; and (8) plant roots also modify the pH in the rhizosphere (Dontsova et al. 2020).

In terms of pH modification, the difference between the pH in the rhizosphere soil and bulk soil can be up to two pH units. The acidification of the rhizospheric soil results from the production of CO_2 during respiration, from the release of organic compounds in the root exudates, and from the balance of ion charges within the roots which depends on the excretion of one H^+ ion for every cation absorbed. These modifications of physicochemical conditions in the rhizosphere make the cations from the minerals more accessible to the plant. In addition, this phenomenon could also be regulated by nutrient deficiencies because plants that grow in potassium- or phosphorus-deficient conditions are able to modify the composition and quantity of root exudates, thus improving the mobilization of these elements (see Calvaruso et al. 2006 and references therein).

7.3.1.2 Animals

There is a modest amount of information about microbially enhanced weathering processes (see Section 7.3.1.3). However, the effect of higher organisms on weathering rates has seldom been addressed. This is somewhat surprising since almost all higher organisms have gut bacterial populations and practically all sedimentary environments contain animals. It seems possible that ingestion and digestion of minerals by higher organisms could produce similar, if not more pronounced, effects on weathering rates as microorganisms.

Experiments show that the combined process of ingestion, digestion, and excretion of fine-grained sediment by annelid worms accelerate weathering and mineral degradation as well as clay mineral authigenesis (formation of minerals in place within sediments or rocks), which was postulated to be as a result of a combination of the low pH and bacteria-rich microenvironment in the guts of animals (Needham et al. 2004). In this study, chlorite was preferentially destroyed during digestion. Both chlorite and muscovite underwent X-ray diffraction (XRD) peak broadening with a skew developing toward higher lattice spacing, characteristic of smectite formation. A newly formed Fe-Mg–rich clay mineral (possibly berthierine) and as-yet undefined clay minerals with very high d-spacing were detected in fecal casts. The gut conditions of certain bivalves have effects on the assimilation of metals from ingested sediment particles (Griscom et al. 2002). Earthworms ingest and digest soil material and thus break down the mineral grains and change the physical properties of soil (Suzuki et al. 2003), manifested as changes in soil aeration, stability, porosity, grain shape, grain size, and water-storage capacity. The copepod *Tigriopus californicus* (the Arthropoda) has been shown to alter ingested tremolite and clinochlore to chamosite and montmorillonite to vermiculite (Syvitski and Lewis 1980).

7.3.1.3 Microbial Communities

Bacteria and other microorganisms are now known to influence not only weathering but also clay mineral precipitation. Generally, colonization of plant roots and microbial communities (e.g. lichens) on rock surface or in soils greatly increases water residence time, which makes minerals more prone to chemical weathering. Barker et al. (1997) summarized several main biological mechanisms that are generally assumed to influence and increase rates of mineral weathering.

Physical disaggregating: Fungal hyphae physically expose mineral surfaces making them more susceptible to chemical weathering processes. Vermiculite has been shown to form from mica via the action of fungi (Weed et al. 1969). Observations by Hall (1990) indicated that expansion and contraction of the mucilage of chasmolithic algae, caused by wetting and drying episodes, resulted in the surface flaking of granitic rocks on nunataks of the Juneau Icefield, Alaska.

Inorganic and organic acids: The hyphal tips of fungi produce acids, which enhance mineral weathering, creating a network of pores in susceptible minerals (Jongmans et al. 1997). Production of both organic and inorganic acids as microbial byproducts can affect the weathering process. At extremes of pH, mineral dissolution rates increase by a power law that is dependent on the hydrogen ion activity. Organic acids in soil, produced as metabolic byproducts from microbes and plants, can thus accelerate silicate mineral dissolution. Oxalic acid is one of many organic acids produced by microorganisms, which, through its ligand-binding capability rather than pH, can dissolve four times the quantity of Al relative to CO_2-rich water (Manley and Evans 1986).

Organic ligands: They affect weathering by forming complexes with mineral ions at the surface thus enhancing dissolution by weakening bonds (Barker et al. 1997). Microbially produced organic ligands include metabolic byproducts, extracellular enzymes, chelates, and both simple and complex organic acids. These substances can enhance silicate dissolution rates by decreasing pH, forming framework-destabilizing surface complexes, or by complexing metals in solution.

Extracellular polymer substances (EPS): They affect weathering in various ways (Barker et al. 1997). Not only do they bind soil and rock fragments, and prevent them from drying out, but they also act as ligands and nucleation sites for mineral authigenesis (Barker et al. 1997). The anti-desiccation qualities of extracellular polymers are particularly important because the residence time of water in soil and fragments is a primary factor that controls mineral weathering rates.

Nutrient absorption: Elements released by weathering processes are utilized as nutrients by microorganisms, enhancing biological activity and stimulating a positive feedback loop. For example, aqueous potassium is often utilized by organisms, of which a common source is the weathering of mica. The potassium released from mica weathering again enhances the growth of weathering organisms. In addition, it is demonstrated that microorganisms benefit from the weathering of silicates that contain trace phosphate mineral inclusions. While organisms assimilate P for biomass production, biomass alone does not accelerate weathering rates. Instead, the key driver is Fe(III) availability in silicates, which stimulates the growth of iron-reducing bacteria and increases their metabolism of mineralogical material (Rogers et al. 2001).

7.3.2 Biological Weathering of Silicate Minerals and Rocks

Biological processes could facilitate the dissolution of silicates by producing organic acids. Chemical weathering of silicate minerals and the growth of authigenic minerals, usually considered in terms of inorganic processes, including oxidation/reduction, dissolution, and hydrolysis, are strongly influenced by weak acids, such as carbonic acid in meteoric water. Such processes can be accelerated by the presence of organic acids, which can be generated as byproducts of metabolism or excreted to extract essential nutrients from minerals and rocks by organisms. Through dissolution reactions, easily exchangeable cations, such as Ca^{2+}, Mg^{2+}, K^+, and Na^+, and variable amounts of aluminum and silica, are leached out from lattice, leaving behind a residual clay phase ($2KAlSi_3O_8$ (K-feldspar) + $2H^+$ + $9H_2O \rightarrow Al_2Si_2O_5(OH)_4$ (kaolinite) + $4Si(OH)_4$ + $2K^+$) or metal oxide ($2KAlSi_3O_8 + 2H^+ + 14H_2O \rightarrow 2Al(OH)_3 + 6Si(OH)_4 + 2K^+$).

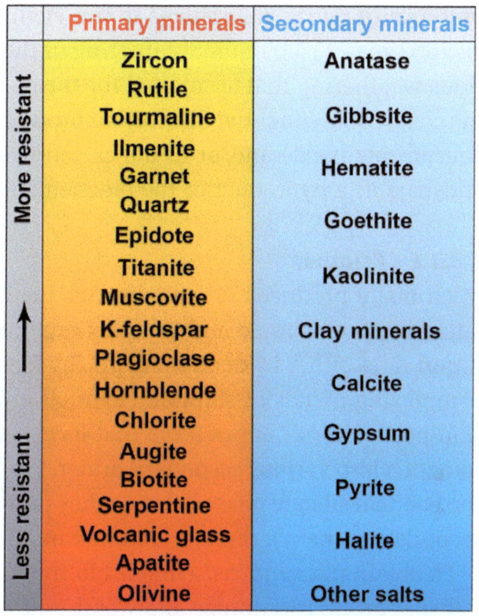

Figure 7.12 Resistance of minerals to weathering. *Source:* Adapted from Perkins (2020) / with permission of Elsevier.

Abiotic weathering patterns may be significantly altered by microorganisms, as they aggressively scavenge limiting nutrients. "Goldich Weathering Sequence" predicts that silicate minerals crystallize at higher temperature (e.g. olivine and pyroxene) in magma are less stable and weather more quickly, while minerals crystallized at lower temperature (e.g. K-feldspar and quartz) are more resistant to weathering on Earth's surface (Figure 7.12; Goldich 1938). The sequence follows a pattern similar to Bowen's Reaction Series, which describes the order of mineral crystallization from magma. However, in an anaerobic and microbially controlled system, the weathering sequence of minerals may be almost reversed, with olivine being more stable than to K-feldspar, because silicate weathering is sometimes driven by the nutrient requirements of the microbial consortium and therefore depends on the trace nutrient content of minerals (Bennett et al. 2001).

There is abundant physical evidence for microbial weathering. For example, colonization of mineral surface by microorganisms leaves surface etching marks that approximate the size and shape of the microorganism. The preferential orientation of etch pits along cleavage planes (see Konhauser 2007; p. 203; fig. 5.6) suggests that dissolution was crystallographically controlled, with the microorganisms taking advantage of structure weak points. Observations carried out with microscopy techniques reveal that fine-sized silicate grains, characterized as quartz, feldspar (mostly K-feldspar), and phyllosilicates (muscovite, illite, biotite, chlorite, smectite), occur coated with biofilms and show different features of alteration (Sanz-Montero and Rodríguez-Aranda 2007). Namely, the quartz grains display surface roughing in the edges; the phyllosilicates, especially the Fe-rich types, have undergone extensive physical disintegration that includes the separation of the layers along the cleavage planes. Besides, these minerals are significantly depleted in Fe, which along with the presence of pyrite precipitates in the vicinity of the altered grains, suggests that Fe(III)-reducing bacteria play a role in the weathering of these minerals. The feldspars have also experienced notorious weathering that accounted for the detachment of fragments and the extensive dissolution of the clasts. Subsequently, dissolution products, basically amorphous masses of silica and aluminum, concentrate inside and/or in the closeness of the minerals. Some Ba-Sr sulfates precipitate in the feldspars as a byproduct of the biochemical dissolution.

7.3.2.1 Feldspar

Microbially produced organic ligands include metabolic byproducts, extracellular enzymes, chelates, and both simple and complex organic acids. These substances can influence feldspar dissolution rates either by decreasing pH, by forming framework-destabilizing surface complexes, or by complexing metals in solution (Bennett et al. 2001). Of importance here are the citric and oxalic acid-producing strains of fungi that have been described in nature as effectively degrading feldspar minerals into various secondary mineral products. Organic acids also influence the composition of the residual silicate phase because they preferentially solubilize Al—O bonds, while in their absence secondary minerals remain enriched in Al (Stillings et al. 1996).

To obtain phosphorus, a major bio-limiting element, microorganisms favor distinct minerals in feldspar group to colonize and weather. *In situ* colonization experiments in the oil-contaminated aquifer, conducted over periods of 3–12 months, demonstrate that Ontario (Canada) microcline and plagioclase surfaces are all barren of microbiota, biofilm development, or chemical weathering because none of these feldspars contain detectable P, although they all contain 3000–6000 ppm Fe(III); oligoclase is lightly colonized with evidence of chemical weathering, which contains a small amount of P (ca.130 ppm); anorthoclase and South Dakota microcline, with 1050 and 1220 ppm P, and 12000 and 2500 ppm Fe(III), respectively, are heavily colonized and weathered (Figure 7.13; Bennett et al. 2001). Initial colonization (observed after 3 months) is by a morphologically diverse consortium, with diversity decreasing over time (12 months) as mineral weathering progresses. On the microcline specimen, secondary clays (smectite) are observed

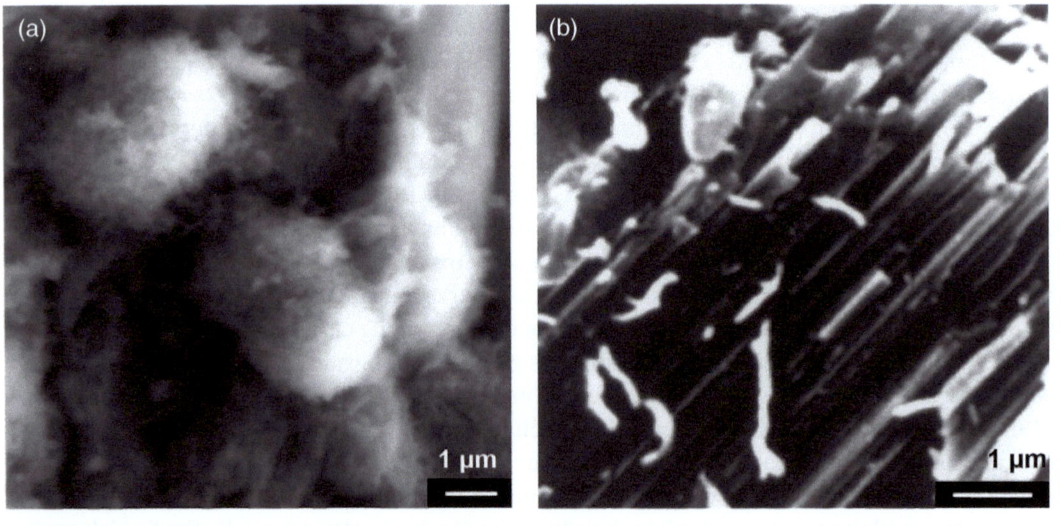

Figure 7.13 (a) SEM image of weathered anorthoclase surface covered with cocci or diplococci, which are coated in a secondary precipitate. (b) SEM image of South Dakota microcline surface with deep etch features. *Source:* Bennett et al. (2001) / Informa UK Limited.

on uncolonized, unweathered mineral surfaces, while clays appear to nucleate and grow on the cell wall surface of organisms colonizing anorthoclase. The P in anorthoclase and microcline occurs as inclusions of bladed apatite. Iron oxides are present in the anorthoclase, while South Dakota microcline does not contain iron-bearing accessory minerals.

When the nutritional requirements are limited by potassium, colonizing bacteria would, on the other hand, attack K-rich feldspars preferentially. In such conditions, surfaces of microcline (a K-rich feldspar) were much more deeply weathered than those of albite (a Na-rich feldspar), after exposure for several months (Bennett et al. 1996).

7.3.2.2 Silica

Silica minerals are resistant to dissolution, but their weathering can be facilitated by diverse microbial processes. The pH value directly influences quartz degradation. At pH values lower than 3.5, no quartz dissolution is possible, while pH values above 9 the dissolution rate of quartz is at a maximum. Since photosynthetic activity and associated CO_2 consumption locally raise pH in selected microenvironments around the cell, microbial leaching and pitting processes may take place in the vicinity of any photosynthesizing cell. Silica weathering is probably enhanced by the presence of organic compounds, especially where microorganisms and their extracellular metabolites (e.g. mucilage) are present. Additionally, silica dissolution is significantly altered by the presence of silica-organic complexes and in NaCl solution, presumably because these substances change the structure of water in a way that increases hydrogen bonding with H_4SiO_4.

There is ample evidence of biologically induced weathering of different types of silica: quartz sand, crystalline quartz, and commercial glass (Brehm et al. 2005). Experiments confirm that diatoms, heterotrophic bacteria, and cyanobacteria from natural biofilms can actively attack quartz and glass. The microbial communities on the grain surface altered the minerals in the vicinity of the living cells, which left their marks on the quartz surface in the form of imprints, depressions, and pits (Gorbushina and Palinska 1999). Subsequently, microbial growth on the surface brought about a general decrease of the grain size in the layer beneath a biofilm. Imprints of diatoms on glass surfaces colonized by diatoms and heterotrophic bacteria confirm the chemical etching activity of these organisms, with depressions corresponding to the shape of individual cells (Figure 7.14). Brehm et al. (2005) conclude that biofilm growth in marine (sub-aquatic) and terrestrial (sub-aerial) conditions can significantly increase the breakdown of silica in the amorphous (glass), sub-crystalline (chert), crystalline, and granular forms of quartz. Microbial growth may therefore substantially modify the processes of transformation of a major rock-forming mineral of great chemical and physical resistance.

7.3.2.3 Mafic Rocks and Minerals

Basalts are relatively rich in P as apatite. Direct evidence of microbial involvement of basalt weathering comes from studies of terrestrial lava flows and pillow basalts. On land, lichen growth facilitates the weathering of the primary rock-forming minerals as a consequence of releasing respiratory CO_2 and excreting organic and lichen acids, which leads to variable stages of etching, mineral fragmentation, and authigenic mineralization (mostly fine-grained calcium oxalate crystals). Microbial communities are also instrumental in altering the kinetics of dissolution. For instance, chemical weathering rates of lichen-colonized lava flows in Hawaii were reported to be up to 100 times that of bare rock (Jackson and Keller 1970) and lichens on Lanzarote lava flows caused a 16 times increase in weathering rates over those found on bare surfaces (Stretch and Viles 2002). Basaltic glass is abundant in the oceanic crust. Experiments demonstrate that bacteria play an important role in the initiation of dissolution reactions, in which they rapidly develop into

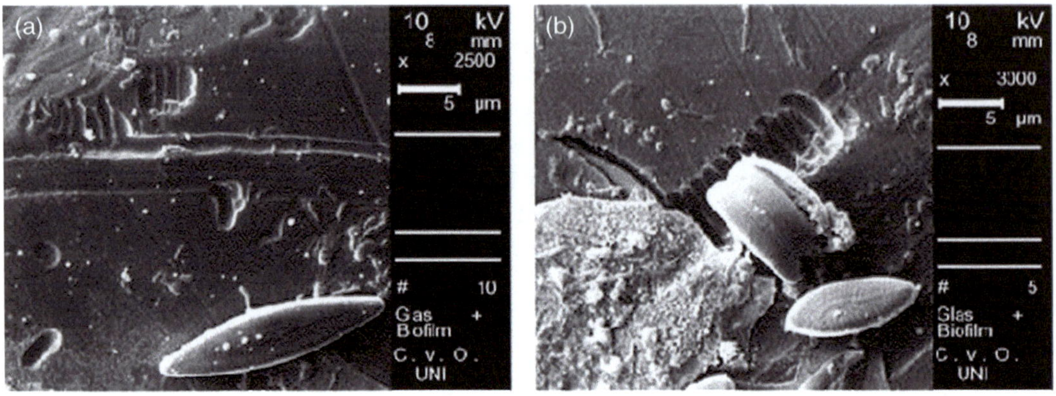

Figure 7.14 (a) SEM micrographs of an etched window glass surface with recognizable traces and cells of bacteria and microscopic fungi or filamentous cyanobacteria. The remains of the biofilm including several bacteria and a diatom in the lower right corner are still present on the surface. (b) SEM micrograph of an etched window glass surface showing pits left by a diatom chain on the surface of glass. The pits are situated in a row characteristic of diatom colonies. Biofilm and several diatoms are clearly visible. *Source:* Brehm et al. (2005) / with the permission of Elsevier.

micrometer-thick biofilms on fresh glass surface. After only weeks to months of colonization, the glasses displayed preferential dissolution at points along fractures and subsequently developed into pronounced etch mark in a year time (Staudigel et al. 1998).

Mafic minerals, such as olivine, pyroxene, amphibole and biotite, are enriched in magnesium and ferrous iron, and weather rapidly in oxic environments as the Fe is initially released through congruent dissolution ([Fe_2SiO_4 (olivine) + $4H^+ \rightarrow 2Fe^{2+}$ + $Si(OH)_4$]), and then oxidized and hydrolyzed to ferric hydroxide ($2Fe^{2+} + 0.5O_2 + 5H_2O \rightarrow 2Fe(OH)_3 + 4H^+$). The congruent dissolution is defined as a complete dissolution of a mineral into its constituent ions without production of second minerals, while the incongruent dissolution is a partially dissolving process, releasing part of components into the solution and leaving behind a new solid phase. During the incongruent dissolution, the original mineral does not dissolve entirely, and it instead transforms into a secondary mineral. However, these dark minerals may be resistant to biological weathering when nutrient contents are limited. A study showed that olivine surfaces were not colonized by biofilms after >12 months, and no indication of weathering was observed (Bennett et al. 2001). This is because the olivine contains no detectable P, although there is Fe and relatively high concentrations of other metals, particularly Ni (>3000 ppm). It is possible that Ni acts as a biocide or in some other way influences the colonization behavior (Bennett et al. 2001). The biotite can undergo two types of biological weathering mechanisms: dissolution by the destruction of the mineral structure (congruent way) or transformation into vermiculite (incongruent way) by the release of interlayer potassium.

7.3.3 Carbonate Weathering

Carbonate minerals are among the most reactive minerals found in abundance on the Earth's surface. Since carbonates dissolve much faster than silicate minerals even at higher pH values, carbonate weathering is commonly attributed to abiotic dissolution. Organisms usually facilitate the erosion of carbonate rocks as macro- and micro-bioeroders (see Section 7.2).

Figure 7.15 Lichens grow on the surface of Cambrian limestone in Kyrshabakty section, in Karatau region of Kazakhstan. *Source:* Photos by Xingliang Zhang.

Carbonate dissolution commonly occurs when liquid water is present, such as dissolution of calcite ($CaCO_3 + H_2CO_3 \rightarrow Ca^{2+}\ 2HCO_3^-$) and dolomite ($CaMg(CO_3)_2 + 2H_2CO_3 \rightarrow Ca^{2+} + Mg^{2+}\ 4HCO_3^-$). However, in arid regions where the availability of water is limited, microbial communities, such as lichens, could facilitate the dissolution of carbonate rocks by creating a suitable environment on their surface and increasing water resistance time (Figure 7.15), although the mechanism is not well-understood. Additionally, calcite is more prone to dissolution and endolithic boring than dolomite (Jones 1989).

7.3.4 Sulfide Mineral Oxidation

As pyrite is the most common sulfide mineral in Earth's crust, we will discuss the pyrite oxidation by microbes as an example for biological weathering of sulfides. The oxidation of pyrite includes indirect and direct mechanisms. The indirect oxidation includes sulfur reactions [Eq. (7.1)] and iron reactions [Eq. (7.2)]. The sulfur reaction can be achieved through several steps and generates various intermediate sulfur products (e.g. thiosulfate, polythionates, sulfite). An essential oxidant in sulfur reactions is Fe^{3+}, which is generated from iron reaction. However, in abiotic conditions, the regeneration of Fe^{3+} is usually slower than its consumption, and therefore, it limits the overall indirect oxidation rate of pyrite (Konhauser 2007). Some acidophilic iron-oxidizing bacteria could obtain energy by oxidizing Fe^{2+} to Fe^{3+} and oxidize Fe^{2+} five to six orders faster than abiotic reactions (Konhauser 2007). As a result, the acidophilic bacteria could provide abundant Fe^{3+} and make the pyrite oxidation no longer limited by the regeneration of Fe^{3+}. On the other hand, pyrite can be oxidized directly (Eq. (7.3)), and iron-oxidizing bacteria can also enhance the rate of direct oxidation.

$$FeS_2 + 8H_2O + 14Fe^{3+} \rightarrow 15Fe^{2+} + 2SO_4^{2-} + 16H^+ \tag{7.1}$$

$$14Fe^{2+} + 3.5O_2 + 14H^+ \xrightarrow{A.f} 14Fe^{3+} + 7H_2O \tag{7.2}$$

$$FeS_2 + H_2O + 3.5O_2 \xrightarrow{A.f} Fe^{2+} + 2SO_4^{2-} + 2H^+ \tag{7.3}$$

In a laboratory incubation experiment, iron-oxidizing bacteria *Acidithiobacillus ferrooxidans* (*A.f*), which increases the oxidation rate of pyrite considerably, leaves distinct pits on the initially smooth surfaces of pyrite (Liu and Zhang 2015). Abiotic reactions generate paralleled and perpendicular cracks on the surface of minerals, and biotic reactions produce distinct, round, and polygonal pits on pyrite surfaces. Such biologically generated pits sometimes are filled with *A.f* cells, suggesting a contact oxidation mechanism. Although the microbes can rarely be preserved through geological processes, such characteristic pits at pyrite surfaces can be found in marine sediment and old sedimentary rocks, such as Ediacaran strata (Zhang et al. 2021). Therefore, like organic molecules (biomarkers) and isotopic fractionations, such pit structures on pyrite surfaces can be used as a biosignature to fingerprint the existence of iron-oxidizing microbes, not only on our Earth but also on extraterrestrial bodies.

7.3.5 Insight into Lichen Weathering

Lichen (e.g. Figure 7.16) is not a single organism, as the way most other living things are, but rather it is a symbiont, consisting of at least two organisms: a fungus (mycobiont) and a photosynthetic partner (photobiont). The photobionts are typically green algae, and other photosynthetic microorganisms such as diatoms and cyanobacteria may also occur. A mass of fungal hyphae or thallus normally forms the greater part of lichen thalli. The photobionts lie just beneath the upper surface, accounting for no more than 20%.

Lichens play an important role in breaking down rock minerals, both physically and chemically, and thus contribute to soil formation. They are ideal microcosms in which to study microbially mediated mineral weathering. Studies show that lichens on lava flows may cause a tremendous increase in weathering rates over those found on bare surfaces (Jackson and Keller 1970; Stretch and Viles 2002).

Lichens colonize ancient monuments and building stones but do not usually cause significant damage in a relatively short time span. Architects and many other people recognize that lichens can give a desirable rustic charm to buildings, adding to local character. If you spray dilute manure on the surface, you can encourage the growth of the colorful yellow *Xanthoria* spp. Lichens can be removed where necessary using fungicides, bleach, and caustic washes.

Figure 7.16 Lichens colonizing on the surface of granite at an altitude of ca. 4000 m, Tianshan Mountain. *Source:* Photo by Xingliang Zhang.

7.3.5.1 Physical Processes

Physical weathering occurs mechanically by hyphae and rhizines that simply grow through the rock and thus result in the breakup of minerals. Rhizines penetrate into the substrate and loosen fragments, and the thallus can also expand and contract depending on moisture availability, pulling and plucking rock fragments from the surface. This seems to be an enormous task for such small organisms. But just consider the amazing ability of mushrooms to push up asphalt and concrete in driveways. Fungal hyphae are able to exert immense turgor pressure, even when they measure only a few microns in diameter. Lichen hyphae may penetrate minerals, especially those with well-developed cleavage planes, such as biotite. We do not know how much weathering is the result of lichen activity. Some people think that lichens may sometimes protect the surface from other kinds of weathering (frost action, wind abrasion, and acidic solutions produced by atmospheric pollutants) by binding together minerals with their hyphae.

7.3.5.2 Biochemical Processes

Biochemical weathering can occur via the production of respiratory carbon dioxide, the secretion of organic acids such as oxalic acid, or the production of weak phenolic acids known as lichen acids, which react with the substrate. Lichens may alter rocks chemically in three ways (Purvis 2000):

1) Simple organic acids, such as oxalic acid secreted by lichen fungi, may react with metals to form metal oxalates. By far the most abundant are the calcium oxalates, weddellite ($CaC_2O_4{\cdot}2H_2O$) and whewellite ($CaC_2O_4{\cdot}H_2O$), which are frequent in and on lichens growing on rocks rich in calcium carbonate, such as limestone. Some lichens, such as *Dirina massiliensis* f. *sorediata*, may form thalli up to 1 cm thick and appear chalk-like in cross-section because of calcium oxalates. Sometimes it is difficult to identify the boundary between lichen and its mineral substrate.
2) The outer layers of hyphae in contact with mineral grains are coated with acidic polysaccharides. These may absorb water and assist in mineral breakdown by extracting particular metal ions from mineral surface.
3) Some lichen substances, such as depsides, are not very soluble in water and, in laboratory experiments, have been found to remove metals from minerals to form metal complexes. We do not know how often this occurs in nature.

7.3.5.3 A Zone Model

Banfield et al. (1999) developed a four-zone model (Figure 7.17) for lichen's weathering based on correlation of different styles of silicate mineral weathering with pore size-controlled microbial distributions. Zone 1 consists of the upper lichen thallus and is devoid of weathering of substrate-derived mineral particles. Zone 2 is a region of extreme mineral weathering, characterized by direct contact between cells, extracellular polymers, associated compounds, and mineral surfaces. Complex, nanometer-scale mixtures of clay products, organic polymers, and primary minerals are common. In some cases, mineral formation on cell surfaces preserves relict cell shapes. Unlike essentially inorganic weathering, the secondary mineral orientations are less controlled by the primary mineral structure and are more determined by the distribution of polymers onto which they nucleate. In Zone 3, weathering reactions are accelerated by dissolved organic molecules (metabolic products, e.g. lichen acid), but not mediated by direct microbial contact. Orientations of secondary minerals are more commonly determined by orientations of the primary minerals, as in essentially inorganic reactions. Zone 4 comprises unweathered rock and rock undergoing early, predominantly inorganic reactions.

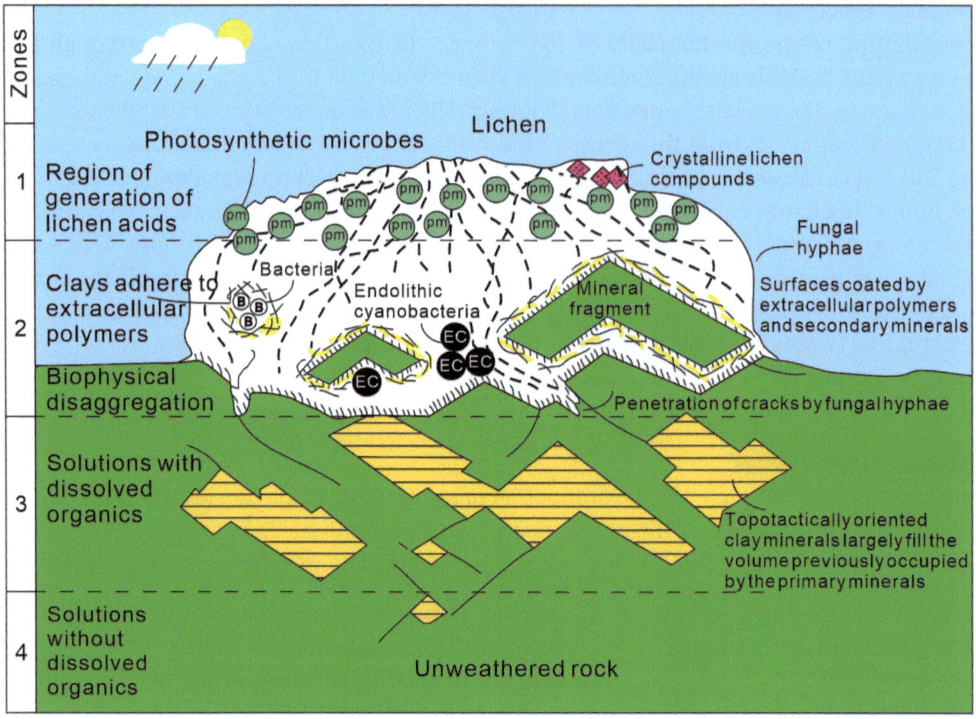

Figure 7.17 Four-zone model cartoon illustrating mineral weathering occurring in zones that are affected by lichen communities to different degrees and in different ways. *Source:* Adapted from Banfield et al. (1999).

7.4 Soil as a Classic System of Biological Weathering

In most places, the land surface is covered by a layer of unconsolidated rock and mineral fragments, regardless of how they were deposited, called **regolith**. It may consist of volcanic ash, sediments transported by wind, streams, or glaciers, or residues of rocking weathering. With the addition of water, air, organic matter, and inorganic nutrients, regolith develops into **soil,** a loose layer on land surface that can support the growth of plants, animals, and microorganisms. The proportions of the soil components are climatically controlled, producing different types of soils, e.g. pedalfers (abundant organic matter, aluminum-rich clays, and iron oxides) in humid regions, pedocals (less organic matter and more unstable minerals, e.g. caliche in horizon B) in arid and semiarid regions and laterite (completely leached, composed largely of aluminum hydroxides, iron oxides, and clay minerals) in the tropics.

The soil matrix is a loose and porous architecture composed of particles of various sizes (e.g. sands and silts). The microenvironment of the matrix contains inorganic minerals (such as clays), organic matter, water, and air. Microorganisms could colonize the surface of organic matter and minerals, interacting with the surrounding environment to obtain essential nutrients and energy (Figure 7.18). The microbial ecosystem of soil holds the largest biodiversity on Earth, at least a quarter of the Earth's total biodiversity; in a single gram of soil, billions of bacterial and archaeal cells, trillions of viruses, tens of thousands of protists, and hundreds of meters of fungal hyphae can be found (Sokol et al. 2022). Such a large biodiversity is associated with a broad range of metabolic

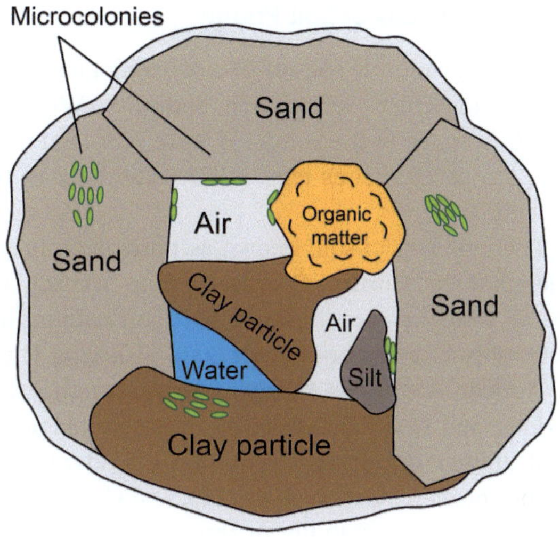

Figure 7.18 A microbial habitat in soil matrix. *Source:* Adapted from Maier et al. (2000).

pathways that would enhance the weathering rate via diverse mechanisms. Therefore, the soil system receives the most extensive biological weathering on Earth.

7.4.1 Soil Development

Soil-forming processes begin at the surface and move gradually downward. Thus, the upper layer of soil is more altered from the parent material than the layers below. The development of soils takes place by the combined action of weathering processes and the growth of successive populations of organisms. Microorganisms play an important role. For example, in rocky areas, the process of soil development usually starts when photosynthetic microorganisms, such as the cyanobacteria, colonize rock crevices. The reproduction of cyanobacteria allows the colonization of the rock by certain other bacteria and fungi. Acids released from the chemical activities of these microorganisms help to dissolve the rock, creating a substrate suitable for the growth of higher organisms. Additionally, lichens and simple plants, such as mosses, also grow on these rocks, and further decompose rocks. Finally, the dead and decaying lichens and mosses begin to create a thin layer of soil in rocky areas, thus allowing larger plants to grow.

The type of soil that develops in a given area is dependent upon complex interactions among several factors, e.g. the type of sediment or bedrock that is weathered, the climate of the region, and the plant species that colonize the soil. For example, soils of mangrove forests in the tropic and subtropic coastal areas of South China, and soils of coniferous forests in the temperate areas of North America, are typically acidic and depleted in nutrients, because microorganisms that decompose the mangrove and coniferous materials release considerable amounts of acid, which together with percolating water from rains in turn leaches the nutrients out of the soil. Hence, soils in mangrove and coniferous forests generally are unsuitable for agriculture, but can be farmed when treated with lime ($CaCO_3$) to raise the pH and with fertilizers to increase the nutrient level. In contrast, indeciduous latifoliate forests or conifer-latifoliate mixed forests in the vast area south to the Yangtze River develop the most productive soil, known as red earth, in South China, because little acid is released to the soil. The forest litter, consisting of fallen leaves, shrubbery, and other plant material, creates a thick humus layer (see Section 7.4.3), from which nutrients are released slowly to fertilize the soil.

7.4.2 Horizons of Soil Profile

Vertically, a mature soil consists of distinct layers or soil horizons, which differ from one another in texture, structure, composition, and color (Wicander and Monroe 2002). The horizons are O, A, B, and C from top to the bottom (Figure 7.19), but their boundaries are transitional rather than sharp. The top two horizons tend to form the bulk of the rhizosphere, the depth to which plant roots extend.

The top layer of soil, known as horizon O, is an organic-rich layer, generally only a few centimeters thick. Horizon O consists of an accumulation of organic litter that is in various stages of decay, from just recently deposited plant remains in the upper part to highly degraded material with refractory humus (Konhauser 2007). Horizon O is very thin or absent in arid regions.

Horizon A consists mostly of resistant minerals such as clay minerals, quartz, and some metal oxides, and is characterized by intense biological activities, because plant roots, bacteria, fungi, and animals (such as worms) are abundant. To extract nutrients from soil minerals, such as P and trace elements (e.g. Zn, Mo, and Co), microbes in the soil use specific weathering strategies to increase the solubility of the elements (Philippot et al. 2024). For example, microbes could enhance the dissolution of nutrient elements via decreasing the pH of soil environment and complex metals through excretion of organic ligands (Philippot et al. 2024). On the other hand, the metabolic reactions that are usually involved with the transfer of electrons would result in a change of redox state in the soil and influence the solubility of some metal elements. For example, the reduced species of some metals [e.g. Fe(II), Mn(II), U(VI), and Cr(VI)] are dissolvable and easily leached. In

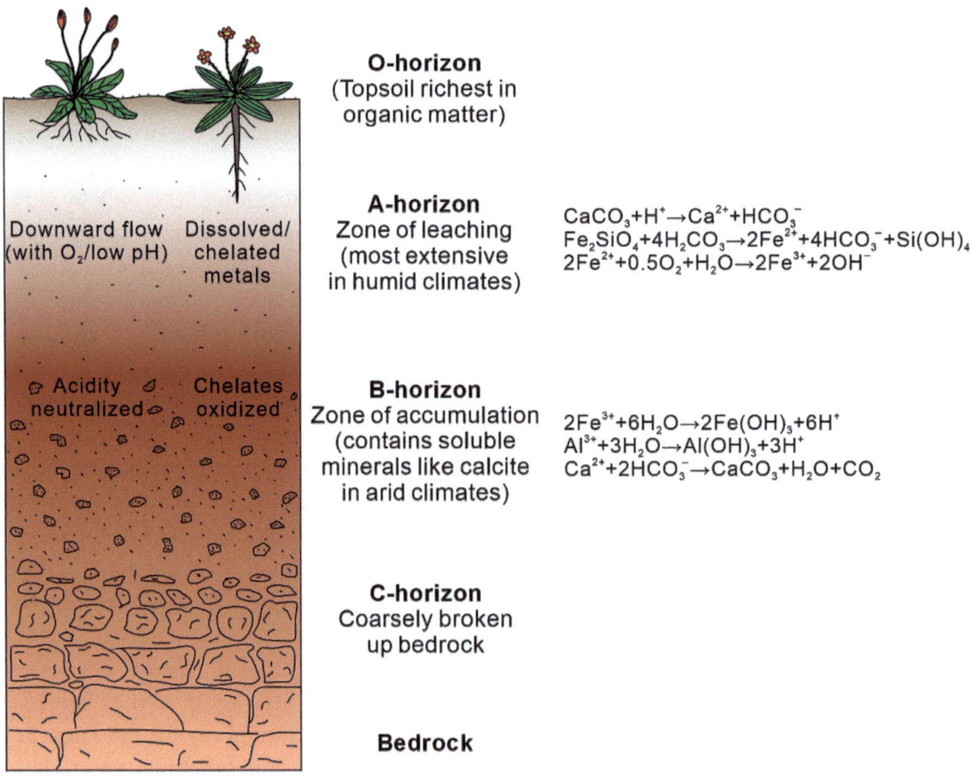

Figure 7.19 A conceptual soil profile showing characteristic zonation. Possible dissolution and precipitation reactions are indicated in each zone.

horizon A, water, acids (generated by biological and abiological processes), and organic chelates percolate through, dissolving soluble minerals (e.g. reactions indicated in Figure 7.19) and carrying them away or downward to lower levels. This process is called leaching, and thus horizon A is also known as the zone of leaching.

Horizon B contains fewer organisms and less organic matter than horizon A and is also called the zone of accumulation. Many soluble cations leached from the horizon A are reprecipitated (e.g. reactions indicated in Figure 7.19) and accumulate as irregular masses. In this horizon, either the acidity is neutralized to facilitate mineral hydrolysis or organic chelates are oxidized by aerobic microbes (Konhauser 2007). In arid soils, calcium carbonate precipitates because the loss of CO_2 and H_2O both increases its saturation state.

Horizon C, the lowest soil layer, consists of partially altered parent material grading downward into unaltered parent material and contains little organic matter. In horizons A and B, the composition and texture of parent material have been so thoroughly altered that the parent material is no longer recognizable. In contrast, rock fragments and mineral grains of parent material retain their original features in horizon C.

7.4.3 Humus in Soils

Areas where sediments accumulate are generally rich in nutrients that stimulate the growth of various populations of microorganisms and higher plants. If the plants grow well and their remains are not rapidly degraded by the soil microorganisms, the soil becomes enriched with organic matter and **humus** from the partial breakdown of roots, stems, and leaves.

Humus consists of poorly degradable residues such as lignin, cellulose, hemicellulose, polypeptides, polysaccharides, nucleic acids, phospholipids, and other materials that have escaped microbial decomposition. These humic substances are characterized by their black to brown color and their very fine-grained size. On the basis of their resistance to degradation and solubility, humic substances have been classified into three groups: (1) fulvic acids that are lowest in molecular weight, lightest in color and soluble in both acid and alkali, (2) humic acids that are medium in molecular weight and color, soluble in alkali but insoluble in acid, and (3) humin that is heaviest in molecular weight, darkest in color, and insoluble in both alkali and acid.

Humus is very important to plants and soil microbial communities. First, it acts as a reservoir of nutrients. For example, as humus is slowly decomposed by microorganisms, it provides plants and soil-dwelling organisms a steady supply of sugars, amino acids, nucleosides, phosphate, ammonia, potassium, and trace elements. Second, humus could hold water when it is abundant and release it back to plant roots during dry conditions, ensuring a steady source of moisture for plants. Last, humus could promote soil aeration by creating a loose texture in soils, improving circulations of air and water, which is essential for the health and growth of land plants and microorganisms in soils.

References

Banfield, J.F., Barker, W.W., Welch, S.A., and Taunton, A. 1999. Biological impact on mineral dissolution: application of the lichen model to understanding mineral weathering in the rhizosphere. *Proceedings of the National Academy of Sciences of the United States of America*, 96, pp. 3404–3411.

Banister, K., and Campbell, A. 1998. *The encyclopedia of aquatic life*, pp. 349. New York, Facts On File, Inc.

Barker, W.W., Welch, S.A., and Banfield, J.F. 1997. Biogeochemical weathering of silicate minerals. *Reviews in Mineralogy*, 35, pp. 391–428.

Benner, J.S., Ekdale, A.A., and de Gibert, J.M. 2004. Macroborings (*Gastrochaenolite*) in Lower Ordovician hardgrounds of Utah: sedimentologic, paleoecologic, and evolutionary implications. *Palaios*, 19, pp. 543–550.

Bennett, P.C., Hiebert, F.K., and Choi, W.J. 1996. Microbial colonization and weathering of silicates in a petroleum-contaminated ground water. *Chemical Geology*, 132, pp. 45–53.

Bennett, P.C., Rogers, J.R., Choi, W.J., and Hiebert, F.K. 2001. Silicates, silicate weathering, and microbial ecology. *Geomicrobiology Journal*, 18, pp. 3–19.

Bolotov, I.N., Aksenova, O.V., Bakken, T., Glasby, C.J., Gofarov, M.Y., Kondakov, A.V., Konopleva, E.S., Lopes-Lima, M., Lyubas, A.A., Wang, Y., Bychkov, A.Y., Sokolova, A.M., Tanmuangpak, K., Tumpeesuwan, S., Vikhrev, I.V., Shyu, J.B.H., Win, T., and Pokrovsky, O.S. 2018. Discovery of a silicate rock-boring organism and macrobioerosion in fresh water. *Nature Communications*, 9, pp. 2882.

Brehm, U., Gorbushina, A., and Mottershead, D. 2005. The role of microorganisms and biofilms in the breakdown and dissolution of quartz and glass. *Palaeogeography, Palaeoclimatology, Palaeoecology*, 219, pp. 117–129.

Calcinai, B., Arillo, A., Cerrano, C., and Bavestrello, G. 2003. Taxonomy-related differences in the excavating micro-patterns of boring sponges. *Journal of the Marine Biological Association of the UK*, 83, pp. 37–39.

Calcinai, B., Bavestrello, G., and Cerrano, C. 2005. Excavating sponge species from the Indo-Pacific Ocean. *Zoological Studies*, 44, pp. 5–18.

Calvaruso, C., Turpault, M.-P., and Frey-Klett, P. 2006. Root-associated bacteria contribute to mineral weathering and to mineral nutrition in trees: a budgeting analysis. *Applied and Environmental Microbiology*, 72, pp. 1258–1266.

Chazottes, V., Le Camion-Alsumard, T., and Peyrot-Clausade, M. 1995. Bioerosion rates on coral reefs: interactions between macroborers, microborers and grazers (Moorea, French Polynesia). *Palaeogeography, Palaeoclimatology, Palaeoecology*, 113, pp. 189–198.

Collins, K.S., Edie, S.M., and Jablonski, D. 2023. Convergence and contingency in the evolution of a specialized mode of life: multiple origins and high disparity of rock-boring bivalves. *Proceedings of the Royal Society B: Biological Sciences*, 290, pp. 20221907.

Dontsova, K., Balogh-Brunstad, Z., and Chorover, J. 2020. Plants as drivers of rock weathering. In: *Biogeochemical cycles: ecological drivers and environmental impact*, Geophysical Monograph Series (eds. K. Dontsova, Z. Balogh-Brunstad, and G. Le Roux), pp. 33–58. Wiley.

Edinger, E.N. 2001. Bioerosion. In: *Palaeobiology II* (eds. D.E.G. Briggs and P.R. Crowther), pp. 273–277. Blackwell Science, Oxford.

Ekdale, A.A. 2006. History of hard habitat homes: the role of bivalves in the evolution of the macroboring niche. In: *The 5th international bioerosion workshop, programme and abstracts* (eds. M. Wisshak, S.B. loffler, C. Schulbert, and A. Freiward), pp. 10–11. Erlangen, Institute of Palaeontology.

Ekdale, A.A., and Bromley, R.G. 2001. Bioerosional innovation for living in carbonate hardgrounds in the Early Ordovician of Sweden. *Lethaia*, 34, pp. 1–12.

Flügel, E. 2004. *Microfacies of carbonates rocks*, pp. 976. Berlin, Springer.

Glaub, I., Balog, S.-J., Bundschuh, M., Gektidis, M., Hofmann, K., Radtke, G., Schmidt, H., and Vogel, K. 1999. Endolithic cyanobacteria/cyanophyta and their traces in earth history. In: *Marine cyanobacteria*, Bulletin de l'Institut océanographique, Monaco (Vol. 19) (eds. L. Charpy and A.W.D. Larkum), pp. 135–142. Institut océanographique, Monaco.

Goldich, S.S. 1938. A study in rock weathering. *Journal of Geology*, 46, pp. 11–58.

Golubic, S., and Schneider, J. 2003. Microbial endoliths as internal biofilms. In: *Fossil and recent biofilm* (eds. W.E. Krumbein, D.M. Paterson, and G.A. Zavarzin), pp. 249–263. Kluwer Academic Publishers, Dordrecht.

Golubic, S., Perkins, R.D., and Lukas, K.J. 1975. Boring microorganisms and microborings in carbonate substrates. In: *The study of trace fossils* (ed. R.W. Frey), pp. 229–259. Springer, Berlin.

Gorbushina, A.A., and Palinska, K.A. 1999. Biodeteriorative processes on glass: experimental proof of the role of fungi and cyanobacteria. *Aerobiologia*, 15, pp. 183–192.

Griscom, S.B., Fisher, N.S., Aller, R.C., and Lee, B.G. 2002. Effects of gut chemistry in marine bivalves on the assimilation of metals from ingested sediment particles. *Journal of Marine Research*, 60, pp. 101–120.

Hall, K. 1990. A note on biological weathering on nunataks of the Juneau Icefield, Alaska. *Permafrost and Periglacial Processes*, 1, pp. 189–196.

Jackson, T.A., and Keller, W.D. 1970. A comparative study of the role of lichens and "inorganic" processes in the chemical weathering of recent Hawaiian lava flows. *American Journal of Science*, 269, pp. 446–466.

James, N.P., Kobluk, D.R., and Pemberton, S.G. 1977. The oldest known macrobores: lower Cambrian of Labrador. *Science*, 197, pp. 980–983.

Jones, B. 1989. The role of microorganisms in phytokarst development on dolostones and limestones, Grand Cayman, Britsh West Indias. *Canadian Jounal of Earth Sciences*, 26, pp. 2204–2213.

Jongmans, A.G., van Breemen, N., Lundström, U., van Hees, P.A.W., Finlay, R.D., Srinivasan, M., Unestam, T., Giesler, R., Melkerud, P.A., and Olsson, M. 1997. Rock-eating fungi. *Nature*, 389, pp. 682–683.

Kazmierczak, J., and Golubic, S. 1976. Oldest organic remains of boring algae from Polish Upper Silurian. *Nature*, 261, pp. 404–406.

Knoll, A.H., Golubic, S., Green, J., and Swett, K. 1986. Organically preserved microbial endoliths from the Late Proterozoic of East Greenland. *Nature*, 321, pp. 856–857.

Konhauser, K. 2007. *Introduction to geomicrobiology*, pp. 425. Oxford, Blackwell.

Liu, W., and Zhang, X. 2015. Experimental study of microbial pyrite oxidation: a suggestion for geologically useful biosignatures. *Geomicrobiology Journal*, 32, pp. 466–471.

Maier, R.M., Pepper, I.L., and Gerba, C.P. 2000. *Environmental microbiology*, pp. 585. San Diego, Academic Press.

Manley, E.P., and Evans, L.J. 1986. Dissolution of feldspars by lowmolecular-weight aliphatic and aromatic acids. *Soil Science*, 141, pp. 106–112.

Needham, S.J., Worden, R.H., and McIlroy, D. 2004. Animal-sediment interactions: the effect of ingestion and excretion by worms on mineralogy. *Biogeosciences*, 1, pp. 113–121.

Neumann, A.C. 1966. Observations on coastal erosion in Bermuda and measurements of boring rate of sponge, *Cliona lampa*. *Limnology and Oceanography*, 11, pp. 92–108.

Perkins, D. 2020. *Mineralogy*. University of North Dakota, Open Educational Resources.

Philippot, L., Chenu, C., Kappler, A., Rillig, M.C., and Fierer, N. 2024. The interplay between microbial communities and soil properties. *Nature Reviews Microbiology*, 22, pp. 226–239.

Purvis, W. 2000. *Lichens*, pp. 112. Washington, DC, Smithsonian Institution Press.

Rogers, J.R., Bennett, P.C., and Choi, W.J. 2001. Enhanced weathering of silicates by subsurface microorganisms: a strategy to release limiting inorganic nutrients? In: *Proceedings of the tenth international symposium on water-rock interaction WRI-10* (ed. R. Cidu), pp. 1461–1464.

Ruppert, E.E., Fox, R.S., and Barnes, R.D. 2004. *Invertebrate zoology, a functional evolutionary approach* (Seventh Edition). Belmont, Brooks/Cole—Thomson Learning.

Rützler, K. 1975. The role of burrowing sponges in bioerosion. *Oecologia*, 19, pp. 205–216.

Rützler, K. 2002. The familay Clionaidae D'Orbigny, 1851. In: *Systema Porifera* (Vol. 1) (eds. J.N.A. Hooper and R.W.M. Van Soest), pp. 173–185. New York, Kluwer Academic/Plenum Publishers.

Rützler, K., and Rieger, G. 1973. Sponge burrowing: fine structure of *Cliona lampa* penetrating calceous substrata. *Marine Biology*, 21, pp. 144–162.

Sajjad, W., Ilahi, N., Kang, S., Bahadur, A., Zada, S., and Iqbal, A. 2022. Endolithic microbes of rocks, their community, function and survival strategies. *International Biodeterioration and Biodegradation*, 169, pp. 105387.

Sanz-Montero, M.E., and Rodríguez-Aranda, J.P. 2007. Microbial weathering of silicates in dolomite-precipitating environments. Miocene lacustrine deposits from the Duero and Madrid Basins, Spain. *Geophysical Research Abstracts*, 9, pp. 06310.

Schönberg, C.H.L. 2002. Substrate effects on the bioeroding demosponge *Cliona orientalis*.1. Bioerosion rates. *Marine Ecology*, 23, pp. 313–326.

Seilacher, A. 1999. Biomat-related lifestyle in the Precambrian. *Palaios*, 14, pp. 86–93.

Sokol, N.W., Slessarev, E., Marschmann, G.L. et al. 2022. Life and death in the soil microbiome: how ecological processes influence biogeochemistry. *Nature Reviews Microbiology*, 20, pp. 415–430.

Staudigel, H., Chastain, R.A.Y., Davies, G.R., Verdurmen, E., Schiffman, P., Bourcier, R., and De Baar, H. 1998. Biologically mediated dissolution of volcanic glass in seawater. *Earth and Planetary Science Letters*, 164, pp. 233–244.

Stillings, L.L., Drever, J.I., Brantley, S.L., Sun, Y., and Oxburgh, R. 1996. Rate of feldspar dissolution at pH 3-7 with 0-8 mmol oxalic acid. *Chemical Geology*, 132, pp. 79–89.

Stretch, R.C., and Viles, H.A. 2002. The nature and rate of weathering by lichens on lava flows on Lanzarote. *Geomorphology*, 47, pp. 87–94.

Suzuki, Y., Matsubara, T., and Hoshino, M. 2003. Breakdown of mineral grains by earthworms and beetle larvae. *Geoderma*, 112, pp. 131–142.

Syvitski, J.P.M., and Lewis, A.G. 1980. Sediment ingestion by *Tigriopus californicus* and other zooplankton: mineral transformation and sedimentological considerations. *Journal of Sedimentary Petrology*, 50, pp. 869–880.

Tapanila, L. 2006. Devonian *Entobia* borings from Nevada, with a revision of *Topsentopsis*. *Journal of Paleontology*, 80, pp. 760–767.

Torunski, H. 1979. Biological erosion and its significance for the morphogenesis of limestone coasts and for near-shore sedimentation. *Seckenbergiana maritima*, 11, pp. 193–265.

Weed, S.B., Davey, C.B., and Cook, M.G. 1969. Weathering of mica by fungi. *Proceedings of the Soil Science Society*, 33, pp. 702–706.

Wicander, R., and Monroe, J.S. 2002. *Essentials of geology* (Third Edition), pp. 523. Pacific Grove, The Wadsworth Group.

Zhang, Y., and Golubic, S. 1987. Endolithic microfossils (cyanophyta) from early Proterozoic stromatolites, Hebei, China. *Acta Micropalaeontologica Sinica*, 4, pp. 1–12.

Zhang, X., Chang, C., Cui, L., and Qiao, Y. 2021. Ecosystem reconstruction during the Cambrian explosion. *Paleontological Research*, 25, pp. 305–314, 310.

8

Co-evolution of Life and Environment

Xingliang Zhang

8.1 Introduction

Cosmologists estimated that the Universe began 13.787 ± 0.020 billion years ago (Planck Collaboration 2020) in what is popularly known as the *Big Bang*, a physical theory that describes how the Universe expanded from an initial state of high density and temperature. According to the *Big Bang* theory, after its initial expansion, the universe cooled sufficiently to allow the formation of subatomic particles. The recombination epoch began after about 379,000 years, when the electrons and nuclei combined into atoms (mostly hydrogen), which were able to emit radiation. The primordial elements in the Universe mostly consisted of hydrogen, with a small quantity of helium and lithium. These light elements are not much good for making up life. However, the original hydrogen atoms perhaps played a very crucial role in the development of the Universe. According to this model, it was nothing more than a dark cloud of hydrogen gas rapidly expanding. The Universe is hard to imagine at this stage because it was dark and had no stars. The solar system accreted from an interstellar cloud of gas and dust, formed after a supernova explosion. This primitive solar nebula was condensed into the Sun and its planets including Earth about 4.567 billion years ago (Halla et al. 2024).

Earth, as a complex and dynamic system, has been changing continually since its origin. These changes and the present-day features we observe resulted from interactions among the various internal and external Earth subsystems and cycles. Earth is unique among the planets of our solar system in that it supports life and has oceans of water, a hospitable atmosphere, and a variety of climates. It supports life because of a combination of factors: it is in the right distance from the Sun, it is protected from harmful solar radiation by its magnetic field, it is kept warm by an insulating atmosphere, and it has the right chemical ingredients for life, including water and carbon. In turn, life processes, over time, have influenced the evolution of Earth's atmosphere, oceans, and to some extent, its crust. These physical factors and the changes they have brought about have also affected the evolution of life. The principal subsystems of Earth are the biosphere, anthroposphere, atmosphere, hydrosphere, lithosphere, mantle, and core. The complex interactions among these subsystems result in a dynamically changing body that exchanges matter and energy, and recycles them into different forms (Figure 8.1).

Living creatures and inanimate worlds they inhabit carry out an intimate interplay. Organisms alter their surroundings as part of their normal activities, and in turn, the changed surroundings

Introduction to Geobiology, First Edition. Xingliang Zhang, Zisheng Guo, Wei Liu, Ruliang He, Luoyang Li, and Weiduo Hao.

Companion website: www.wiley.com/go/Zhang_1e

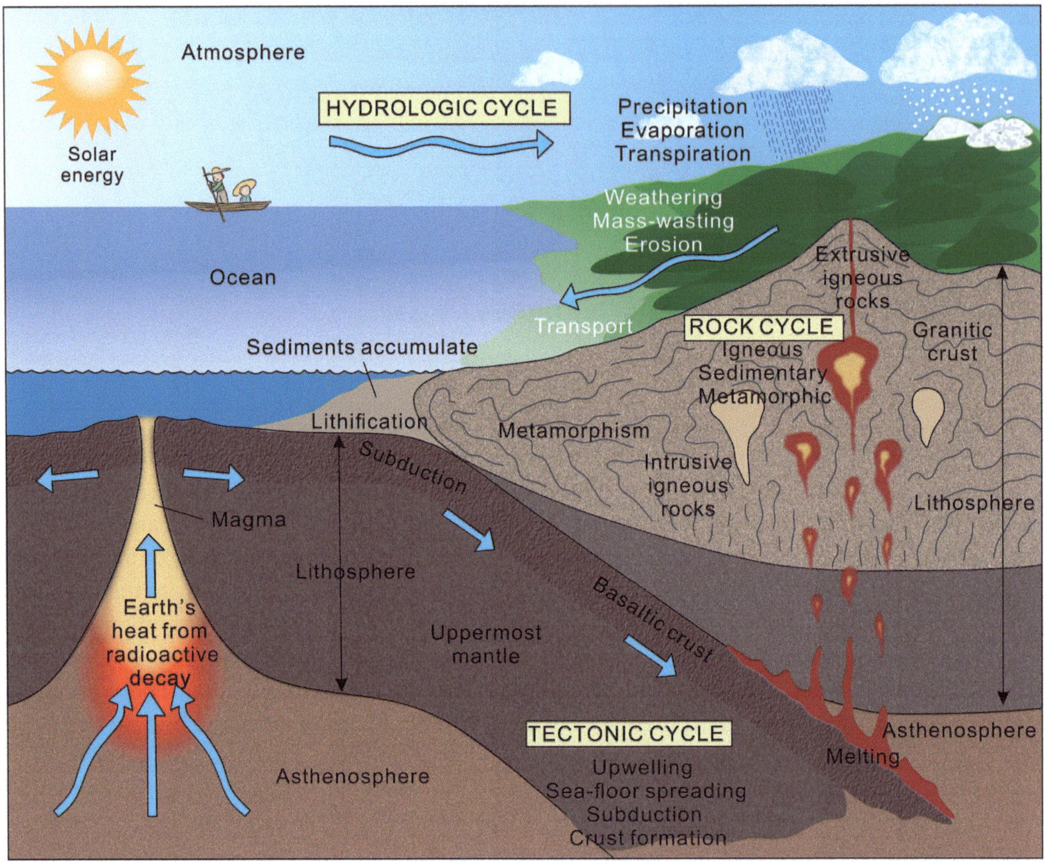

Figure 8.1 The interactions among the subsystems of the modern Earth system—such as the atmosphere, hydrosphere, lithosphere, biosphere, and anthroposphere—are the result of billions of years of evolution since Earth's formation 4.567 billion years ago. These interactions will persist and continue to evolve indefinitely.

nurture different types of life and altered modes of living. This interplay between life and environment has evolved through time. Often complex feedbacks operate. For example, loss of vegetation promotes erosion, which then alters the distribution of plants and changes the climate, leading to further modification of erosion patterns. In another domain, organisms exert a powerful influence on the chemistry of the atmosphere and the oceans by modifying concentrations of components such as oxygen, carbon dioxide, and methane; biogenic shifts in atmospheric chemistry can affect life in ways that feed changes back into the atmosphere.

This chapter addresses the evolution of the interactions amongst the solid Earth, the fluid Earth, and the living Earth on a global scale from Hadean to Phanerozoic (~4.567 Ga to the present). First, we have short reviews on Hadean–Archean geological records and surface environment deduced from these records, followed by separate reviews on the evolution of atmosphere, hydrosphere, and biosphere, and finally a summary on the co-evolution of Earth and life. Specifically, this chapter primarily focuses on early life and its environment, a major topic in Geobiology. Knowledge about Phanerozoic Earth–life evolution can be found in traditional Paleontology and Historical Geology books. By studying the co-evolution of life and the environment, students will gain a better understanding of how the coupled Earth–life system evolved into today's world.

8.2 Earth's Earliest Records and Habitability

8.2.1 Hadean (~4.567 to 4.031 Ga)

The Hadean is the first of the four eons of Earth history, enclosing the accretion and differentiation of the Earth and other planets, the Moon-forming Giant Impact (Theia collided with the Earth), the beginning of the suggested Late Heavy Bombardment (LHB), and the formation of the Earth's protocrust. Its lower boundary age 4567.30 ± 0.16 Ma, ratified as the base of the International Commission on Stratigraphy (ICS) Geological Time Chart, was obtained from the absolute date of calcium–aluminum-rich inclusions (CAIs) in chondritic meteorites, which are the first solids known to have formed in the Solar System (Halla et al. 2024). This age is currently the best-constrained older limit for the age of the Solar System. Astronomical, chemical, physical, and chronological lines of evidence point to a very fast formation of the Solar System and a rapid accretion and differentiation of the proto-Earth in only a few million years. There are several age estimates for the Earth's core segregation, all implying that Earth existed, and its core segregated within 10–30 million years after the CIAs were formed (Kleine et al. 2002; Kleine and Walker 2017). Hence, it is reasonable to set the age of the Earth at the beginning of the formation of the Solar System.

Geological records of the Hadean are very rare, largely consisting of granular zircons from one locality in Western Australia. Our knowledge on the Hadean mostly comes from the very limited suite of ancient geological samples and the interpretations are fraught with controversy. Life may have emerged during the Hadean. While the LHBs, which may have occurred frequently during ~4.1 to 3.8 Ga and heated the oceans and the atmosphere as to make the Earth briefly uninhabitable, sterilized at several hundred degrees Celsius, could obliterate any life forms that emerged during the Hadean.

8.2.1.1 Hadean Zircons and Implications

The Hadean Earth is characterized by a lack of preserved rocks. However, geologists have carefully sorted out more than 100,000 microscopic, detrital zircons in metamorphosed rocks at Mt. Narryer and Jack Hills in the Narryer Gneiss Terrane, Yilgarn Craton, Western Australia. The host rock is >3.3 Ga, with the majority of the Narryer Gneiss Terrane >3.6 Ga, while Jack Hills zircons range from >3.0 to nearly 4.4 Ga in age. The zircon crystals contain trace elements and isotopic signatures as well as microscopic inclusions, such as graphite and gas bubbles, that provide a unique window into conditions on Earth as life arose and the first continents formed. Just three of the very oldest zircons have been found, ones that dated back to almost 4.4 billion years ago (Valley et al. 2014; Wilde et al. 2001; Figure 8.2).

There are several implications from analyses of the zircons and their host rocks. First, by 3.3 Ga ago, in the mid-Archean, there was already old continental crust being eroded and redeposited by water. Second, abundant zircons are typical of rocks broadly characteristic of continental crust. This line of reasoning thus suggests granitoid rocks and continental crust were present in the Hadean. Intuitive reasoning would suggest komatiitic and basaltic rocks would be expected to be typical of 4.4 Ga crust, rather than granitoids, but the existence of Hadean zircon implies otherwise, at least locally in what is now Western Australia, there was continental crust. Moreover, to form granitoid nowadays, subduction of an old hydrated oceanic plate is needed: water is needed to make granites, and subduction is needed to supply the water. Oxygen isotope ratios in 4.4 Ga old zircons suggest that liquid water was indeed present at that time (Mojzsis et al. 2001). The zircons contain isotopically heavy oxygen: suggesting derivation from liquid surface water (Nisbet and Fowler 2005). Bell et al. (2015) reported carbon isotopic measurements for graphite inclusions in a 4.10 ± 0.01 Ga

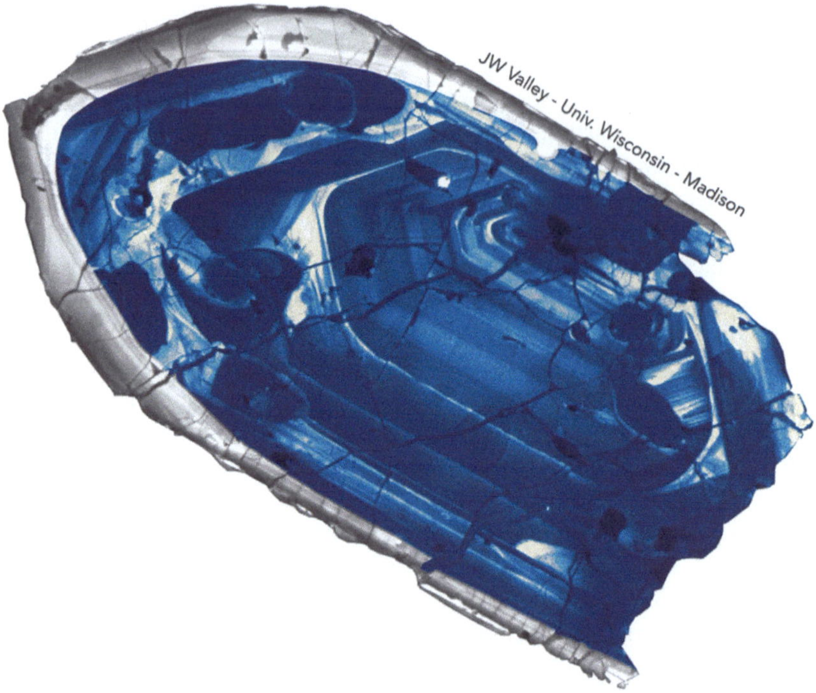

Figure 8.2 False colored cathodoluminescence image of a 4.4-billion-year-old detrital zircon crystal (~399 μm long, $ZrSiO_4$) from the Jack Hills region of Australia, dated as the oldest known mineral on our planet. *Source:* Contesy of John Valley.

zircon crystal and interpreted it as evidence that a terrestrial biosphere had emerged by 4.1 Ga for their $\delta^{13}C$ of $-24 \pm 5‰$ consistent with a biogenic origin.

8.2.1.2 Hadean Environment

The very old Jack Hills zircons provide evidence for the existence of continental crust and liquid water by 4.4 Ga, though they both might be limited in extent. Oxygen isotope evidence of zircons further supports the deduction that oceans of liquid water were in place by 4.4 Ga (Wilde et al. 2001). However, these are clues from a very limited geological record, and just as "one swallow does not make a summer, one zircon does not make either a continent or an ocean of water" (Moorbath 1983). Hadean geophysical models and the surface environment remain uncertain and controversial among scientists. Presently, there are more imaginations than facts.

The Hadean Earth may have been characterized by stagnant lid and mantle drip tectonics. Continental crust was formed by differentiation of initially basaltic crust to produce a stratified ensemble comprising underlying cumulates, mid-crustal intrusions, and an upper crust of volcanic rocks and intrusions. The collision of oceanic arcs and continental blocks was supposed to produce fractionated, more felsic magma. The early granites were typical tonalite-trondhjemite-granitoids (TTGs) that were formed predominantly through intracrustal melting. The earlier, hotter crust would not have been able to support massive continents and high mountains as does the modern crust, and landmasses were small, likely resembling volcanic islands today. The surface was subject to heavy bombardment by extraterrestrial material.

The Sun was ca. 30% less luminous than today and emitted more high-energy radiation, with high UV fluxes to the surface of the Earth because of the lack of oxygen and ozone in the Hadean atmosphere, though its composition is uncertain. Sleep et al. (2014) proposed that the initial atmosphere, formed after Moon-forming impact, comprised hydrogen, helium, water vapor, methane, and ammonia, while a secondary atmosphere was rich in CO_2, molecular nitrogen, and water vapor. The steamy atmosphere with approximately 100 bars of CO_2 led to the surface temperature approaching 100 °C. Only when atmospheric CO_2 decreased to about 25 bars and surface temperatures reached <122 °C could significant, life-producing, prebiotic chemistry initiate (Sleep et al. 2014; Zahnle et al. 2015). On the contrary, in an earlier publication, Sleep and Zahnle (2001) imagined that degassing of CO_2 would have been vigorous but so would have been the return of carbon to the interior, and it is likely that the mantle cycle would have dominated. Moreover, frequent meteorite impacts would have created vast quantities of basalt ejecta that would also have reacted with CO_2 to precipitate carbonate. Therefore, they concluded that so much CO_2 would have been held in the mantle that the greenhouse warming would have been small: the Earth was probably heavily glaciated; the Hadean was probably a Norse ice-hell.

The ocean may have appeared very early. As the accretionary energy input waned, surface temperatures dropped to below 100 °C and water condensed to form the primordial oceans. However, during the Hadean, the accretionary impact by planetesimals took place episodically. Therefore, the early atmospheres and oceans may have been reprocessed several times by episodic bombardments, with lost volatiles continuously replenished by mantle outgassing and addition of volatile-rich meteoritic and cometary material. It was suggested that the Earth may have acquired as much as 10^{21} L of water between 4.5 and 3.8 Ga from cometary collisions alone, essentially equivalent to all the water in modern oceans (Chyba 1987). The early seawater was supposed to be enriched in transition metals, such as soluble Fe^{2+}, Ni^{2+}, Co^{2+}, and Mn^{2+}, the salinity was estimated as either being similar to that of the present-day Earth or higher, and the estimates of the seawater temperatures range from a lower value of 26 °C up to between 50 and 70 °C (see Westall and Xiao 2024 for a review).

8.2.2 Archean (~4.031 to 2.5 Ga)

The Archean is the second of the four geological eons of Earth's history, subdivided into four eras including Eoarchean, Paleoarchean, Mesoarchean, and Neoarchean. Its lower boundary, automatically coincident with the upper boundary of the Hadean and lower boundary of the Eoarchean, was officially defined as 4.031 ± 0.003 Ga by the ICS, which is the age of the oldest known rock on Earth, i.e. Acasta Gneiss Complex Formation, Slave Province, Canadian Shield. Geological material from the preceding Hadean is therefore restricted by definition to either non-rock or non-terrestrial sources such as individual zircon grains and lunar samples. The Archean upper boundary, the base of the Proterozoic, is defined chronometrically as 2.5 Ga by the Global Standard Stratigraphic Age (GSSA) but it will be replaced by a Global Boundary Stratotype Section and Point (GSSP). In comparison with the Hadean, geological records of the Archean are substantial in number, widely distributed in most continents (Figure 8.3), and informative for deciphering the origin and evolution of the Earth–life system.

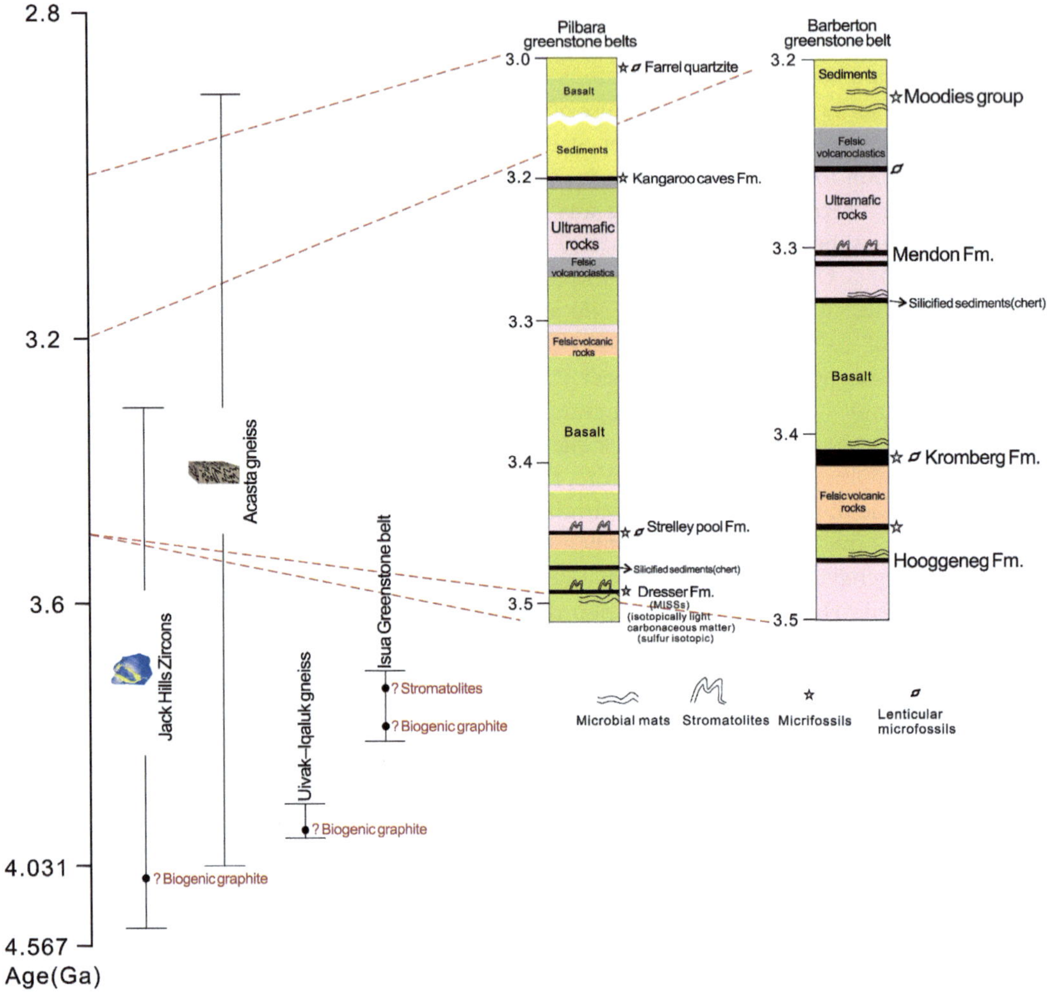

Figure 8.3 Earliest records of life on Earth. Simplified stratigraphy and reported traces of early life from the Barberton Supergroup of South Africa in comparison with the Pilbara Supergroup of Western Australia.

8.2.2.1 Geological Record of Eoarchean (~4.031–3.6 Ga)

The Eoarchean is characterized by frequent LHBs, the formation of the planet's first solid crust, and the emergence of life. Some of the oldest preserved rock formations date back to this era, including the Isua Greenstone Belt in Greenland, the Acasta Gneiss in Canada, and the Nuvvuagittuq Greenstone Belt in Quebec.

Acasta gneiss complex (AGC): It is a ~2400 km^2 terrane that contains primary magmatic zircons in metaigneous host rocks, allowing for dating the oldest known zircon-bearing rocks on Earth, up to 4031 ± 3 Ma in age (Reimink et al. 2019). The AGC is a highly deformed, metamorphically recrystallized, and heterogeneous rock assemblage of foliated to gneissic tonalites, trondhjemites, granodiorites, and granites, which contains minor quartz-diorites, diorites, gabbros, and ultramafic rocks. This rock assemblage primarily formed during discrete pulses of magmatic activity from ~4.031 to 2.9 Ga and constitutes the northwestern edge of the Slave Craton in the

Northwest Territories, Canada. There are various such terrains worldwide. Examples include the Nain Province in Labrador (containing the oldest metasedimentary rocks, Uivak–Iqaluk Gneiss, ~3.9 Ga), Canada; the Napier complex in Antarctica (up to ~3.7 Ga); and the Narryer complex, Australia (host rock up to ~3.7 Ga, hosting the older zircons). Some of these terrains are up to several thousand square kilometers, though the datable older rocks may only be a small proportion of the whole. The implication is that massifs of continental crust at least up to the size of Luxembourg or Rhode Island existed in the Eoarchean.

The Isua Greenstone Belt (IGB): It is an Eoarchean greenstone belt in southwestern Greenland, aged between 3.7 and 3.8 Ga. It comprises two major sequences of metamorphosed mafic volcanic and sedimentary rocks, the southern terrane approximately 3.8 Ga and the northern terrane approximately 3.7 Ga. The belt contains the oldest known sedimentary rocks being 3.8 Ga old, which consists of volcanoclastic material, clastic turbidites, and chemical precipitates such as banded iron formations (BIFs) and chert. This informative old sequence is reminiscent of material deposited today in volcanic island arcs. Many profound implications come from this rock unit. There was clearly an ocean present, and land masses as well (at least volcanoes, possibly other older crust); erosion occurred and sediments were deposited; volcanic eruptions must have been normal features of the geological setting. There is also good evidence for a meteoritic component in sediments (Schoenberg et al. 2002). Therefore, at Eoarchean, the Earth was still under heavy bombardment by meteorites. Volcanoes and hydrothermal systems were well-being because many primary features were preserved in this greenstone belt, including mafic lavas with fine-grained cooling rims, and quartz globules in pillow breccias (Appel et al. 2001). These globules are interpreted as former gas vesicles, infilled with quartz from hydrothermal veins that formed during and immediately after volcanism. These quartz infillings contain rare fluid inclusions, remnants of two independent fluid/mineral systems, comprising pure methane and highly saline (25% NaCl) aqueous fluids, and co-precipitating calcite. These fluids strongly resemble modern sea-floor hydrothermal fluids. Appel et al. (2001) concluded that methane-brine hydrothermal systems operated in the Eoarchean. More importantly, the emitted methane was useful for metabolism.

Evidence for Eoarchean life: Because of its age, the IGB has long been the focus of studies seeking to identify signs of early terrestrial life. Mojzsis et al. (1996) reported isotopic data of graphite inclusion from the Akilia association (tentatively dated at 3.85 Ga), $\delta^{13}C$ values ranging from −21‰ to −49‰. The data was, therefore, cited as evidence for very early life. However, Fedo and Whitehouse (2002) showed that the rock studied by Mojzsis et al. (1996) was not sedimentary but an ultramafic igneous rock. They further considered that the isotopic ratios of the carbon particles at Akilia were indicative of high-temperature metamorphic processes rather than evidence of life, resulting in the formation of abiotic hydrocarbons. At the second Greenland location, carbon microparticles (2–5 μm graphite globules) from turbiditic sediments and pelagic mud of the Isua Formation are also strongly depleted in ^{13}C, with $\delta^{13}C$ value around −19‰ (Rosing 1999). Apart from this possible biogenic graphite, Hassenkam et al. (2017) described very tiny inclusions of lightly metamorphosed organic material inside garnets from 3.7-billion-year-old rocks in Greenland. The combined analyses of atomic force microscopy and infrared spectroscopy demonstrated occurrences of various chemical bonds including carbonyl, anhydride, phosphate, and nitrile, which were interpreted as biogenic carbon relics. Additionally, Nutman et al. (2016) described stromatolites from ~3.7 Ga metamorphosed rocks of the IGB in Greenland. However, the interpretation of the Isua stromatolites (Nutman et al. 2016) has elicited much debate and they were later reinterpreted as abiotic structures (see Westall and Xiao 2024). Outside the IGB, Eoarchean graphite micro-granules have been discovered. Tashiro et al. (2017) reported the total organic carbon contents and carbon isotope values of graphite and carbonate in the oldest metasedimentary rocks, known as Uivak-Iqaluk Gneiss, in the Saglek Block of northern Labrador, Canada. The significant fractionation between the $\delta^{13}C_{carb}$ and $\delta^{13}C_{org}$ values, up to 25‰, was interpreted as evidence for life existing more than 3.95 billion years ago. Nevertheless, the abiotic formation of organic matter and graphite with a light carbon isotope composition during metamorphism challenges the biogenicity of these earliest graphite particles (Guo et al. 2024). In addition, hematite filaments and tubes were described from the >3.75 Ga jasper-carbonate BIFs in the Nuvvuagittuq Craton of Canada, and interpreted as possible life forms around hydrothermal vents (Dodd et al. 2017; Papineau et al. 2022), while both their biogenicity and age remain controversial (Westall and Xiao 2024).

8.2.2.2 Geological Record of Paleoarchean (3.6–3.2 Ga)

Rocks of relatively low metamorphic grade from the Paleoarchean have been found in several places on Earth, particularly the Pilbara Craton in Western Australia and the Kaapvaal Craton in Africa, where multiple lines of evidence indicate that a diverse range of prokaryotic life forms existed during the Paleoarchean (Figure 8.3).

Pilbara craton in Western Australia: The Pilbara Craton is subdivided into three principal components: (1) The East Pilbara Granite–Greenstone Terrane comprising 3655–2850 Ma domical, multicomponent granitic complexes and ~3515–3000 Ma greenstones of the Pilbara Supergroup; (2) The West Pilbara Granite–Greenstone Terrane is composed of two temporally and compositionally distinct greenstone belts, ~3270 Ma Roebourne Group and ~3120 Ma Whundo Group; (3) The Kurrana Granite–Greenstone Terrane consists of ~3190 Ma orthogneisses with enclaves of older greenstones (Van Kranendonk 2006). The ~3515–3320 Ma Warrawoona Group of the Pilbara Supergroup in the Pilbara Craton is composed of a wide range of rock types, including both lavas and sedimentary rocks, which have received significant geobiological attention for exploring the earliest life on Earth. Among formations of this group, the ~3.5 Ga Dresser Formation is composed of between one and four units of bedded "chert±barite±carbonate" associations interbedded with pillowed mafic volcanic rocks and contains an abundance of well-preserved evidence for the existence of microbial life, including the microbially induced sedimentary structures (MISSs) of phototrophic microbial mats (Noffke et al. 2013), stromatolites of both photoautotrophs and anaerobic chemoautotrophs (Van Kranendonk 2006), isotopically light carbonaceous matter (Figure 8.4) with aromatic and aliphatic components (Weimann et al. 2024), and sulfur isotopic evidence for microbial sulfate reduction (Shen et al. 2001). Weimann et al. (2024) proposed that the black bedded barite of the Dresser Formation was deposited in a hydrothermal environment. The report of pyritic filamentous microfossils from a 3.235 Ga volcanogenic massive sulfide deposit also implies the presence of hyperthermophile microbial life during Paleoarchean (Rasmussen 2000). Numerous bedded chert units are interbedded with volcanic rocks of the Warrawoona Group. Putative microfossils and biological markers have been reported from six of these chert units (Figure 8.3), e.g. biomarker from the ~3.5 Ga Dresser Formation (Derenne et al. 2008), Awramik microfossil locality (Awramik et al. 1983), and Apex chert microfossil locality (Schopf 1993). Carbonaceous filaments from the Dresser Formation cross-cutting chert veins have a $\delta^{13}C$ of −56‰, suggestive of microbial methanogenesis (Ueno et al. 2006). However, the biogenicity of putative microfossils from these chert veins in the Warrawoona Group remains controversial and requires further detailed study because such structures have only been observed from hydrothermal veins and similar in both morphology and physicochemical setting to the abiogenic structures (Brasier et al. 2002; Van Kranendonk 2006).

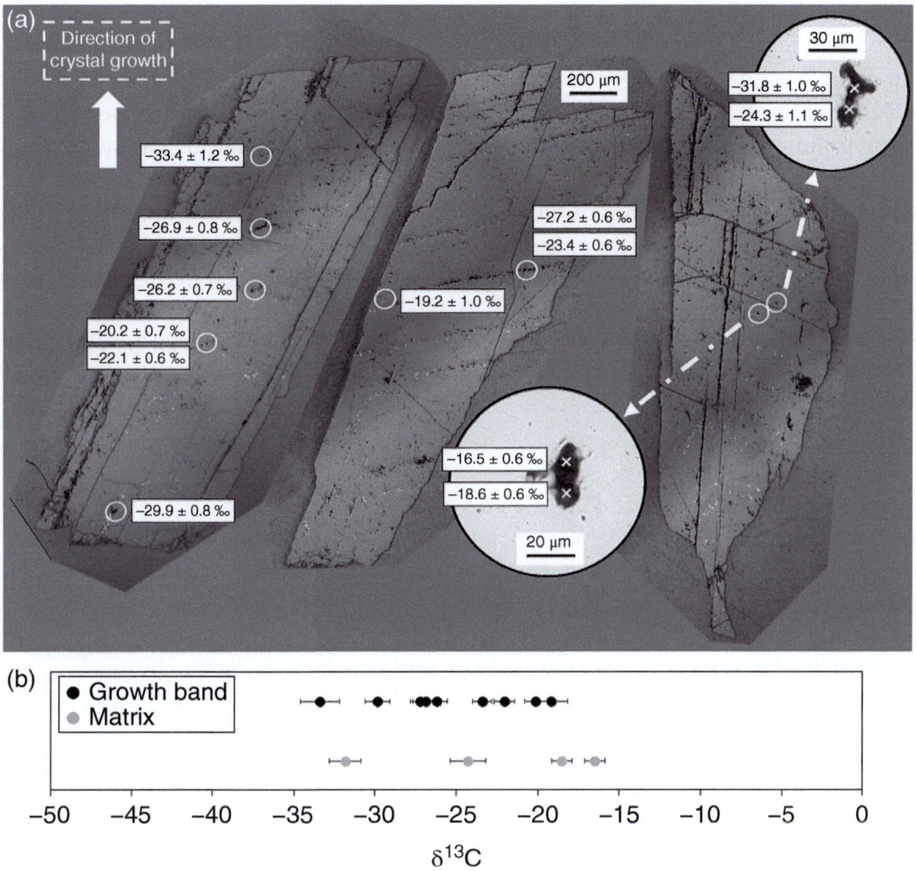

Figure 8.4 Micro-scale $\delta^{13}C$ measurements of solid carbonaceous matter (CM) preserved in barites of the ~3.5 Ga Dresser Formation via Secondary ion mass spectrometry (SIMS) showing $\delta^{13}C$ values ranged from −33.4 ± 1.2‰ to −16.5 ± 0.6‰ in accordance with a biogenic origin. (a) SEM image showing barite fragments, including CM inclusions along growth bands or within the barite matrix. (b) Summary of $\delta^{13}C$ values from CM along growth bands (n = 9) and inside the barite matrix (n = 4). *Source:* Contesy of Lena Weimann.

Barberton greenstone belt in the Kaapvaal Craton of South Africa and Swaziland: The Barberton Supergroup (the former Swaziland Supergroup) is a volcanic-sedimentary succession approximately 15 km thick and subdivided into three stratigraphic units, i.e. Onverwacht (~3.55–3.26 Ga), Fig Tree (~3.25 Ga), and Moodies (~3.22 Ga) groups in ascending order. Geological and geochemical evidence from the Onverwacht group suggests that hydrothermal systems were active in both deep oceans and shallow seas, discharging hot vents into cold waters, and local glacial conditions may have existed around 3.5 Ga. The whole succession, ranging from 3.55 to 3.20 Ga, hosts a variety of macro- and microscopic biological remains, as well as geochemical evidence for early microbial life (Homann 2019). The biogenicity of stromatolites and laminated silica crusts from the upper Onverwacht Group and Fig Tree Group is elusive for lacking evidence of microfossils. While microbial mats and biofilms occurring as MISSs, and filamentous, spheroidal, and lenticular microfossils have been described from a number of horizons of the Onverwacht Group, widespread from the bottom 3.472 Ga Hoogenoeg Formation to the top 3.26 Ga Mendon Formation. The microbial mats and microfossils from the Buck Reef

Chert (3.416 Ga) of the Kromberg Formation represent a particularly widespread and well-documented record of the Paleoarchean life (Homann 2019). The ca. 3.22 Ga Moodies Group consists of quartz-rich sandstones with subordinate conglomerates, mudstones, siltstones, thin volcanic tuff beds, minor BIFs, and basaltic lava. These rocks were deposited in marine and terrestrial paleoenvironments, representing the oldest well-preserved alluvial to shallow-marine tidal deposits. Large spheroidal microfossils, shallow-marine tufted microbial mats, and remnants of cavity-dwelling microbial communities have been reported from this group. Terrestrial microbial mats interbedded with fluvial sandstones and conglomerates show a significant difference in organic carbon and nitrogen isotopes compared to those of the tidal marine mats, and were thus interpreted as the oldest macroscopically visible traces of life on land (Homann et al. 2018). The presence of a Paleoarchean terrestrial biosphere is further supported by paleosols reported from the Moodies Group, which carry signals of biogenic sulfur isotope fractionation (Nabhan et al. 2016).

8.2.2.3 Geological Record of Mesoarchean (3.2–2.8 Ga)

During the Mesoarchean era, modern-style plate tectonics may have begun and the Earth experienced the expansion of microbial life and first glaciation. Geological records with evidence for life are preserved in many crotons. The following examples are from three classic areas.

Pongola Supergroup of the Kaapvaal Craton in South Africa and Swaziland: This supergroup (2.99 to 2.87 Ga) is an exceptionally well-preserved succession of volcanic and sedimentary rocks deposited in a volcano-sedimentary continental or epicontinental setting and comprises two groups. The lower Nsuze Group is dominated by volcanic rocks ranging from basalt to basalt-andesite through intermediate to rhyolite compositions, and the upper Mozaan Group is dominated by sedimentary units (Luskin et al. 2019). The diamictite of the Mozaan group (Figure 8.5) in KwaZulu-Natal and Eswatini was interpreted as the oldest known glacial deposits on the planet, known as the Pongola glaciation, which was constrained within the interval 2985 ± 1 and 2837 ± 5 Ma in age. In the Nsuze Group (2.99 Ga), a wide variety of microbial buildups were

Figure 8.5 Diamictite of the ~2.9 Ga of the Mozaan Group cropping out in Vryheid, KwaZulu-Natal of South Africa, interpreted as glacial deposits during the Pongola glaciation. *Source:* Photo by Xingliang Zhang.

found in the carbonate facies, including stromatolites, ooids, and oncoids, while MISSs occur in siliciclastic facies. The Mozaan Group sandstones contain wrinkle structures, which in thin section reveal filamentous textures forming carpet-like microbial mat fabrics. The textures resemble the trichomes of some modern cyanobacteria. Results of mineralogical, geochemical, and isotopic analyses are consistent with a biological origin of the filament-like texture. Carbonaceous filaments with biogenic isotopic signatures ($\delta^{13}C = -24.2‰$) are closely associated with hematite, goethite, and chert minerals, which may derive from the former presence of oxygen within the microbial mats (Noffke et al. 2003). At the Nhlazatse section, more diverse and complex MISSs have been found, including erosional remnants and pockets, multidirected ripple marks, polygonal oscillation cracks, and gas domes (Noffke et al. 2008).

Pilbara Craton in Western Australia: A convergent margin with a modern-style oceanic arc may have existed at the boundary between West and East Pilbara approximately 3.12 Ga. The West Pilbara Terrane converged with and accreted onto the East Pilbara Terrane by 2.97 Ga. Therefore, it is speculated that modern-style plate subduction occurred during the Mesoarchean. The 3.0 Ga Farrel Quartzite hosts abundant carbonaceous acritarchs demonstrating a significant expansion of microbial life in diversity and complexity. There is also evidence for the presence of methanotrophs in the ~2.8 Ga old Mount Roe paleosol. The paleosol contains highly isotopically fractionated organic carbon, probably recording the activity of methanotrophs living near ephemeral ponds, which implies that significant biological methane sources existed in Mesoarchean (Rye and Holland 2000).

Carbonate Platform in Northwest Ontario of Canada: Earth's oldest thick, marine carbonate platform facies are well-preserved in the Mesoarchean sequence of the North American Craton. The 2930 Ma Red Lake–Wallace Lake Carbonate Platform in the Western Superior Province of Canada contains laterally linked and isolated domal stromatolites in the peritidal zone as well as large low relief domes in deeper water regions (McIntyre and Fralick 2017). The high Mn concentration (mean 1.3 wt%) and positive Ce anomalies in the iron formation directly below the carbonate platform as well as the enrichment of redox-sensitive elements in the chemical sedimentary rocks indicate the presence of some O_2 in the depositional sites ~2.93 billion years ago (Afroz et al. 2023). The ~2.8 Ga Mosher Carbonate Formation of the Steep Rock Group at Steep Rock Lake in south central Canada was thought to be Earth's oldest thick (up to 500 m), well-preserved marine carbonate platform facies and contains a variety of stromatolites, ranging from small domes and finger-like columns, to giant domes 5 m or more across (Fralick and Riding 2015). Isotopic evidence from the Steep Rock Group indicates that Rubisco-mediated carbon capture (i.e. oxygenic photosynthesis) controlled the global carbon partitioning between carbon dioxide and carbonate: this is some of the oldest evidence for global oxygenic photosynthesis (Nisbet and Fowler 2005). These very large stromatolitic domes together with negative Ce anomalies, positive Gd anomalies, and reduced positive Eu indicate significant phototrophic biomass development and oxygen production.

8.2.2.4 Geological Record of Neoarchean (2.8–2.5 Ga)

This era is marked by the formation and stabilization of large continental landmasses. The Kaapvaal and Pilbara cratons may have united into a supercontinent known as Vaalbara. Apart from the classic Kaapvaal and Pilbara cratons, where carbonate platform facies were well developed, notably large domical stromatolites up to tens of meters in height and diameter, sedimentary rocks are also substantially exposed in other regions, e.g. North China Craton. Geochemical evidence of Neoarchean sediments indicates a range of autotrophic carbon fixation mechanisms and heterotrophic processes including fermentation and methanotroph, as well as the emergence of oxygenic ecosystems around this time.

Kaapvaal Craton: The Ngesi Group of the Belingwe Greenstone Belt, Zimbabwe, is composed of Manjeri, Reliance, Zeederbergs, and Cheshire formations in ascending order. The Manjeri (2.7 Ga) and Cheshire (2.65 Ga) formations have a wide and diverse array of evidence for Neoarchean life, including domical, columnar, and tabular stromatolites. The base of the Group includes shallow-water facies locally rich in carbon and sulfur that is highly fractionated isotopically, suggesting the original presence of methanogens, as well as the operation of complex sulfur fractionating processes. Close to the contact with the lavas, the uppermost sediments of the Manjeri Formation are very rich in kerogen and sulfides, with highly variable fractionated carbon and sulfur isotopes. The complex isotopic fractionation implies a consortium of prokaryotes, some reducing sulfate and some perhaps oxidizing sulfur, others generating methane, some photosynthesizing and capturing carbon by using Rubisco, and perhaps carrying out other microbial biochemistry (using metal enzymes). The limestones of the Cheshire Formation also have extensive and very well-preserved stromatolites, showing evidence for Rubisco fractionation of kerogen and carbonate carbon (Nisbet and Fowler 2005). In South Africa, the Ventersdorp Supergroup, an undeformed, sub-greenschist to greenschist facies of volcano-sedimentary succession, deposited on the Kaapvaal Craton between 2729 and 2665 Ma and hosts small lobate, domical, columnar, and linked pseudo-columnar stromatolites formed during periods of clastic input quiescence in a high-energy environment.

Pilbara Croton: In the Neoarchean of Western Australia, there is much more evidence for life. For example, 2.7 Ga stromatolites from the Tumbiana Formation and the Fortescue Group have diverse morphology and formed in ephemeral intercontinental basin lakes, demonstrating the gradual extension of life inland, away from continental margins. The 2.7-billion-year-old shales in the Pilbara Craton of Australia contain biomolecules thought to be evidence of eukaryotes and photosynthetic cyanobacteria, e.g. 2-methylhopanes (Summons et al. 1999). More recent work by Rasmussen et al. (2008) suggested that the organic biomarkers are not of Archean age and must have entered the rocks later, after peak metamorphism approximately 2.2 Ga. For a long time, 2-methylhopanes were considered definitive evidence of cyanobacteria. However, recent studies have shown that these lipids can also be produced by Alphaproteobacteria, which has led to some debate about their exclusivity as cyanobacterial biomarkers.

North China Craton: Neoarchean BIFs crop out in Anshan-Benxi, Wutai, and Jidong areas of North China. The metasedimentary rocks associated with the BIFs were deposited in arc or back-arc basins on an ancient continental margin during 2.6 to 2.5 Ga. A craton-wide geochemical study suggested an anoxic ocean–atmosphere system during the deposition of North China BIFs (Wang et al. 2024). Sulfur isotopic analyses on the diagenetic pyrite minerals from the BIFs show $\delta^{34}S$ varying in a large range from −22.0‰ to +11.8‰, which indicates that the biological sulfate reduction contributes to the precipitation of the pyrite, and the sulfate concentration in the oceans, at least locally, is above 1 mmol/L (Hou et al. 2007).

8.2.2.5 Archean Environment—Habitable for Prokaryotes

Solar luminosity would have been significantly lower than today but would have gradually increased over time. Meanwhile, solar UV radiation would have remained at a high level detrimental to eukaryotes, albeit slowly decreasing. The Earth rotated faster and was much closer to the moon as indicated by Milankovitch cycles recorded in ancient sediments. The Earth's precession cycle had a significantly higher frequency than it does today, signaling shorter day lengths and a closer Earth–Moon distance. It was estimated that the astronomical precession frequency was 108.6 ± 8.5 arcsec/yr, corresponding to an Earth–Moon distance of $321{,}800 \pm 6500$ km and a day length of 16.9 ± 0.2 hours at 2.46 Ga (Lantink et al. 2022). The LHB persisted into the early Eoarchean, potentially triggering interactions between solids and fluids, increasing reaction rate by shock waves and heat, and facilitating the transport of essential organic molecules necessary for the origin of life on Earth. Geodynamics transitioned from the Hadean's stagnant lid to horizontal tectonics, with modern-style plate tectonics potentially emerging in the Neoarchean. The continental crust, initially formed during the Hadean, experienced significant growth throughout the Archean.

The geological record suggests the presence of rain, subaerial exposure, weathering, erosion, and deposition, as well as extensive oceans capable of evaporating rainwater since the Eoarchean. However, the atmospheric composition remains controversial, with a notable absence of free oxygen throughout the Archean. Atmospheric CO_2 and CH_4 levels ranged ~10–2500 and 10^2–10^4 PAL, respectively. The greenhouse gas concentrations were sufficient to offset a fainter Sun (Catling and Zahnle 2020). Oceans were generally anoxic and unable to support eukaryotes, though transient oxic seawater may have existed locally due to abiotic processes generating small amounts of oxygen, such as UV radiolysis of water and reactions between water and surface-bound radicals on abraded quartz surfaces (He et al. 2021). Volcanic activity was more violent than today, emitting gases possibly including CO_2, H_2S, SO_2, and N_2. Vigorous and abundant subseafloor hydrothermal systems leached crust and exhaled ions into the oceans. Before the Great Oxidation Event (GOE) at ~2.3 Ga, local whiffs of oxygen or oxygen oases may have existed since the origin of oxygenic photosynthesis in the Mesoarchean (Sánchez-Baracaldo et al. 2022). Surface temperatures are poorly constrained, with estimates ranging from moderately hot (40 °C) to very hot (55–80 °C), within the temperature limits of thermophiles, despite regional glaciation in the Archean. Overall, the environmental conditions during the Archean were capable of hosting prokaryotes with diverse metabolisms, as evidenced by the geological record discussed above.

8.3 Evolution of Atmosphere

The exact composition of the earliest atmosphere remains uncertain, but it was vastly different from today's atmosphere. It likely originated from Earth's interior as the planet experienced melting and segregation into a core, mantle, and crust. During this process, volatile materials were released, probably containing significant amounts of nitrogen gas, considerable carbon dioxide, some methane and ammonia, and almost no molecular oxygen.

So, where did the oxygen in our current atmosphere come from? There are two primary natural processes that produce oxygen gas: (1) abiotic processes, such as the breakdown of water into hydrogen gas and oxygen gas by ultraviolet (UV) radiation; (2) oxygenic photosynthesis, which uses sunlight to convert carbon dioxide and water into carbohydrates, releasing free oxygen gas as a by-product. The latter is much more efficient at producing oxygen. According to one study, oxygenic photosynthesis generates an estimated 357 billion tons of oxygen annually (Borisov and Björn 2018). Many scientists believe that the high oxygen levels in our atmosphere today are largely due to photosynthetic organisms releasing O_2 over the past three billion years. As atmospheric oxygen levels increased, organisms evolved to utilize oxygen in an efficient form of metabolism known as aerobic respiration. These changes ultimately enabled the evolution of complex, multicellular life forms.

In this section, we start with a short review on the earliest atmosphere and the faint young sun paradox (FYS), and then we discuss the evolution of O_2 and CO_2 in atmosphere through geological time, which are two fundamental gases to life and climate.

8.3.1 The Prebiotic Atmosphere

The Earth was formed by accretion of solid material that was condensed from the solar nebula. Any primary captured atmosphere must have been lost. The present atmosphere is secondary, which was generated from volatile compounds contained within the solid planetesimals. There are various theories about the composition of the primitive atmosphere from which the present atmosphere evolved.

In the early 20th century, scientists theorized that the primordial atmosphere was primarily composed of carbon dioxide, methane (commonly known as natural gas), ammonia, and hydrogen gas. This perspective on atmospheric composition was essential to support various hypotheses of life origin. However, in the 1950s, the understanding of the primordial atmosphere began to shift, focusing on the origins of the gases. Geologists increasingly recognized that the atmospheric gases originated from within Earth itself, primarily through volcanic activity. Geochemists determined that nearly the entire atmosphere had formed by the time Earth was just one million years old, consisting mostly of carbon dioxide and nitrogen, with very small amounts of methane and ammonia, and almost no free hydrogen gas. This composition was markedly different from previous assumptions. In the 1960s, it was believed that Earth formed relatively slowly, with a cold interior, and that most of its volatiles were initially trapped inside the planet. Over time, the Earth's interior was heated by radioactive decay, gradually releasing these trapped gases through volcanic outgassing. Initially, these volcanic gases were highly reduced, containing H_2, CH_4, and NH_3. However, after the Earth's core formed, the composition of volcanic gases became similar to modern volcanic emissions, consisting of H_2O, CO_2, and N_2, with traces of H_2 and CO.

More recent models of planetary accretion suggest that Earth formed over a period of 10–100 million years, with its interior initially hot due to LHBs, including one that may have led to the formation of the moon. The Earth's core likely formed simultaneously with accretion, allowing metallic iron to be removed from the upper mantle. Consequently, volcanic gases could have been relatively oxidized as early as 4.5 billion years ago.

Moreover, many of Earth's volatiles were probably released on impact, and this process may have formed a transient steam atmosphere during at least part of the accretionary period. Hydrodynamic escape became difficult in the post-accretionary period (after ~4.5 Ga) because the solar extreme UV flux would have been lower and the energy available to fuel the escape process would have been reduced. Furthermore, the escape rate would have become limited by diffusion once H_2 became a minor component of the atmosphere. H_2 would have been a major component of an accretionary, steam atmosphere as a consequence of reduction of water by infalling, metallic iron-rich planetesimals. Once the main accretionary phase ended about 3.8 Ga ago, the surface heat flux would have dwindled, and the steam atmosphere would have rained out to form an ocean. The remaining atmosphere would probably have been dominated by carbon and nitrogen compounds, primarily CO_2, CO, and N_2. Kasting (1993) estimated that the early atmosphere contained around 10 bars of $CO_2 + CO$ and approximately 1 bar of N_2 during the first several hundred million years of Earth's history. Climate models suggest that the mean surface temperature of such an atmosphere would have been around 85 °C. Impacts of LHB could have also influenced atmospheric composition by introducing sources of CO and NO. CO would have been produced by the oxidation of organic carbon in carbonaceous impactors or by the reduction of atmospheric CO_2 by iron-rich impactors. NO would have been generated by the shock heating of atmospheric CO_2 and N_2. Consequently, CO could have potentially reached high concentrations in the atmosphere (Kasting 1993).

Clearly, there was no consensus on the composition of the primitive atmosphere, with opinions ranging from strongly reducing to relatively oxidized. There is no geological evidence to definitively support either view, although it is generally accepted that O_2 was absent. While it is beyond our knowledge to delve deeply into this question, it is worth noting that atmospheric chemists mostly favor a composition high in CO_2 and N_2, whereas prebiotic chemists tend to favor more reducing conditions, which are necessary for the efficient synthesis of amino acids, purines, pyrimidines, and sugars.

8.3.2 The Atmosphere at the Time of Life Origin

Life likely existed by the earliest Eoarchean, as suggested by isotopic data from potentially biogenic graphite particles (see Section 8.2). Before the LHB, large impacts would have repeatedly vaporized the uppermost layers of the ocean. Impact events with objects exceeding 440 km in diameter could have vaporized the entire ocean, making such events fatal to any life form. Consequently, although life could have originated multiple times during and before the LHB, it probably did not survive until the end of the LHB at 3.8 Ga (Kasting 1993).

According to theories on the origin of life, an atmosphere composed of CH_4, NH_3, H_2, and H_2O is preferable. Experiments, such as the Miller-Urey experiment, demonstrated that biologically important organic compounds, including sugars and amino acids, could be formed by spark discharge (simulating lightning) in such an atmosphere. However, CH_4 and NH_3 may not have been present in early Earth's atmosphere. Their presence in the primitive atmosphere likely depended on the oxidation state of the upper mantle at that time, which is challenging to determine. Without a volcanic source of CH_4 and NH_3, the post-heavy bombardment atmosphere was probably dominated by CO_2 and N_2, with traces of CO, H_2, and reduced sulfur gases, resulting in a weakly reduced atmosphere. In an atmosphere lacking CH_4 and NH_3, formaldehyde (H_2CO), which is necessary for the synthesis of sugars, could not be synthesized and thus should have been not readily available. However, the formation of hydrogen cyanide (HCN), essential for the synthesis of amino acids and nucleotides, is much more challenging because it requires breaking both an $N{\equiv}N$ and a $C{\equiv}O$ triple bond (if starting from N_2 and CO_2). Explaining the formation of HCN remains one of the major challenges for theories of life origin that rely on the atmosphere as the source of starting materials. Consequently, the source of biological precursor molecules may not lie in the primitive atmosphere. Indeed, alternative theories suggest that biologically important organic compounds were delivered by impacting comets or other planetesimals, or were synthesized within submarine hydrothermal vents (See Section 8.5.1).

8.3.3 Atmospheric Greenhouse Gases and Long-term Climate Change

Certain gases in the atmosphere, such as water vapor, carbon dioxide, and methane, act like a trap by absorbing and reflecting infrared radiation emitted by Earth. These gases, known as greenhouse gases, conserve heat in a manner similar to the glass of a greenhouse. The concentration of greenhouse gases is crucial in regulating Earth's climate and surface temperature. Without the natural greenhouse effect, Earth's surface would be a frozen mass, with average global temperatures around −18 °C, much less than the 20th-century average of about 15 °C. Conversely, higher levels of greenhouse gases can lead to catastrophic global warming.

Among the various greenhouse gases (water vapor, CH_4, CO_2, O_3, NO_x, and chlorofluorocarbons), CO_2 and CH_4 are naturally occurring gases that may have been present on Earth's surface since its formation. Currently, CH_4 is not found in concentrations anywhere near those of CO_2 in the atmosphere, but it is a powerful greenhouse gas. In the first few days after emitting, CH_4 is approximately 80 times more powerful than CO_2 in trapping heat, but it is removed from the atmosphere much more quickly—after about a decade, whereas CO_2 remains in the atmosphere for centuries. In the distant past, before the rise of O_2, CH_4 levels could have been significantly higher. Therefore, it is believed that variations in atmospheric CH_4 concentrations were just as important as variations in CO_2 in determining past surface temperatures and regulating climate stability (Kasting 2005).

8.3.3.1 The Faint Young Sun Paradox

The fundamental issue of climate during the Precambrian is often referred to as the "faint young Sun (FYS) paradox." Standard models of solar luminosity predict that the Sun was about 30% less luminous when it first formed around 4.567 billion years ago and has increased in luminosity steadily since then. A useful formula to describe this is: $S = S_0/(1 + 0.4\,t/t_0)$, where S_0 represents the current solar luminosity ($\sim$1370 W/m^2), $t_0 = 4.6$, and t is the time in billions of years before the present. This increase in luminosity results from the gradual conversion of hydrogen into helium in the Sun's core, which increases the core density, causing it to contract and heat up, thereby accelerating nuclear fusion reactions.

The reduced solar luminosity implies a colder climate in the past. If the early atmosphere had the same composition as today's atmosphere, Earth's mean surface temperature would have been below the freezing point of water before approximately 2.0 Ga (Figure 8.6). This would likely have resulted in constant global glaciation for more than two billion years. However, evidence from sedimentary rocks and the evolution of life since the Eoarchean indicates that liquid ocean teeming with life existed on Earth's surface at least since 3.8 Ga. This issue is known as the FYS paradox (Feulner 2012). There are speculations on some astrophysical solutions to the FYS paradox. For instance, various astronomers have suggested that the Sun could have been more massive when it first formed and lost its mass by way of a greatly enhanced solar wind, while others have suggested

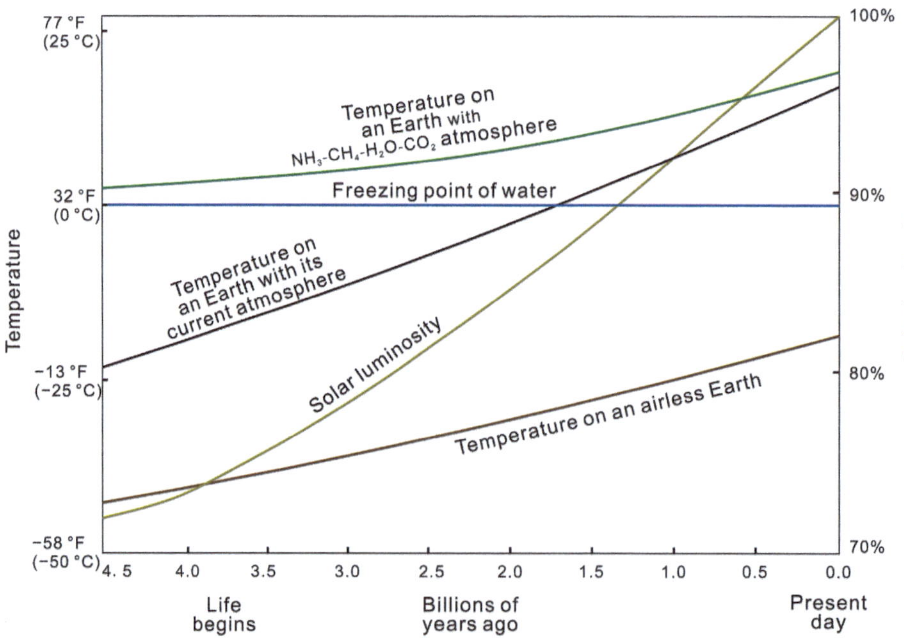

Figure 8.6 Evolution of average surface temperature as a function of increasing solar luminosity through Earth history. *Source:* Modified from Sagan and Mullen (1972) and Feulner (2012).

that a lower Earth's albedo in the distant past could have warmed the climate to some extent. However, Kasting (2005) argued both astrophysical solutions were unlikely and restated that greenhouse warming is required in any tenable solution to the FYS paradox, with CO_2 and CH_4 being the most likely candidates of greenhouse gases in the early atmosphere.

8.3.3.2 Atmospheric CO_2 Concentrations in the Precambrian

Multiple lines of evidence, including paleontological, sedimentological, and isotopic records, indicate that Earth has maintained surface pressures and temperatures conducive to the presence of liquid water and oceans, thereby supporting a global habitat for life since the Eoarchean. Models of solar luminosity suggest that without a much greater greenhouse effect early in Earth's history, global glaciation would have dominated throughout the Archean. Therefore, higher concentrations of greenhouse gases, primarily CO_2 (but also water vapor and methane), would have been necessary to maintain normal surface temperatures to overcome the FYS paradox in early Earth history. However, atmospheric CO_2 concentration in the Precambrian is poorly constrained. Several constraints are based on geochemical analyses of paleosols and organic microfossils, or inferred from micrometeorite oxidation, whilst the secular trend of atmospheric CO_2 concentration is derived from climate models.

Figure 8.7 illustrates the atmospheric CO_2 concentrations required to offset the reduced solar luminosity in the past, based on a climate model that assumes no changes in surface or cloud albedo. This calculation presumes that CO_2 and water vapor were the only significant greenhouse gases and that variations in cloudiness can be disregarded. The 10-bar upper limit at 4.5 Ga is derived from Walker's model of an ocean-covered Earth. CO_2 levels during the glacial periods

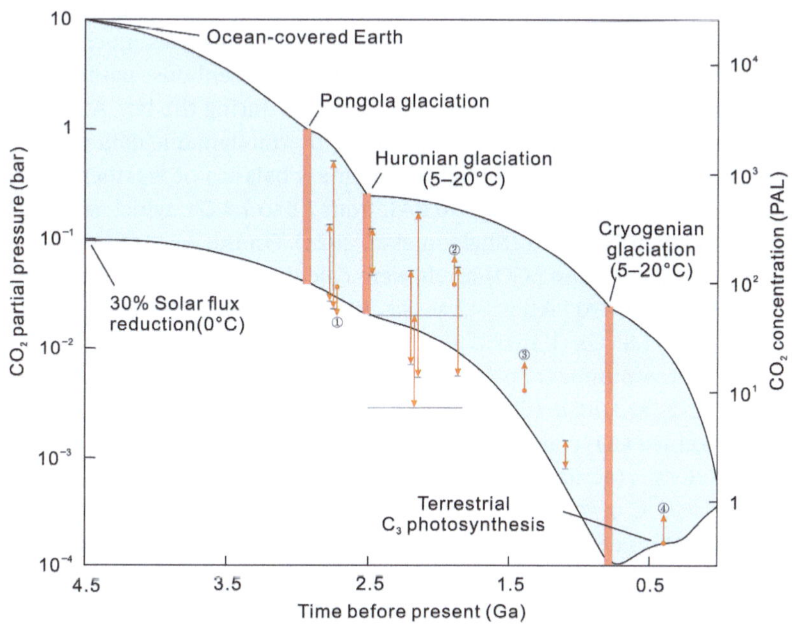

Figure 8.7 Atmospheric P_{CO_2} levels in Precambrian based on a climate model. *Source:* Adapted from Kasting (1993). Indicated constraints are based on descriptions in Section 8.3.3. constrains ① Mt. Roe paleosol (Rye et al. 1995), ② massive siderite-rich beds in pre-1.8-Gyr sedimentary sequences and carbon isotope ratios (Ohmoto et al. 2004), ③ microfossil analyses (Kaufman and Xiao 2003), and ④ terrestrial C_3 photosynthesis.

in the Paleoproterozoic (Huronian glaciation) and 0.72–0.635 Ga (Cryogenian glaciations) were calculated using a radiative convective climate model, assuming a mean surface temperature between 5 and 20 °C (the current mean surface temperature is 15 °C). CO_2 concentrations at other times were estimated by assuming that Earth remained ice-free throughout most of the Precambrian (Kasting 1993).

A negative feedback mechanism involving changes in the rate of CO_2 consumption by silicate weathering could explain high CO_2 levels in the past. The argument is as follows: If other factors affecting climate remain unchanged, a dimmer Sun would have led to colder surface temperatures. This would have decreased the silicate weathering rate, causing volcanically emitted CO_2 to accumulate in the atmosphere. Higher atmospheric CO_2 levels would have increased the greenhouse effect, warming the climate. Atmospheric CO_2 concentrations would have risen to the level needed to prevent the oceans from freezing and to balance the global carbon budget. To achieve this, atmospheric CO_2 concentration must have declined from over 300 PAL (pre-Industrial Atmosphere Level, 280 ppm) at 4.5 Ga to around 1–10 PAL near the end of the Proterozoic. Declining atmospheric CO_2 concentrations over the past 2.5 Ga are qualitatively consistent with the observed increase in the $^{13}C/^{12}C$ ratio of marine kerogens.

There are several paleo-CO_2 indicators that can be used to test the CO_2 concentrations based on the model discussed above. Soils form at the Earth's surface in direct contact with the atmosphere. Thus, paleosols serve as archives of Earth's atmospheric composition and are widely used to reconstruct pCO_2 and pO_2 (Sheldon et al. 2021). Rye et al. (1995) used well-studied paleosol profiles located north of Thessalon, Ontario, Canada, to constrain atmospheric pCO_2 between 2.75 and 2.2 Ga. The observation that iron lost from the tops of these profiles was reprecipitated lower down as iron silicate minerals, rather than as iron carbonate, indicates that atmospheric pCO_2 must have been less than $10^{-1.4}$ atm—about 100 times of the 1995 level 360 ppm, and at least 5 times lower than that required in one-dimensional climate models to compensate for lower solar luminosity at 2.75 Ga. Accordingly, Rye et al. (1995) suggested that either Earth's early climate was much more sensitive to increases in pCO_2 than previously thought, or that one or more greenhouse gases other than CO_2 contributed significantly to the atmosphere's radiative balance during the late Archean and early Proterozoic eons. More recent recalculations using update thermodynamic data or new approaches generated divergent results. The model based on the mass balance of weathering in paleosols gave consistent results of pCO_2 values about 8–70 PAL from 2.5 to 1.8 Ga, which is insufficient to overcome the "faint young Sun" paradox (Sheldon et al. 2021). On the basis of the modeling approach of mass dissolution kinetics, the pCO_2 levels were calculated to be 85–510 PAL at ~2.77 Ga, 78–2500 PAL at ~2.75 Ga, 160–490 PAL at ~2.46 Ga, 30–190 PAL at ~2.15 Ga, 20–620 PAL at ~2.08 Ga and 23–210 PAL at ~1.85 Ga (Kanzaki and Murakami 2015). There are no paleosols described between 1.8 and 1.2 Ga. A number of paleosols from the Midcontinent Rift have consistently indicated a 4–6 PAL of pCO_2 at 1.1 Ga. (Sheldon et al. 2021 and reference therein).

Kaufman and Xiao (2003) constrained CO_2 concentration by comparing the carbon isotope compositions of eukaryotic algal microfossils (acritarchs) extracted from a Proterozoic (~1.4 Ga) shale in North China. Analyses of the $^{13}C/^{12}C$ of individual acritarchs, coupled with calculated magnitudes of the carbon isotope fractionation depending on the CO_2 levels of the growth medium and volume/surface-ratio estimates of the original algal forms, suggest that, around 1400 million years ago, pCO_2 levels were at least 10–200 PAL. Additional studies based on data from microbialite gave estimates of ≤10 PAL at 1.2 Ga and ≤ 20 PAL at 1.1 Ga. These results appear to be close to the limits required in atmospheric models to overcome diminished solar luminosity (Figure 8.7).

Generally speaking, regardless of the approach taken, reconstructed pCO_2 levels during the Precambrian show a long-term decreasing trend and were consistently many times greater than pre-Industrial atmospheric levels. These pCO_2 levels were still insufficient to compensate for the FYS, suggesting that other greenhouse gases, such as methane, were present and provided enough additional greenhouse to maintain habitable conditions at Earth's surface throughout its history. However, a recent study based on model of micrometeorite oxidation inferred a very high pCO_2 concentration of >70% by volume at 2.7 Ga, which is alone sufficient to keep the Archean Earth warm (Lehmer et al. 2020).

8.3.3.3 Methane-rich Atmosphere in the Precambrian

The discussions above suggest that pCO_2 levels during most of the Precambrian were insufficient to maintain surface temperatures near the modern global average temperature of 15 °C. Additionally, Sleep and Zahnle (2001) argued that CO_2 was primarily removed from the Archean atmosphere-ocean system through seafloor carbonatization rather than continental weathering, as it is today. If this was the case, the negative feedback loop between atmospheric CO_2 levels and climate may not have functioned effectively at that time, potentially resulting in a colder Archean climate, which contradicts the lack of evidence for extensive glaciation during that period. Therefore, additional greenhouse warming was necessary throughout most of the Precambrian.

Pavlov et al. (2000) proposed that a combination of gases, including methane, prevented the early Earth from being permanently frozen (Figure 8.8). The net methane flux from the oxygen-poor Proterozoic oceans could have been 10–20 times higher than the present biological methane flux, which would have been sufficient to maintain methane concentrations at 100–300 ppm and keep the surface warm throughout the Proterozoic without high CO_2 levels (although CO_2 abundance could have been higher as well). The second oxygenation event at the end of the Proterozoic would have resulted in a decrease in methane flux and could have caused the first Neoproterozoic "Snowball Earth" glaciation (Pavlov et al. 2003).

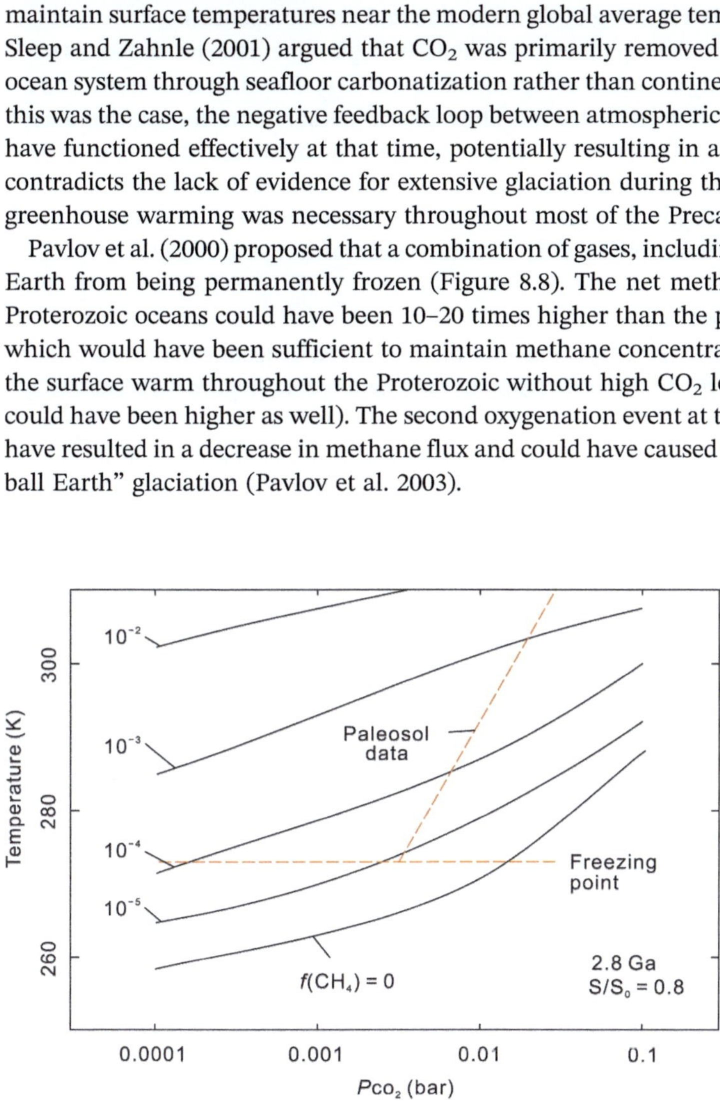

Figure 8.8 Global mean surface temperature as a function of Pco_2 and CH_4 mixing ratio. Calculations are for a solar luminosity of 80% of the present value, assumed to be valid at ~2.8 Ga. The dashed curves show the freezing point of water and the upper limit on Pco_2 derived from paleosols. *Source:* Modified from Pavlov et al. (2000).

Kasting (2005) argued that CH_4 was a significant atmospheric constituent and greenhouse gas during much of the Precambrian. Small amounts of CH_4 (1–10 ppmv) likely formed abiotically on the Hadean Earth from impacts and serpentinization of ultramafic rocks on the seafloor. Much larger CH_4 concentrations (~1000 ppmv or more) may have been present during the Late Archean/Paleoproterozoic, prior to the rise of O_2. This CH_4 would have been produced by methanogens metabolizing CO_2 and H_2 from the atmosphere or acetate from the fermentation of photosynthetically produced organic matter. The rise of O_2 in the atmosphere around 2.3 Ga likely caused a drastic decrease in atmospheric CH_4 concentrations, potentially triggering the Paleoproterozoic glaciation (Huronian glaciation). CH_4 concentrations could have remained elevated throughout much of the Proterozoic (10–100 ppmv) due to low concentrations of dissolved O_2 and sulfate in the deep oceans, leading to increased organic matter recycling by fermentation and methanogenesis.

However, in modern anoxic and sulfate-poor environments, biological methane production is relatively low efficient, converting only about 5% of organic carbon to CH_4. Based on this restriction, a simple model of marine carbon cycling estimated that atmospheric methane concentrations did not exceed 1 ppm (about twice the pre-industrial value) throughout the Proterozoic and $\leq$100 ppm in the Archean (Laakso and Schrag 2019). Alternatively, Zhao et al. (2018) proposed that terrestrial methanogenic mats supplied a globally significant flux of methane to a relatively low O_2 atmosphere during the Proterozoic, buffering the atmospheric CH_4 inventory and warming the climate.

8.3.3.4 Phanerozoic CO_2 Concentrations

Following the Precambrian, atmospheric CO_2 concentrations continued to decline in the first-order trend but experienced significant fluctuations throughout the Phanerozoic (Figure 8.9). The pCO_2

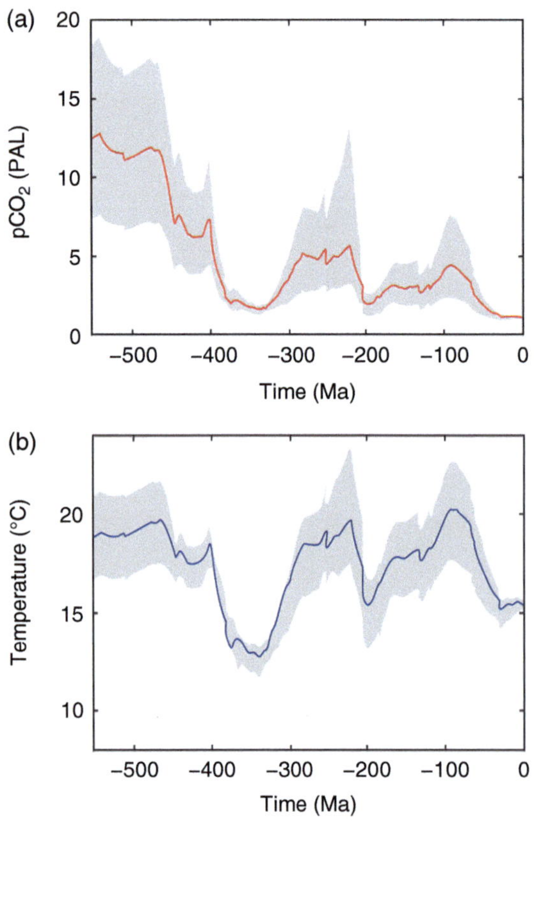

Figure 8.9 COPSE model of Earth's climate over Phanerozoic time. *Source:* Lenton et al. (2018) / with permission of Elsevier. (a) Atmospheric pCO_2; (b) global temperature.

levels remained high during the early Paleozoic, often exceeding 4000 ppm and contributing to a warm climate; dropped significantly in the late Paleozoic, leading to the Carboniferous and Permian glaciations; rose again in the Mesozoic, reaching around 2000 ppm during the Triassic and Jurassic periods, contributing to a warm and stable climate; gradually declined during the Cenozoic, with significant drops during the Oligocene and Miocene epochs, leading to cooler climates and the development of polar ice sheets. These fluctuations have played a crucial role in shaping Earth's climate and the evolution of life throughout the Phanerozoic.

Numerous proxies for evaluating Phanerozoic atmospheric CO_2 concentrations have been developed and refined. The most reliable method involves determining the composition of air trapped in glacial ice. To date, this method has only been useful for the past 0.8 million years (Figure 8.10) due to the absence of older ice. Consequently, other methods have been applied to study the older geological record. Proxy records, such as leaf stomata, pedogenic carbonate $\delta^{13}C$ and boron isotopes, extend back to ~420 Ma, showing pCO_2 peaking above 2000 ppm during two greenhouse episodes (Silurian and early Mesozoic), each followed by declines to near-modern levels associated with icehouse climates (Lu et al. 2024). Pre-Silurian pCO_2 was poorly constrained. Carbon isotope measurements from Ordovician paleosols suggest that CO_2 levels were 14–16 PAL.

The updated COPSE (Carbon, Oxygen, Phosphorus, Sulfur, and Evolution) biogeochemical model predicts a two-phase drawdown of atmospheric CO_2 with the rise of land plants and associated cooling phases during the Late Ordovician to Early Carboniferous (~465–350 Ma), followed by broad peaks in atmospheric CO_2 and temperature during the Triassic and mid-Cretaceous. The model reasonably captures the long-term trends in Phanerozoic geochemical proxies, although some of the features in the CO_2 proxy record are missed (Figure 8.9). Prior to the rise of land plants,

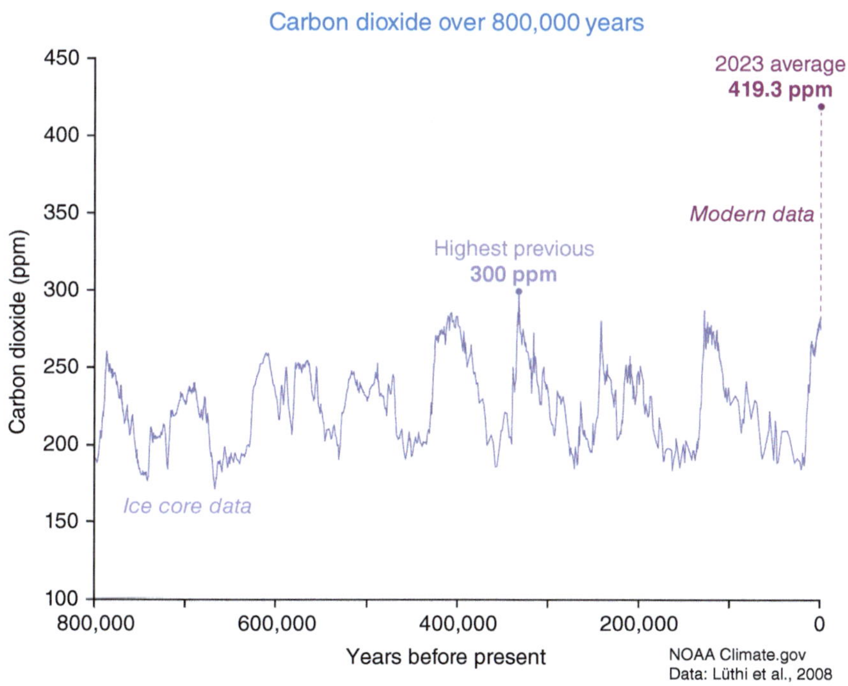

Figure 8.10 Atmospheric carbon dioxide (CO_2) in parts per million (ppm) for the past 800,000 years based on ice-core data (light purple line) compared to 2022 concentration (bright purple dot). *Source:* Lüthi et al. (2008) / NOAA / Public Domain.

from the Cambrian to the Early Ordovician (~540–465 Ma), atmospheric CO_2 levels were generally uncertain due to a lack of proxies. Subsequently, atmospheric CO_2 was well regulated, particularly against decreases, but there were peaks during the Triassic and mid-Cretaceous. The rising through the later Carboniferous and Permian to a Triassic peak corresponds to the assembly of the super-continent Pangea. However, predicted atmospheric CO_2 levels are particularly uncertain because the model cannot reproduce the marked Permian drop in $^{87}Sr/^{86}Sr$, which may indicate uncaptured changes in degassing or lithology. The Jurassic to Neogene interval (~200–0 Ma) includes the rise of calcareous phytoplankton and the break-up of Pangea—with increasing uplift and a peak then decline in degassing. The model captures an overall peak and then decline of CO_2 driven by the peak in degassing (and to some degree the rise of uplift), but it does not capture the finer-scale structure suggested by the proxy compilation (Lenton et al. 2018).

The overall decreasing trend was likely due to the silicate weathering-negative feedback and the expansion of land plants, which negated the slow increase of solar irradiance over the past 500 million years, ensuring Earth's long-term habitability. However, if fossil fuel use continues unabated, atmospheric CO_2 levels could rise to values not seen since the early Eocene (50 million years ago) by the middle of the 21st century. If CO_2 levels continue to rise into the 23rd century, coupled with the increase in radiative forcing, the Earth's system response would likely be without geological precedent in the last half billion years (Foster et al. 2017).

8.3.4 The Rise of Atmospheric O_2

Oxygen gas is the most important component of the present atmosphere. Organisms, especially phytoplankton and green plants, are responsible for producing nearly all of the oxygen we breathe. Over the past decades, data from various fields, including biology, geology, geochemistry, oceanography, and atmospheric science, have revealed two significant increases in atmospheric O_2 levels at approximately 2.3 and 0.8–0.6 billion years ago, respectively (see review by Catling and Claire 2005), which broadly corresponds to the three-stage model of atmospheric O_2 evolution (Figure 8.11, see Kasting 1993).

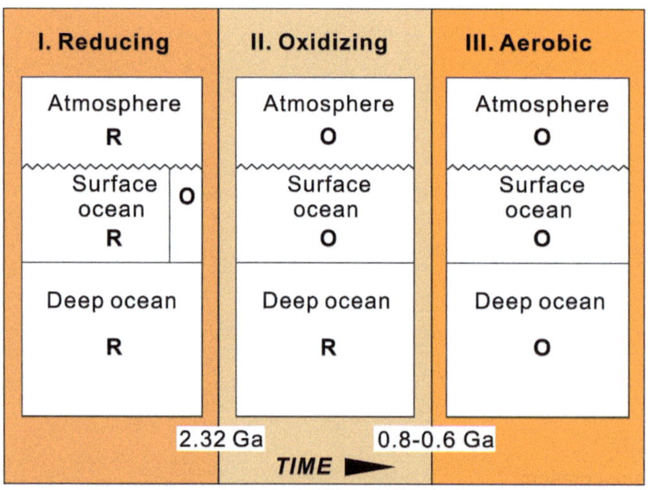

Figure 8.11 Three-box model for the oxygenation of Earth's surface. O, Oxidizing conditions; R, reducing conditions. The small box in the surface ocean during stage I represents a local oxygen oasis. *Source:* Adapted from Kasting (1993).

8.3.4.1 Net Accumulation of O_2 in Atmosphere

It has long been recognized that oxygenic photosynthesis, followed by the burial of organic carbon, is the major source of atmospheric O_2. Additionally, abiotic photodissociation of H_2O followed by the escape of H_2 to space, and microbial sulfate reduction followed by pyrite burial, can produce small amounts of O_2 but contribute little to the atmosphere.

Currently, photosynthesis on land is carried out by green plants but terrestrial plants have limited effect on the long-term atmospheric O_2 evolution because it is largely balanced by the reverse processes of respiration and emerged much later than marine photosynthesis. In contrast, marine photosynthesis by microorganisms evolved early in the Archean (Sánchez-Baracaldo et al. 2022) and significantly contributes to atmospheric O_2 because a small fraction of the organic matter synthesized in the oceans is buried in sediments. This small leak in the marine organic carbon cycle is responsible for most of atmospheric O_2 (Kasting and Siefert 2002).

Respiration and decay on a time scale of $\sim 10^2$ years consume >99% of the O_2 produced by photosynthesis. Only a small fraction (0.1–0.2%) of organic carbon escapes oxidation through burial in sediments. The burial of 1 mol of organic carbon liberates 1 mol of O_2. The fate of the O_2 depends on the kinetics of the various O_2 sinks (see Section 4.3.4). O_2 will not, in fact, accumulate in the atmosphere if the flux of kinetically rapid sinks exceeds the slow flux of organic burial, which is important when considering the lack of atmospheric O_2 in the Archean. The oxygen cycle is complicated by the burial of other redox-sensitive elements besides carbon. The quantitatively important elements are sulfur and iron. During weathering, O_2 dissolved in water oxidizes sulfur and iron of pyrite, making soluble SO_4^{2-}, which is carried by rivers to the ocean. Then, in the ocean, microbes reduce SO_4^{2-} and Fe^{3+} to pyrite. The reducing power of photosynthesized organic carbon is essentially transferred to pyrite. Therefore, the reduction of sulfate, followed by the burial of pyrite also adds a minor flux of O_2, whereas the burial of sulfate minerals in sediments removes O_2 (Catling and Claire 2005). Hence, pyrite deposits also represent an oxygen source for the atmosphere (Canfield 2014). The sum reaction equation is $2Fe_2O_3 + 8SO_4^{2-} + 16H^+ \leftrightarrow 15O_2 + 4FeS_2 + 8H_2O$.

Geological evidence indicates that the transition of atmospheric oxygenation occurred at $\sim$2.4–2.3 Ga (see below), while oxygenic cyanobacteria evolved long before, at least 300 Ma earlier than the transition (Sánchez-Baracaldo et al. 2022). This apparent delay could be explained by the fact that excess reductants scavenged photosynthetic O_2 from the early atmosphere. Given that all contemporary cyanobacteria are capable of oxygenic photosynthesis, can live anaerobically and aerobically, and have a deep origin as early as Archean, they are universally believed to have been responsible for the initial rise of atmospheric O_2 in the Paleoproterozoic (Kasting and Siefert 2002).

8.3.4.2 Constraints on Precambrian Atmospheric O_2 Levels

Kasting (1987, 1991) proposed a three-stage model of O_2 evolution, also known as a "three-box" model (Figure 8.11), with atmosphere, surface ocean, and deep ocean constituting three distinct O_2 reservoirs. In stage I, which probably lasted until red beds began to form at $\sim$2.0 Ga, the Earth's surface environment was generally devoid of free O_2, except possibly in localized regions of high productivity in the surface ocean. In stage II, which may have lasted for less than 150 Ma or much longer, the atmosphere and surface ocean were oxidizing, while the deep ocean remained anoxic. In stage III, which has lasted until the present, all three surface reservoirs contained appreciable amounts of free O_2. While the beginning of stage III is controversial, it may have commenced with the disappearance of BIFs at $\sim$1.85 Ga (Kasting 1991), the appearance of Ediacaran macro-biotas at $\sim$575 Ma (Canfield et al. 2007), or the ecological expansion of land plants at $\sim$420 Ma (Stockey et al. 2024). According to this "three-box" model, the atmospheric O_2 level during stage II could not have exceeded 3% PAL because O_2 that down welled into the deep ocean would have been lost faster than it was produced and could not have been less than 0.2–0.5% PAL during stage III because the deep ocean would have remained anoxic due to the influx of reductants from hydrothermal vents.

There are more stringent constraints on pO_2 in the Precambrian. Eukaryotic organisms likely require at least 1% PAL of dissolved O_2 in the water column to satisfy their metabolic requirements. This level is below the upper limit for stage II and could even have occurred during stage I. The upper limit for dissolved O_2 in localized oxygen oases is approximately 8% PAL. Thus, the existence of eukaryotic organisms alone does not necessarily indicate that the atmosphere contained free O_2. However, photochemical model calculations predict that a biologically effective UV screen (ozone layer) would have been established at an O_2 concentration of 1–10% PAL. Before atmospheric O_2 concentrations reached this level, eukaryotic phytoplankton would have been excluded from the surface oceans due to their intolerance to high levels of UV radiation. Multiple lines of evidence indicate that eukaryotes diversified at ~1.8 Ga (Westall and Xiao 2024), implying that pO_2 must have been above 1% PAL by that time to develop a protective ozone layer.

Additional constraints can be derived from analyses of various geological records, some of which are listed here and indicated in Figure 8.12. (1) The presence of detrital pyrite and uraninite in Archean sediments reveals that the atmosphere in the earliest Archean contains no free O_2 (Cloud 1972). (2) Bekker et al. (2004) measured sulfur isotopic composition of syngenetic pyrite in organic-rich shales of the 2.32-Ga Rooihoogte and Timeball Hill formations of South Africa and found no evidence of mass-independent fractionation (MIF), indicating $pO_2 > 10^{-5}$ PAL at ~2.32 Ga. (3) Qualitative calculations of pO_2 from Precambrian paleosols are many. Sheldon et al. (2021) compiled pO_2 constraints from paleosols from ~3.0 Ga to the present. The results generally match model and metal-redox derived estimates both in terms of the timing and magnitude of pO_2 changes from below to above 1‰ PAL at 2.32 Ga. These data also support a prolonged low pO_2 (<10% PAL) state throughout much of the Proterozoic until a second rise in the Neoproterozoic, although possible transient increases in pO_2 were suggested by other geochemical proxies (Lyons et al. 2021). (4) Canfield and Teske (1996) used sedimentary sulfur isotope ratios to argue

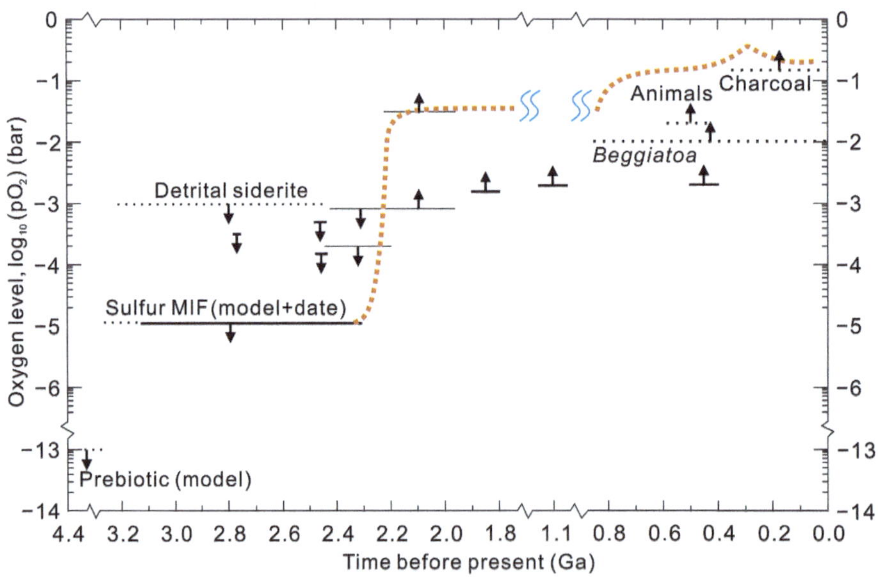

Figure 8.12 Change in atmospheric O_2 level during geological history based on multiple constraints described in 8.3.4. *Source:* Adapted from Catling and Claire (2005).

that pO_2 remained below 5–18% PAL until 1.05–0.64 Ga. They suggested that non-photosynthetic sulfide-oxidizing bacteria were absent until this time because continental shelf sediments were overlain by anoxic waters. (5) Canfield et al. (2007) provided evidence for oxygenation of the deep ocean at 580 Ma, which requires that the minimum pO_2 was about 15% PAL.

8.3.4.3 Evidence for the Paleoproterozoic Rise of Atmospheric O_2

The oxidation state of paleosols, the history of red beds and detrital grains of reduced minerals, the age distribution of BIFs, isotopic signals of carbon and sulfur, and the evolution of eukaryotes are all consistent with the hypothesis that the O_2 content of the atmosphere increased dramatically during of the early Paleoproterozoic (Figure 8.13), known as the GOE, which began 2.460–2.426 Ga and ended 2.060 Ga (Gumsley et al. 2017). A rise in O_2 would promote the rapid destruction of the methane, cooling the Earth by a few tens of degrees Celsius, which could explain the ~2.3–2.2 Ga Makganyene "Snowball Earth" (Huronian glaciation). Oxygen also has another important repercussion.

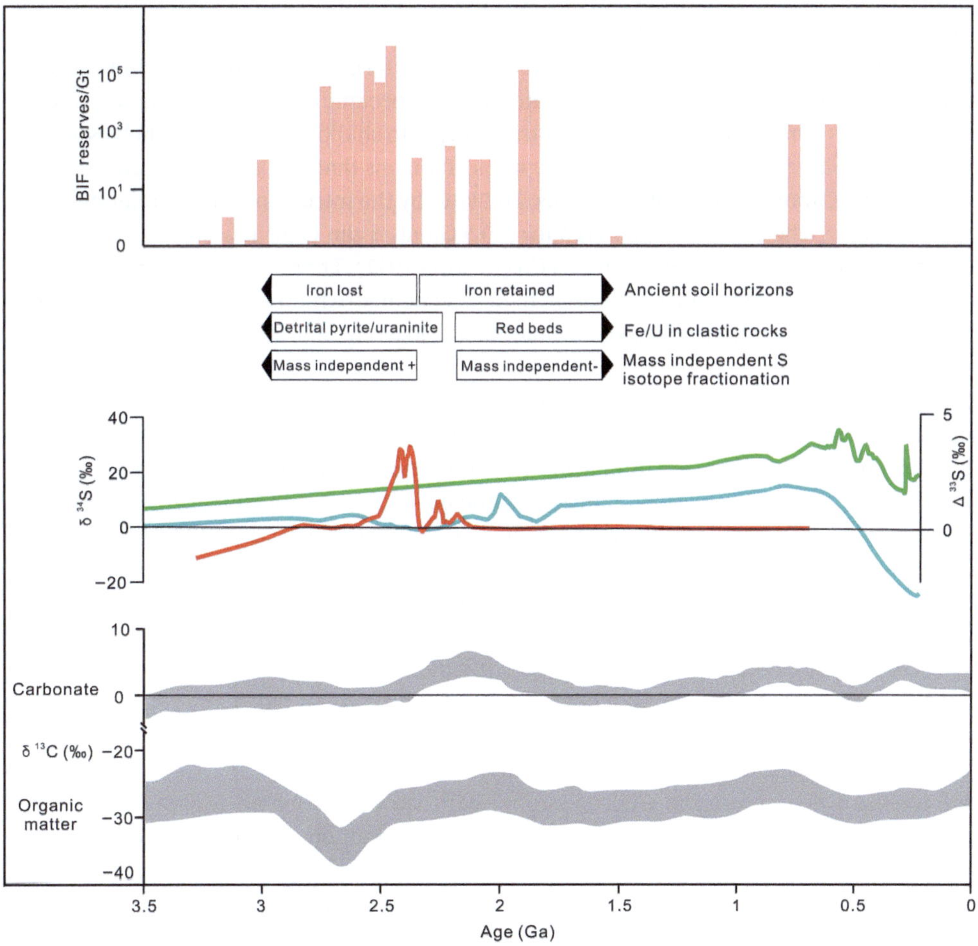

Figure 8.13 Geological evidence for the rise of atmospheric O_2 in the Paleoproterozoic based on multiple sources described in the text (Section 8.3.4) and mentioned below. Carbon isotope records adapted from Garcia et al. (2021), sulfur isotope records from Fakhraee et al. (2025), and BIF record from Bekker et al. (2010).

Ozone (O_3) derives from O_2. Once O_2 rose to ~1% PAL, an effective stratospheric ozone screen for UV radiation emerged, shielding the surface biosphere (Catling and Claire 2005).

Iron was leached from paleosols before the ~2.4 Ga oxic transition but not afterward because anoxic rainwater flushes soluble ferrous iron through soil, whereas oxygenated rainwater produces insoluble and immobile ferric iron. Therefore, the iron lost or retained in paleosols is a good indicator of the atmospheric oxidative state. Detrital grains in pre-2.4 Ga riverbeds commonly contain reduced minerals (pyrite, siderite, and uraninite) that would only survive at low pO_2. Detrital grains of pyrite, uraninite, and siderite place upper bounds on Archean pO_2 of ~0.1, ~0.01, and ~0.001 bar, respectively. The roundness of such grains shows that they were transported long distances in aerated waters. Such grains of reduced minerals are not present in sedimentary rocks younger than 2.4 Ga. Red beds are sedimentary sandstones derived from windblown or river-transported particles coated with red-colored iron oxides, usually hematite, which have long been considered evidence for the presence of free O_2 in the atmosphere. Red beds are common only in sedimentary successions deposited after ~2.2 Ga.

BIFs appear from the start of the sedimentary record (3.8 Ga) but decline in abundance through the Paleoproterozoic and disappear after ~1.8 Ga (Section 6.2.4). No BIFs have been found in rocks between ~1.8 and ~0.8 Ga. The brief Neoproterozoic renaissance of BIFs ended well before the beginning of the Phanerozoic. Trace and rare earth element patterns indicate that iron in BIFs owes its origin to the hydrothermal circulation of seawater through oceanic crust, perhaps largely at mid-oceanic ridges. Fe^{2+} is hypothesized to have been transported to shallow continental shelves, where Fe^{2+} oxidation precipitated ferric iron by a variety of mechanisms that probably included oxidation by solar UV radiation, anoxygenic photosynthesis, and later by the intrusion of atmospheric O_2 into near-surface seawater. The persistence of BIFs for ~0.5 Ga after the GOE has been used to argue that the deep ocean was not oxygenated until 1.8 Ga (Kasting 1991). Therefore, the disappearance of BIFs at ~1.8 Ga may be related more to a change in ocean chemistry than to a rise in pO_2. If so, the reappearance of BIFs in the late Neoproterozoic may be due to a return of severe ocean stratification rather than to a severe drop in pO_2.

Isotopes of carbon and sulfur in marine sediments also indicate a major change of atmospheric O_2 concentration in the Paleoproterozoic. A very positive, longest durable $\delta^{13}C_{carb}$ excursion (Figure 8.13) occurred between ~2.3 and 2.1, known as the Lomagundi-Jatuli Carbon Isotope Excursion, is possibly associated with enhanced organic burial that is crucial for O_2 accumulation in the atmosphere. The sulfur isotope record indicates that there was <0.2 mM sulfate in the Archean oceans and sulfate levels increased in the Proterozoic. For concentrations exceeding ~0.05–0.2 mM, sulfate-reducing bacteria preferentially use ^{32}S rather than ^{34}S, producing sulfides that are enriched in ^{32}S relative to coexisting seawater sulfate. Sulfides with significant ^{32}S enrichment only become widespread after ~2.3 Ga, suggesting more available sulfate and hence oxygen. MIF of sulfur isotopes was recorded in Archean sedimentary rocks but faded after ~2.45 Ga. This signature independently suggests that oxygen began to accumulate in the atmosphere early in the Paleoproterozoic, regardless of other lines of evidence.

Evidence for the Paleoproterozoic GOE is strong, with various proposed causes (Catling and Claire 2005). Some argue that O_2 levels rose due to a cascade of abiotic and biotic processes, such as an increase in phosphorus supply and an elevated rate of primary production accompanied by a high rate of organic and pyrite burial, either as a pulse or a secular trend. Others suggest that a decrease in the O_2 sink was the key factor, simply because the flux of volcanic reducing gases dwindled. Still others consider that the flux of H_2 from Earth's interior and its escape to space played a significant role in Earth's oxygenation. In any case, it is the excess of supply over demand that causes O_2 accumulation in the atmosphere. Holland's (2009) modeling suggests that rates of O_2

liberation first exceeded rates of oxygen demand between 2.7 and 2.2 Ga. Early Earth was tectonically active and released reduced gases (mostly H_2), which reacted with O_2 and may titrate all O_2 liberated from burial of organic carbon and pyrite. As Earth cooled, tectonic activities slowed, and this resulted in a reduction in H_2 flux from the mantle (Canfield 2014). Therefore, it is the slowdown of H_2 flux coupled with the increase of O_2 liberation that ultimately caused the GOE.

8.3.4.4 Second Rise of Atmospheric O_2

The second major episode in the rise of atmospheric O_2 levels occurred in the late Neoproterozoic, between 800 and 540 Ma, known as the Neoproterozoic Oxygenation Event (NOE), during which oxygen possibly accumulated to the levels required to support complex life forms (Och and Shields-Zhou 2012), critically metazoans that diversified around the Ediacaran–Cambrian boundary (~539 Ma).

Generally high $\delta^{13}C$ values in marine carbonates formed at late Neoproterozoic indicate a marked increase in the rate of organic carbon burial and hence in the rate of O_2 liberation. A demonstrably global and primary record of extremely negative $\delta^{13}C$ values, particularly Shuram–Wonoka (equivalent to DOUCE and EN3 in South China) excursion between ~573 and 562 Ma, strongly suggests the oxidation of a large dissolved organic carbon (DOC) pool. The increase of $^{87}Sr/^{86}Sr$ isotope ratio toward the Neoproterozoic–Cambrian transition indicates that the O_2 sink from hydrothermal reductant was diminishing in magnitude and continental weathering was enhancing, which may have fueled higher organic production and burial during the late Neoproterozoic. Some of the excess O_2 was probably released to the atmosphere.

Rise of benthic communities of complex multicellular life after the Cryogenian Snowball Earth period indicates a widespread seafloor oxygenation on shallow shelves, while the deep oceans may largely remain anoxic till the mid-Paleozoic (Stockey et al. 2024). In comparison with overwhelming data on ocean oxygenation, quantitative constraints on the atmospheric pO_2 are sparse and the balance of O_2 liberation and O_2 demand remains uncertain. On the basis of Cr isotope data from across the globe, Planavsky et al. (2014) found evidence for inhibited oxidation of Cr at Earth's surface during 1.8 to 0.8 Ga, suggesting a $pO_2 < 0.1\%$ PAL. In contrast, pO_2 was inferred to have increased from 1% to 3% PAL to greater than 5–18% PAL in the Neoproterozoic, based on sedimentary sulfides with ^{32}S enrichment exceeding the isotope discrimination of sulfate-reducing bacteria (Catling and Claire 2005). Krause et al. (2022) provided a continuous quantitative reconstruction of pO_2 over the past 1.5 Ga using an isotope mass balance approach that combines bulk geochemistry and tectonic recycling rate calculations. The result demonstrated that pO_2 oscillated between ~1% and 50% PAL, there was no simple unidirectional rise in pO_2 during the Neoproterozoic (Figure 8.14), and the first animals evolved against a backdrop of extreme O_2 variability. This extreme variability in pO_2 was assumed to be likely a response to the global tectonic regime during the Cryogenian to Cambrian periods, notably the breakup of supercontinent Rodinia (~800 to 600 Ma) and assembly of Gondwana (~650 to 500 Ma) (Figure 8.14).

8.3.4.5 Phanerozoic Variations of Atmospheric O_2

Much attention has been paid to the pO_2 rise in the Precambrian and its association with biological evolution. However, there are also multiple lines of evidence for constraints and variations in pO_2 during the Phanerozoic.

Fossil record demonstrates that invertebrate metazoans were initially diversified during the Ediacaran–Cambrian transition and have a continuous fossil record from then on (Zhang and Shu 2021), which requires at least 2 mg/L dissolved oxygen to support their metabolism and maintain their diversity. If this concentration is extrapolated to equilibrium of the atmosphere with

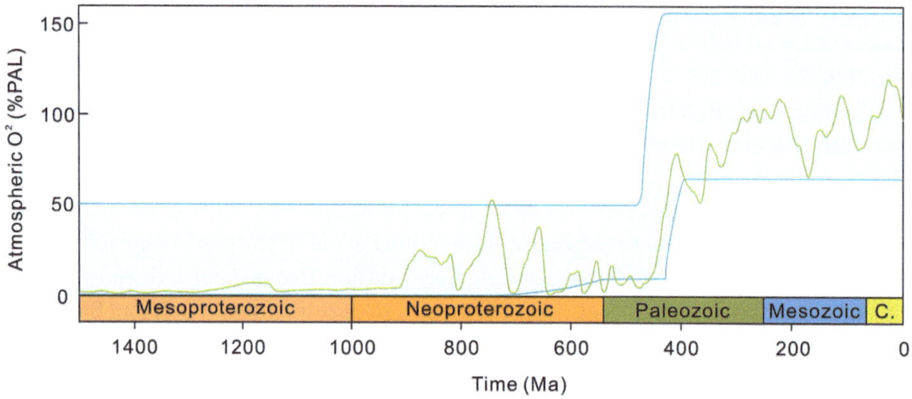

Figure 8.14 Variability of atmospheric oxygen levels during the Neoproterozoic to Paleozoic transition estimated from the NEOCARBSULF model. *Source:* Adapted from Krause et al. (2022). The green line is the mean result of 5000 model runs and the shaded band between the two blue lines is the O_2 best guess estimate.

well-mixed 15 °C surface waters with 35‰ salinity, it corresponds to an O_2 concentration of ~5%. Therefore, the atmospheric O_2 level could be at or above 25% PAL since the Cambrian to maintain the evolution of marine biodiversity (Zhang and Cui 2016). There are many examples of hypoxia-tolerant animals that are able to survive in low or even oxygen-free environments for extended periods by slowing their metabolism and using anaerobic respiration. While an oxygen concentration of 2 mg/L is necessary for any aquatic ecosystem to maintain the taxonomic and metabolic diversity, complete life cycles, active behaviors, and complex ecology, such as skeleton-making and predation, that are oxygen-consuming processes, all of which have been in operation since the Cambrian.

Charcoal is the carbon-rich residue left after the incomplete combustion of plant wood. The presence of charcoal in the geological record indicates the occurrence of wildfires, which require a minimum atmospheric oxygen concentration of about 16%, or roughly 75% PAL, to sustain combustion (Glasspool and Gastaldo 2022). Charcoal has a nearly continuous record since mid-Silurian (~430 Ma), which indicates that pO_2 has reached the threshold of 75% PAL since that time. The existence of terrestrial forests themselves also provides a crude upper bound on pO_2, ~0.3–0.35 bar or ~150–170% PAL, because excess of O_2 could result in photorespiration outcompetes photosynthesis and plants experience zero or negative net growth. Also, higher pO_2 may have caused more extensive, possibly catastrophic, wildfires. The persistence of terrestrial plant communities from the Silurian through the present does imply that O_2 concentrations have not risen too high during the past 430 Ma.

The COPSE modeling indicates an increasing first-order trend of pO_2 during the Phanerozoic, in contrast to the decreasing trend of pCO_2, albeit with considerable temporal variations and uncertainties (Figure 8.15). The model robustly predicts a mid-Paleozoic oxygenation event due to the rise of land plants, with pO_2 rising from ~24% to >80% PAL and oxygenating the deep ocean. Thereafter, pO_2 was effectively regulated with remaining fluctuations including a Carboniferous–Permian O_2 peak ~124% PAL linked to burial of terrestrial organic matter in coal swamps, a Triassic–Jurassic O_2 minimum ~100% PAL linked to low uplift, a Cretaceous O_2 peak ~124% PAL linked to high degassing and weathering fluxes, and a Cenozoic pO_2 decline (Lenton et al. 2018).

There is also geological evidence that pO_2 was greater than 100% PAL during the late Paleozoic and dropped at the end of the Permian. Coals from the Carboniferous and Permian contain a

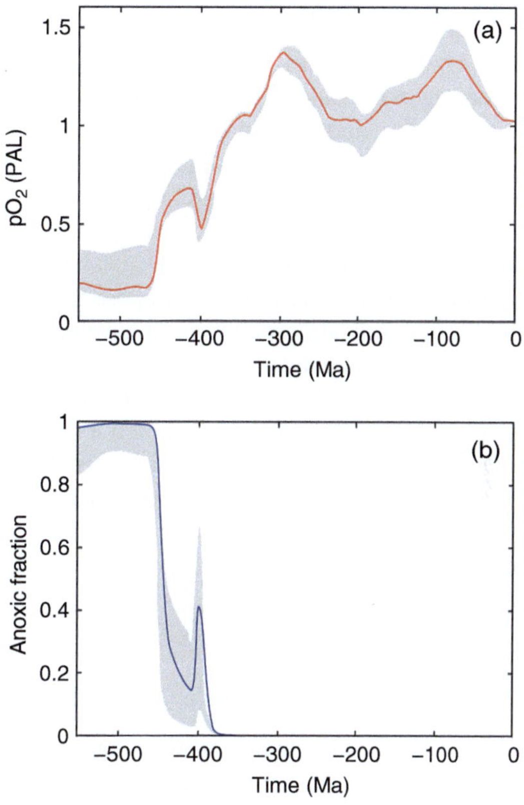

Figure 8.15 COPSE model of Earth's surface oxygenation over Phanerozoic time. *Source:* Lenton et al. (2018) / with permission of Elsevier. (a) Atmospheric pO_2; (b) anoxic fraction of the ocean.

greater abundance of fusain, a product of woody tissue combustion and charring, than those from any period of the subsequent geological past, suggesting more abundant forest fires and possibly higher pO_2 at this time. The most obvious paleontological support for a late Paleozoic peak in pO_2 is the contemporaneous expression of gigantism within numerous arthropod lineages. Terrestrial arthropods rely on tracheal networks for O_2 diffusion to support their metabolism, which sets upper limits on body size for a given pO_2. In comparison with arthropod communities of the Carboniferous and Permian, modern terrestrial arthropods are rather small. Dragonflies at that time reached 70 cm wingspans, mayflies reached 45 cm wingspans, and millipedes reached 1 m or more in length. These large body sizes could not be supported by today's 21% O_2 atmosphere and instead require elevated pO_2 between 350 and 250 Ma. The largest Phanerozoic mass extinction occurred at the end of the Permian. A noticeable decrease in the burial of organic matter in marine sediments across the Permian–Triassic boundary may be associated with a global decline in primary productivity and O_2 supply. The gigantic terrestrial insect lineages did not survive across this boundary, further suggesting a global drop in pO_2.

8.4 Evolution of the Ocean

The temperature, oxygenation, and salinity of the oceans are major environmental variables relevant to the course of life evolution on Earth. This section describes the chemical and physical evolution of the ocean in relation to life in the sea.

8.4.1 Origin of the Oceans

By ~4.4 Ga, the Earth probably had liquid water. We can assign a minimum age for the ocean according to the oldest known sedimentary rocks (~3.8 Ga) that show evidence for water weathering, erosion, and deposition. But where did the water come from? Various hypotheses have been proposed. Many scientists believe that during the early accretion of the Earth, much water (initially in the form of water vapor) and large amounts of other gases were released by volcanoes. Modern volcanoes also spew out water, although much of their output may actually be recycled seawater. A significant amount of water may also have been deposited by icy comets that collided with the primitive Earth. Some scientists propose that chemical reactions between hydrogen and oxygen in the Earth's early atmosphere could have produced water, which then condensed and formed the oceans. Volcanic outgassing, comet and asteroid impacts, and chemical reactions could all have played a role in delivering and creating the water that makes up Earth's oceans, while the exact proportion of each contribution is still a subject of research.

Over time, the oceans have accumulated dissolved materials from volcanic input and weathered material transported from continental rocks. The oceans contain ions, or salts, of elements such as sodium (Na), magnesium (Mg), and chlorine (Cl), along with smaller amounts of calcium (Ca), silicon (Si), potassium (K), and various dissolved gases, including oxygen, carbon dioxide, and sulfur dioxide.

8.4.2 Cooling of the Surface Ocean Temperature

Modern open ocean temperatures range from −2 to 30 °C, with a mean surface temperature of about 17.2 °C in 2023. Seawater temperatures below 0 °C (32 °F) are possible because saltwater freezes at a lower temperature than pure water. In the geological past, ocean temperatures have varied over time. Three independent lines of evidence reveal a general cooling trend (Figure 8.16): from hot Archean oceans with surface temperatures ~50–80 °C, through warm Paleo- and Mesoproterozoic oceans of ~40 °C, to normal Neoproterozoic to Phanerozoic oceans with surface seawater temperatures below 40 °C, dropping below 0 °C during Cryogenian Snowball Earth intervals. However, climate moderation by the carbon cycle suggested average surface temperatures between 0 and 40 °C (Catling and Zahnle 2020), and oxygen isotope compositions of phosphates in Barberton sediments support temperate Archean oceans, with surface temperatures of 26–35 °C during 3.2–3.5 Ga (Blake et al. 2010). Cold environments may have been present locally because shallow-water glacial diamictites were reported from the Paleoarchean sequences of South Africa (de Wit and Furnes 2016).

The ratios of $^{18}O/^{16}O$ provide a method for measuring past temperatures. Lower $^{18}O/^{16}O$ ratios in seawater reflect higher vaporization and, consequently, higher surface temperatures. Therefore, measurements of $\delta^{18}O$ in uncrystallized, original carbonate precipitates, such as seashells or fibrous aragonitic marine cements, can yield the actual temperature of the seawater. Measurements across a single shell may reveal seasonal variations during the organism's life. However, it is rare to find original precipitates in strata older than the Cenozoic, let alone in deep Precambrian rocks, because all marine precipitates eventually undergo diagenesis. During this process, the originally deposited microcrystalline high-Mg calcite, aragonite, and calcite dissolve and reprecipitate into interlocking crystals of calcite and/or dolomite, altering their isotopic composition. Therefore, oxygen isotope paleo-thermometry of carbonates, as practiced on Cenozoic material, usually cannot be applied to older samples. Alternatively, scientists choose cherts to measure the oxygen and silicon isotopic composition. Cherts are composed of microcrystalline quartz, which is very common in

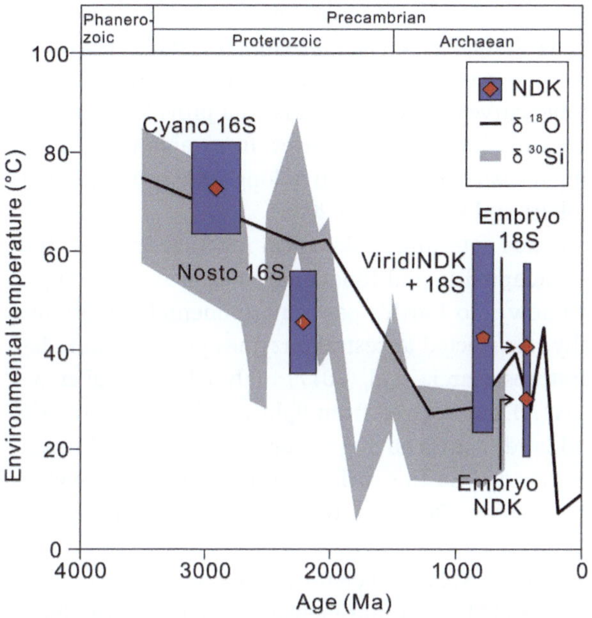

Figure 8.16 A cooling trend of ocean temperatures over geological time based on the $\delta^{18}O$ and $\delta^{30}Si$ values from cherts incorporated with temperature ranges inferred from reconstructed ancestral NDK Tms plotted against fossil-record-indicated first appearance of the various groups. *Source:* Modified from Robert and Chaussidon (2006) and Garcia et al. (2017).

sedimentary rocks of all ages and typically forms during early diagenesis of sedimentary rocks. Quartz itself is one of the most resistant minerals to later isotopic alteration. Masses of chert have essentially no measurable permeability and strongly resist the later dissolution/reprecipitation episodes. Recrystallization can occur during metamorphic heating but water/rock ratios at that point are usually so low that isotope composition is largely unchanged. The isotopic composition of cherts thus provides a possible approach for determining past diagenesis temperatures. If diagenesis occurs during shallow burial (as is generally the case for shallow water deposits), such temperatures approximate the climatic temperature during diagenesis (Knauth 2005).

Knauth (2005) plotted all available oxygen isotopic data of cherts from Archean to Phanerozoic, showing an overall trend of elevated $\delta^{18}O$ through geological time. Archean cherts have $\delta^{18}O$ values about 10‰ lower than those of Phanerozoic cherts. This 10‰ increase corresponds to an overall decrease in temperature of ∼40 °C (1‰ decrease corresponding to a 4.2 °C increase). If global temperatures in the Paleo- and Neogene were on the order of 10–20 °C, then the lower $\delta^{18}O$ values at ∼3.5 Ga correspond to temperatures on the order of 50–80 °C (Knauth and Lowe 2003). A striking feature of the chert $\delta^{18}O$ record is that values jump to as high as 24.5‰ by 2.45 Ga and never return to values similar to those of the Archean. This jump was interpreted to correspond to an approximate general temperature decrease on the order of 20 °C, suggesting Paleoproterozoic temperatures on the order of ∼40 °C. Data are scarce during much of the Proterozoic, especially at the important Neoproterozoic–Phanerozoic transition period. $\delta^{18}O$ values appear to rise progressively throughout the Proterozoic and become more similar to Phanerozoic values by ∼1.2 Ga. The interval between ∼1.2 and 0.685 Ga yields $\delta^{18}O$ values >30‰, similar to post-Ordovician cherts, suggesting that climatic temperatures during the non-glacial periods of the Neoproterozoic were not much different from most of the early Paleozoic (Knauth 2005).

The silicon isotopic composition of cherts from the pre-550 Ma sedimentary rocks demonstrates systematic variations with age and is correlated with the oxygen isotope composition. This means that the isotopic composition of cherts is likely to record their formation temperature. Using numerical models of the silicon cycle, the isotopic curve derived from this data provides a unique record of Precambrian seawater temperature variations, with changes from about 70–20 °C between 3.5 and 0.8 billion years ago, supporting the earlier conclusion of long-term ocean cooling based on the oxygen isotope data of cherts (Robert and Chaussidon 2006).

Based on the assumption that molecular sequence determines biological function and that the ancient molecules would have necessarily been adapted to and functional in their surroundings, the reconstructed molecules should offer a window into Earth's past environmental conditions. Therefore, the thermostability of experimentally resurrected ancestral enzymes provides an independent indicator to assess paleo-ocean temperatures. Garcia et al. (2017) analyzed ancestral enzymatically active nucleoside diphosphate kinases (NDKs) derived from light-requiring prokaryotic and eukaryotic phototrophs (cyanobacteria and land plants) having widely separated fossil-based divergence ages. Their results indicate a cooling trend of Earth's surface temperatures from ~65 to 80 °C in the Archean (~3.0 Ga) to ~35 °C at the Silurian–Devonian transition (~420 Ma), consistent with previous geological results.

The oxygen isotope compositions of carbonate and phosphatic skeletons have been widely used to reconstruct surface ocean temperatures in the last 500 Ma. A comprehensive compilation of Phanerozoic $\delta^{18}O$ data for carbonate and phosphate skeletons shows rapid shifts that can be explained by temperature change. Average low-latitude (30°S–30°N) temperatures of surface seawater declined from 42.0 ± 3.1 °C during the late Cambrian (Furongian) to Middle Ordovician to 35.6 ± 2.4 °C for the Late Ordovician through the Devonian, then varied between 19 and 35 °C from the Mississippian to today, including relatively cool icehouse temperatures from Carboniferous to Middle Permian (19–24 °C), and Early Triassic and Middle Cretaceous hothouse intervals (Grossman and Joachimski 2022). Correlations between pCO_2 forcing and temperature curves support the view that pCO_2 was a climate driver in the Paleozoic (Figure 8.9). Notably, the doubling of atmospheric CO_2 concentration from ~410 to 860 ppm was supposed to cause a >10 °C increase of surface temperature at a rate of 100–1000 °C per million years in the latest Permian (Song et al. 2021). This rapid warming that might be accompanied with intensified El Niños led to deforestation, reef demise, and a plankton crisis, all of which resulted in a positive feedback cycle that led to catastrophic but diachronous terrestrial and marine biodiversity losses (Sun et al. 2024).

8.4.3 Salinity History of Ocean

Salinity is defined as the total amount of salt dissolved in seawater and is typically about 35‰ in today's oceans. Due to problems with making accurate measurements of total salts, salinity is now normally expressed as a ratio of the electrical conductivity of the seawater to that of a standard concentration of KCl solution. The concentration of the standard solution is set such that the ratio of the approximate average salinity of the ocean today is equal to 35. The previous concentration value (‰) is thus unnecessary, but it is still widely used to highlight the basic meaning of the term. The relative proportions of the dissolved constituents are approximately constant, so the amount of Cl^-, the major component, is a proxy for salinity. Salinity history refers to concentration changes of dissolved salts in seawater over geological time.

Seawater contains a small amount of almost everything, but most solutes, or dissolved materials, are made up of a surprisingly small group of ions. In fact, only six ions (Cl^-, Na^+, $SO^2{}_4{}^-$, Mg^{2+}, Ca^{2+}, K^+) compose over 98% of seawater (Table 8.1). Na^+ and Cl^- account for about 85% of the

Table 8.1 The composition of seawater of 35‰ salinity. Although the concentration varies slightly from place to place in the ocean, the percentage of total salinity of each ion remains constant.

Ion	Concentration (‰)	Percentage of total salinity
Chloride (Cl^-)	19.345	55.03
Sodium (Na^+)	10.752	30.59
Sulfate (SO_4^{2-})	2.701	7.68
Magnesium (Mg^{2+})	1.295	3.68
Calcium (Ca^{2+})	0.416	1.18
Potassium (K^+)	0.390	1.11
Bicarbonate (HCO_3^-)	0.145	0.41
Bromide (Br^-)	0.066	0.19
Borate ($H_2BO_3^-$)	0.027	0.08
Strontium (Sr^{2+})	0.013	0.04
Fluoride (F^-)	0.001	0.003
Other dissolved material	<0.001	<0.001

salinity, which is why seawater tastes like table salt. Salinity is an important environmental factor that controls and limits the nature of metabolic activity, especially for marine animals. For example, no marine animals can survive in the restricted lagoons of the Persian Gulf, where salinities approach twice the open ocean values. Although halophiles thrive at higher salinities (up to ten times ocean salinity), populations of marine bacteria generally decrease with increasing salinity, and the diversity is exceptionally low in hypersaline water masses. Salinity can also strongly affect the solubility of atmospheric oxygen and thus impact metazoans. Throughout geological history, the composition of seawater has changed over time. Furthermore, changes in concentrations of ions like Ca^{2+}, Mg^{2+}, and SO_4^{2-} have affected the dominant minerals precipitating from seawater through geological time. This section discusses the available studies showing variations in total salinity and concentrations of Ca^{2+}, Mg^{2+}, and SO_4^{2-} ions, and trace elements in seawater in the geological past.

8.4.3.1 Total Salinity

Ancient seawater may be preserved in carbonates or other minerals as fluid inclusions. Analysis of primary, one-phase fluid inclusions precipitated directly from ancient seawater can provide direct salinity measurements. However, such samples are extremely rare because initial precipitates typically undergo dissolution/reprecipitation episodes during subsequent burial and metamorphism. There are rare cases of ancient seawater investigations. Analyses of primary, one-phase fluid inclusions in Cambrian and Ordovician primary marine cements yield salinities of 31–47‰ at about 500 Ma (Johnson and Goldstein 1993), within the salinity range of modeling estimates by Hay et al. (2006). Sanford et al. (2013) determined high-salinity groundwater (70‰) found more than 1000 m deep under the Chesapeake Bay as remnant seawater from the Early Cretaceous North Atlantic Sea (~100–145 Ma). The preservation of ancient seawater was supposed to be aided by the impact of a massive comet or meteorite that struck the area about 35 million years ago, creating Chesapeake Bay. Models of geochemical cycles yield no constraints on past salinity levels, and reliable geochemical proxies, isotopic records, or other possible indicators of ancient oceanic salinity are under development.

Since Na^+ and Cl^- constitute about 85% of the salts in the sea, the geological record of halite deposits allows insights into how ocean salinity may have varied over time. Enormous salt beds (saline giants) are found to be not evenly distributed in the geological record. The episodic nature of the deposition and subsequent episodic erosion of these units during tectonic uplift has surely led to major salinity variations in oceans over time. Two large time intervals in which most of the known salt deposits were deposited have been recognized (Figure 8.17). About 40% of the

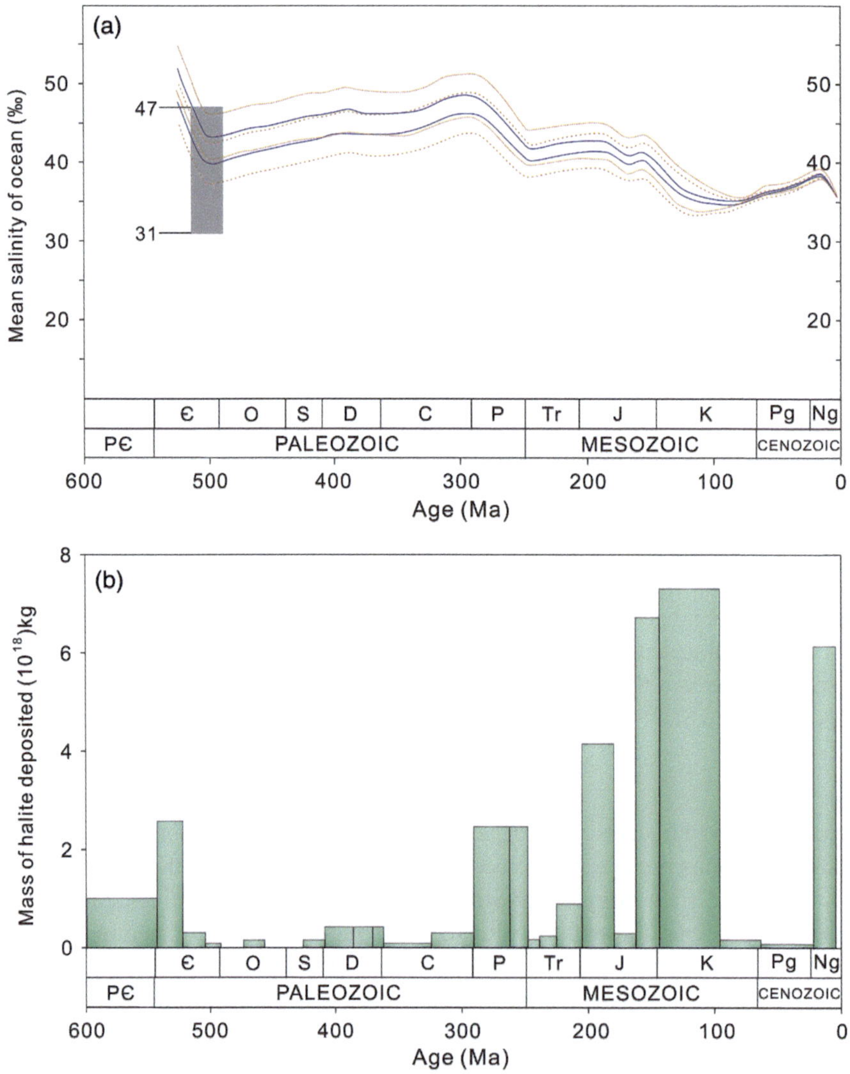

Figure 8.17 Abundance of halite deposits and estimates of salinity variations in the Phanerozoic. (a) Four models reconstructing the mean salinity of the ocean during the Phanerozoic, with fluid inclusion estimate indicated; the short-dashed lines are the maximum and minimum estimates based on the model that no water is lost through subduction, and the upper solid line is the average for this model; the long-dashed lines are the maximum and minimum estimates based on the model that water is steadily lost through subduction, and the lower solid line is the average for this model. (b) Distribution of existing halite deposits through the Phanerozoic showing the sporadic nature of deposition. *Source:* Adapted from Hay et al. (2006).

Phanerozoic inventory was sequestered in the interval 180–250 Ma. Much of the rest was deposited in the Neoproterozoic (Knauth 2005).

Knauth (2005) argued that the initial salinity was 1.5–2 times of the modern value, or even higher, if all salt and brine currently sequestered on the continents were returned to the ocean from which it was extracted by sedimentation on continents, and remained high throughout the Archean in the absence of long-lived continental cratons required to sequester giant halite beds and brine derived from evaporating seawater. The first great decline of oceanic salinity probably occurred in the latest Precambrian when enormous amounts of salt and brine were sequestered in giant Neoproterozoic evaporite basins. Since most forms of modern macroscopic life cannot tolerate salinities above 50‰, this "delayed" decline of ocean salinity was hypothesized to explain why the evolution of complex life took so long (Knauth 1998).

In the Phanerozoic, the deposition of the saline giants is strikingly episodic. Over half of the salt currently identified was deposited between 100 and 300 Ma following the breakup of the supercontinent Pangea. Using maximal and minimal estimates of volumes of existing evaporite deposits, combined with constant and declining volumes of ocean water through the Phanerozoic, Hay et al. (2006) developed models to reconstruct ocean salinity through the Phanerozoic (Figure 8.17). The modeling suggests that there was a major salinity decline during the Precambrian–Cambrian transition. The mean salinity of the ocean changes significantly, showing a general decline throughout the Phanerozoic (Figure 8.17a). There were a few major extractions of salt during the Paleozoic, a time of relatively stable but slowly increasing salinities. The Mesozoic was a time of generally declining salinity associated with the deep-sea salt extractions of the North Atlantic and Gulf of Mexico (Middle to Late Jurassic) and South Atlantic (Early Cretaceous). The decline in salinity corresponds closely to the evolution of both planktonic foraminifera and calcareous nannoplankton. Both groups were restricted to shelf regions in the Jurassic and Early Cretaceous, but spread into the open ocean in the mid-Cretaceous. Their availability to inhabit the open ocean may be directly related to the decline in salinity (Hay et al. 2006). Estimated salinities during the Early Miocene were between 37‰ and 39‰.

8.4.3.2 Secular Variations of SO_4^{2-}, Ca^{2+}, and Mg^{2+} Concentrations

Sulfate is the second most abundant anion in modern seawater, with a concentration of ~29 mM. Microbial sulfate reduction, which always couples with the oxidation of organic matter, plays a central role in the global cycles of carbon and sulfur. In some marine settings, sulfate reduction may account for more than 50% of organic carbon oxidation in sediments. Seawater sulfate concentrations can vary in response to oxidative weathering, massive evaporite deposition, reduced sediment ventilation, intensive microbial sulfate reduction, increased pyrite burial, or other geobiological processes. Concentrations of dissolved ions (e.g. Ca^{2+}, Mg^{2+}, and SO_4^{2-}) in ancient seawater can be determined by analyzing fluid inclusions in marine halite and using quantitative geochemical methods coupled with modeling (Algeo et al. 2015; Berner 2004; Lowenstein et al. 2001).

Quantitative geochemical studies have revealed that the sulfate concentration of seawater has varied significantly through geological time, increasing episodically in line with the pace of oxygenation (Figure 8.27). It showed an increasing trend from less than 200 μM during the Archean and Paleoproterozoic before the GOE, elevated to ~5 mM or less during most of the Proterozoic, rose sharply across the Ediacaran–Cambrian boundary to ~5–10 mM, and rose again during the Permian to near-modern levels of ~10–30 mM (Algeo et al. 2015). The data from fluid inclusions suggest a negative couple of Ca^{2+} against Mg^{2+} and SO_4^{2-}. Seawater had high Mg^{2+}/Ca^{2+} ratios (>2.5) and relatively high Na^+ concentrations during the Late Ediacaran, Permian, and mid-Eocene to the present (40 to 0 Ma), when aragonite and $MgSO_4$ salts were the dominant marine precipitates.

Conversely, seawater had low Mg^{2+}/Ca^{2+} ratios (<2.3) and relatively low Na^+ concentrations during the early Cambrian (540 to 520 Ma), Silurian, and Cretaceous, when calcite was the dominant non-skeletal carbonate and K, Mg, and Ca-bearing chloride salts were the only potash evaporites (Lowenstein et al. 2001). Berner (2004) used a theoretical model to constrain concentrations of SO_4^{2-}, Ca^{2+}, and Mg^{2+} in seawater over the past 550 Ma. The result agrees well with the data of salt fluid inclusions for variations of Ca^{2+} and SO_4^{2-} (Figure 8.18). Hardia (2003) extended the previous seawater secular-variation model to predict the secular variations in Precambrian seawater chemistry (concentrations of Ca^{2+}, Mg^{2+}, K^+, and SO_4^{2-}) back to the Late Archean. The strong correlation in Figure 8.19 supports the predictions: (1) Precambrian seawater was a saline NaCl water with $Ca^{2+} > HCO_3^-$ since at least the Late Archean. (2) The major ion compositions of Precambrian seawater chemistry and their secular variations are in the same ranges as those of Phanerozoic seawater. (3) The Mg/Ca mole ratio in seawater has controlled the types of $CaCO_3$ polymorphs (Hardie 2003).

Abiotic $CaCO_3$ polymorphs are dependent on saturation state, Mg/Ca mole ratio, and temperature of seawater. In general, aragonite precipitation is favored when the Mg/Ca mole ratio is $> \sim 2$ and water temperature > 20 °C. No abiotic calcite precipitation occurs when temperature > 20 °C. No calcite or aragonite nucleation occurs when super saturation index <18 and Mg/Ca mole ratio > 2, while calcite nucleation is possible with Mg/Ca mole ratio < 2 in a range of saturation states. Calcite nucleation barrier surpasses that of metastable aragonite in solutions with Mg/Ca ratios consistent with modern seawater (Mg/Ca = 5.2) and hence only aragonite is preferred to nucleate (Figure 8.20; Sun et al. 2015). Under these constraints, biotic aragonite precipitation is possible throughout the Phanerozoic. By using Mg/Ca = 2 as the only boundary criterion for calcite seas (Mg/Ca > 2) versus aragonite seas (Mg/Ca ≤ 2), calcite is predicted to have been precipitated in the intervals 40–170, 330–520, 925–1105, 1310–1425, 1675–1875, 2025–2125, and 2375–2630 Ma, and aragonite (±Mg-calcite) in the intervals 0–40, 170–330, 550–925, 1105–1310, 1425–1675, 1875–2025, 2125–2375, and 2630–2750 Ma (Figure 8.21). However, the mineralogy of biogenic carbonate is not strictly regulated by Mg/Ca ratio of seawater. Mineralogy of carbonate skeletons, especially those formed through BCM, are largely biologically controlled (see Chapter 5). Biominerals

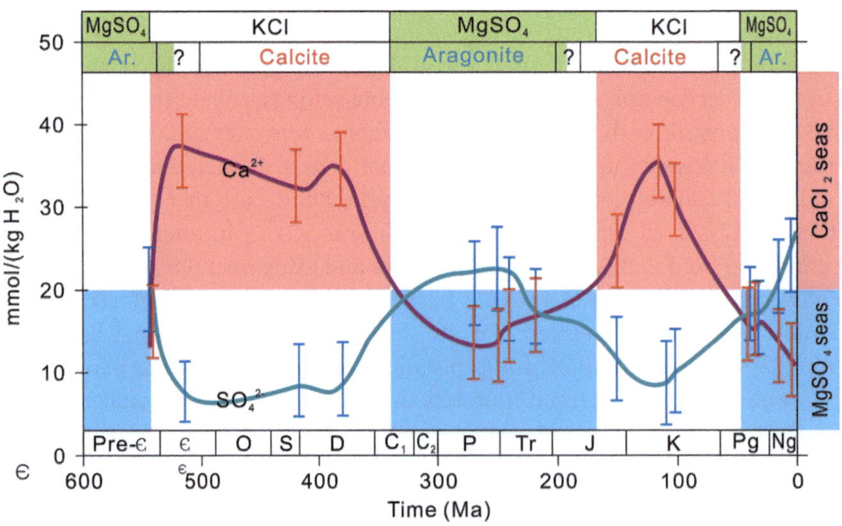

Figure 8.18 Secular variation in Ca^{2+} and SO_4^{2-} concentrations of seawater during Phanerozoic time, determined by the analysis of fluid inclusions in marine halite. Also plotted are temporal distributions of primary Phanerozoic non-skeletal carbonate minerals. *Source:* Modified from Lowenstein et al. (2003).

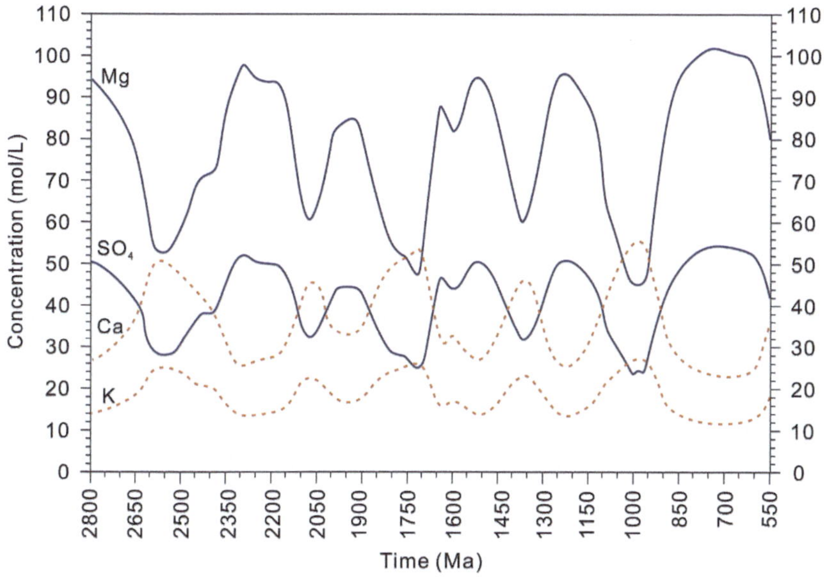

Figure 8.19 Predicted secular variations in concentrations of Mg^{2+}, $SO_4{}^{2-}$, Ca^{2+}, and K^+ ions in seawater during the Precambrian. *Source:* Adapted from Hardie (2003).

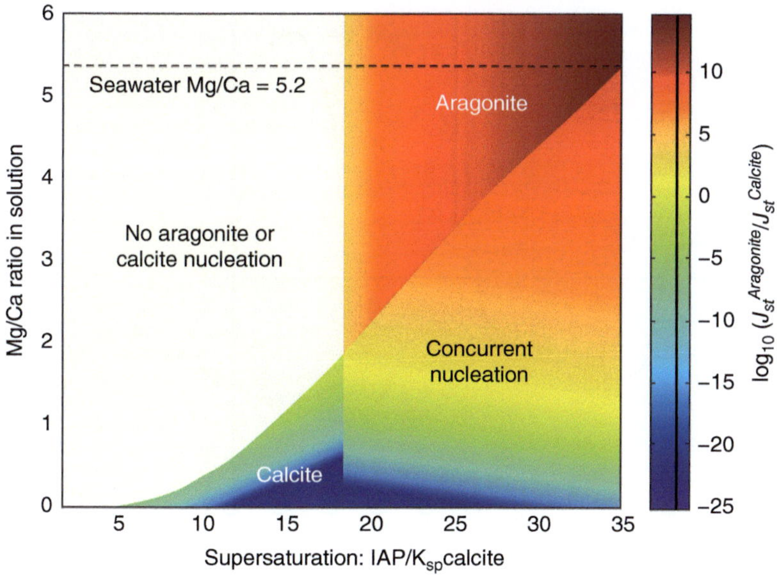

Figure 8.20 Kinetic phase diagram of the relative nucleation rate between calcite and aragonite (color-coded) as a function of solution Mg:Ca ratio and the supersaturation. For Mg:Ca = 5.2 (modern seawater), only aragonite is preferred to nucleate. Concurrent nucleation of calcite and aragonite occurs for a broad span of supersaturations near Mg:Ca = 2. *Source:* Sun et al. (2015) / National Academy of Sciences.

do not form in seawater but in compartments never completely isolated from seawater. The organism regulates the concentration and speciation of inorganic ions, polysaccharides, and proteins, all of which can select the mineral polymorph in the final and can vary activation barriers for particle formation, phase transitions, and growth (Gilbert et al. 2022).

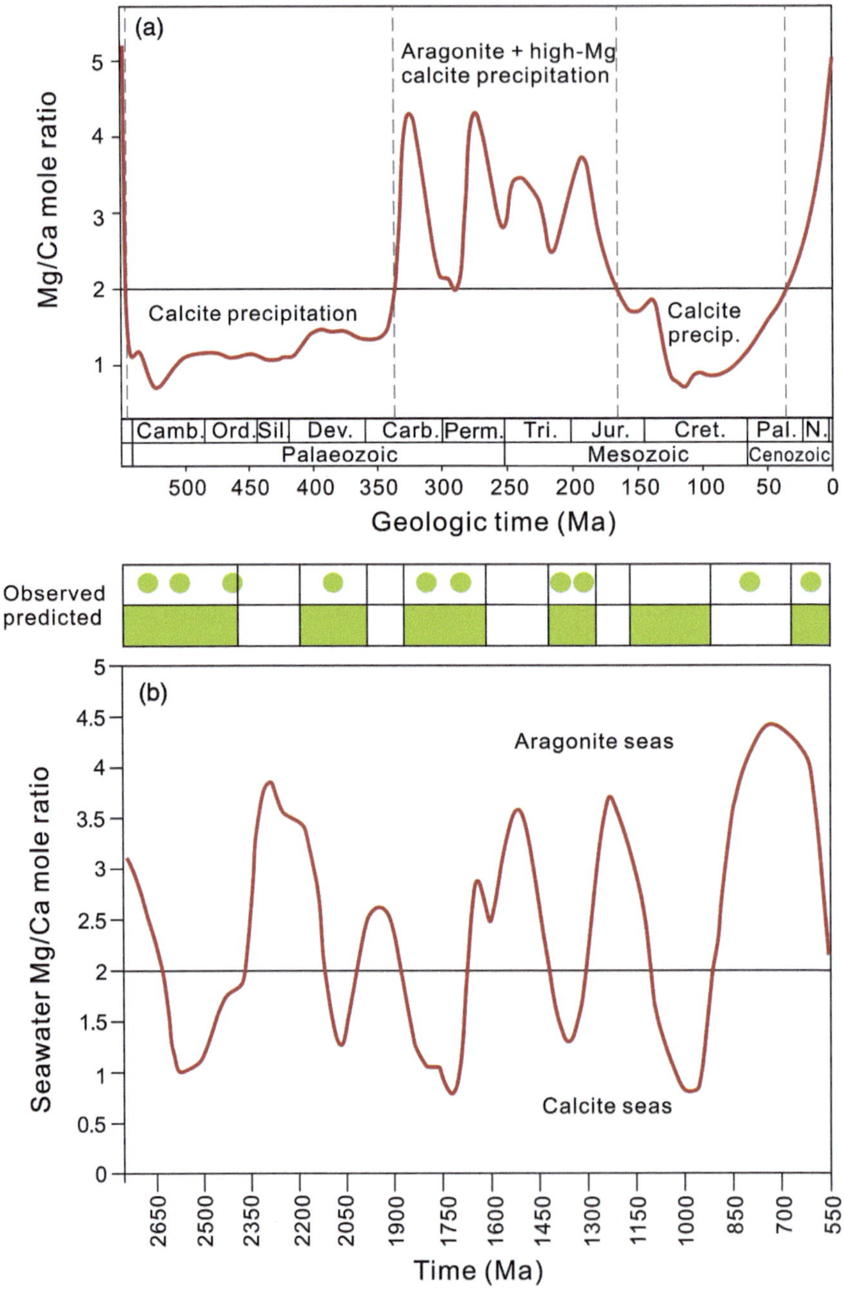

Figure 8.21 Secular variation in Mg/Ca mole ratio of seawater and the timing of aragonite seas and calcite seas. (a) The ratio of Mg/Ca in seawater over Phanerozoic time, black vertical bars represent data of fluid inclusions in marine halite. *Source:* Adapted from Lowenstein et al. (2001), Horita et al. (2002) and the gray line represents model calculations by Berner (2004). (b) Mg/Ca mole ratio of Precambrian seawater. *Source:* Adapted from Hardie (2003).

8.4.3.3 Bioavailability of Phosphorus and Nitrogen

Phosphorus and nitrogen are primary components of many biomolecules. They are key limiting nutrients for terrestrial and marine primary production. Therefore, their biogeochemical cycles control ecosystem growth and are closely linked to organic carbon burial, oxygenation, and life evolution.

The total phosphorus inventory quickly grew to a modern level after Earth's formation, and has been redistributed into different reservoirs during major geological events (Walton et al. 2023). From a geobiological standpoint, bioavailable phosphorus is derived from weathering, a vital fertilizing agent for biological productivity, and considered as an ultimate limiting nutrient in the ocean on timescales exceeding 1000 years. Variations in the marine phosphorus cycle have certainly occurred over time and determining these variations is challenging due to a lack of effective marine proxies for the P cycle. No useful phosphorus isotope exists to constrain the phosphorus cycle in deep time, nor are fossil-based proxies adequate for determining past changes in the phosphorus cycle. The geological record indicates an increasing trend in the phosphorus content of marine sediments, with iron-bound phosphorus likely playing an important role in the increased phosphorus levels in continental margin sediments. This trend generally mirrors the oxygenation of the Earth and could have been driven by shifts in the atmosphere-ocean redox state (Zhao et al. 2024), which is supported by the fact that both modern and ancient sediments deposited in anoxic settings contain much less phosphorus than those deposited under oxygenated water columns (Canfield 2014). The redox-dependent recycling capacity of phosphorus in oceanic environments increased episodically in line with the pace of oxygenation (Walton et al. 2023). The modern ocean has a phosphorus concentration of about 2.5 μM, whereas seawater in the Archean and early Proterozoic likely contained 0.04–0.13 μM phosphorus on average (Jones et al. 2015).

Nitrogen cycle has co-evolved with life and redox conditions over geological time. Sedimentary nitrogen isotope record, together with other proxies, demonstrates that the nitrogen cycle has responded to and perhaps modulated events such as the emergence of life, oxygenation events, major climatic perturbations, and mass extinction events. Quantitative estimates of bioavailable nitrogen through geological time are available (Stüeken et al. 2024). Nitrogen fixation in modern ecosystems generates 10^{13} mol N/yr. During the Hadean and Eoarchean before the evolution of N_2 fixation, abiotic processes such as lightning, photochemical reactions, meteorite impacts, and hydrothermalism, might have converted $\sim 10^9$–10^{12} mol N/yr into bioavailable nitrogen oxides and ammonium, sufficient to sustain a small biosphere. Biologically controlled nitrogen cycle has been established since the origin of N_2 fixation from the Paleoarchean to Mesoarchean onward but operated under anoxic conditions. $\delta^{15}N$ records of the mid-Archean varying between $\sim$0‰ and 1‰ suggest a predominance of N_2 fixation and rapid recycling of ammonium, with dissolved ammonium concentrations on the order of a few μmol/L or less. Organic carbon burial and localized ammonium oxidation to N_2 coupling to reduction of ferric iron were the main sinks of nitrogen from the Mesoarchean oceans. Nitrate has probably begun to accumulate in oxygenated surface seawaters since the GOE that might cause the onset of nitrification and denitrification. Therefore, aerobic processes became more established in the N cycle, possibly enabling the diversification of eukaryotes in the mid-Proterozoic, while nitrate might have been limiting for the emergence of complex life until the Neoproterozoic to Paleozoic oxygenation of the deep ocean. During the Phanerozoic, transitions in the nitrogen cycle over 100 Ma timescales were associated with icehouse-to-greenhouse conditions. Short-lived perturbations occurred during mass extinctions and anoxic events, impacting nitrogen availability and cycling.

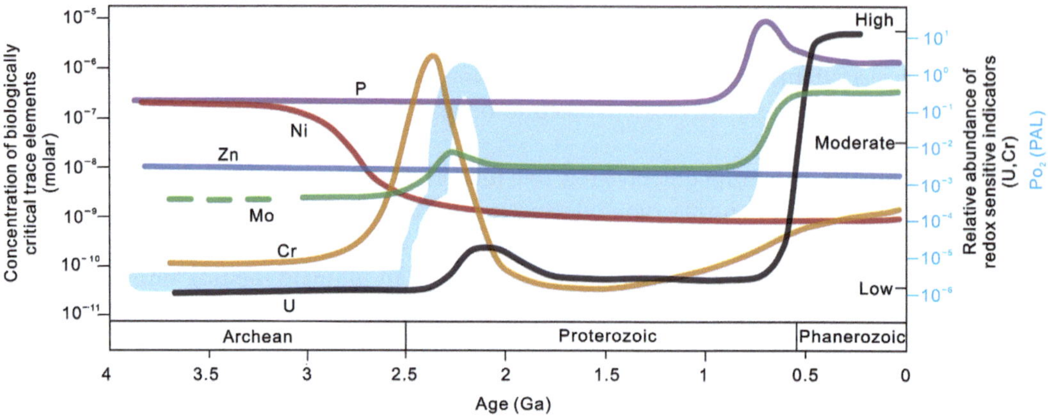

Figure 8.22 Secular variations of marine nutrients (e.g., P) and bio-essential trace metals in the background of Earth's oxygenation through geological time based on data described in Section 8.4.3.

8.4.3.4 Bio-essential Trace Metals

Life requires a wide variety of trace metals (e.g. Fe, Co, Ni, Zn, and Cu) to act as structural components and reactive centers in metalloenzymes. Availability of bio-essential trace metals in marine environments has been closely linked to exhalations from the Earth's interior, redox state of oceans, and evolution of life. The temporal changes in concentrations of these elements have profound implications for the evolution of ocean chemistry and development of biological systems through Earth's history.

A comprehensive sedimentary pyrite–trace element database of the Precambrian demonstrated how changing availability and scarcity influenced early life. The data show relatively elevated concentrations of Ni, Co, Cu, and Fe in the Archean, and enriched Cr in the Paleoproterozoic. Abundances of Ni, Co, Fe, and Cu in the Archean potentially facilitated their widespread utilization by prokaryotes. The Paleoproterozoic–Mesoproterozoic saw increases in Zn and Mo but a marked decline in Ni, Co, Cu, Se, and Fe. An overall increase in a multitude of elements (Ni, Co, Cu, Cr, Se, V, Mo, and P) in the Neoproterozoic and Cambrian was considered to benefit the rise of complex life forms from then on (Mukherjee and Large 2020). Robbins et al. (2016) compiled records of bio-essential trace elements (Ni, Mo, P, Zn, Co, Cr, Se, and I) in sedimentary archives. The redox-sensitive elements either increase (e.g. U and Mo) or decrease (e.g. Ni) following the pace of ocean oxygenation. Cr is redox-sensitive but not bio-essential, showing an enrichment in the Paleoproterozoic and Phanerozoic. In contrast, Zn displays less redox-sensitive behavior, maintaining low content throughout Earth's history (Figure 8.22).

Fe and Mo are important elements for biological nitrogen fixation and nitrate assimilation. Fe is effectively removed from solution in both oxic and sulfidic conditions. In the anoxic Archean oceans, the Fe concentration may have been as high as 50 μM, more than three orders of magnitude higher than modern oxygenated seawater and sulfidic deep waters of the chemically stratified Black Sea (the closest modern analog to a sulfidic ocean). In contrast, Mo forms the highly mobile molybdate anion (MoO_4^{2-}) under oxidizing conditions, and thus Mo is the most abundant transition metal in modern oceans, with a concentration of 105 nM and an ocean residence time of ~8–10^5 years. Under sulfidic conditions, Mo is readily removed to sediments by reduction to insoluble sulfides or conversion to particle-reactive thiomolybdate (MoS_4^{2-}). Therefore, in the Black Sea, Mo concentrations fall from ~40 nM in oxygenated surface waters to ~3 nM below the chemocline.

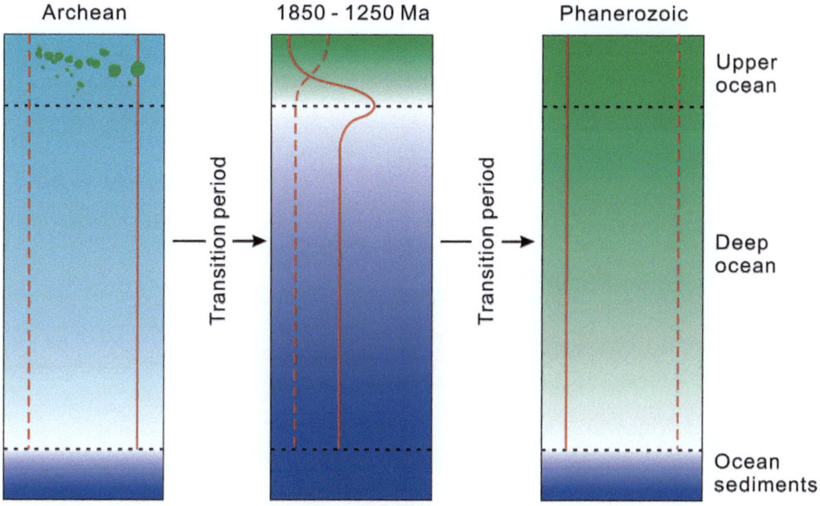

Figure 8.23 Schematic depiction of effects of changing ocean redox conditions on the depth distributions of Mo (dashed lines) and Fe (solid lines). During the Archean, oceans were anoxic but not sulfidic. O_2 may be present only in cyanobacterial "blooms," Mo was scarce because it is not readily mobilized from crustal rocks during weathering under low Po_2, and Fe was abundant in the absence of O_2 and H_2S. From 1850 to 1250 Ma, moderate Po_2 oxygenates surface waters but sulfidic deep waters develop (Canfield Ocean); Mo was scarce because of rapid removal in sulfidic deep waters but somewhat elevated at the surface because of upper ocean oxygenation and enhanced oxidative weathering; Fe was depleted in sulfidic deep waters, severely depleted in oxic surface waters, and enriched near the redoxcline where both O_2 and H_2S were scarce. During the Phanerozoic, O_2 penetrates to the sediment-water interface, Mo and Fe distributions were similar to today's ocean. *Source:* Adapted from Anbar and Knoll (2002).

Both elements are redox-sensitive but opposite in chemical behaviors, which resulted in different responses to the dynamic oxygenation of stratified oceans (Figure 8.23). Geochemical data indicate that for much of the Proterozoic, Earth's oceans were moderately oxic at the surface and sulfidic at depth. Under these conditions, Fe concentrations declined in surface ocean after the GOE, while deep ocean remained ferruginous in the Archean and shifted to be sulfidic in the post-GOE Proterozoic. It was inferred that Mo surface concentrations in the late Paleoproterozoic and early Mesoproterozoic oceans were less than 10% of present levels if sulfidic conditions covered >10% of the seafloor. Since the oxygenation of the deep ocean, Mo concentrations have become similar to today's ocean (Anbar and Knoll 2002).

8.4.4 Evolution of Ocean Oxygenation

As discussed above, the early ocean, with higher temperature and salinity, was significantly different from the modern ocean in many physical and chemical properties, including oxygen solubility. Knauth (2005) examined the solubility of oxygen in seawater through geological time as a function of salinity and temperature under the present (pre-industry) atmospheric level (PAL) of oxygen (Figure 8.24). At the current salinity of 35‰ and a surface mean temperature below 25 °C, dissolved O_2 is greater than 5 mL/L. However, during the Archean, surface ocean temperatures exceeding 55 °C and salinity about 50–70‰, dissolved O_2 would be less than 3 mL/L (<4 mg/L) even if atmospheric O_2 were at modern levels. If the Archean atmosphere had just 30% less oxygen than today, the oceans would have been anoxic due to the high temperature and salinity alone. In other words,

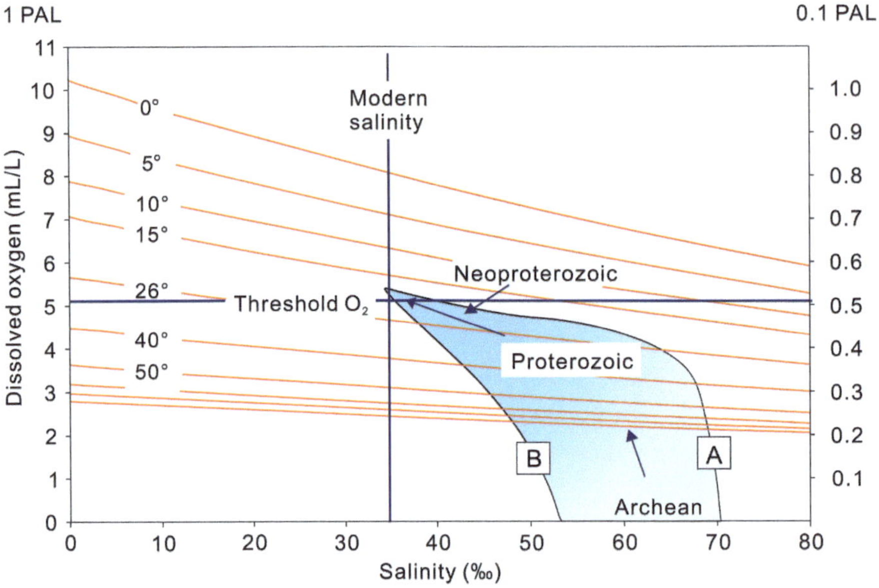

Figure 8.24 Dissolved oxygen in seawater as a function of salinity and temperature for the present atmosphere (pO_2 = 1 PAL) and an ancient atmosphere (pO_2 = 10% PAL). An example threshold level of 0.5 mL/L for dissolved O_2 required for Neoproterozoic metazoans is shown together with example trajectories for the evolution of dissolved O_2 in seawater with time. Curve B invokes a minimum salinity in the Archean and a relatively steady decrease throughout the Precambrian. Curve A is for high initial salinity with little decline until the deposition of the great Neoproterozoic saline giants. Under an atmosphere of 10% PAL O_2 and temperatures of 15–25 °C, reduction of ocean salinity by deposition of these salt deposits could shift dissolved O_2 to values above that required for metazoan respiration, as shown by the arrow labeled "Neoproterozoic." *Source:* Knauth (2005) / with permission of Elsevier.

high salinity and temperatures alone would ensure a predominantly anoxic ocean environment as long as pO_2 was below 70% PAL.

The three-box model of oxygenation (Kasting 1987, 1991) proposed a stratified ocean throughout most of Earth's history. It stated that the Earth's surface environment was generally devoid of free O_2 before the appearance of redbeds at ~2.0 Ga, except for oxygen whiffs in localized regions of high productivity in the surface ocean. Thereafter, the surface ocean became oxidizing, while the deep ocean remained anoxic. In the later stage, from the disappearance of BIFs around 1.85 billion years ago to the present, both the surface and deep ocean contained appreciable amounts of free O_2. This three-box model and three-stage evolution are generally supported by numerous geochemical data based on multiple proxies, except for the timing of deep ocean oxygenation. Multiple lines of evidence support the oxygenation of the surface ocean during and after the GOE. Geochemistry of Fe speciation suggested that the deep ocean remained anoxic and ferruginous during pre-Gaskiers Neoproterozoic and became oxic after the Gaskiers glaciation (580 Ma), which requires a pO_2 at least 15% PAL (Canfield et al. 2007). However, analyses of a large geochemical dataset coupled with Earth system modeling reveal a moderate long-term increase in atmospheric oxygen and marine productivity from the Ediacaran to the mid-Paleozoic, with no evidence for wholesale ocean oxygenation during the NOE. The increase in pO_2 and marine productivity during the Ediacaran–Cambrian might be insufficient to oxygenate the deep ocean. The global ocean did not become persistently oxygenated to the near-modern level until the Silurian–Devonian transition, when land plants expanded (Stockey et al. 2024). The COPSE model (Figure 8.15), δ^{98}Mo record, and appearance of giant sea scorpions also suggest a major rise of atmospheric oxygen level at around 420 Ma (Canfield 2014).

The disappearance of BIFs since 1.85 Ga was previously cited as evidence for deep ocean oxygenation. However, Anbar and Knoll (2002) argued that the low solubility of Fe-sulfides could alternatively indicate that the deep oceans became sulfidic rather than oxic after 1.85 Ga. This argument aligns with the Canfield Ocean model proposed by Canfield (1998), which describes ocean oxygenation during the Proterozoic, particularly during the Boring Billion (1.8 to 0.8 billion years ago). During this period, the ocean was characterized by an anoxic and sulfidic condition known as euxinia (similar to the Black Sea). Thereafter, the euxinic water bodies dwindled, while ferruginous bottom water expanded. The size of sulfidic water bodies may have varied from place to place and from time to time and also may have rimmed the global ocean as "dynamic wedges" at mid-water depths between the overlying oxic water and underlying ferruginous water (Li et al. 2010) and persisted into the Phanerozoic, at least locally.

In summary, the ocean redox state on a global scale has shown spatial heterogeneity and temporal instability throughout Earth's history. Archean oceans were generally anoxic but not sulfidic, though O_2 whiffs associated with oxygenic cyanobacterial "blooms" may have existed. The Paleoproterozoic rise in atmospheric O_2 caused moderate oxygenation of Earth's surface oceans. However, at depth, oceans remained anoxic and sulfidic for much of the Proterozoic. Persistent oxygenation of deep oceans occurred during the late Ediacaran or later at the Silurian–Devonian transition, while transient, local anoxic and/or sulfidic water bodies may have been present at different times and in different places. The modern ocean is generally oxygenated, with cold, oxygen-rich bottom water constantly replenished by the flow of Arctic water through global circulation, while oxygen minimum zone (oxygen at its lowest level) exists at depths of about 200–1500 m.

8.5 Evolution of Biosphere

This section provides a concise overview of the history of life, emphasizing major evolutionary events that made Earth. It is presented in a chronological framework, beginning with a brief introduction to the origin of life, followed by a framework of life and its metabolic evolution through geological time, and concluding with a summary of the geobiological evolution of Earth on a global and long-term scale.

8.5.1 The Origin of Life

The Earth was probably inhabited by microbial life at some point between late Hadean and Eoarchean, when Earth's crust began to solidify and surface temperature dropped to a life tolerant range from −20 to 121 °C, currently known temperature extremes for psychrophiles and hyperthermophiles capable of completing their life cycles, respectively. The life origin is one of the most fascinating and critical questions in science. According to the classic theoretical model, the first life was a product of chemical evolution in four stages, which can be found in many textbooks as the following: (1) abiotic synthesis of small organic molecules, such as amino acids and nucleotides; (2) joining of small molecules into polymers, including proteins and nucleic acids; (3) origin of self-replicating molecules that eventually made inheritance possible; and (4) packaging of all these molecules into "protobionts," droplets with membranes that maintained an internal chemistry different from the surroundings. These are all speculative, of course, but the first two stages can be tested in the laboratory. The famous Miller-Urey experiment effectively synthesized organic monomers, and laboratory simulations of early-Earth conditions have produced organic polymers. While the exact process remains a mystery, several hypotheses and theories have been proposed to explain how life might have arisen on Earth, which are summarized below.

The panspermia hypothesis suggests that life did not originate on Earth but was instead brought here from elsewhere in the universe. If life had existed near the surface of Mars, or any terrestrial planet in a continuously habitable zone, it could feasibly have been ejected as part of a meteorite and delivered intact as a living organism to Earth, seeding the development of life on our planet. This theory circumvents the challenging task of elucidating how life began on Earth by transferring the problem to another planet far away and long ago. However, many compounds of biological significance, such as acetaldehyde (CH_3CHO), cyanamide (NH_2CN), and HCN, have been spectroscopically identified in interstellar regions.

The RNA world hypothesis posits that RNA, a molecule capable of storing genetic information and catalyzing chemical reactions, was a key player in the origin of life. RNA molecules could have formed spontaneously and then replicated themselves, eventually leading to the development of more complex life forms. Therefore, it is hypothesized that the first hereditary material was not DNA, but RNA, which means the molecular biology today was preceded by an "RNA world."

The primordial soup hypothesis speculated that life began in a "primordial soup" of organic molecules that had diverse origins from chemical reactions in early oceans or on the surface of the Earth, or from extraterrestrial material. Potential energy source for organic synthesis comes from UV radiation, electrical discharges, and the shock waves of incoming extraterrestrial bolides. Synthesis of amino acids was successfully accomplished by subjecting a reduced gaseous mixture of CH_4, H_2, H_2O, and NH_3 to electric discharges (Miller-Urey experiment). The shock waves of meteorites have been considered more efficient than UV radiation in producing organic chemicals. Hypersonic meteorites can generate high temperatures up to 10,000 K that is sufficient to ionize atmospheric gases and yield a variety of secondary organic compounds. Simulated experiments produced small amounts of HCN, aldehydes, and amino acids from shock vaporization of meteorites in a reduced gas mixture of CH_4, C_2H_6, NH_3, and H_2O (see Konhauser 2007 and references therein). Under reducing conditions, organic material can be synthesized in the Miller-Urey experiment, while under non-reducing, CO_2-rich atmospheric conditions, the generation of organic molecules may be much lower, which has led to the suggestion that extraterrestrial material may instead have provided a more suitable source of organic compounds.

The hydrothermal vent hypothesis proposes that life originated at hydrothermal vents on the ocean floor. These vents release hot, mineral-rich water that could provide energy and chemical precursors for the origin of protobionts. Molecular phylogenetic analyses indicate that the ancestors of modern prokaryotes thrived in hot conditions and may have lived on inorganic sulfur compounds that are common in deep-sea vent environments. Laboratory studies also point to the environments of deep-sea vents as abiotic sources of some of the organic compounds, e.g. acetyl coenzyme A (acetyl-CoA), a crucial molecule essential for energy production in cells. Sulfides of iron and nickel, common in deep areas, catalyze the formation of acetic acid and precursors of acetyl coenzyme from CO and H_2S. In addition, abiotic formation of hydrocarbons may occur today in mid-ocean ridge systems. Strong selection for hyperthermophiles would also have existed on early Earth as a result of episodic meteorite bombardment. In the absence of ozone, the Earth's surface was probably bathed in lethal UV radiation. This problem would not apply to the hydrothermal habitat, which favors abiotic synthesis of ammonia as well as the formation of key elements of the citric acid cycle from CO and H_2S in the presence of metal catalysts (Line 2002).

The alkaline hydrothermal vent hypothesis suggests that life originated at deep-sea alkaline hydrothermal vents, such as the Lost City Hydrothermal Field, located on the Atlantis Massif at the intersection between the Mid-Atlantic Ridge and the Atlantis Transform Fault in the Atlantic Ocean. Unlike common hydrothermal vents, such as black smokers and methane seeps, Lost City hydrothermal vents are generated when seawater encounters hot ultramafic rocks (such as olivine and peridotite) from the Earth's mantle. The water-rock reaction known as serpentinization produces alkaline, less hot fluids (40–90 °C) filled with hydrogen and methane gas. The calcium-rich fluids then react with carbon in the seawater to form Lost City's iconic carbonate chimneys (white smokers). Serpentinization requires only water and olivine but is able to create an alkaline condition with unlimited energy and matter source that a cell needs to synthesize biomolecules. Additionally, interconnected micropores in the walls of Lost City chimneys provide an ideal place for chemical reactions. Taken together, it's plausible that the high-pH water swirling into chimney micropores could generate favorable conditions for life to begin. The hydrogen gas and other molecules mixing together in the micropores may create a precursor to a cell, known as a protocell. Currently, the Lost City Hydrothermal Field is a prime location for investigating the origin of life (Kusmer 2018).

The Iron–Sulfur World Hypothesis suggests that the first life was a chemoautotroph that may have originated on the surface of iron and nickel sulfide minerals, particularly in hydrothermal vent environments (Wächtershäuser 2010). These minerals could have catalyzed the formation of organic molecules, providing a surface for chemical reactions to occur and leading to the development of early life forms. Both the primordial soup and hydrothermal models reveal that bio-essential organic molecules could form abundantly in early oceans. However, synthesis of amino acids and other organics, and their further polymerization into a peptide, require condensation reactions. In the absence of enzymes, it is difficult to conceive how these reactions may have occurred in early oceans with hopelessly diluted concentration of organic molecules. A critical step of life origin requires a means to select and concentrate right molecules from the watery prebiotic soup (Hazen 2005). Accordingly, organic compounds can be synthesized by autocatalytic reactions between simple inorganic components (e.g. CO_2, $H_2PO_4^-$, and NH_4^+) on mineral sorbing surfaces. For instance, in the presence of Fe^{2+}, and at pH values up to 5, pyrite has a positively charged surface capable of adsorbing inorganic anions and negatively charged organic molecules from the fluids. These ions and molecules could accumulate and polymerize into more complex compounds directly on the mineral surface. Sowerby et al. (1996, 1998) exposed ideally flat cleavage faces of graphite (C) and molybdenite (MoS_2) to solutions of adenine, guanine, and other biologically interesting organic species. These cyclic molecules were adsorbed to the mineral surfaces and organized into two-dimensional structures (Figure 8.25). Interestingly, some everyday minerals like quartz and calcite are able to select and concentrate on specific biological amino acids from an equimolar mix. In addition, certain crystal faces of these and other minerals display an aptitude for adsorbing handed or "chiral" molecules (Hazen 2005). Therefore, minerals are able to provide organizing templates for chemical reactions.

8.5.2 The Timing of Life Origin

Three independent approaches have been used to constrain the timing of life origin on Earth or elsewhere in the universe. The results range from 4.2 to 3.6 billion years ago, challenging the assumption that the LHB was inimical to the origin and continuity of life. Serpentinization, a universal geological process, might have occurred on the Hadean Earth or other planets. Critically,

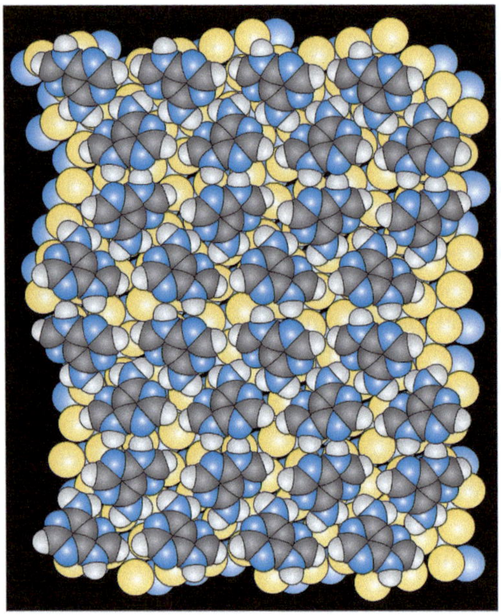

Figure 8.25 Adenine ($C_5H_5N_5$), a biomolecule that contributes to both metabolic and genetic processes (illustrated as clusters of 15 smaller silver, gray, and blue spheres, which represent hydrogen, carbon, and nitrogen atoms, respectively) forms a periodic two-dimensional structure on molybdenite (MoS_2, illustrated as larger yellow and blue spheres, which represent sulfur and molybdenum atoms, respectively). Similar molecular adsorption and organization on mineral surfaces may have played a role in the origin of life. *Source:* Adapted from Sowerby et al. (1996).

serpentinization is known to generate sufficient energy and matter, ideal pH conditions, and Goldilocks compartments needed for spontaneous life. Thus, it has significant implications for the potential habitability of early Earth and other planetary bodies, such as Mars and icy moons like Europa and Enceladus.

The presence of isotopically depleted graphite microparticles from Hadean zircons and Eoarchean metasedimentary rocks (Figure 8.3) provides suggestive but not conclusive evidence for life existing since 4.1 Ga. The $\delta^{13}C$ values of these earliest graphite microgranules, ranging from −19‰ to −50‰, are consistent with the results of biological carbon fractionation, probably through photosynthesis or methane metabolism. However, abiotic processes, such as Fischer-Tropsch-type (FTT) synthesis, are known to convert carbon monoxide (CO) and hydrogen (H_2) into hydrocarbons that are depleted in ^{13}C under hydrothermal conditions. This indicates that carbon isotopic composition may not be a particularly effective diagnostic means to differentiate between biological and non-biological sources (McCollom and Seewald 2006). Therefore, geological evidence for the origin of life in the Hadean remains elusive.

Nano magnetite grains (average 15–20 nm in diameter), polycyclic aromatic hydrocarbons (PAHs), and fossil nanobacteria were reported from carbonate minerals of ALH84001, a Martian meteorite dating back to 3.9–4.1 Ga. Specifically, the tiny magnetite grains were initially interpreted as magnetofossils (comparable to intracellular magnetite crystals formed by terrestrial magnetotactic bacteria) and cited as evidence for life existing on Mars as early as 4.1–3.9 billion years ago. However, ultrastructural features (spiral defects) discovered in some of the ALH84001 magnetite suggest they were formed by vapor deposition at high temperatures. Additionally, these magnetites exhibit epitaxial growth relationships with the host carbonate, making it unlikely that they formed within bacterial cells (Bradley et al. 1998). Similarly, other supposed evidence for Martian life from ALH84001, such as nanofossils and PAHs, is also untenable (Farmer 2001). Early Martian life remains to be discovered.

The third approach uses molecular-clock techniques to estimate the age of the Last Universal Common Ancestor (LUCA), the hypothetical common ancestral cell from which all life on Earth descended. LUCA is imagined as a complex life form with a lipid bilayer, genetic code, and ribosomes for translating DNA or RNA into proteins. It is not the origin of life itself but represents a point in evolutionary history where the three domains of life—Bacteria, Archaea, and Eukarya—diverged. Early work dated LUCA back to approximately 4.0 billion years ago (Hedges et al. 2001). A more recent study inferred that LUCA lived ~4.09–4.33 Ga through divergence time analysis of pre-LUCA gene duplicates. The age is generally in agreement with the earliest known graphite with a light carbon isotope composition. Phylogenetic reconciliation suggests that LUCA had a genome of at least 2.5 Mb, encoding ~2600 proteins, comparable to modern prokaryotes (Moody et al. 2024).

8.5.3 The Framework of Life History

8.5.3.1 Prokaryotes Dominated the Biosphere for 2.0 Billion Years

The long-term trend of life evolution is characterized by a progression from simple to complex, from small to large, from few to many, and from monotonous to diverse life forms. Given that prokaryotic cells are generally smaller and much simpler than eukaryotic cells, it is reasonable to presume that the earliest life forms were prokaryotes. This presumption is particularly supported by the fossil record dating back to the Precambrian. The presence of life in the Hadean and Eoarchean is only evidenced by ambitious biogenic graphite microgranules and suggested by the ancient age of LUCA. In contrast, multiple lines of evidence indicate that diverse life forms were in place in a range of ecosystems in the Paleoarchean, including anoxygenic photobionts, sulfate reducers, methanogens, hydrothermal microbial communities, and even the earliest life on land as well (Westall and Xiao 2024). They are all invisible prokaryotes but were recorded in rocks as biogenic sedimentary structures such as stromatolites (Figure 8.26) and MISSs, microfossils, organic matter, and isotope signatures (Figure 8.4). The combination of these observations indicates a Paleoarchean radiation of prokaryotes. This prokaryote-dominated biosphere may have lasted for more than two billion years and extended to the late Paleoproterozoic. Eukaryotes have become taxonomically diverse and ecologically important since the Mesoproterozoic, though they might have evolved in the Mesoarchean (Westall and Xiao 2024).

Figure 8.26 A polished slab of a stromatolite specimen from the late Archean of South Australia. *Source:* Courtesy of Mike Reich.

The two domains of prokaryotes, Bacteria and Archaea, may have diverged several hundred million years before the diverse life forms found in Paleoarchean rocks, such as the ∼3.5 Ga Dresser Formation in Western Australia. An obvious phenomenon of evolution on a temporal macroscale is evolutionary addition and metabolic flexibility. All these early prokaryotes, whether as taxonomic or functional groups, have descendants thriving in various environments today. They were neither eliminated by the emergence of more complex life forms nor did their evolution cease due to environmental threats such as extreme climate conditions during the Snowball Earth periods. Two groups of prokaryotes, cyanobacteria and sulfate-reducing microbes, are particularly noteworthy. Both rose early and are still thriving in modern ecosystems. They were likely the major players of the Archean biosphere, functioning as primary producers and decomposers, respectively, and driving biogeochemical cycles of the ecosystems.

Cyanobacteria are a monophyletic group of the Bacteria domain and have a Paleoarchean origin. They are the only prokaryotes to have evolved oxygenic photosynthesis that oxygenated the Earth. Interestingly, some extant cyanobacteria can switch to anoxygenic photosynthesis when alternative electron donors (e.g. H_2S and Fe^{2+}) are available (Sánchez-Baracaldo et al. 2022). This metabolic flexibility could have been important on the early Earth, partly explaining the early rise of cyanobacteria and the delayed oxygenation of Earth's surface.

Sulfate-reducers are a functional group of microorganisms, including bacteria and archaea. They are ecologically and metabolically opposite cyanobacteria, as they reduce sulfate to decompose organic matter produced by cyanobacteria, thereby recycling carbon and sulfur. Many sulfate reducers are known to oxidize methane. They were likely the major decomposers before the GOE. Sulfur isotopic evidence suggests that microbial sulfate reduction was in place during the Paleoarchean (Shen et al. 2001). Molecular clock analyses inferred that sulfate-reducing bacteria diverged when cyanobacteria oxygenated the Earth's surface and enriched sulfate concentrations in the oceans during the GOE (Cui et al. 2024; Figure 8.27).

8.5.3.2 Eukaryotic Life Emerged Early and Diverged Late

The Eukarya have long been considered as one of the three domains of life. More recent studies are challenging this three-domain phylogeny of life. Many eukaryogenic theories concur that an archaeal host cell and an alphaproteobacterial endosymbiont were essential for initiating eukaryogenesis (Nobs et al. 2022). Eukaryotic cells are generally larger and much more complex than prokaryotic cells. They are believed to have evolved from a symbiotic community of prokaryotes living within larger prokaryotic cells. The mitochondria in our cells, as well as those in all other eukaryotes, are descendants of these "endosymbionts" (symbiotic cells living within larger host cells). Similarly, the chloroplasts found in plants and certain algae also originated from these endosymbiotic relationships.

Three-stage evolution of eukaryotes in the Precambrian has been proposed (see Westall and Xiao 2024 and references therein). They are (1) evolution of stem group eukaryotes during the Neoarchean to early Mesoproterozoic, (2) divergence of major eukaryote clades during the late Mesoproterozoic to early Neoproterozoic, and (3) the radiation of multicellular eukaryotes after the Cryogenian Snowball Earth during the Ediacaran. Stage 3 is well supported by fossil record (see Westall and Xiao 2024 and references therein), while the fossil record of stage 1 remains uncertain and the taxonomic diversity of eukaryotes during the pre-Cryogenian Proterozoic is still very low, less than 100 genera and species in each time period on a global scale (Tang et al. 2025). The Archean origin of eukaryotes has been suggested by molecular clock studies, but no convincing eukaryotic fossil or biomarker has been reported from Archean rocks. Several Paleoproterozoic macrofossils offer suggestive but inconclusive evidence for early eukaryotes. For example, the

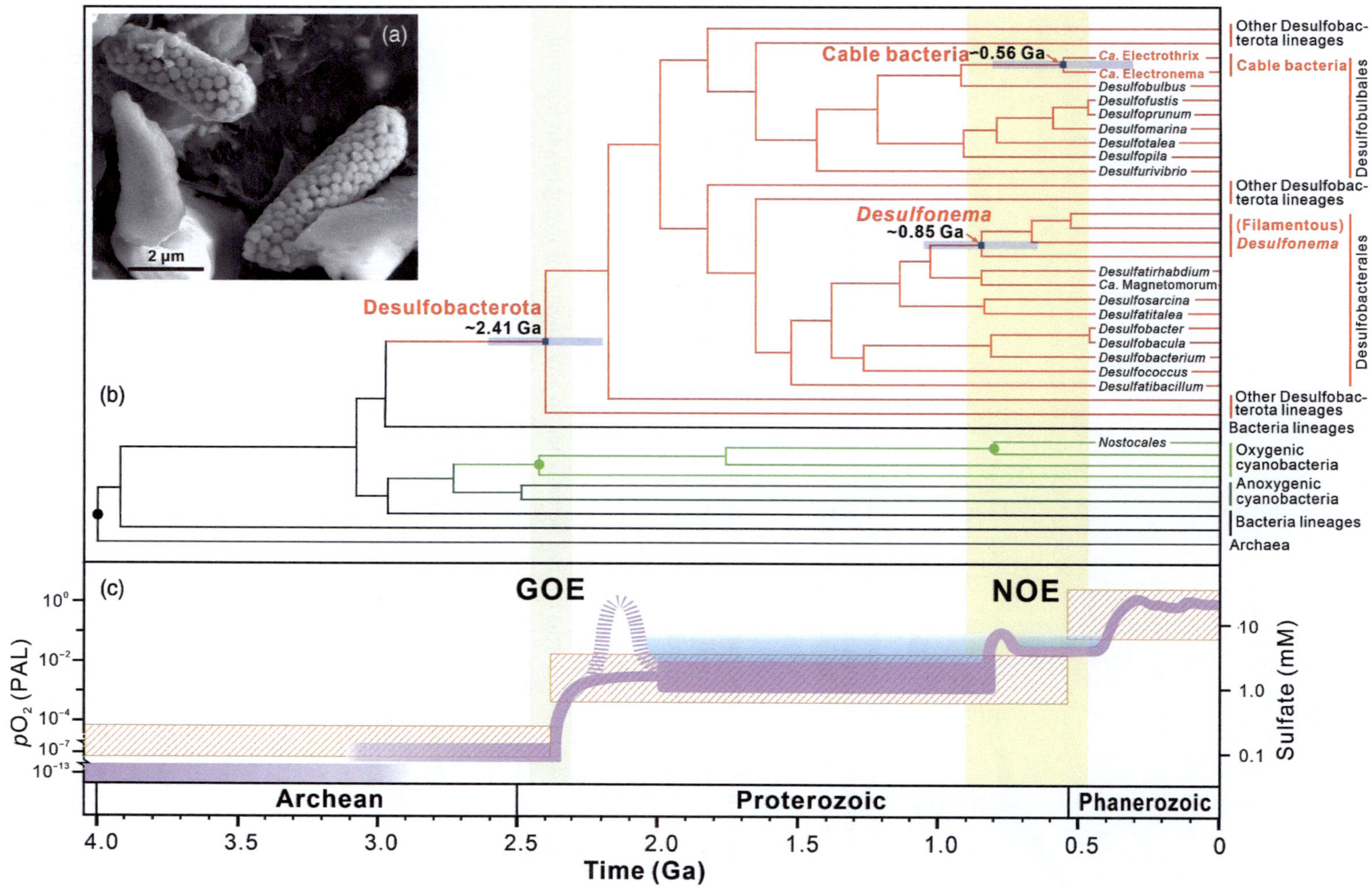

Figure 8.27 Evolution of major sulfate-reducing lineages (b) and temporal association with geological events, oxygenation and sulfate concentration of ocean (c). The pink bands represent changes in atmospheric oxygen levels, while the yellow diagonal bands indicate variations in oceanic sulfate concentrations. (a), *Qingjiangonema*, a sulfate-reducing microfossil from the ~518 Ma Qingjiang biota of South China. *Source:* Cui et al. (2024) / Elsevier / CC BY 4.0.

Figure 8.28 A ribbon-like fossil, *Grypania* from the Gaoyuzhuang Formation (~1500–1400 Ma) of North China. *Source:* Photo courtesy of Hong Hua. Similar fossil from 2.1 billion-year-old rocks was interpreted as remains of early eukaryotes (Han and Runnegar 1992).

2.1 Ga carbonaceous fossil *Grypania*, preserved as spirally coiled and millimeter-wide ribbons, was once considered as the oldest fossil of eukaryotes (Han and Runnegar 1992). However, these ribbons (Figure 8.28) were alternatively interpreted as remnants of single-celled tubes, more or less like the hollow stalks of a living eukaryotic seaweed (*Acetabularia*), or colonies of the prokaryotic cyanobacterium *Nostoc* (Schopf 1999). More convincing candidates are represented by those multicellular filaments and acritarchs with vesicle-wall sculptures from 1.7 to 1.6 Ga rocks (e.g. Miao et al. 2024; Zhang 1986). These forms cannot be assigned to any lineage of living eukaryotes and hence were considered as stem-group eukaryotes (Westall and Xiao 2024).

8.5.3.3 Multicellularity Across Lineages of the Life Tree

The emergence of more complex cells paved the way for the diversification of eukaryotic life. A wide range of unicellular forms evolved, leading to a diversity of protists and their single-celled descendants, which continue to thrive today. Multicellularity also evolved, giving rise to various multicellular algae, plants, fungi, and animals. In multicellular organisms, such as animals, development typically begins from a single cell—the fertilized egg, or zygote, in the case of sexual reproduction. Through cell division and differentiation, this single cell transforms into a multicellular adult with many specialized cell types. Multicellularity enables sexual reproduction, which requires the differentiation of cells into germ cells and soma, and allows for the development of large, complex organisms. This complexity permits selection to operate at various levels, including genes, cells, individuals, and populations. Notably, despite the diversity of animal phyla, they are all based on a relatively limited basic body plan determined by a surprisingly small number of genes, varying from several to 30,000 among phyla. The evolution of multicellular bodies broke through the threshold in structural organization and provided the stock for new waves of evolutionary diversification.

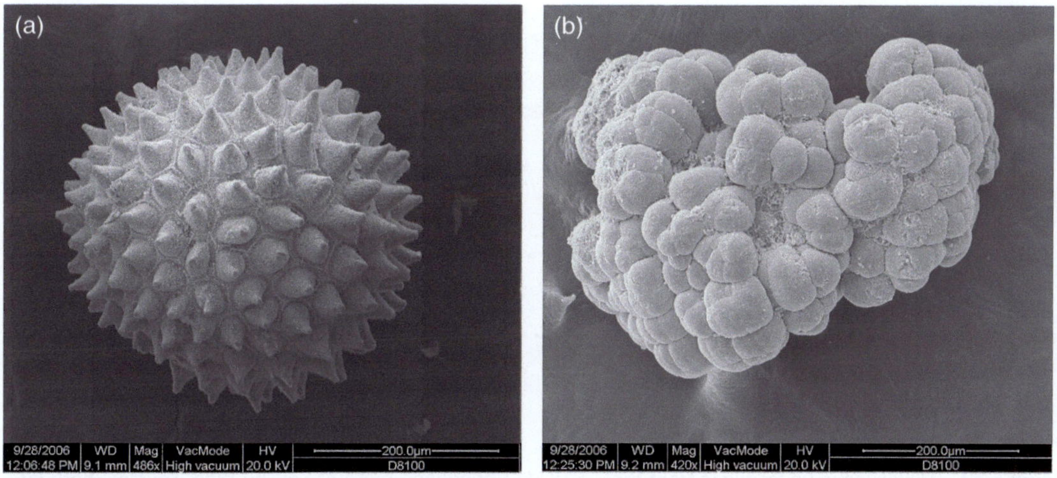

Figure 8.29 Phosphatized microfossils from the ~600 Ma Weng'an biota of South China. (a) *Tianzhushania spinosa*. (b) Multicellular algae. *Source:* Photos by Xingliang Zhang.

Multicellularity is present in the life cycles of many organisms across domains and kingdoms in the tree of life and has evolved independently multiple times in different lineages at different times. Protists are primarily unicellular, but some can form colonies or multicellular structures (Figure 8.29, megasphere). Ferns have multicellular fronds but they develop from single-celled spores. Filamentous multicellular cyanobacteria have contributed to building microbial mats in the Archean, and similarly multicellular sulfate-reducing microbial fossils were discovered in 518 Ma black shales (Cui et al. 2024). The evolution of multicellularity within the Eukarya gave rise to plants, animals, fungi, and various groups of algae. Multicellular eukaryotes may have evolved in the latest Paleoproterozoic. Septate filaments (*Rafatazmia chitrakootensis*) and multicellular thalli (*Ramathallus lobatus*) from the 1.6 Ga Tirohan Dolomite of the Semri Group in central India were interpreted as possible crown-group red algae (Bengtson et al. 2017). The septate filamentous microfossil *Qingshania magnifica* from the 1.63 Ga Chuanlinggou Formation of the Changcheng Group in North China shows intriguing evidence of cellular differentiation, with its expanded clavate terminal cell possibly representing a specialized reproductive cell akin to spores (Miao et al. 2024). These three taxa are among the oldest known candidates for multicellular eukaryotes.

8.5.3.4 The Cambrian Explosion of Animals

The Cambrian Explosion initially stemmed from a paleontological phenomenon, characterized by the relatively sudden appearance of skeletal remains of diverse animals in the lowest known fossiliferous rocks of the Cambrian. This event is a three-phased explosion of animal body plans, including the oldest known fish (Figure 8.30), alongside episodic biomineralization, pulsed changes in generic diversity, body size variation, and a progressive increase in ecosystem complexity. Animals may have emerged in the Ediacaran or earlier, becoming taxonomically divergent and ecologically significant since the late Ediacaran, and eventually dominating marine ecosystems by the end of the Cambrian Explosion. Early animals shared seafloors with Ediacarans (vendobionts and tubular forms of uncertain affinity) for the last 20 million years of the Ediacaran, although their ecological relationships are less understood. During the Ediacaran, animals might have been ecologically

Figure 8.30 *Myllokunmingia fengjiaoa* from the ~518 Ma Chengjiang biota of South China. *Source:* Photo by Xingliang Zhang, which was described as the earliest known jawless fish (Shu et al. 1999). Scale bar = 20 mm.

modest, but they evolved organs and systems, developed orderly repetition of body parts (e.g. segmentation), and explored other possibilities, enabling the evolution of morphological, physiological, and ecological variations and complexity. In contrast, Ediacarans maintained their less differentiated body designs, tissue-grade organization, and likely osmotic physiology. Consequently, Ediacarans died off at the end of their era for unknown reasons. Following this, animals rapidly diversified, generating numerous phylum-rank stem or crown lineages with different fates in the first 20 million years of the Cambrian Period (Figure 8.31). Some lineages immediately diverged in ecology and morphology and persisted to the present day (e.g. arthropods and molluscs), some temporarily succeeded in diversity but became extinct sooner or later, and some have remained relatively lonely but have survived to the present day.

The Cambrian Explosion is not purely a biotic process but a polythetic event in natural history that requires many sets of necessary conditions, from molecules to forms, and from isotopes to habitability. The explosive appearance and ecological expansion of animal forms occurred when the world oceans became habitable for various forms of animals, the developmental gene regulatory networks (GRNs) were sufficiently complex for constructing complex forms, and resource supply was less restricted. What sparked the Cambrian Explosion? Intrinsic and extrinsic causes have been extensively discussed, but they are merely prerequisites for the Cambrian Explosion. Without molecular evolution, there could be no Cambrian Explosion. However, the developmental system alone is insufficient to explain it. Time-equivalent environmental changes are often considered extrinsic causes, but the time coincidence is also insufficient to establish causality. Like any other evolutionary event, it is the ecology that made the Cambrian Explosion possible, though ecological processes failed to cause a burst of new body plans in subsequent evolutionary radiations. The Cambrian Explosion is a polythetic event in natural history, manifested in many aspects. No simple, single cause can explain the entire phenomenon (Zhang and Shu 2021).

8.5.3.5 The Development of Life on Land

On land, wherever aquatic systems or moisture are available, either permanently or intermittently, life can be found. Therefore, it is reasonable to assume that organisms may have occupied land since the earliest history of life. In particular, prokaryotes are flexible in metabolism and have been

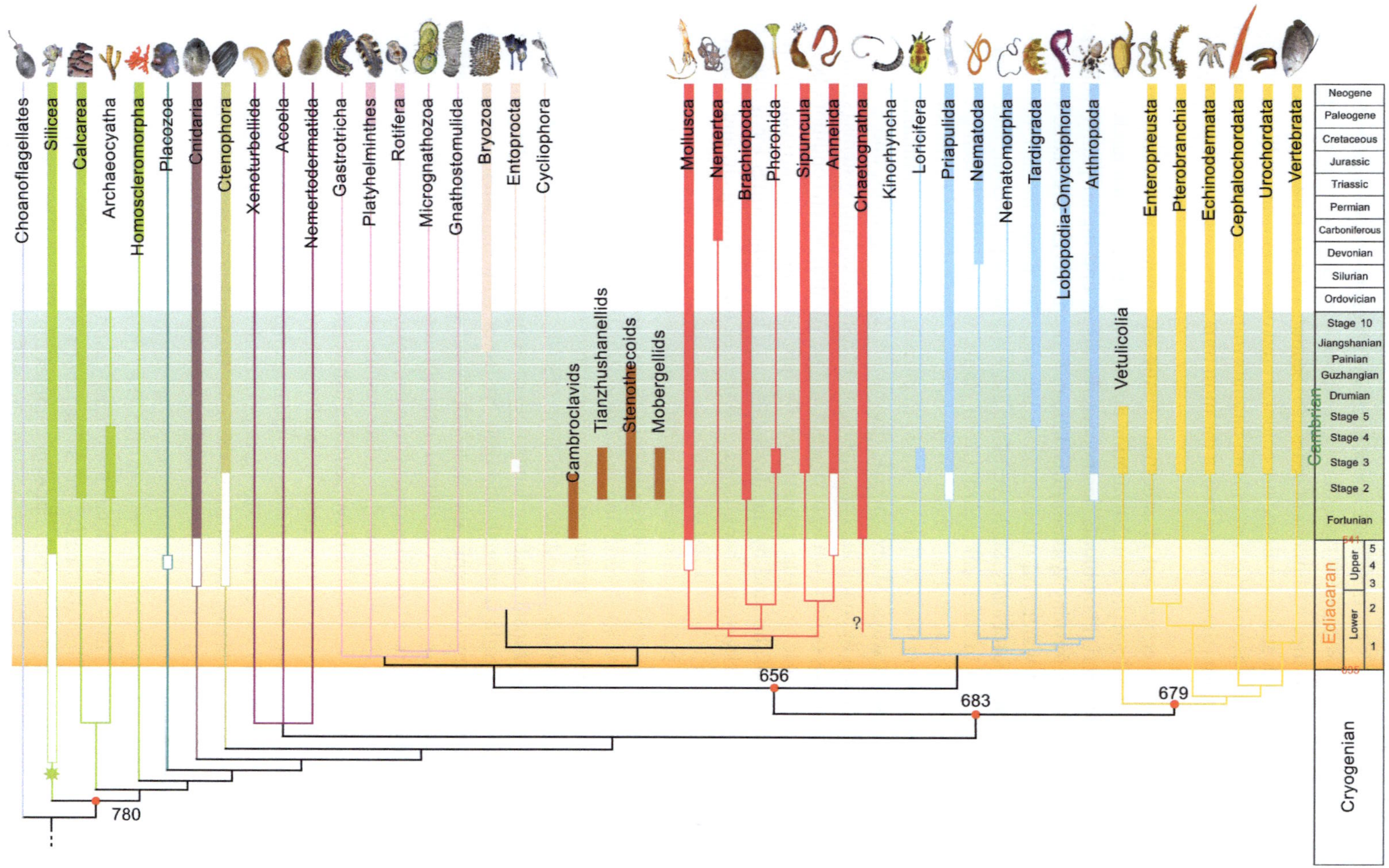

Figure 8.31 Phylogeny and fossil first appearances of animal phyla during the geological periods. The Ediacaran–Cambrian occurrences are resolved to geological stages. This tree is a summary of diverse sources, red nodes with numbers indicating the time of divergence between lineages in millions of years based on molecular clock data. *Source:* Zhang and Shu (2014) / Springer Nature.

found everywhere, many of which are able to thrive in extreme environments. This assumption is supported by the report of ~3.5 Ga terrestrial hot springs with biosignatures (Djokic et al. 2017). The presence of Archean paleosols and MISS in terrestrial facies also indicates that microbial communities colonized terrestrial surfaces billions of years ago (Section 8.2.2.2). However, macroscopic life in the form of plants, fungi, and animals did not colonize land until about 460 million years ago, during the Late Ordovician. The colonization of land habitats from the sea by plants and animals, known as terrestrialization, was thought to be one of the most important events in the history of life on Earth (Selden 2005).

This gradual evolutionary venture out of ancestral aquatic environments was associated with numerous adaptations that helped prevent dehydration and respire the air, which made it possible to grow and reproduce on land. Photobionts (cyanobacteria, algae, and higher plants) colonized land in the company of fungi, forming symbiont-like lichen. Photobionts are associated with fungi that aid in the absorption of water and minerals from the soil. The fungi, in turn, obtain their organic nutrients from the photobionts. Such symbiotic associations of photobiont and fungi are dated back to the Ordovician in the fossil record. Green plants, which evolved from green algae, developed structures like roots to anchor them in the soil, vascular tissues to transport water and nutrients, and a waterproof waxy cuticle on leaves to prevent desiccation. Animals adapt terrestrial ecosystems in several ways. Many animals (freshwater sponges, fishes, and soil animal nematodes) avoid the problem by living in lakes, rivers, or interstitial aquatic systems (soils). Poikilohydric animals can tolerate desiccation by drying out and rehydrating when conditions become favorable again. Homoiohydric animals, represented by tetrapods and land arthropods, have achieved the true conquest of the land by the development of limbs and wings for walking on the ground and flying in the sky, lungs, book lungs, and trachea for breathing air, and protective coverings to prevent water loss (Selden 2005).

The terrestrialization had a profound impact on Earth's surface conditions and ecosystems. The evolution of early land plants correlates with a dramatic decline of pCO_2 (Figure 8.9) and a rise of pO_2 in the atmosphere (Figure 8.15), testifying to a first-order disturbance in the global carbon cycle. Terrestrialization created new resources and new habitats (such as forests and grasslands), contributed to the cycling of nutrients and the formation of soils, and hence led to the diversification of life on Earth. Plants transformed the landscape, creating new opportunities for all life, including herbivorous (plant-eating) animals and their predators. Although many animal groups are represented in terrestrial environments, the most widespread and diverse land animals are certain arthropods and vertebrates. The first terrestrial invertebrates are myriapods that occurred in the fossil record in the late Silurian. The terrestrial vertebrates are called tetrapods, a reference to the four walking limbs that distinguish them from the various aquatic vertebrates we call fishes. An excellent fossil record documents that amphibians evolved from fish during the Devonian. Evolution of the amniotic egg expanded the success of vertebrates on land: reptiles evolved from amphibians during the Carboniferous and flourished in the Mesozoic era, and mammals and birds evolved from reptiles during the Triassic and Jurassic, respectively. Most groups of modern mammals were established by ~50–60 Ma before present, including the primates, the order that includes apes and humans.

8.5.3.6 Phanerozoic Events

The billions of years of evolution during the Precambrian, the explosive appearances of animal body plans in the early Cambrian, Great Ordovician Biodiversification Event (GOBE), and the rise of green plants on land laid the foundations for the subsequent evolution of life. The Phanerozoic era witnessed the rise and fall of numerous clades of complex life. In most cases, a clade emerged cryptically and accidentally, while its taxonomical diversification and ecological expansion required the availability of resources, conditions, and space in the environment. For example, mammals have a long and relatively inconspicuous history, and their evolutionary flourishing in the Cenozoic is often attributed to the available niches released by the environment-driven mass extinction at the end of the Mesozoic, which eliminated non-bird dinosaurs. Looking back to the Triassic or earlier, it would have been hard to predict that little mammals, e.g. *Brasilodon quadrangularis* (Cabreira et al. 2022; Figure 8.32) would achieve great success millions of years later.

Life history is periodically interrupted by mass extinctions, a classic topic for the co-evolution of life and the environment. Only in the Phanerozoic, with the appearance of animals and the rise of land plants, is there a sufficient fossil record to reveal these great biotic crises. Statistics of taxonomic diversity changes through time led to the recognition of five great mass extinctions ("big five"), which occurred during the Late Ordovician, the Late Devonian, the Late Permian, at the end of the Triassic, and at the end of the Cretaceous, respectively. These events had significant impacts on the course of evolution. For instance, the Late Ordovician extinction led to the replacement of the trilobite-rich Cambrian fauna by the brachiopod-rich Paleozoic fauna, which again was replaced by the mollusc-rich modern fauna after the two pulses of the end-Permian mass extinction (Figure 8.33).

The causes of mass extinctions have been widely debated, leading to numerous theories. Each extinction event is driven by an ultimate cause—a trigger that initiates various climatic and environmental changes—as well as one or more proximate causes, which are the specific environmental shifts responsible for increased biotic mortality (Algeo and Shen 2023). Commonly cited drivers of mass extinctions include meteoritic impacts, large igneous province (LIP) activity, tectonic and oceanic transformations, climatic warming, ocean acidification, and anoxic conditions. However, in the cases of the Phanerozoic, there is a remarkable degree of similarity between the empirical record of sedimentary rock quantity-age in continental crust and the first-order history of all of life and of molecular oxygen in the atmosphere (Figure 8.34). Therefore, the possibility of an ultimate

Figure 8.32 *Brasilodon quadrangularis* is a small, shrew-like creature that lived approximately 225 million years ago during the Late Triassic period and proposed as the earliest known mammal. *Source:* Cabreira et al. (2022) / John Wiley & Sons.

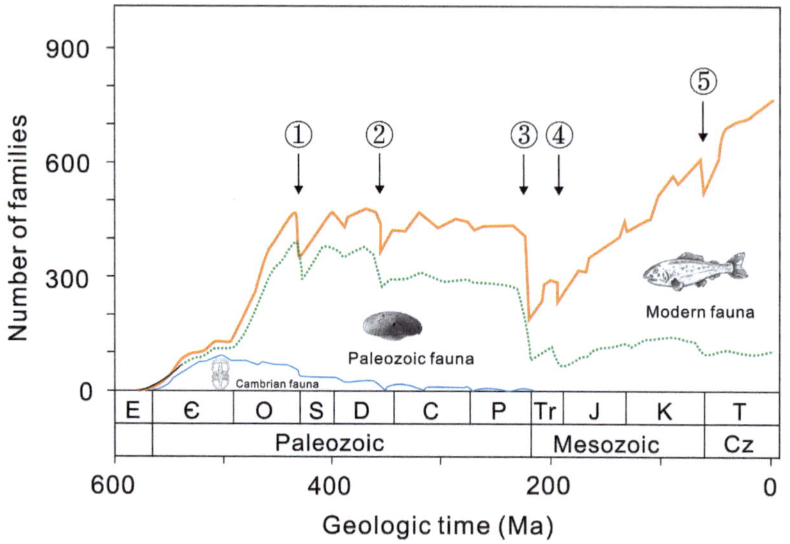

Figure 8.33 Diversity of marine faunas through the Phanerozoic. The thick orange line shows the total number of marine families in each of 83 geological stages. The circled numbers correspond to the five major mass extinctions: ① Late Ordovician, ② Late Devonian, ③ Late Permian, ④ end-Triassic, and ⑤ end-Cretaceous. The lower blue line shows the proportion of family diversity assignable to the Cambrian evolutionary fauna (trilobite rich); the middle green line shows the proportion of family diversity assignable to the Paleozoic evolutionary fauna (brachiopod rich), and the area between the green and orange lines is the proportion of family diversity assignable to the modern evolutionary fauna (mollusc rich). *Source:* Adapted from Sepkoski (1984).

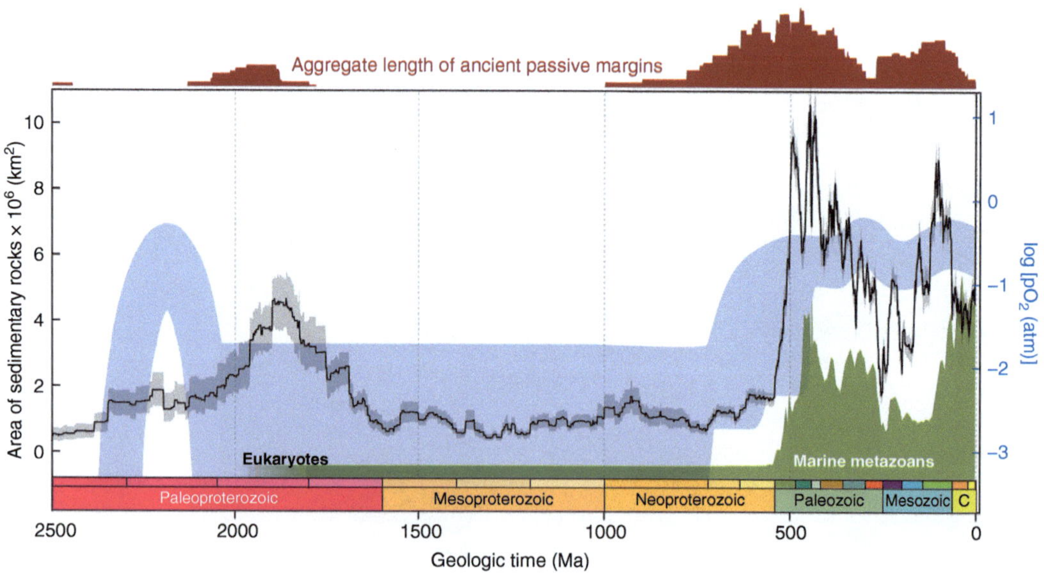

Figure 8.34 A representation of the coevolution of life, the environment, and the rock record. A time series of continental sedimentary rock area in North America (left axis) is compared to plausible levels of atmospheric oxygen throughout Earth history (right axis) and the generic diversity (scaled) of eukaryotes in the Proterozoic and marine metazoans in the Phanerozoic. The red histogram at the top shows a normalized record of the global aggregate length of ancient passive margins. *Source:* Peters et al. (2022) / with permission of Annual Review.

common-cause mechanism linking patterns in the rock record to the first- and second-order history of life and environment was reconsidered (Peters et al. 2022). This similarity is unlikely to be the result of an overarching bias in the geological record that passively shaped the evolution of life and the environment to mirror the quantity-age distribution of sedimentary rocks in the continental crust. Rather, the primary patterns observed in the sedimentary rock record likely reflect genuine shifts in the fundamental boundary conditions governing the surface environment. These shifts have played a crucial role in regulating the timing of biological evolution and atmospheric oxygenation.

8.5.4 Evolution of Metabolism

Reconstructing the evolution of metabolism is a challenging task. Theoretically, genomic and molecular genetic approaches hold potential for elucidating the evolution of life, particularly some metabolic details. However, these approaches have many limitations and should be used with great caution. As DeLong (1998) stated, "Despite utility for identification, molecular sequence comparisons do not provide much detail on phenotype," and as paraphrased by Nealson and Rye (2005), "between rapid and unpredictable rates of mutational substitution on one hand, and horizontal gene transfer on the other, trying to use molecular data to specify ancient metabolic evolution is at best an effort fraught with many intellectual traps, and at worst, impossible." Currently, genomic and molecular genetic approaches offer limited insights into the precise timing and exact sequence of events that led to the metabolic diversity that life enjoys today. Recent analyses using molecular clocks, machine learning, and phylogenetic reconciliation to present a reconstruction of the evolution of microbial metabolisms over the past 4 billion years. Their results showed that the last common ancestor of cellular life existed 4.4 to 3.9 billion years ago and the majority of metabolic pathways were established within the first 2 billion years of Earth history (Davín et al. 2025; Moody et al. 2025). Physical fossils of past life have a deep record and are valuable for reconstructing a picture of organismal evolution. However, this record does little to elucidate the evolution of metabolism. Nevertheless, various geological and geochemical data (e.g. signature organic molecules, inorganic chemistry, and isotopic fractionations of ancient sediments) provide fragmentary records on the evolution of metabolism.

Both **carbon fixation** and **nitrogen fixation** were likely present before the divergence of the three or two domains of life. The ~4.2 Ga LUCA is proposed to have been an anaerobic autotroph using the Wood–Ljungdahl pathway (WLP) for carbon fixation and capable of nitrogen fixation using a nitrogenase (Moody et al. 2024). While the metabolic completeness and individual enzyme analyses suggested that nitrogen fixation evolved in the mid-late Archean, around 3 billion years ago (Moody et al. 2025). The oldest known graphite with light carbon isotope composition suggests that carbon fixation might have been in place around ~4.1 Ga. Siliciclastic sedimentary rocks of low to moderate metamorphic grade deposited between 3.5 and 2.9 Ga show a mean $\delta^{15}N_{bulk}$ values of 1.3‰ (±3.6‰, 1σ, $n = 191$) and $\delta^{15}N_{ker}$ values of −0.12‰ (±3.6‰, 1σ, $n = 107$) across marine and non-marine depositional settings, suggesting the presence of biological nitrogen fixation in the Paleoarchean (Stüeken et al. 2024). Cyanobacteria that emerged in the Paleoarchean are able to fix nitrogen. Trait evolution analyses indicate that the acquisition of nitrogenase, the key enzyme in biological nitrogen fixation, occurred near the root of the cyanobacterial tree (Sánchez-Baracaldo et al. 2022). Therefore, cyanobacteria might be the major fixers of carbon and nitrogen in the earliest biosphere.

Photosynthesis can proceed via anoxygenic pathways (photosystem I: M00598; photosystem II: M00597) or through oxygenic pathways (photosystem I: M00163; photosystem II: M00161).

Comparative phylogenomic analyses indicate that oxygenic photosynthesis expanded rapidly between 3.25 billion years ago (Ga) and 2 Ga, reaching full completeness in both photosystems by the latter time. In contrast, components of anoxygenic photosystem I appear earlier, preceding the rise of oxygenic systems, whereas anoxygenic photosystem II shows a marked increase in completeness during the Mesoproterozoic (Moody et al. 2025).Carbon isotopes have been extensively used as evidence for the appearance of life on Earth. Negative $\delta^{13}C$ values connote isotopically lighter carbon and are taken to indicate the possibility of biological activity. Organic compounds produced by chemo- and photosynthesis have $\delta^{13}C$ values roughly ranging from −10‰ to −35‰. When the $\delta^{13}C$ values fall below −40‰, the involvement of methane-producing and consuming microbes is suspected. The very light carbon isotopic composition ($\delta^{13}C = -56‰$) of carbonaceous filaments from the Dresser Formation coupled with other geological evidence is suggestive of microbial **methanogenesis** and **methanotrophy** at ~3.5 Ga (see Westall and Xiao 2024). Phylogenomic studies inferred the presences of the mcrD subunit at 3.77 Ga, mcrB at 3.05 Ga, the rest of the modern mcr complex by 2.54 Ga (The MCR complex is the key enzyme in archaeal methanogenesis, including subunits such as mcrD and mcrB), indicating that modern methanogenesis had evolved before the GOE (Moody et al. 2025).

Isotopes of sulfur are likewise extremely valuable for tracing sulfur metabolism and have been extensively used for the study of ancient sediments. The fractionations of −20‰ between microscopic sulfide and coexisting sulfate from ancient sediments indicate that microbial **sulfate reduction** started on Earth as early as ~3.5 Ga (Shen et al. 2001), which is also supported by molecular clock analyses (Cui et al. 2024). However, phylogenomic results indicate assimilatory sulfate reduction predates dissimilatory sulfate reduction by almost 1.0 Ga, with dsrA and dsrB originated at 2.17 Ga, in the aftermath of the GOE (Moody et al. 2025).

Comparative structural biology and phylogenetic analyses suggest that anoxygenic and oxygenic photoautotrophs coexisted in time but were separated by environment for most of the Archean, prior to the rise of cyanobacteria to ecological dominance. Molecular clock estimates for the gene family that includes photosystems I and II align with evidence showing traces of oxygen throughout the Archean. Therefore, it is probable that oxygenic photosynthesis was already established by 3.0 Ga. Early forms of water oxidation, carried out by ancestral homodimeric photosystems, could have originated a billion years before the GOE. The highly sophisticated photosystem II found in extant cyanobacteria can be traced back to the Mesoarchean (Sánchez-Baracaldo et al. 2022). Stromatolites from 2.7 Ga lacustrine sediments indicate the presence of oxygenic photosynthesis, as the lake lacked sulfate and hydrothermal reductants needed for anoxygenic photosynthesis, while the stromatolites display a photosynthetic habit (Buick 1992). Data of lead isotopes suggest that oxygenic photosynthesis existed as early as 3.7 Ga (Rosing and Frei 2004). Comparison of ribosomal RNA from cyanobacteria with portions of the DNA inside chloroplasts implies that all eukaryotes, including algae and higher plants, derived their photosynthetic capabilities from cyanobacteria through endosymbiosis. The cyanobacterium *Gloeomargarita* and the protist *Paulinella* may be living descendants of ancestral organisms involved in this event (Sánchez-Baracaldo et al. 2017). It appears that oxygenic photosynthesis—an extremely complex biochemical process—was "invented" only once, and a primitive cyanobacterium was the organism responsible.

Massive accumulations of oxidized iron in BIF deposits (~3.8–1.85 Ga) indicate that iron oxidation and reduction were important processes in early Earth history. Biological models, such as those proposed by Konhauser et al. (2002, 2005), suggest that microbial Fe(II) oxidation and Fe(III) reduction were involved in BIF deposition and diagenesis. If this proposal is correct, metabolic Fe(II) oxidation and Fe(III) reduction might have been present on early Earth. Anoxygenic photosynthesis, which can oxidize Fe, may have contributed to BIF deposition before ocean oxygenation.

As for biomineralization, biologically influenced and induced mineralization may have evolved very early because a number of metabolic processes, e.g. photosynthesis, sulfate reduction, and iron oxidation, are able to induce the precipitation of minerals like calcite, pyrite, and hematite, which may have been responsible for the extensive microbial carbonates and BIFs in the Precambrian. Magnetotactic bacteria, capable of producing nanoscale magnetite crystals via biologically controlled process, may have evolved early in the deep Precambrian. In eukaryotes, biologically controlled mineralization (BCM) evolved many times in different lineages (see Section 5.5.2). It seems likely that the invention of the BCM in eukaryotes is rather late though it took place multiple times in different lineages. Scale microfossils from the 810.7 ± 6.3 Ma Fifteenmile Group of Yukon, Canada, constructed a hierarchically organized interwoven network of fibrous hydroxyapatite crystals, each elongated along the [001] direction. Accordingly, the scales were interpreted to form through the BCM (Cohen et al. 2017). At the beginning of the Cambrian, many metazoan phyla evolved the ability to make their skeletons and thus left extensive fossil records (see Section 5.5). Notably, calcification of animals is a major mechanism of carbonate genesis in the Phanerozoic.

8.6 A Summary: Four Billion Years Co-evolution to One Intelligent Species

The co-evolution of life and environment is based on the notion that interplays between biological and geological processes have driven the development of Earth's environment, climate, and ecosystems, with each influencing the other in a dynamic and interconnected way, and the interplays themselves have been evolving through time as well. Figure 8.35 presents a summary of the co-evolution of the Earth–life system, with geological and biological events put in the chronological framework.

The accretion of Earth with a metal core was completed by ~4.5 Ga. The atmosphere and hydrosphere almost certainly originated from the Earth's interior as the planet underwent melting and segregation into a core, mantle, and crust, simultaneously degassing volatile materials. Evidence provided by 4.3–4.4 Ga old zircon crystals and their isotopic composition (having a high ^{18}O abundance, which is characteristic of interactions between molten rock and liquid water), suggests the possible presence of Hadean oceans. The existence of Hadean zircons also implies, at least locally, there was continental crust at that time.

During the Hadean and Eoarchean transition, about 4.1 to 3.8 billion years ago, Earth experienced intense LHBs that could have been powerful enough to convert all the oceans to steam, possibly on multiple occasions. These LHBs may have influenced the chemical pathways leading to the origin of life. The shock waves and heat generated by impacts might have increased reaction rates, and extraterrestrial bodies (meteoroids and asteroids) could have introduced elements and simple organic compounds necessary for life. Coincidentally, both biological and geological evidence suggest that life emerged during this time. Another intriguing possibility is that the LHBs selected for hyperthermophilic biota by periodically boiling the global oceans, with perhaps two lineages surviving the selection, which are identified today as the deepest branches in the Bacteria and Archaea. Conversely, the LHBs could have been catastrophic, destroying early life forms.

As discussed in Section 8.2.2, during the Archean, the Sun emitted less energy, the moon was closer to Earth, the climate was hot, and the oceans were saltier and anoxic. A methane-rich atmosphere, along with the greenhouse gas CO_2, helped overcome the FYS paradox, keeping Earth's surface warm enough to be habitable. The Earth's surface conditions were similar to extreme

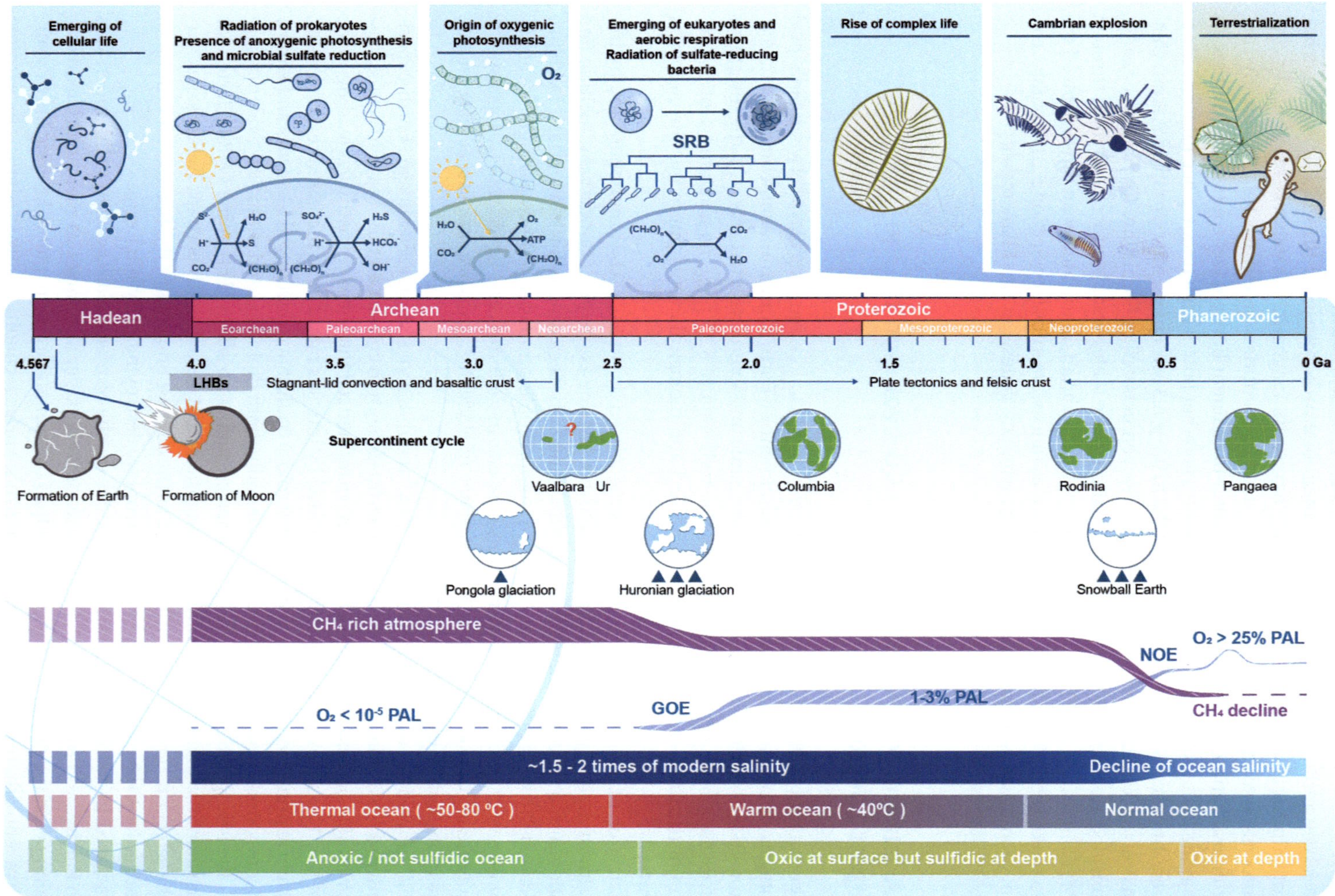

Figure 8.35 Summary for co-evolution of biosphere and geosphere, with a timeline showing major events in Earth history.

environments found today, making it an ideal habitat for prokaryotes. Multiple lines of evidence suggest a radiation of prokaryotes in the Paleoarchean.

A striking example of the co-evolution of life and the environment is the interaction between autotrophs (including chemo- and photoautotrophs) and geological processes, which led to the oxygenation of Earth and had a profound impact on the history of life. Photoautotrophic prokaryotes include both anoxygenic and oxygenic forms. Anoxygenic forms utilize reductants such as Fe^{2+} and H_2S instead of water as electron donors. In environments where alternative electron donors are present, anoxygenic photosynthetic bacteria often outcompete oxygenic cyanobacteria. Additionally, some modern cyanobacteria can switch to anoxygenic photosynthesis when H_2S is available (Sánchez-Baracaldo et al. 2022). This metabolic flexibility and diversity could have been crucial on early Earth. It is clear that an excess of reductants limits oxygenic photosynthesis. Anoxygenic phototrophs, along with chemoautotrophs, helped oxidize excess reductants on early Earth's surface, facilitating the ecological rise of oxygenic cyanobacteria. Another limitation for the success of oxygenic cyanobacteria is the availability of phosphorus. Many Archean environments were likely depleted in phosphorus (Jones et al. 2015). Therefore, the increase of phosphorus and the decrease of reductants such as Fe^{2+} and H_2S in oceans provided physical drivers for the rise of cyanobacteria to ecological prominence and oxygen production.

Geologically, before 2.7 Ga, Earth was characterized by stagnant-lid convection and continents were basaltic; at ~2.5 Ga before the GOE, plate tectonics and subduction were initiated, shifting continental crust-forming magmatism from basaltic to felsic compositions and decreasing the flux of mantle-derived reductants (Fe^{2+}, H_2S, and H_2) via magmatism (Lee et al. 2016). Growing geological evidence indicates the presence of large low-latitude continental landmasses in the early Paleoproterozoic, including the Kaapvaal Craton and supercraton Superia. These landmasses experienced a series of LIP events between 2510 and 2426 Ma. Collectively, these LIPs on extensive low-latitude continental landmasses likely triggered global glaciations via enhanced chemical weathering of aerially extensive, nutrient-rich continental flood basalts. The weathering increased carbon dioxide drawdown and the flux of phosphorus and other essential nutrients to continental margins and intracratonic basins. The enhanced nutrient flux would have significantly boosted photosynthetic activity and oxygen production, leading to higher net burial of organic carbon in sediments, as reflected by the Lomagundi-Jatuli $\delta^{13}C$ excursion. Even a small rise in free atmospheric oxygen would have rapidly oxidized atmospheric methane, causing catastrophic climate change and global glaciation. LIPs also contributed CO_2, helping to rebuild the greenhouse atmosphere and terminate the first Snowball Earth state of the early Paleoproterozoic (Gumsley et al. 2017).

After the early Paleoproterozoic glaciation that is broadly coeval with the onset of the GOE, the Earth's surface was oxygenated, sheltered by an ozone layer from UV radiation, and had a mild temperature at the order of 40 °C. The Paleoproterozoic Earth became much milder and well habitable for aerobic organisms including eukaryotes. It may not be a mere coincidence that diverse potential eukaryotic fossils were reported from Paleoproterozoic rocks (see Westall and Xiao 2024).

During the interval between GOE and NOE, often referred to as the "Boring Billion," the carbon cycle was less perturbed, tectonic activity was minimal, atmospheric CH_4 and CO_2 concentrations could have still remained elevated and the O_2 concentrations were estimated to be below 5% PAL. The ocean was stratified, oxic at surface and euxinic at depth, and its salinity probably remained at a high level intolerable for most forms of modern macroscopic life.

The links of LIPS, oxygenation, glaciation, and biological events in the Paleoproterozoic were compared with a similar sequence of events in the Neoproterozoic (Gumsley et al. 2017). Multiple LIPs on the low-latitude supercontinent Rodinia during incipient rifting and breakup immediately preceded the Sturtian glaciation (~717 to 660 Ma). The Neoproterozoic is characterized by the NOE linked with Snowball Earth glaciations and accompanied by high rates of organic carbon burial. The triggering mechanisms for the initial global glaciations might be different because methane was probably a more important greenhouse gas before the Makganyene glaciation than before the Sturtian glaciation. Similar events include low-latitude supercraton or supercontinent capped by LIPs, incipient rifting and/or breakup. All of these factors enhanced chemical weathering and increased the flux of bio-limiting nutrients to depositional basins, thus leading to a biotic response of higher primary production and organic carbon burial. The growth of a continental carbon reservoir and a sharp increase of sedimentary record during the Precambrian–Cambrian transition were correlated to the NOE (Lee et al. 2016; Peters et al. 2022). Coincidently, macrostratigraphical data also revealed an abrupt increase in net continental sedimentation during the NOE (Peters et al. 2022). However, whether the NOE took place before or after the "Snowball Earth" is controversial.

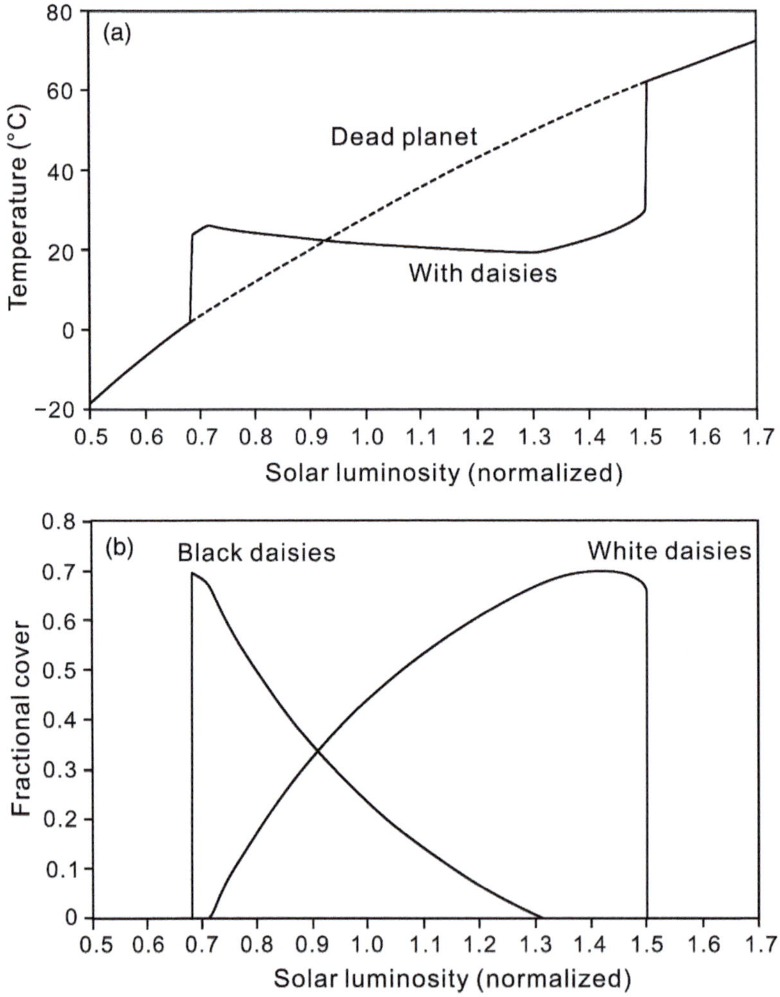

Figure 8.36 Rein control of the Daisyworld that is planted with two types of daisies. Black daisies help warm the planet by absorbing sunlight and they thrive particularly in cooler conditions. White daisies cool the planet by reflecting sunlight but they flourish in warmer environments. (a) Planetary temperature with and without daisies; (b) fractional coverage of black and white daisies. *Source:* Modified from Lenton and Watson (2011).

Canfield et al. (2007) suggested that glacial melting increased the nutrient load to the ocean. This stimulated primary production and carbon burial and thus increased atmospheric oxygen levels without a major change in global deep-ocean oxygenation (Stockey et al. 2024).

Following the Neoproterozoic snowball climate and the oxygenation during the Ediacaran and subsequent periods, the Earth's oceans became suitable for accommodating complex life like Ediacara-type biotas and metazoans, which evolved against a backdrop of extreme O_2 variability. Physical drivers of terrestrialization are presently uncertain, while the origin and initial diversification of terrestrial ecosystems were occurring as the supercontinent Pangaea was coalescing. In addition, the onset of the end-Devonian glaciation after a protracted period of "greenhouse" climatic conditions was frequently discussed and causally linked to the invasion of the continents by land plants. Conversely, terrestrialization created niches for land animals and the development of vegetation is able to regulate the habitability of the Earth system, as demonstrated by the Daisyworld hypothesis (Figure 8.36). The evolution of early land plants correlates with a rise in pO_2 and a decline in pCO_2, indicating a significant disturbance in the global carbon cycle. The invasion of land plants increased the atmospheric oxygen concentration to levels exceeding the minimum requirement (>15% PAL) for healthy human brain function, reproduction, and civilization (Lenton and Watson 2011), making it possible for the subsequent emergence of the only known intelligent species after four billion years of evolution.

References

Afroz, M., Fralick, P.W., and Lalonde, S.V. 2023. Sedimentology and geochemistry of basinal lithofacies in the Mesoarchean (2.93 Ga) Red Lake carbonate platform, northwest Ontario, Canada. *Precambrian Research*, 388, pp. 106996.

Algeo, T.J., and Shen, J. 2023. Theory and classification of mass extinction causation. *National Science Review*, 11, pp. nwad237.

Algeo, T.J., Luo, G.M., Song, H.Y., Lyons, T.W., and Canfield, D.E. 2015. Reconstruction of secular variation in seawater sulfate concentrations. *Biogeosciences*, 12, pp. 2131–2151.

Anbar, A.D., and Knoll, A.H. 2002. Proterozoic ocean chemistry and evolution: aA bioinorganic bridge? *Science*, 297, pp. 1137–1142.

Appel, P.W.U., Rollinson, H.R., and Touret, J.L.R. 2001. Remnants of an early Archean (>3.75 Ga) seafloor hydrothermal system in the Isua greenstone belt. *Precambrian Research*, 112, pp. 27–49.

Awramik, S.M., Schopf, J.W., and Walter, M.R. 1983. Filamentous fossil bacteria from the Archean of Western Australia. *Precambrian Research*, 20, pp. 357–374.

Bekker, A., Holland, H., Wang, P.L., Rumble, D., III, Stein, H.J., Hannah, J.L., Coetzee, L.L., and Beukes, N.J. 2004. Dating the rise of atmospheric oxygen. *Nature*, 427, pp. 117–120.

Bekker, A., Slack, J.F., Planavsky, A., Krapež, B., Hofmann, A., Konhauser, K.O., and Rouxel, O.J. 2010. Iron formation: tThe sedimentary product of a complex interplay among mantle, tectonic, oceanic, and biospheric processes. *Economic Geology*, 105, pp. 467–508.

Bell, E.A., Boehnke, P., Harrison, T.M., and Mao, W.L. 2015. Potentially biogenic carbon preserved in a 4.1 billion-year-old zircon. *Proceedings of National Academy of Science of United States of America*, 112, pp. 14518–14521.

Bengtson, S., Sallstedt, T., Belivanova, V., and Whitehouse, M. 2017. Three-dimensional preservation of cellular and subcellular structures suggests 1.6 billion-year-old crown-group red algae. *PLoS Biology*, 15, pp. e2000735.

Berner, R.A. 2004. Model for calcium, magnesium and sulfate in seawater over Phanerozoic time. *American Journal of Science*, 304, pp. 438–453.

Blake, R., Chang, S., and Lepland, A. 2010. Phosphate oxygen isotopic evidence for a temperate and biologically active Archaean ocean. *Nature*, 464, pp. 1029–1032.

Borisov, A.Y., and Björn, L.O. 2018. On oxygen production by photosynthesis: aA viewpoint. *Photosynthetica*, 56, pp. 44–47.

Bradley, J.P., McSween, H.Y., and Harvey, R.P. 1998. Epitaxial growth of nanophase magnetite in Martian meteorite Allan Hills 84001: implications for biogenic mineralization. *Meteoritics Planetary Science*, 33, pp. 765–773.

Brasier, M.D., Green, O.R., Jephcoat, A.P. et al. 2002. Questioning the evidence for Earth''s oldest fossils. *Nature*, 416, pp. 76–81.

Buick, R. 1992. The antiquity of oxygenic photosynthesis: evidence from stromatolites in sulphate-deficient Archaean lakes. *Science*, 255, pp. 74–77.

Cabreira, S.F., Schultz, C.L., da Silva, L.R. et al. 2022. Diphyodont tooth replacement of *Brasilodon*-A Late Triassic eucynodont that challenges the time of origin of mammals. *Journal of Anatomy*, 241, pp. 1424–1440.

Canfield, D.E. 1998. A new model for Proterozoic ocean chemistry. *Nature*, 396, pp. 450–453.

Canfield, D.E. 2014. *Oxygen: a four-billion-year history*, pp. 1–196. Princeton, Princeton University Press.

Canfield, D.E., and Teske, A. 1996. Late Proterozoic rise in atmospheric oxygen concentration inferred from phylogenetic and sulfur-isotope studies. *Nature*, 382, pp. 127–132.

Canfield, D.E., Poulton, S.W., and Narbonne, G.M. 2007. Late-Neoproterozoic deep-ocean oxygenation and the rise of animal life. *Science*, 315, pp. 93–95.

Catling, D.C., and Claire, M.W. 2005. How Earth''s atmosphere evolved to an oxic state: aA status report. *Earth and Planetary Science Letters*, 237, pp. 1–20.

Catling, D.C., and Zahnle, K.J. 2020. The Archean atmosphere. *Science Advances*, 6, pp. eaax142.

Chyba, C.F. 1987. Cometary contribution to the oceans of primitive Earth. *Nature*, 330, pp. 632–635.

Cloud, P.A. 1972. Working model of the primitive Earth. *American Journal of Science*, 272, pp. 537–548.

Cohen, P.A., Strauss, J.V., Rooney, A.D., Sharma, M., and Tosca, N. 2017. Controlled hydroxyapatite biomineralization in an ~810 million-year-old unicellular eukaryote. *Science Advances*, 3, pp. e1700095.

Cui, L.H., Zhu, K., Li, R., Chang, C., Wu, L., Liu, W., Fu, D., Liu, P., Qiu, H., Tang, G., Li, Q., Gaines, R.R., Tao, Y., Wang, Y., Li, J., and Zhang, X. 2024. The Cambrian microfossil *Qingjiangonema* reveals the co-evolution of sulfate-reducing bacteria and the oxygenation of Earth''s surface. *Science Bulletin*, 69, pp. 1486–1494.

Davín, A.A., Woodcroft, B.J., Soo, R.M. et al. 2025. A geological timescale for bacterial evolution and oxygen adaptation. *Science*, 388, eadp1853.

DeLong, E. 1998. Microbiology—Archaeal means and extremes. *Science*, 280, pp. 542–543.

Derenne, S., Robert, F., and Skrzypczak-Bonduelle, A. 2008. Molecular evidence for life in the 3.5 billion year old Warrawoona chert. *Earth Planetary Science Letters*, 272, pp. 476–480.

Djokic, T., Van Kranendonk, M., Campbell, K. et al. 2017. Earliest signs of life on land preserved in ca. 3.5 Ga hot spring deposits. *Nature Communications*, 8, pp. 15263.

Dodd, M., Papineau, D., Grenne, T. et al. 2017. Evidence for early life in Earth''s oldest hydrothermal vent precipitates. *Nature*, 543, pp. 60–64.

Fakhraee, M., Crockford, P.W., Bauer, K.W. et al. 2025. The history of Earth''s sulfur cycle. *Nature Reviews Earth & Environment*, 6, pp. 106–125.

Farmer, J.D. 2001. Exploring for a fossil record of extraterrestrial life. In: *Palaeobiology II* (eds. D.E.G. Briggs and P.R. Crowther), pp. 8–13. London, Blackwell Science.

Fedo, C.M., and Whitehouse, M.J. 2002. Metasomatic origin of quartz-pyroxene rock, Akilia, Greenland, and implications for Earth"s earliest life. *Science*, 296, pp. 1148–1152.

Feulner, G. 2012. The faint young Sun problem. *Reviews of Geophysics*, 50, pp. RG2006.

Foster, G., Royer, D., and Lunt, D. 2017. Future climate forcing potentially without precedent in the last 420 million years. *Nature Communications*, 8, pp. 14845.

Fralick, P., and Riding, R. 2015. Steep Rock Lake: Sedimentology and geochemistry of an Archean carbonate platform. *Earth Science Reviews*, 15, pp. 132–175.

Garcia, A.K., Schopf, J.W., Yokobori, S., Akanuma, S., and Yamagishi, A. 2017. Reconstructed ancestral enzymes suggest long-term cooling of Earth"s photic zone since the Archean. *Proceedings of National Academy of Science of United States of America*, 114, pp. 4619–4624.

Garcia, A.K., Cavanaugh, C.M., and Kacar, B. 2021. The curious consistency of carbon biosignatures over billions of years of earth-life coevolution. *The ISME Journal*, 15, pp. 2183–2194.

Gilbert, P.U.P.A. et al. 2022. Biomineralization: iIntegrating mechanism and evolutionary history. *Science Advances*, 8, pp. eabl9653.

Glasspool, I.J., and Gastaldo, R.A. 2022. Silurian wildfire proxies and atmospheric oxygen. *Geology*, 50, pp. 1048–1052.

Grossman, E.L., and Joachimski, M.M. 2022. Ocean temperatures through the Phanerozoic reassessed. *Scientific Reports*, 12, pp. 8938.

Gumsley, A.P., Chamberlain, K.R., Bleeker, W. et al. 2017. Timing and tempo of the Great Oxidation Event. *Proceedings of National Academy of Science of United States of America*, 114, pp. 1811–1816.

Guo, Z., Papineau, D., O"Neil, J. et al. 2024. Abiotic synthesis of graphitic carbons in the Eoarchean Saglek-Hebron metasedimentary rocks. *Nature Communications*, 15, pp. 5679.

Halla, J., Noffke, N., Reis, H., Awramik, S. et al. 2024. Ratification of the base of the ICS Geological Time Scale: the Global Standard Stratigraphic Age (GSSA) for the Hadean lower boundary. *Episodes*, 47, pp. 381–389.

Han, T.M., and Runnegar, B. 1992. Megascopic eukaryotic algae from the 2.1 billion-year-old Negaunee Iron-Formation, Michigan. *Science*, 257, pp. 232–235.

Hardie, L.A. 2003. Secular variations in Precambrian seawater chemistry and the timing of Precambrian aragonite seas and calcite seas. *Geology*, 31, pp. 785–788.

Hassenkam, T., Andersson, M., Dalby, K. et al. 2017. Elements of Eoarchean life trapped in mineral inclusions. *Nature*, 548, pp. 78–81.

Hay, W.W., Migdisov, A., Balukhovsky, A.N., Wold, C.N., Flögel, S., and Söding, E. 2006. Evaporites and the salinity of the ocean during the Phanerozoic: iImplications for climate, ocean circulation and life. *Palaeogeography Palaeoclimatology Palaeoecology*, 240, pp. 3–46.

Hazen, R.M. 2005. Genesis: rocks, minerals, and the geochemical origin of life. *Elements*, 1, pp. 135–137.

He, H., Wu, X., Xian, H. et al. 2021. An abiotic source of Archean hydrogen peroxide and oxygen that pre-dates oxygenic photosynthesis. *Nature Communications*, 12, pp. 6611.

Hedges, S.B., Chen, H., Kumar, S., Wang, D.Y., Thompson, A.S., and Watanabe, H. 2001. A genomic timescale for the origin of eukaryotes. *BMC Evolutionary Biology*, 1, pp. 4.

Holland, H.D. 2009. Why atmosphere became oxygenated: a proposal. *Geochimica et Cosmochimica Acta*, 73, pp. 5241–5255.

Homann, M. 2019. Earliest life on earth: evidence from the Barberton Greenstone Belt. South Africa. *Earth-Science Reviews*, 196, pp. 102888.

Homann, M., Sansjofre, P., Van Zuilen, M., Heubeck, C., Gong, J., Killingsworth, B., Foster, I.S., Airo, A., Van Kranendonk, M.J., Ader, M., and Lalonde, S.V. 2018. Microbial life and biogeochemical cycling on land 3,220 million years ago. *Nature Geoscience*, 11, pp. 665–671.

Hou, K.-J., Li, Y.-H., and Wan, D.-F. 2007. Constraints on the Archean atmospheric oxygen and sulfur cycle from mass-independent sulfur records from Anshan-Benxi BIFs, Liaoning Province, China. *Science in China Series D-Earth Sciences*, 50, pp. 1471–1478.

Johnson, W., and Goldstein, R. 1993. Cambrian sea water preserved as inclusions in marine low-magnesium calcite cement. *Nature*, 362, pp. 335–337.

Jones, C., Nomosatryo, S., Crowe, S.A., Bjerrum, C.J., and Canfield, D.E. 2015. Iron oxides, divalent cations, silica, and the early Earth phosphorus crisis. *Geology*, 43, pp. 135–138.

Kanzaki, Y., and Murakami, T. 2015. Estimates of atmospheric CO_2 in the Neoarchean-Paleoproterozoic from paleosols. *Geochimica et Cosmochimica Acta*, 159, pp. 190–219.

Kasting, J.F. 1987. Theoretical constraints on oxygen and carbon dioxide concentrations in the Precambrian atmosphere. *Precambrian Research*, 34, pp. 205–229.

Kasting, J.F. 1991. Box models for the evolution of atmospheric oxygen: an update. *Palaeogeography Palaeoclimatology Palaeoecology*, 97, pp. 125–131.

Kasting, J.F. 1993. Earth"s early atmosphere. *Science*, 259, pp. 920–926.

Kasting, J.F. 2005. Methane and climate during the Precambrian era. *Precambrian Research*, 137, pp. 119–129.

Kasting, J.F., and Siefert, J.L. 2002. Life and the evolution of Earth"s atmosphere. *Science*, 296, pp. 1066–1068.

Kaufman, A.J., and Xiao, S. 2003. High CO_2 levels in the Proterozoic atmosphere estimated from analyses of individual microfossils. *Nature*, 425, pp. 279–282.

Kleine, T., and Walker, R.J. 2017. Tungsten isotopes in planets. *Annual Review of Earth and Planetary Science*, 45, pp. 389–417.

Kleine, T., Münker, C., Mezger, K., and Palme, H. 2002. Rapid accretion and early core formation on asteroids and the terrestrial planets from Hf–W chronometry. *Nature*, 418, pp. 952–955.

Knauth, L.P. 1998. Salinity history of the Earth"s early ocean. *Nature*, 395, pp. 554–555.

Knauth, L.P. 2005. Temperature and salinity history of the Precambrian ocean: implications for the course of microbial evolution. *Palaeogeography Palaeoclimatology Palaeoecology*, 219, pp. 53–69.

Knauth, L.P., and Lowe, D.R. 2003. High Archean climatic temperature inferred from oxygen isotope geochemistry of cherts in the 3.5 Ga Swaziland Supergroup, South Africa. *GSA Bulletin*, 115, pp. 566–580.

Konhauser, K.O. 2007. *Introduction to geomicrobiology*, pp. 425. Oxford, Blackwell Science.

Konhauser, K.O., Hamade, T., Morris, R.C., Ferris, F.G., Southam, G., Raiswell, R., and Canfield, D. 2002. Did bacteria form Precambrian banded iron formations? *Geology*, 30, pp. 1079–1082.

Konhauser, K.O., Newman, D., and K. and Kappler, A. 2005. The potential significance of microbial Fe(III) reduction during deposition of Precambrian banded iron formations. *Geobiology*, 3, pp. 167–177.

Krause, A.J., Mills, B.J.W., Merdith, A.S., Lenton, T.M., and Poulton, S.W. 2022. Extreme variability in atmospheric oxygen levels in the late Precambrian. *Science Advances*, 8, pp. eabm8191.

Kusmer, A. 2018. Diving deep to reveal the microbial mysteries of Lost City. Smithsonian Magazine, Smithsonian Institution. https://learninglab.si.edu/q/r/3183938

Laakso, T.A., and Schrag, D.P. 2019. Methane in the Precambrian atmosphere. *Earth and Planetary Science Letters*, 522, pp. 48–54.

Lantink, M.L., Davies, J.H., Ovtcharova, M., and Hilgen, F.J. 2022. Milankovitch cycles in banded iron formations constrain the Earth–moon system 2.46 billion years ago. *Proceedings of National Academy of Science of United States of America*, 119, pp. e2117146119.

Lee, C.T., Yeung, L., McKenzie, N. et al. 2016. Two-step rise of atmospheric oxygen linked to the growth of continents. *Nature Geoscience*, 9, pp. 417–424.

Lehmer, O.R. et al. 2020. Atmospheric CO_2 levels from 2.7 billion years ago inferred from micrometeorite oxidation. *Science Advances*, 6, pp. eaay4644.

Lenton, T.M., and Watson, A. 2011. *Revolutions that made the Earth*, pp. 423. Oxford, Oxford University Press.

Lenton, T.M., Daines, S.J., and Mills, B.J.W. 2018. COPSE reloaded: aAn improved model of biogeochemical cycling over Phanerozoic time. *Earth-Science Reviews*, 178, pp. 1–28.

Li, C., Love, G.D., Lyons, T.W., Fike, D.A., Sessions, A.L., and Chu, X.-L. 2010. A stratified redox model for the Ediacaran ocean. *Science*, 328, pp. 80–83.

Line, M.A. 2002. The enigma of the origin of life and its timing. *Microbiology*, 148, pp. 21–27.

Lowenstein, T.K., Timofeeff, M.N., Brennan, S.T., Hardie, L.A., and Demicco, R.M. 2001. Oscillations in Phanerozoic seawater chemistry: evidence from fluid inclusions. *Science*, 294, pp. 1086–1088.

Lowenstein, T.K., Hardie, L.A., Timofeeff, M.N., and Demicco, R.V. 2003. Secular variation in seawater chemistry and the origin of calcium chloride basinal brines. *Geology*, 31, pp. 857–860.

Lu, Z., Rosalind, E.M.R., Jonathan, L.P., and Ashley, N.P. 2024. Phanerozoic co-evolution of O_2-CO_2 and ocean habitability. *National Science Review*, 11, pp. nwae099.

Luskin, C., Wilson, A., Gold, D., and Hofmann, A. 2019. The *Pongola Supergroup*: Mesoarchaean deposition following Kaapvaal Craton Stabilization. In: *The Archaean Geology of the Kaapvaal Craton, Southern Africa* (eds. A. Kröner and A. Hofmann), pp. 225–254. Springer, Cham, Regional Geology Reviews.

Lüthi, D., Le Floch, M., Bereiter, B. et al. 2008. High-resolution carbon dioxide concentration record 650,000–800,000 years before present. *Nature*, 453, pp. 379–382.

Lyons, T.W. et al. 2021. Oxygenation, life, and the planetary system during Earth's middle history: aAn oOverview. *Astrobiology*, 21, pp. 906–923.

McCollom, T.M., and Seewald, J.S. 2006. Carbon isotope composition of organic compounds produced by abiotic synthesis under hydrothermal conditions. *Earth and Planetary Science Letters*, 243, pp. 74–84.

McIntyre, T., and Fralick, P. 2017. Sedimentology and geochemistry of the 2930 Ma Red Lake-Wallace Lake carbonate platform, Western Superior Province. Canada. *The Depositional Record*, 3, pp. 258–287.

Miao, L., Yin, Z., Knoll, A.H., Qu, Y., and Zhu, M. 2024. 1.63-billion-year-old multicellular eukaryotes from the Chuanlinggou Formation in North China. *Science Advances*, 10, pp. eadk3208.

Mojzsis, S.J., Arrhenius, G., Mckeegan, K.D., Harrison, T.M., Nutman, A.P., and Friend, C.R.L. 1996. Evidence for life on Earth before 3.800 million years ago. *Nature*, 384, pp. 55–59.

Mojzsis, S.J., Harrison, T.M., and Pidgeon, R.T. 2001. Oxygen-isotope evidence from ancient zircons for liquid water at the Earth's surface 4,300 Myr ago. *Nature*, 409, pp. 178–180.

Moody, E.R.R., Álvarez-Carretero, S., Mahendrarajah, T.A. et al. 2024. The nature of the last universal common ancestor and its impact on the early Earth system. *Nature Ecology Evolution*, 8, pp. 1654–1666.

Moorbath, S. 1983. The most ancient rocks. *Nature*, 304, pp. 585–586.

Mukherjee, I., and Large, R.R. 2020. Co-evolution of trace elements and life in Precambrian oceans: The pyrite edition. *Geology*, 48, pp. 1018–1022.

Nabhan, S., Luber, T., Scheffler, F., and Heubeck, C. 2016. Climatic and geochemical implications of Archean pedogenic gypsum in the Moodies Group ($\sim$3.2 Ga), Barberton Greenstone Belt, South Africa. *Precambrian Research*, 275, pp. 119–134.

Nealson, K.H., and Rye, R. 2005. Evolution of metabolism. In: *Biogeochemistry. Treatise on Geochemistry* (Vol. 8) (ed. W.H. Schlesinger), pp. 41–61. Oxford, Elsevier-Pergamon.

Nisbet, E.G., and Fowler, C.M.R. 2005. The early history of life. In: *Biogeochemistry. Treatise on Geochemistry* (Vol. 8) (ed. W.H. Schlesinger), pp. 1–39. Oxford, Elsevier-Pergamon.

Nobs, S.-J., MacLeod, F.I., Wong, H.L., and Burns, B.P. 2022. Eukarya the chimera: eukaryotes, a secondary innovation of the two domains of life? *Trends in Microbiology*, 30, pp. 421–431.

Noffke, N., Hazen, R.M., and Nhleko, N. 2003. Earth's earliest microbial mats in a siliciclastic marine environment (2.9 Ga Mozaan Group, South Africa). *Geology*, 31, pp. 673–676.

Noffke, N., Christian, D., Wacey, D., and Hazen, R.M. 2013. Microbially induced sedimentary structures recording an ancient ecosystem in the ca. 3.48 billion-year-old Dresser Formation, Pilbara. Western Australia. *Astrobiology*, 13, pp. 1103–1124.

Noffke, N., Hazen, R.M., Beukes, N., Bower, D., and Swift, J.P. 2008. An actualistic perspective into Archean worlds – (cyano-)bacterially induced sedimentary structures in the siliciclastic Nhlazatse Section, 2.9 Ga Pongola Supergroup, South Africa. *Geobiology*, 6, pp. 5–20.

Nutman, A.P., Bennett, V.C., Friend, C.R., Van Kranendonk, M.J., and Chivas, A.R. 2016. Rapid emergence of life shown by discovery of 3,700-million-year-old microbial structures. *Nature*, 537, pp. 535–538.

Och, L.M., and Shields-Zhou, G.A. 2012. The Neoproterozoic oxygenation event: eEnvironmental perturbations and biogeochemical cycling. *Earth-Science Reviews*, 110, pp. 26–57.

Ohmoto, H., Watanbe, Y., and Kumazawa, K. 2004. Evidence from massive siderite beds for a CO2-rich atmosphere before ~1.8 billion years ago. *Nature*, 429, pp. 295–399.

Papineau, D., She, Z., Dodd, M.S., Iacoviello, F., Slack, J.F., Hauri, E., and P. and Little, C. 2022. Metabolically diverse primordial microbial communities in Earth"s oldest seafloor hydrothermal jasper. *Science Advances*, 8, pp. eabm2296.

Pavlov, A.A., Hurtgen, M.T., Kasting, J.F., and Arthur, M.A. 2003. Methane-rich Proterozoic atmosphere. *Geology*, 31, pp. 87–90.

Pavlov, A.A., Kasting, J.F., Brown, L.L., Rages, K.A., and Freedman, R. 2000. Greenhouse warming by CH_4 in the atmosphere of early Earth. *Journal of Geophysical Research*, 105, pp. 11981–11990.

Peters, S.E., Quinn, D.P., Husson, J.M., and Gaines, R.R. 2022. Macrostratigraphy: insights into cyclic and secular evolution of the Earth–life system. *Annual Review of Earth and Planetary Sciences*, 50, pp. 419–449.

Planavsky, N.J., Reinhard, C.T., Wang, X., Thomson, D., McGoldrick, P., Rainbird, R.H., Johnson, T., Fischer, W.W., and Lyons, T.W. 2014. Low Mid-Proterozoic atmospheric oxygen levels and the delayed rise of animals. *Science*, 346, pp. 635–638.

Planck Collaboration. 2020. Planck 2018 results VI Cosmological parameters. *Astronomy & Astrophysics*, 641(A6), pp. 1–67.

Rasmussen, B. 2000. Filamentous microfossils in a 3,235-million-year-old volcanogenic massive sulfide deposit. *Nature*, 405, pp. 676–679.

Rasmussen, B., Fletcher, I., Brocks, J., and Kilburn, M.R. 2008. Reassessing the first appearance of eukaryotes and cyanobacteria. *Nature*, 455, pp. 1101–1104.

Reimink, J.R., Baue, A.M., and Chacko, T. 2019. The Acasta Gneiss Complex. In: *Earth's Oldest Rocks* (eds. M.J. Van Kranekdonk, R.H. Smithies, and V.C. Bennett), pp. 329–347. Amsterdam, Elsevier.

Robbins, L.J. et al. 2016. Trace elements at the intersection of marine biological and geochemical evolution. *Earth-Science Reviews*, 163, pp. 323–348.

Robert, F., and Chaussidon, M. 2006. A palaeotemperature curve for the Precambrian oceans based on silicon isotopes in cherts. *Nature*, 443, pp. 969–972.

Rosing, M.T. 1999. ^{13}C-depleted carbon in >3700 Ma seafloor sedimentary rocks from West Greenland. *Science*, 283, pp. 674–676.

Rosing, M.T., and Frei, R. 2004. U-rich Archaean Sea-floor sediments from Greenland-indications of >3700 Ma oxygenic photosynthesis. *Earth Planetary Science Letters*, 217, pp. 237–244.

Rye, R., and Holland, H.D. 2000. Life associated with a 2.76 Ga ephemeral pond? *Geology*, 28, pp. 483–486.

Rye, R., Kuo, P.H., and Holland, H.D. 1995. Atmospheric carbon dioxide concentrations before 2.2 billion years ago. *Nature*, 378, pp. 603–605.

Sagan, C., and Mullen, G. 1972. Earth and Mars: evolution of atmospheres and surface temperatures. *Science*, 177, pp. 52–56.

Sánchez-Baracaldo, P., Raven, J.A., Pisani, D., and Knoll, A.H. 2017. Early photosynthetic eukaryotes inhabited low-salinity habitats. *Proceedings of National Academy of Science of United States of America*, 114, pp. E7737–E7745.

Sánchez-Baracaldo, P., Bianchini, G., Wilson, J.D., and Knoll, A.H. 2022. Cyanobacteria and biogeochemical cycles through Earth history. *Trend in Microbiology*, 30, pp. 143–157.

Sanford, W., Doughten, M., Coplen, T., Hunt, A.G., and Bullen, T.D. 2013. Evidence for high salinity of Early Cretaceous sea water from the Chesapeake Bay crater. *Nature*, 503, pp. 252–256.

Schoenberg, R., Kambeer, B.S., Collerson, K.D., and Moorbath, S. 2002. Tungsten isotope evidence from 3.8 Gyr metamorphosed sediments for early meteorite bombardment of Earth. *Nature*, 418, pp. 403–405.

Schopf, J.W. 1993. Microfossils of the Early Archaean Apex Chert: new evidence of the antiquity of life. *Science*, 260, pp. 640–646.

Schopf, J.W. 1999. *Cradle of life*. Princeton and Oxford, Princeton University Press.

Selden, P.A. 2005. Terrestrialization (Precambrian–Devonian). In: *Encyclopedia of Life Sciences*. Chichester, Wiley.

Sepkoski, J.J., Jr. 1984. A kinetic model of Phanerozoic taxonomic diversity III. Post-Paleozoic families and mass extinction. *Paleobiology*, 10, pp. 246–267.

Sheldon, N.D., Mitchell, R.L., and Dzombak, R.M. 2021. *Reconstructing Precambrian* pCO_2 *and* pO_2 *using Paleosols, pp. 1–28. Cambridge, Cambridge University Press.*

Shen, Y., Buick, R., and Canfield, D.E. 2001. Isotopic evidence for microbial sulfate reduction in the early Archean era. *Nature*, 410, pp. 77–81.

Shu, D., Luo, H., Conway Morris, S., Zhang, X., Hu, S. et al. 1999. Lower Cambrian vertebrates from south China. *Nature*, 402, pp. 42–46.

Sleep, N.H., and Zahnle, K.J. 2001. Carbon dioxide cycling and implications for climate on ancient Earth. *Journal of Geophysics Research*, 106, pp. 1373–1399.

Sleep, N.H., Zahnle, K.J., and Lupu, R.E. 2014. Terrestrial aftermath of the moon-forming impact. *Philosophical Transactions of the Royal Society London A*, 372, pp. 20130172.

Song, H., Kemp, D.B., Tian, L. et al. 2021. Thresholds of temperature change for mass extinctions. *Nature Communications*, 12, pp. 4694.

Sowerby, S.J., Heckl, W.M., and Petersen, G.B. 1996. Chiral symmetry breaking during the self-assembly of monolayers from achiral purine molecules. *Journal of Molecular Evolution*, 43, pp. 419–424.

Sowerby, S.J., Edelwirth, M., and Heckl, W.M. 1998. Self-assembly at the prebiotic solid-liquid interface: structure of self-assembled monolayers of adenine and guanine bases formed on inorganic surfaces. *Journal of Physical Chemistry B*, 102, pp. 5914–5922.

Stockey, R.G., Cole, D.B., Farrell, U.C., Agić, H., Boag, T.H. et al. 2024. Sustained increases in atmospheric oxygen and marine productivity in the Neoproterozoic and Palaeozoic eras. *Nature Geoscience*, 17, pp. 667–674.

Stüeken, E.E., Pellerin, A., Thomazo, C., Johnson, B.W., Duncanson, S. et al. 2024. Marine biogeochemical nitrogen cycling through Earth"s history. *Nature Reviews Earth & Environment*, 5, pp. 732–747.

Summons, R.E., Jahnke, L.L., Hope, J.M., and Logan, G.A. 1999. 2-Methylhopanoids as biomarkers for cyanobacterial oxygenenic photosynthesis. *Nature*, 400, pp. 554–557.

Sun, W., Jayaraman, S., Chen, W., and Ceder, G. 2015. Nucleation of metastable aragonite $CaCO_3$ in seawater. *Proceedings of National Academy of Science of United States of America*, 112, pp. 3199–3204.

Sun, Y., Farnsworth, A., Joachimski, M.M., Wignall, P.B., Krystyn, L. et al. 2024. Mega El Niño instigated the end-Permian mass extinction. *Science*, 385, pp. 1189–1195.

Tang, Q., Zheng, W., Zhang, S. et al. 2025. Quantifying the global biodiversity of Proterozoic eukaryotes. *Science*, 386, pp. eadm9137.

Tashiro, T., Ishida, A., Hori, M., Igisu, M., Koike, M. et al. 2017. Early trace of life from 3.95 Ga sedimentary rocks in Labrador, Canada. *Nature*, 549, pp. 516–518.

Ueno, Y., Yamada, K., Yoshida, N., Maruyama, S., and Isozaki, Y. 2006. Evidence from fluid inclusions for microbial methanogenesis in the early Archaean era. *Nature*, 440, pp. 516–519.

Valley, J., Cavosie, A., Ushikubo, T., Reinhard, D.A., Lawrence, D.F. et al. 2014. Hadean age for a post-magma-ocean zircon confirmed by atom-probe tomography. *Nature Geoscience*, 7, pp. 219–223.

Van Kranendonk, M.J. 2006. Volcanic degassing, hydrothermal circulation and the flourishing of early life on Earth: aA review of the evidence from c. 3490-3240 Ma rocks of the Pilbara Supergroup, Pilbara Craton, Western Australia. *Earth-Science Reviews*, 74, pp. 197–240.

Wächtershäuser, G. 2010. Chemoautotrophic origin of life: the Iron–Sulfur World Hypothesis. In: *Geomicrobiology: Molecular and Environmental Perspective* (eds. L. Barton, M. Mandl, and A. Loy), pp. 1–35. Dordrecht, Springer.

Walton, C.R., Ewens, S., Coates, J.D., Blake, R.E., Planavsky, N.J. et al. 2023. Phosphorus availability on the early Earth and the impacts of life. *Nature Geoscience*, 16, pp. 399–409.

Wang, C., Dong, Z., Robbins, L., Zhang, B., Peng, Z. et al. 2024. A craton-wide geochemical survey of late Archean banded iron formations in China. *Earth and Planetary Science Letters*, 642, pp. 118879.

Weimann, L., Reinhardt, M., Duda, J.-P., Mißbach-Karmrodt, H., Drake, H. et al. 2024. Carbonaceous matter in ~3.5 Ga black bedded barite from the Dresser Formation (Pilbara Craton, Western Australia) – Insights into organic cycling on the juvenile Earth. *Precambrian Research*, 403, pp. 107321.

Westall, F., and Xiao, S. 2024. Precambrian Earth: Co-evolution of life and geodynamics. *Precambrian Research*, 414, pp. 107589.

Wilde, S., Valley, J., Peck, W., and Graham, C.M. 2001. Evidence from detrital zircons for the existence of continental crust and oceans on the Earth 4.4 Gyr ago. *Nature*, 409, pp. 175–178.

Moody, E.R.R., Williams, T.A., Álvarez-Carretero, S., Szöllősi, G.J., Pisani, D., Lenton, T.M., and Donoghue, P.C.J. 2025. The emergence ofmetabolisms through Earth history andimplications for biospheric evolution. *Philosophical Transection of Royal Society B*, 380, pp. 20240097.

de Wit, M.J., and Furnes, H. 2016. 3.5-Ga hydrothermal fields and diamictites in the Barberton Greenstone Belt-Paleoarchean crust in cold environments. *Science Advances*, 2, pp. e1500368.

Zahnle, K.J., Lupu, R., Dobrovolskis, A., and Sleep, N.H. 2015. The tethered moon. *Earth Planetary Science Letters*, 427, pp. 74–82.

Zhang, Z.Y. 1986. Clastic facies microfossils from the Chuanlinggou Formation (1800 Ma) near Jixian, North China. *Journal of Micropalaeontology*, 5, pp. 9–16.

Zhang, X., and Cui, L. 2016. Oxygen requirements for the Cambrian explosion. *Journal of Earth Science*, 27, pp. 187–195.

Zhang, X., and Shu, D. 2014. Causes and consequences of the Cambrian explosion. *Science China-Earth Sciences*, 57, pp. 930–942.

Zhang, X., and Shu, D. 2021. Current understanding on the Cambrian Explosion: questions and answers. *PalZ*, 95, pp. 641–660.

Zhao, M., Reinhard, C., and Planavsky, N. 2018. Terrestrial methane fluxes and Proterozoic climate. *Geology*, 46, pp. 139–142.

Zhao, M., Mills, B.J.W., Poulton, S.W. et al. 2024. Drivers of the global phosphorus cycle over geological time. *Nature Review Earth & Environment*, 5, pp. 873–889.

9

Exploring Extraterrestrial Life

Ruliang He and Xingliang Zhang

9.1 Introduction

Astrobiology is a subject to understand the origin, evolution, and distribution of life in the past, present, and future of the Universe. The only known instance of life is on Earth. In the previous chapters, we have discussed interactions between life and environments on Earth and demonstrated how various life forms adapt to and modify the forever-changing environments throughout Earth's history. In this chapter, we seek to understand life in the Universe and how the indigenous intelligent species is searching for life elsewhere. All the ingredients required by life, including stars, planets, elements, and energy, have formed since the Big Bang, but they do not define life only until being cooked at the right conditions. Therefore, we discuss the physical and chemical boundaries that may harbor life. At this stage, we have figured out the conditions that favor the origin and persistence of life in the Universe, and we can define the habitability and search for habitable zones in the Universe. We take a closer look at our own solar system, studying the environmental conditions on Mars and in subsurface oceans of icy moons. We try to answer the possibility of biological processes on those planets or moons in the past, right now, and in the future.

9.2 Habitability and Habitable Zone

For astrobiologists to investigate the origin, evolution, and distribution of life in the extraterrestrial space, understanding the term "habitability" is essential. Habitability is defined as the ability of an environment to support the metabolic activity that allows for survival, maintenance, growth, or reproduction for at least one known organism (Cockell et al. 2016). Although it remains unknown whether terrestrial life represents a universal norm, the conservative definition of habitability here constrains it to a known life-form based on the current biology knowledge (Cockell et al. 2016). It should be noted that the habitability can be decoupled from the presence of life; for example, the ancient Mars can be considered as an"uninhabited habitat" where habitable conditions have developed but no life signature has been found.

Introduction to Geobiology, First Edition. Xingliang Zhang, Zisheng Guo, Wei Liu, Ruliang He, Luoyang Li, and Weiduo Hao.

Companion website: www.wiley.com/go/Zhang_1e

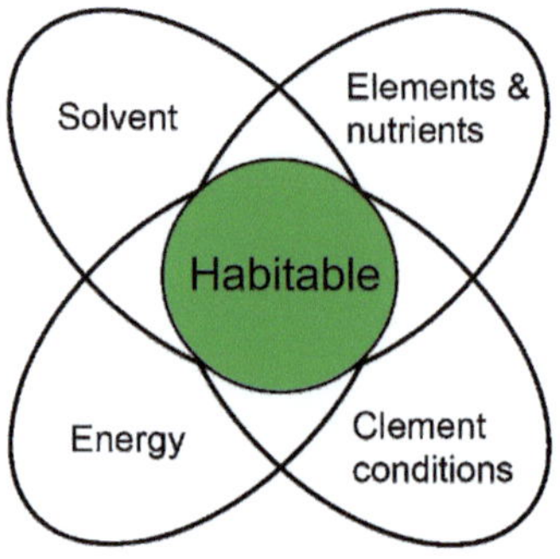

Figure 9.1 Four key requirements for habitability. *Source:* Adapted from Hoehler (2007).

9.2.1 Key Requirements for Habitability

There are four key requirements to keep an environment habitable, including a solvent, enough raw materials, available energy, and clement physicochemical conditions (Hoehler 2007). The boundary for each separate requirement can be relatively broad, but the habitability must be a consequence as an intersection of the four requirement sets with narrower boundary conditions (Figure 9.1).

9.2.1.1 A Solvent

All the building blocks (e.g. proteins, carbohydrates, lipids, and nucleic acids) of life have to be assembled in a liquid medium, which means metabolic reactions are not efficient without a solvent (Cockell 2016). Water, or H_2O molecule, is the only known solvent prevalently used by organisms on Earth, because of its abundance and chemical properties. Hydrogen and oxygen atoms of water molecule are bound by polar covalent bonds and this polarity enables water to easily dissolve both salts as a source of ions and organic molecules as basic units for important redox reactions and synthesis of macromolecules. The polarity also helps the formation of hydrogen bonds between water molecules, which results in a wide temperature range for liquid water to exist (e.g. on the surface of Earth and interiors of icy moons) (Cockell 2016). The lower density of ice than liquid water could prevent a complete frozen condition of lakes and oceans, because ice would float at the surface when freezing. Such unique properties would provide a refuge for life in extreme cold environments (e.g. during Snowball Earth events and in Jovian moons). Alternative solvents have also been considered by astrobiologists. A suitable candidate should be abundant in the universe or capable of remaining liquid in a wide temperature range, including ammonia (NH_4), sulfuric acid (H_2SO_4), formamide (CH_3NO), and methane (CH_4) (Grefenstette et al. 2024). However, water is still at the front for human to search life beyond Earth.

9.2.1.2 Elements and Nutrients

For terrestrial life, C, H, O, N, P, and S are common in all life forms, and compose fundamental biomolecules, such as nucleotides, amino acids, lipids, and sugars. Carbon builds basic frameworks of most biomolecules, and its covalent bonds enable the formation of various structures and functions with other elements. Hydrogen is usually bound with C, N, and S in almost all organic compounds, and oxygen can be contained in carbohydrates, alcohol, and other molecules (e.g. ring structures). More importantly, the redox potential of H- and O-containing compounds provides energy via biogeochemical reactions (Colón-Santos et al. 2024). Nitrogen is a critical component of proteins, DNA, chlorophyll, and other biomolecules, while the abundant N_2 gas in the Earth's atmosphere is biologically inaccessible. Phosphorus is a fundamental component in nucleic acids, energy transfer molecules (e.g. ATP), and cellular membranes (e.g. phospholipid), occurring as orthophosphate (PO_4^{3-}) (Canfield et al. 2005). Sulfur can form the disulfide bond as protein bridges and influence the folding and stability of proteins; it also forms iron-sulfur clusters in electron transfer proteins and provides energy for microbial life due to its diverse oxidation states (Colón-Santos et al. 2024). When C, H, O, N, P, and S are combined together, their versatility builds up the complexity of organic compounds and expands the function of life. Elements of minor content in organisms are also important, such as Na, K, Cl, Ca, Mg, and Fe. For example, they serve to balance differences in ion concentration and charge between two sides of the cellular membrane; in addition, transition metals (Fe, Mn, Cu, Zn, Mo, V, Co, Ni) help to catalyze biochemical reactions by promoting reaction rate without being used up (Colón-Santos et al. 2024).

9.2.1.3 Energy

Life needs to turn raw nutrient materials and energy from environments into those available for cellular use via metabolism. Energy is especially essential for organisms to keep a low-entropy state and to power chemical reactions that are not spontaneous (Thweatt et al. 2024). For organisms on Earth, although they may have different metabolic pathways (Section 1.3), the external energy (e.g. light or chemical sources) is ultimately converted for cell use via redox reactions that involve the electron flow from electron donors to electron acceptors (Figure 9.2). Finally, energy from redox reactions can be stored temporarily within phosphate bonds of ATP (adenosine triphosphate) and used by subsequent cellular processes. Therefore, the availability of external energy sources and redox couples (i.e. electron donor and electron acceptor) in the environment is important for the evaluation of habitability beyond Earth (Figure 9.2). The energy budget of life can be assessed using thermodynamics of different redox couples (e.g. Gibbs free energy). Apart from solar and chemical sources, extraterrestrial life could make use of thermal gradients, mechanical energy, gravitational energy, other radiation energy, and osmotic gradients as alternative energy sources (Grefenstette et al. 2024).

9.2.1.4 Clement Physical and Chemical Conditions

While available liquid water, nutrients, and energy sources are fundamental requirements for the presence of life, a suitable surrounding condition is also indispensable. Physical and chemical stressors commonly include, but not limited to, temperature, pH, pressure, salinity, radiation, aridity,

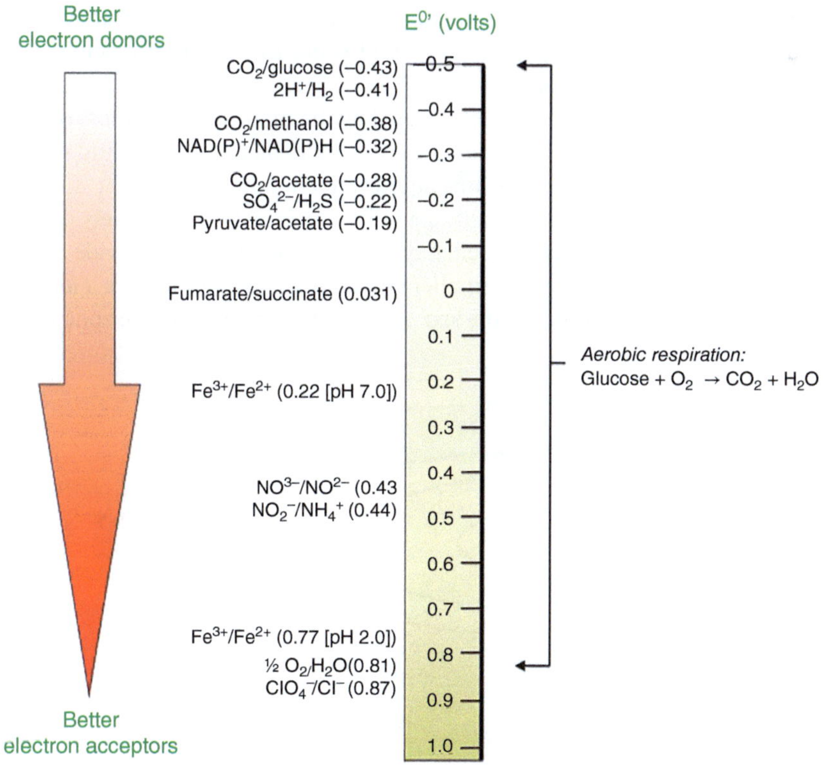

Figure 9.2 Redox couples and their electrode potentials at the standard condition. Energy can be produced via a full redox reaction, consisting of two half reactions (redox couples) that either give up electrons or accept electrons. The aerobic respiration of glucose by molecular oxygen is shown as an example. *Source:* Cockell (2016) / John Wiley & Sons.

and toxic metal concentration. Since we are still learning how Earth life could tolerate extreme environments, the edges of biosphere on Earth and extraterrestrial cannot be well constrained currently. Studying extremophiles is helpful for understanding how life could adapt to extreme environments beyond our Earth (Section 2.3–10).

9.2.2 The Habitable Zone

Astrobiologists define the habitable zone, also called the Goldilocks zone, as a range of orbits surrounding a star where liquid water can be stably sustained on the planetary surfaces. The definition would underestimate the habitability of those extraterrestrial bodies with liquid water below the surface (e.g. icy moons of Jupiter and Saturn), and overlook other requirements of habitability (e.g. essential nutrient elements). However, defining a habitable zone would considerably help to narrow down our scope while searching for life beyond Earth. The habitable zone is a circular band, with clement temperature to sustain liquid water (i.e. neither too hot nor too cold) (Figure 9.3). Within the inner edge of the habitable zone, planets and moons would receive strong radiation and consequently any liquid water would evaporate from their surfaces; beyond its outer edge, low temperature on the surface is unable to sustain liquid water. The habitable zone of our solar system is commonly thought between Venus and Mars (Figure 9.3), although the exact position is dependent on different assumptions used in the models (e.g. Kasting et al. 1993; Kopparapu et al. 2013). Venus might have had oceans at a very early age, but solar radiation finally caused a runaway greenhouse effect, and only a tiny amount of water is now preserved in its atmosphere. For Mars, although there is evidence of liquid water on its surface in the past, it is almost frozen currently on or near the surface. Actually, many environments with liquid water in the solar system have been found beyond the habitable zone; for example, liquid water can be present below the surfaces of icy moons related to the tidal energy.

The location and stability of the habit zone can be determined by the mass and brightness of the star. A massive star generates stronger UV radiation, and thereby its habitable zone would move outward to stay in a suitable temperate range that sustains liquid water; on the contrary, for smaller and dimmer stars, habit zones shift inward. Since the lifespan of a star is usually negatively correlated with its mass, the habitable zone of a massive stellar system may be unable to provide sufficient time for life and even intelligence to develop. A smaller and dimmer star would have a longer time span of habitability; due to its close distance to the star, a planet in such a habitable zone may

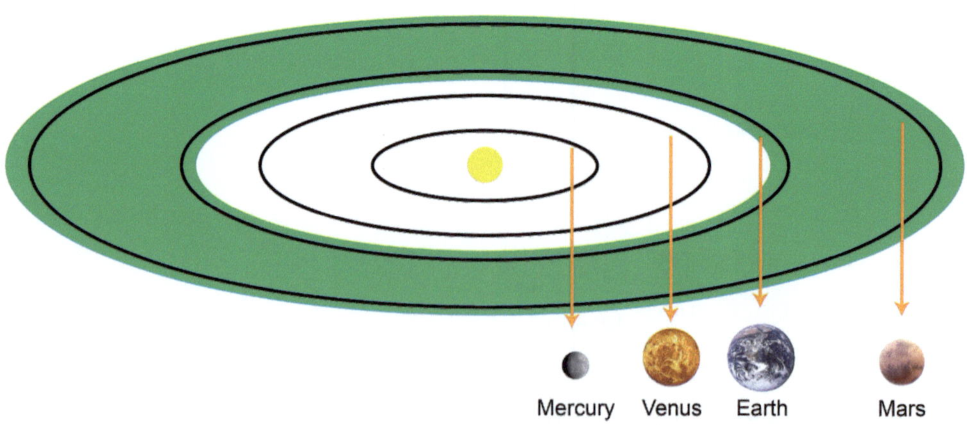

Figure 9.3 The habitable zone of the solar system. Inner and outer boundaries. *Source:* Data of inner and outer boundaries are based on Kopparapu et al. (2013).

face a tidal locking problem, which means one face of the planet is permanently locked toward the star and the other face would always be dark and too cold to support life. Additionally, such stellar systems would generate ion radiation flares more frequently and consequently endanger potential life on planetary surfaces. Since the luminosity of stars increases with their ages, the habitable zone would shift gradually outward due to the evolution of a stellar system. A concept of a continuously habitable zone is proposed to represent a zone that supports habitability during the entire lifespan of a stellar system. A continuously habitable zone is normally in a narrower range than a common habitable zone.

9.3 Potentially Habitable Worlds Beyond Earth

Our knowledge of habitability beyond Earth is mostly from the solar system. Venus has a strong greenhouse effect, and thereby the extremely high surface temperature cannot sustain any liquid water or potential life. Although Mars is less possible to host life at present, there is ample evidence for liquid water and a clement climate on the early Mars. We have made many direct observations of Mars by sending different space missions. Planetary systems beyond the "frost line" were initially thought to be impossible to hold life due to the extremely low temperature and a lack of rocky crust, but recent discovery of liquid water beneath the icy shells of their moons suggests icy moons of the solar system would be a major target for future astrobiology research (e.g. Europa, Enceladus, Titan). For exoplanets outside the solar system, since we are unable to send spacecraft for direct observation, most information comes from remote observation. However, considering the large number of exoplanets, they are of great potential for astrobiology study from the statistical perspective.

9.3.1 Venus

Venus is the second planet from the Sun and located near the inner edge of the habitable zone of the solar system. It may have witnessed a deterioration of habitability and availability of liquid water in the past. As the brightness of the Sun increased in the past 4.6 billion years, the habitable zone of our solar system gradually moved outward. At the same time, Venus, which may have been initially within the habitable zone, would move to the inner edge and lose liquid water when the temperature of the planet increased during the stellar evolution.

Currently, Venus has a very dense atmosphere; the atmospheric pressure at the surface of Venus is estimated to be over 90 times greater than that of Earth, equaling to the pressure at ~1 km depth in the ocean of Earth. The atmosphere contains ~96% CO_2 and ~ 3% N_2, with sulfuric acid occurring in the cloud deck. Therefore, the dense atmosphere is nearly opaque for the outgoing thermal radiation from the surface of Venus, resulting in a strong greenhouse effect. The temperature on the surface of Venus could reach 462 °C and exceed the upper temperature limit for life forms (e.g. Strain 121) we know on Earth. The surface of Venus is desiccated due to the extremely high temperature. Water vapor content in its atmosphere is at parts per million (ppm; 1×10^{-6}) level, three orders lower than that of Earth. During the runaway greenhouse stage, water may have been evaporated to the upper atmosphere and subsequently been photolyzed to hydrogen and oxygen (Kasting 1988).

However, Venus may have had liquid water in the past. The current deuterium to hydrogen ratio (D/H or $^2H/^1H$) of the Venus atmosphere is considerably higher than that of seawater of Earth. After photolysis, lighter H_2 molecules are easier to escape to the space than the heavier DH, and thereby the atmosphere would be enriched in deuterium gradually. Assuming Venus and Earth had the same D/H ratio when they were formed, higher D/H in modern Venusian atmosphere would suggest the planet may have contained water in the past (Kasting and Pollack 1983). Without such a strong greenhouse effect and with a faint young Sun in the past, the temperature of Venus may have been cool enough to condense liquid water. Therefore, it is possible that Venus may have been habitable and harbored life in the past.

9.3.2 Mars

9.3.2.1 A Brief Introduction to Mars

Mars is the fourth planet from the Sun and a neighbor of Earth. Compared to other extraterrestrial bodies, Mars has always been attracting astrobiologists' interests and imaginations. Mars has a rusty red appearance and is also referred to as "the Red Planet," due to abundant ferric oxides on its surface (Figure 9.4). The atmosphere of Mars is considerably thinner than that of Earth (e.g. 0.6% of Earth's atmospheric pressure on Mars surface), and the average temperature of Mars' atmosphere is normally below −60 °C due to the further distance to the Sun and a weaker greenhouse effect (Haberle 2015). Carbon dioxide (CO_2) is a primary component (>95%) of the Martian atmosphere, and it also contains argon (Ar), nitrogen (N_2), trace amount of oxygen (O_2), and other components. The atmosphere usually contains tiny particles, and therefore, dust storms occur very frequently on Mars. The internal structure of Mars is similar to Earth. The uppermost layer is crust, and the mantle is also composed of silicates. The core of Mars is in a melted state and made of metal iron and nickel. However, the present Mars is tectonically inactive.

History of Climate and Aqueous Activity on Mars

Mars is the same age as Earth and formed around 4.567 billion years ago (Ga) (Figure 9.5). Its geological evolution history can be divided into three eons, including Noachian (~4.1 to 3.7 Ga), Hesperian (~3.7 to 3.1 Ga), and Amazonian (~3.1 Ga to present). During Noachian, Mars had a

Figure 9.4 The surface feature of Mars, captured by Zhurong rover. *Source:* CNSA / GRAS / Public domain.

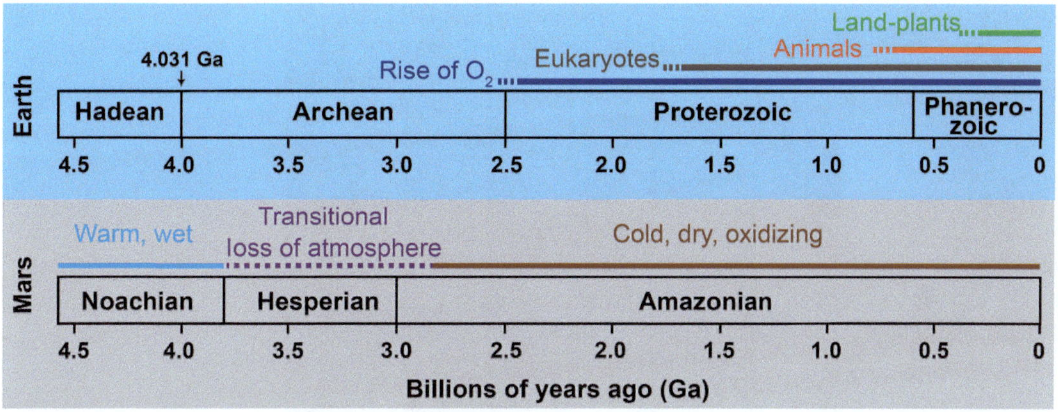

Figure 9.5 Geological timescales of Earth and Mars and key environmental and biological evolution trends.

relatively warm and humid climate, with evidence for aqueous activity (e.g. liquid water and rock interaction) and a denser atmosphere, which might be habitable to life. Transitioning into the Hesperian, the surface environment started to deteriorate, showing signs of desiccation, loss of both atmosphere and aqueous activity. During Amazonian, as we see on the present Mars, its surface finally became cold, dry, and desert-like, which seems to be much less habitable for biological activities. The history of Mars witnesses a long-term deterioration of liquid water activity and therefore the habitability (Figure 9.5).

Astrobiology Research on Mars

The exploration of Mars started from remote observations using telescopes that simulated a trend of imaging civilization on Mars, to recently advances of sending spacecrafts (e.g. orbiters, landers, and rovers) that finally revealed barren landscapes on Mars. Many agencies are able to send spacecraft to Mars, including those of the United States, Soviet Union, Europe, China, India, and the UAE, to study the climate, geology, and astrobiology of Mars. The early explorations since the 1960s, such as Mariner flyby and orbiter missions, started human's first close-up look at the Red Planet. In the 1970s, Viking orbiters and landers started to send back detailed images of Mars surface and to conduct a series of biological experiments on Mars surface to detect any signs of life there. Later, following missions to Mars have been providing detailed topographic maps, chemical and mineralogical compositions of Martian rocks, details of "marsquakes" and other features of Mars. In the future, we are planning to return samples from Mars to Earth for more sophisticated investigation and even transport humans to the Mars surface. Mars is thus the most extensively studied body in terms of astrobiology beyond Earth.

9.3.2.2 Water on Mars

As multiple lines of evidence suggest liquid water activities may have altered the mineralogical composition of the Martian surface and reformed its surface morphology during the Noachian and Hesperian eons, Mars may have contained liquid water on its surface at its early age. However, there is little evidence of liquid water to occur on the surface of Mars at present. Most of the water on Mars has now been converted to ice on its polar surface or beneath dust and rocks (e.g. ice caps and ground ice), or has been locked into crystal structures of minerals as hydroxyl (−OH) of clay minerals or as water of crystallization in evaporites (e.g. sulfate).

Figure 9.6 False color image of the Martian Jezero Crater showing channel, fan, and delta features. Sediments containing phyllosilicates are presented in green. *Source:* NASA / JPL-Caltech / MSSS / JHU-APL / NASA / Public Domain.

Mineralogical Evidence

When mafic volcanic rocks, such as basalt (a major rock type of Mars crust), are exposed on Mars surface, the mafic minerals (e.g. plagioclase feldspar, pyroxene, and olivine) would react rapidly with water and form secondary minerals—phyllosilicates (e.g. smectites or chlorites). Phyllosilicates can be found in many Pre-Noachian and Noachian terrains, and thus an alternative name for this time interval is Phyllocian, suggesting the early Mars may have had ample liquid water on its surface (Figure 9.6). In hydrothermal systems, when ultramafic or basaltic rocks react with fluids, serpentine minerals would be formed, which, as a phyllosilicate mineral, could also retain water as hydroxyl. They have also been found on Mars surface, even being less common. Water molecules can also be stored in crystal structures of minerals, such as sulfates, opaline silica, goethite, and kieserite, which can be formed under evaporating, hydrothermal, and other aqueous conditions on Mars. Other minerals, which do not contain water molecules or hydroxyl, but require a liquid water environment to form, are also found on Mars, such as carbonate, anhydrite, and hematite. Such mineralogical evidence can be found on Mars surface and Martian meteorites, suggesting Mars had aqueous activity in the past.

Geomorphological and Sedimentary Evidence

Geomorphology suggests liquid water has carved and scoured the surface of Mars in the past. For example, branched network of valleys is an important geomorphologic feature on Noachian terrains that resembles terrestrial river drainage basins on Earth. Additionally, catastrophic outflow channels could extend from tens to thousands of kilometers on the planet, and they may have developed since Hesperian, suggesting activities of outburst floods through time (Figure 9.7a). Deltas have also been identified on Mars (e.g. Eberswalde Delta captured by the Mars Global Surveyor; Figure 9.7b), suggesting deposition from a liquid and flowing medium, most likely in lotic water

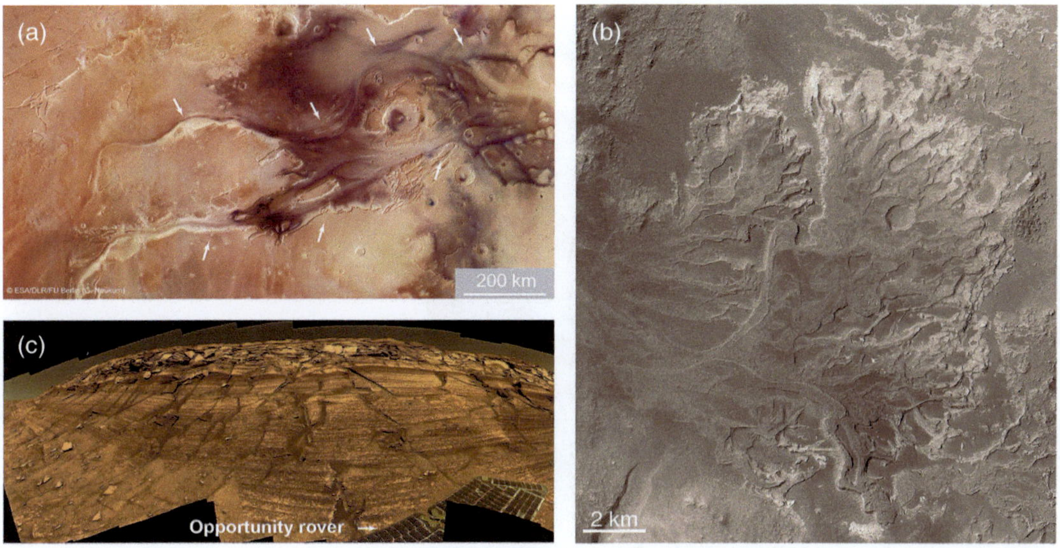

Figure 9.7 Morphological and sedimentary features related to aqueous activities on Mars. (a) Catastrophic outflow channels in Kasei Valles denoted by white arrows; (b) Eberswalde delta; (c) Burns Cliff in Endurance Crater. *Sources:* (a) ESA/DLR/FU Berlin (G. Neukum)/European Space Agency/CC BY SA 3.0, (b) NASA/JPL/MSSS/NASA/Public Domain, (c) NASA/JPL/Cornell/NASA/Public Domain.

systems. Some layered stratigraphic structures in craters that are featured by cementation of aeolian dunes may be related to the seeping of groundwater and redeposition of cementing minerals (e.g. Burns Cliff in Endurance Crater; Figure 9.7c). It is hypothesized that the low-elevation regions in the northern hemisphere may have been a vast ocean that occupied one-third of Martian surface, as suggested by some potential shoreline structures. However, such hypothesis is still under debate.

Evidence for potential recent flows of brine water on Mars has been found, although they are still controversial. The recurrent slope lineae (RSL), featured as dark streaks occurring downslopes in the summer of Mars every year, are considered as flows of brine water, although we cannot rule out that they may be related to flows of dry grains (Figure 9.8). In addition, the young gully deposits may also be produced by liquid brine, although there could be alternative explanations.

Water Ice on Mars

From the perspective of thermodynamics, most places on Mars surface favor the formation of either solid or vapor phases of water, instead of liquid water, because the average atmosphere pressure on Mars surface is close to the triple point of water (Figure 9.9). In other words, on the Mars surface, water ice would be sublimated directly, skipping the liquid phase, except in some basins where the atmospheric pressure is slightly higher (Figure 9.9). Therefore, brine with a high concentration of salt and water ice could be more common forms of water on Mars surface.

Water ice can be found as ice caps in Mars' polar regions, whose size could be comparable to that of the Greenland ice sheet on Earth (Figure 9.10). Ice caps on Mars may have even been larger in the past (e.g. Hesperian). Glacier-like features (e.g. lobate debris apron) also suggest the occurrence of glaciers on Mars and periglacial features suggest frozen water may be below the surface that is similar to the permafrost regions on Earth. Such evidence indicates considerable water ice resources beneath dust and debris on Mars.

Using neutron and gamma ray spectrometers aboard Mars Odyssey spacecraft, scientists measured hydrogen content that was subsequently converted to water concentration by stoichiometry.

Figure 9.8 Recurring slope lineae (RSL) in Newton crater of Mars. *Source:* NASA/JPL-Caltech/Uarizona/NASA/ Public Domain.

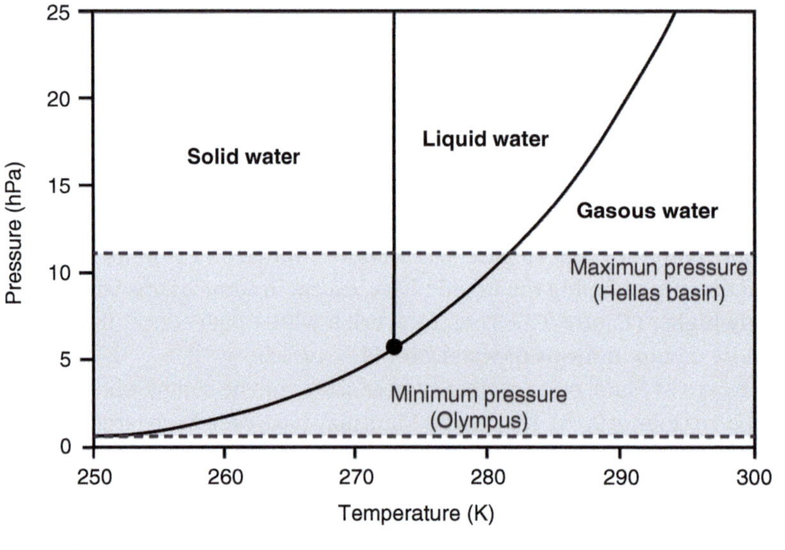

Figure 9.9 Phase diagram of water. The gray box shows the upper and lower boundaries of atmospheric pressure on Mars surface. *Source:* Modified from Davila et al. (2020).

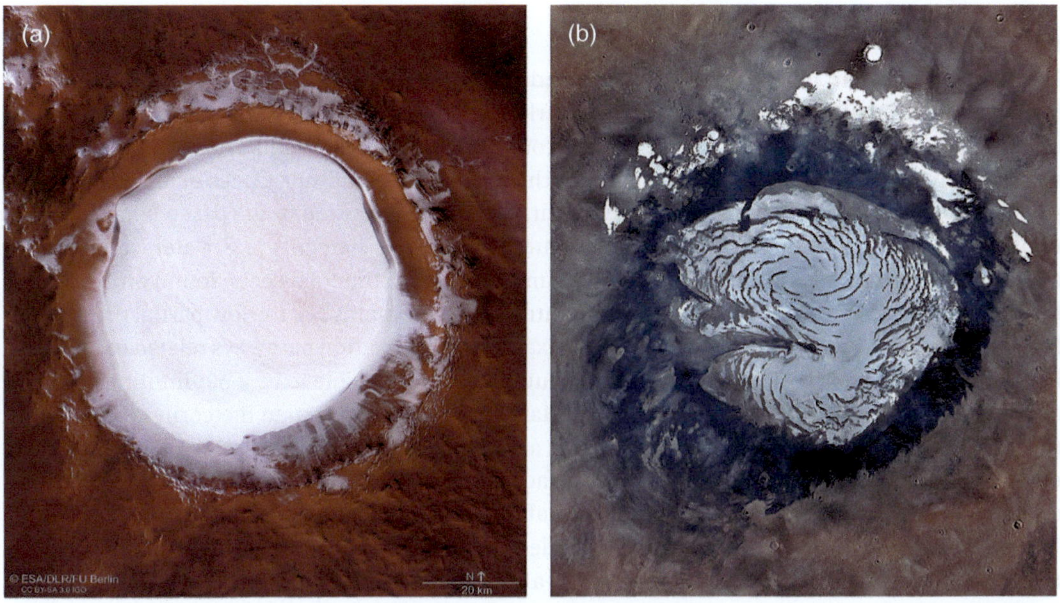

Figure 9.10 Water ice caps on Mars. (a) Korolev crater cap; (b) North polar cap. *Sources:* (a) ESA / DLR / FU Berlin / European Space Agency/CC BY SA 3.0, (b) CNSA / GRAS / Public domain.

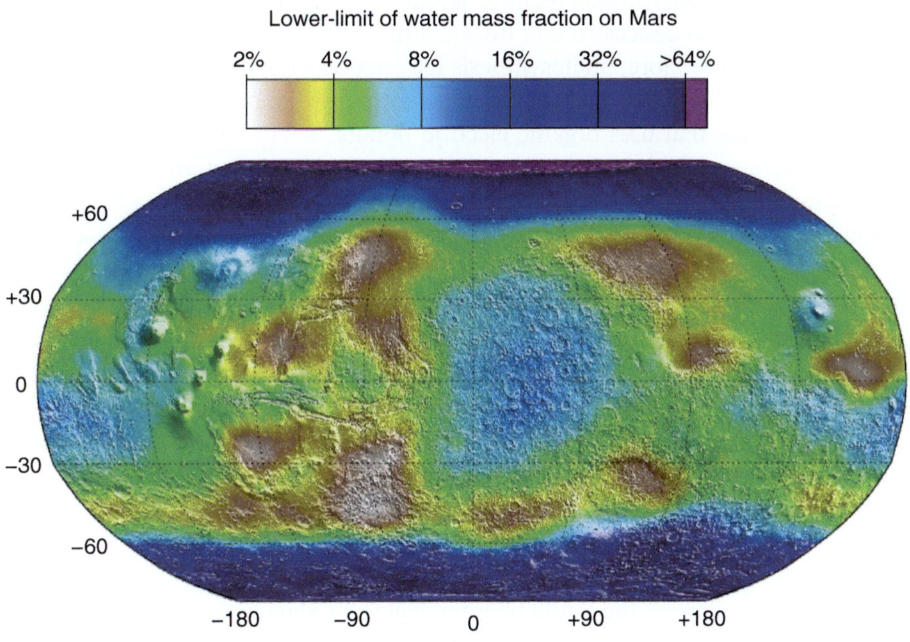

Figure 9.11 Lower limit of water content of the uppermost Martian soil. *Source:* NASA / JPL / Los Alamos National Laboratory / NASA / Public Domain.

The overall global distribution of water has been mapped and water content varies from 2% to 100% from low latitudes to polar regions. Such water may be buried as ice phase under a thin rocky or dusty layer (Figure 9.11). A recent study also found potential evidence for groundwater that might be retained in pores and cracks in the outer crust of Mars, which could be 10–20 km deep.

9.3.2.3 Habitability of Mars

Essential Nutrients

Nutrient elements are important for building and functioning of life. Carbon atoms for life to use could be from Mar atmosphere as it contains nearly 95% CO_2, while the content and distribution of organic carbon, a major energy source for heterotrophic organisms, is still unclear on Mars. The source for hydrogen could be water that exists both on the surface and subsurface of Mars, or hydrogen could be generated as a byproduct of serpentinization that occurred in the past when Mars had liquid water. Oxygen could form various compounds with other elements (e.g. water, sulfate, perchlorates, and iron oxides). Fixed nitrogen (e.g. ammonia and nitrate) has been found on Mars surface and in Martian meteorites. Although Mars' atmosphere contains N_2, the low partial pressure of N_2 on present Mars may not allow its biological fixation. Other fixation pathways related to impacts, lightning, and volcanism are possible on Mars, but nitrogen is possible to be a bio-limiting element on Mars. Phosphorous has been found both on Mars surface and in Martian meteorites (e.g. as apatite found by Mars Exploration Rovers). Sulfur is also present on Mars as sulfate, sulfide, and other compounds, although its distribution on subsurface is unknown. Other important major and trace elements are also contained in ultramafic and mafic rocks and their weathering products on Mars. For example, Fe and Mg are abundant in olivine and pyroxene, Ca, Na, and K are contained in feldspar minerals, and other trace elements are also abundant in igneous rocks.

Energy Sources

Energy available on Mars for biological use could be from solar radiation (phototrophy) and chemical reactions (chemotrophy). Photosynthesis requires liquid water, iron or sulfur compounds as electron donors, but at present it is desiccated and frozen at the surface and near-surface of Mars, where sunlight can penetrate. Therefore, photosynthesis may not be a dominant energy pathway for Martian life, if it exists. Chemotrophy includes chemoautotrophy that uses inorganic electron donors and chemoheterotrophy that uses organic electron donors. There are various metabolic pathways that require different redox couples, such as anaerobic iron oxidation, methanogenesis, sulfate reduction, and nitrate reduction (Section 1.3). Investigating possible reaction pathways and complete redox couples on Mars is important for understanding energy limitations on Mars.

Environmental Conditions

Although a lack of liquid water is a primary limitation of habitability on Mars, harboring life also requires clement physical and chemical conditions. Mars does not have an ozone layer that could absorb short-wavelength radiation; ultraviolet (UV) radiation could directly reach the surface of Mars and cause damage to DNA, the fundamental life-building block. Subsurface spaces cannot be impacted by UV radiation and some extremophiles (e.g. radiodurans in Section 2.9) that are able to withstand a high dose of radiation could live on the surface.

Liquid Water

Currently, although still under debate, the liquid phase of water on Mars is possibly brine rather than pure water, and such a condition may have developed since the Noachian to Hesperian transition. Therefore, low water activity and high chaotropicity in brines would decrease the habitability on Mars. Overall, the surface of Mars has an inclement condition to support life at present, but we cannot rule out that Mars had a habitable environment and harbored life in the past.

9.3.2.4 Search for Extant Life in Subsurface of Mars

The surface of Mars might have contained a large amount of liquid water in the past, but lost the reservoir after the late Noachian. However, it is possible that part of the water resource could now be preserved as ground ice and groundwater below the surface, sustaining a habitable environment.

For example, an estimation, based on Martian geothermal gradient, suggests that the thickness of permafrost (or cryosphere) on Mars could be 0–9 km at equator and 10–22 km at poles; groundwater could exist below the permafrost as temperature increases (Clifford et al. 2010). In addition, radar measurement on the Mars Express spacecraft indicates that liquid water may exist below the ice of the South Polar Layered Deposits (i.e. a subglacial lake) (Orosei et al. 2018), which should be much shallower than the sub-permafrost groundwater. Compared to the Martian surface environment, the subsurface would not be impact by a series of disadvantages, including strong oxidant (e.g. perchlorate), desiccation, UV-radiation, and extremely low temperatures (Carrier et al. 2020).

In analogous subsurface environments on Earth, there is ample evidence for life. For example, life has been detected 4–5 km below the surface on continental crust, and detected 2.5 km below the surface of marine sediments on Earth; terrestrial life could also survive spanning a large range of geothermal gradient from −54 to 122 °C (Onstott et al. 2019). Life could live in even deeper regions where we are unable to explore at present. The Earth's subsurface biosphere is estimated to contain around 1×10^{30} cells, nearly 10% of the surface biomass (Onstott et al. 2019). Therefore, the subsurface environment of Mars also has the potential to harbor life.

The metabolism of potential life in the subsurface of Mars could be similar to that of Earth's subsurface. Similar environments on Earth include ice and ice-rock interfaces, deep basaltic crust, deep subsurface continental crust in mafic and siliceous igneous and metamorphic rocks, sedimentary aquifers, and impact crater hydrothermal systems (Onstott et al. 2019). In the absence of sunlight, the primary producers in the deep environments on Mars should be chemoautotrophic. Energy could be derived from various ways, such as H_2 generated via water-rock interactions (e.g. serpentinization) or oxidation of ferrous minerals (e.g. in mafic rocks) (Carrier et al. 2020). If the subsurface liquid water has existed stably since the geological past, such an environment might have had a chance to witness the origin and evolution of Martian life.

On a billion-year timescale, the subsurface of Mars could also be considered as a refugia for Martian life. When the environment started to deteriorate in the late Noachian, Martian life harbored on the surface of Mars may have retreated to deep subsurface that became more habitable than the average condition of Mars. Other refugia on Mars also include caves, ice, and salts, which have all the requirements for habitability, develop widely on Mars, and have adapted life forms on Earth (Carrier et al. 2020). Astrobiology studies of Mars' subsurface in the future may be helped by sample return that would enable the use of more advanced methodologies than those used on Mars, as well as by studying analog conditions in laboratories and field sites on Earth (Carrier et al. 2020).

9.3.3 Europa

Europa, the second moon of Jupiter, is slightly smaller than our Moon, but has a reservoir of liquid water even larger than the Earth's, which is essential for life survival. Europa has a global icy crust, a liquid ocean beneath the crust, a silicate mantle, and a potentially metal core (Figure 9.12). The ice crust is distributed globally and its thickness ranges from 3 to 30 km. The ocean beneath the icy crust is estimated to be over 100 km deep (compared to ~4 km of Earth), and the liquid water volume may be the twice of ocean of the Earth (Table 9.1).

The magnetometer on NASA's Galileo spacecraft provides the most compelling evidence for liquid water. It found that Jupiter's magnetic field was disrupted around Europa and a special magnetic field is induced in the interior of Europa, which must be related to a conducting layer within Europa (Styczinski et al. 2024). The most plausible candidate for the material is a salty liquid ocean. Why can liquid water be maintained in such a cold environment? The heat is probably from tidal heating. The orbit of Europa is non-circular, and the gravitational force of the moons influences each other differently and results in orbital resonances. Therefore, the surface of Europa distorts at different orbiting periods and heat can be generated under such tidal flexing process (Tobie et al. 2003).

Figure 9.12 Artists' illustration of Europa's Interior, with an icy crust, subsurface ocean, and hydrothermal activities on the rocky seafloor. *Source:* NASA / JPL-Caltech / NASA / Public Domain.

Table 9.1 Comparison of liquid oceans of Earth and Europa.

	Earth	**Europa**
Diameter	12,742 km	3120 km
Average ocean depth	4 km	100 km
Ocean volume	1.4×10^9 km^3	3.0×10^9 km^3
Surface	29% land and 71% liquid water	Global water ice crust of 3–30 km thick
Sea floor	Rocky	Rocky

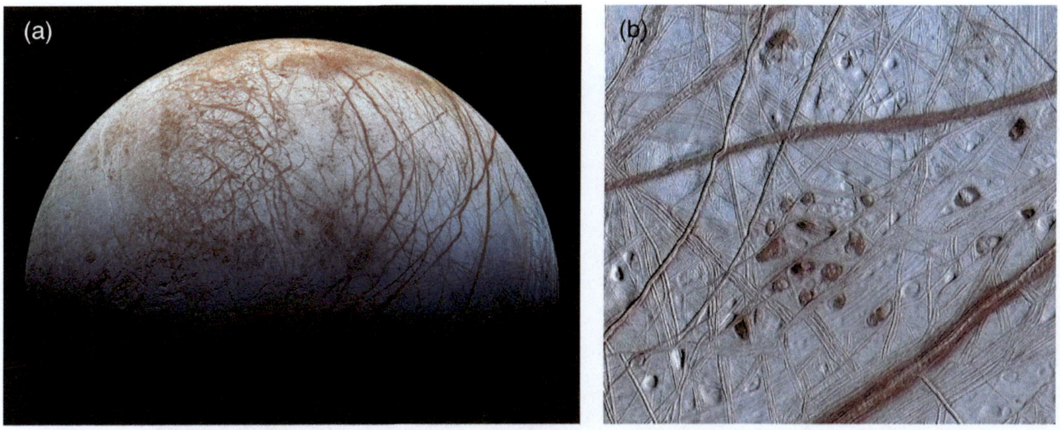

Figure 9.13 The general surface feature of Europa. (a) The icy surface featured by few impacts; (b) lineae and lenticulae on Europa's surface. *Sources:* (a) NASA / JPL-Caltech / SETI Institute / NASA / Public Domain; (b) NASA / JPL / University of Arizona/University of Colorad / NASA / Public Domain.

The surface of Europa has few impact craters, suggesting the surface is relatively young and its interior (Figure 9.13), especially the subsurface ocean, can be an active world. The classic surface features on Europa, such as *lineae*, *lenticulae*, and chaos terrains, could be related to deformation of icy crust. The *lineae* is a series of crisscrossing stripes of dark and light colors. The line structures might be the spreading of icy crust and upwelling of warmer ice, which is similar to the mid-ocean ridge on Earth (Figure 9.13a,b). The deformation and motion of the icy crust may be related to the tidal force when Europa moves to different positions in its orbit. The *lenticulae* are dark spots, domes, or patches of various sizes that may be related to the release of meltwater or rising of warm ice to the surface (similar to the motion of magma on Earth) (Figure 9.13b). Chaos terrain is a surface feature that combines ridges, fractures, and plains in a specific area. It may be related to the breaking up of icy crust that has already generated cracks but has not yet split the ice apart. Although we are unable to study Europa's interior currently, such surface features suggest the icy crust of Europa may be floating on a subsurface ocean that could still be active to support biological activities.

The surface of Europa may be inclement due to a frozen condition and high ion radiation, but the subsurface ocean may have a more habitable environment for life. The tidal heating provides consistent energy source to maintain water as liquid phase and to drive metabolic reactions. Hydrothermal activities and water-rock reactions at the bottom of the ocean could leach essential elements for life. Energetic particles and photons arriving at Europa's surface could generate oxidant, which, if combined with water-rock reaction that generates reductant, could produce completed redox couples that are necessary for chemotrophs. With temperature and pressure ranges still near the boundary for extremophiles on Earth, the subsurface ocean on Europa is of great potential for future astrobiology study. The recent Europa Clipper by NASA and JUICE mission by ESA have been launched to search for signs of life on the icy moon.

9.3.4 Enceladus

Enceladus is a moon of Saturn. It has a small body with a diameter of 500 km, ~15% size of the Moon. The surface of Enceladus is covered by water ice, beneath which is a global ocean that is in contact with a rocky core. Enceladus was first visited by Voyager 1 and 2 in the 1980s. The Cassini spacecraft flew by the moon in 2005, and the mission attracted great interest from astrobiologists on

Figure 9.14 Plumes in the south polar region of Enceladus. *Source:* NASA / JPL-Caltech / Space Science Institute / NASA / Public Domain.

the small icy moon because of the discovery of plumes vented from its south polar terrain (Figure 9.14). The data suggest that, like Europa, Enceladus also has a global ocean beneath the floating icy crust.

In the south polar region of Enceladus, the most intriguing feature is "tiger stripes," which are four nearly parallelled ridges, extending over 100 km, with fractures in the center. The Cassini mission found above a hundred jets from hot spots along the tiger stripes (Figure 9.14). The jets converge as the plume that extends for thousands of kilometers and finally merges into the E ring surrounding the Saturn. The Cassini *in situ* collected plume materials that were able to reach the altitude of the spacecraft during the flyby. Analyses indicate the plume is composed of gases and grains. The major component of gas in the plume is water vapor (>95%), while CO_2, CH_4, NH_3, and H_2 are also contained near or below 1% (Cable et al. 2020) (Table 9.2). There are three types of grain in the plume, including type I which is nearly water ice, type II which is either organic rich or silicate rich and type III which contains Na and K salts (Cable et al. 2020) (Table 9.3). Throughout the Cassini flies, the mass of output plume is in a range from 100 to 1000 kg/s, with an average of 200 kg/s (Cable et al. 2020). The plume is probably generated from the subsurface ocean and emitted through fissures and channels of the icy crust.

Enceladus is the only planetary body where we are able to obtain the subsurface materials, without drilling it. It contains a subsurface ocean and thereby has fulfilled the first requirement for

Table 9.2 Major composition of Enceladus' plume gas.

Constituent	Mixing ratio (%)
H_2O	96–99
CO_2	0.3–0.8
CH_4	0.1–0.3
NH_3	0.4–1.3
H_2	0.4–1.4

Source: Cable et al. (2020) / University of Arizona Press.

Table 9.3 Major composition of Enceladus' plume grains.

Grain type	Plume mixing ratio	Composition
Type I	30	Almost pure water ice
Type II	>30	Organic-rich or silicate-rich
Type III	<40	0.5–2% by mass Na, K salts

Source: Adapted from Cable et al. (2020).

habitability. Major nutrient elements, such as C, H, O, N, and S, have been detected in its plume materials. Phosphorus, once considered as a limitation for the biosphere on Enceladus, has been proven to exist in the plume gas and subsurface ocean. Other major and trace elements required by life could be generated from ocean and rocky core interactions. The heat generated from tidal flexing could be a potential energy source. Chemical disequilibrium detected in the plume suggests potential energy that can be used for metabolism in the subsurface ocean (e.g. CO_2 and H_2), such as H_2-based methanogenesis. Finally, the presence of salts as antifreeze materials at a specific concentration in the plume and subsurface ocean suggests the temperature of the ocean is supposed to be above −1 °C. The pH of the ocean is estimated to be 9–11, slightly alkaline, based on the components in the plume and E ring of Saturn. Therefore, Enceladus appears to be habitable. If the subsurface ocean has existed and stayed active for a long time, it may have a chance to host life independent of Earth.

Besides studying the plume of Enceladus, sending landers to investigate more details of its surface and even sending a probe visiting the subsurface ocean to investigate potential microbial ecosystems would be future directions.

9.3.5 Titan

Titan is the largest moon of Saturn, and the second-largest moon in the solar system. The size of Titan is 40% of that our Earth and greater than Mercury. This moon is named after Titans, giants in Greek mythology.

Titan has been closely studied by the Voyager 1 and Cassini-Huygens spacecraft. Titan is the only moon in the solar system known to contain substantial atmosphere; the atmospheric pressure on the surface of Titan is estimated to be 1.6 bar. The upper atmosphere of Titan contains complex hydrocarbons, and the atmosphere of the moon looks to be opaque and surrounded by yellow haze (Hörst 2017). Similar to that of Earth, the atmosphere of Titan is stratified with thermosphere, mesosphere, stratosphere, and troposphere, but each layer is stretched due to the lower gravity on the moon (Figure 9.15).

The Titan's atmosphere is dominated by nitrogen gas (98%) with minor methane (1–2‰), and the content of methane increases when approaching the surface. Nitrogen and methane gases in the upper atmosphere of Titan can be photolyzed by UV irradiation, producing radicals that recombine to form complex organic molecules (Figure 9.16). Titan has a methane and hydrocarbon cycle similar to the water cycle on Earth. Methane clouds in the atmosphere could form methane rains, which deposit as rivers and streams; such liquid methane would finally flow into lakes and seas (Figure 9.16).

The *Cassini-Huygens* mission revealed detailed surface features of Titan. The low latitude of the moon is featured by extensive sand dunes or "sand seas," and the mid latitude is dominated by plains; the polar region is featured by seas, lakes, and rivers filled by liquid hydrocarbons (e.g. methane and ethane) (Lunine et al. 2020) (Figure 9.17a). On Titan, methane behaves like water on Earth,

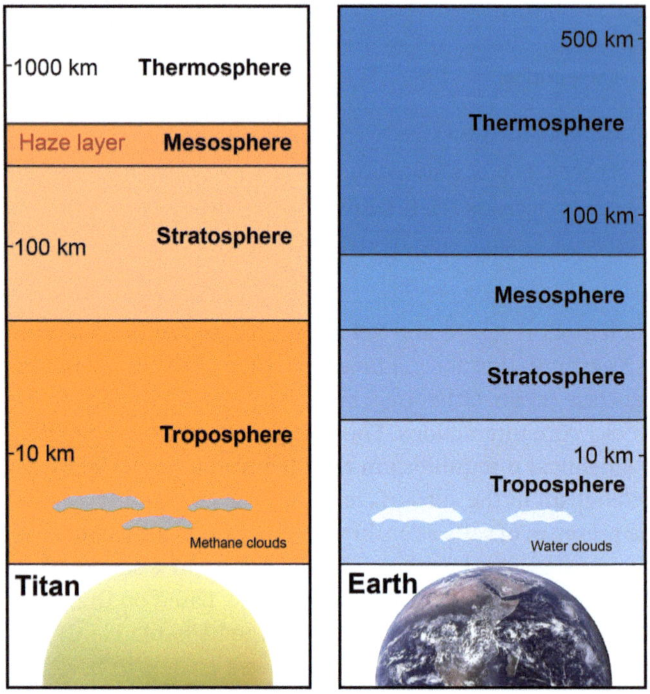

Figure 9.15 Atmospheric structures of Titan and Earth. *Source:* Adapted from Cockell (2016).

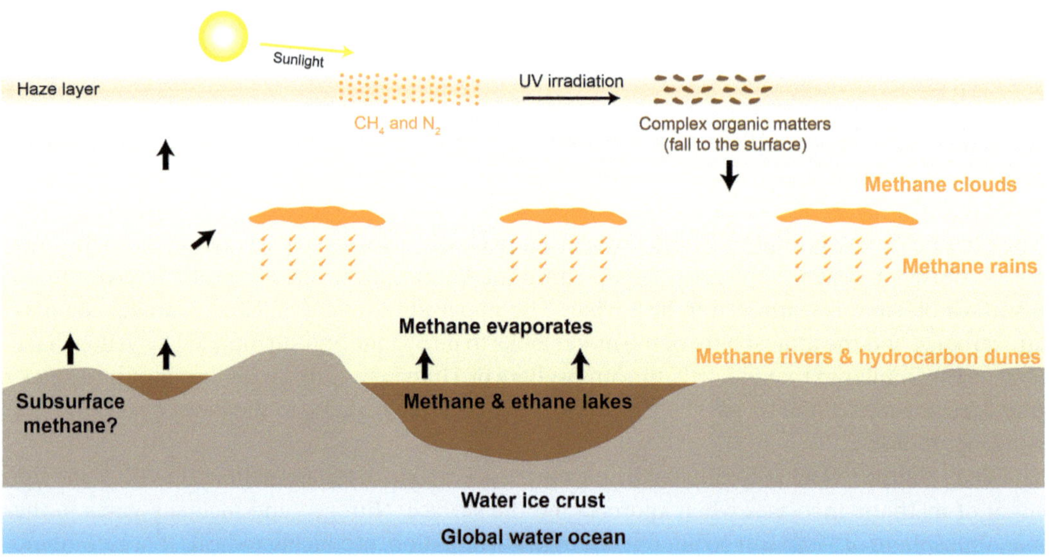

Figure 9.16 UV radiation-induced organic reactions and cycle of methane on Mars.

such as precipitating from atmosphere, carving the surface, flowing as rivers; on the other hand, water ice on the surface of Titan behaves like rocks on Earth, such as forming boulders (Figure 9.17b); additionally, extensive dunes on the equatorial region are composed of complex hydrocarbons rather than silicate sands on Earth (Cockell 2016).

Like Europa and Enceladus, Titan may bear another liquid environment, the subsurface liquid water ocean. Titan also has a rocky core that is in direct contact with the subsurface ocean,

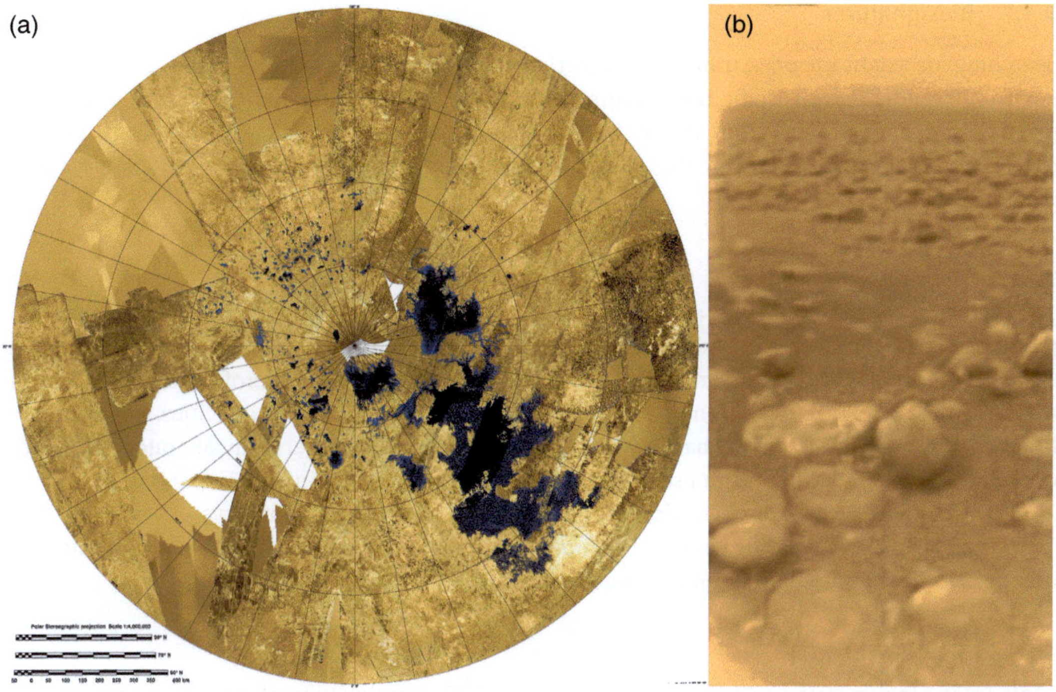

Figure 9.17 Surface features of Titan. (a) Tian's north polar lakes and sea (in blue); (b) Titan's surface captured by Huygens probe. *Sources:* (a) NASA / JPL-Caltech / ASI / USGS / NASA / Public Domain; (b) ESA / NASA.

supporting hydrothermal activities. The UV radiation-induced organic reactions in Titan's atmosphere and the large amount of organic matter on Titan make it as a giant natural laboratory for the formation of prebiotic compounds, such as ammonia acids. If the subsurface ocean is habitable, the organic materials might be able to feed a potential biosphere there. Such hypothesis would be tested by future space missions to Titan.

9.4 Detecting Life Beyond Earth

9.4.1 Life Detection

Besides understanding habitable conditions in extraterrestrial bodies, we also wonder if life exists or existed there. It can be studied in several ways, including *in situ* analysis of samples, remote observation of planets and moons, simulating extraterrestrial conditions in laboratory, and comparing analog conditions on Earth. *In situ* analysis can be achieved by loading instruments on spacecraft and conducting analysis after landing, or analyzing samples directly on Earth after sample return missions or from meteorites. Remote observation investigates morphological features and chemical composition of the surface and atmosphere, via orbiting spacecraft or using telescopes either on Earth or in space. In the laboratory, simulating physical and chemical environments similar to extraterrestrial environments helps to understand possible ongoing biogeochemical processes on other planets and icy moons. Finding analogous conditions is to search for extreme conditions on Earth and understand how life could develop, and if signs of life are preserved in such environments (e.g. Atacama Desert and Qaidam Basin for Mars, and Antarctica subglacial lakes for icy moons).

9.4.2 Biosignatures

Searching for extant life on extraterrestrial worlds may still be challenging, but life could alter the environment and leave specific biosignatures that are traceable. Biosignatures can be a series of physical or chemical signals that are related to essential attributes of life (e.g. metabolism, growth, reproduction) and distinct from the abiotic environment background, such as biofabrics and biomarkers.

9.4.2.1 Physical Signals

The organism itself, its biological activities, and interactions with the environment can all generate physical imprints; if the organism becomes extinct, such physical biosignatures could be preserved as fossils. The remains of organism bodies can be preserved as macro- or microfossils, depending on their original size. Trace fossils, on the other hand, would record biological activities. For example, leaching marks of iron-oxidizing bacteria were found in/on sulfide minerals in laboratory cultured, natural environments, and ancient sediments (Liu and Zhang 2011; Liu and Zhang 2015; Liu et al. 2020) (Figure 9.18). For microbes, biofilms and microbial mats can be preserved as microbialites (e.g. stromatolites). Microbially induced sedimentary structures (MISS) are able to record interaction between life and environments as well (Section 6.4).

9.4.2.2 Chemical Signals

Molecular biosignature is an important indicator for life. Finding biomolecules that are essential to biological processes (such as proteins, carbohydrates, lipids, and nucleic acids) or their oligomers and monomeric units (e.g. nucleobases and amino acids) usually suggests a life source. These molecules could degrade rapidly, making them difficult to be detected once the organism is dead. However, even after the degradation, the main structure of some biomolecules can be preserved and the remaining molecule is called biomarker, sometimes considered as "molecular fossil." Biomarkers can be used to track the occurrence of biomolecules. For example, degradation product of membrane lipids, hopanes, and steranes can be used to study the evolution of cyanobacteria and eukaryotes through geological history.

Structural and compositional biases are indicative of biological processes, including chirality, stereoisomerism, and even-over-odd preference. All known terrain organisms favor a specific chiral structure of biomolecules, which is known as homochirality, while abiotic processes tend to form molecules of equal enantiomers. The homochirality is common for amino acids and carbohydrates.

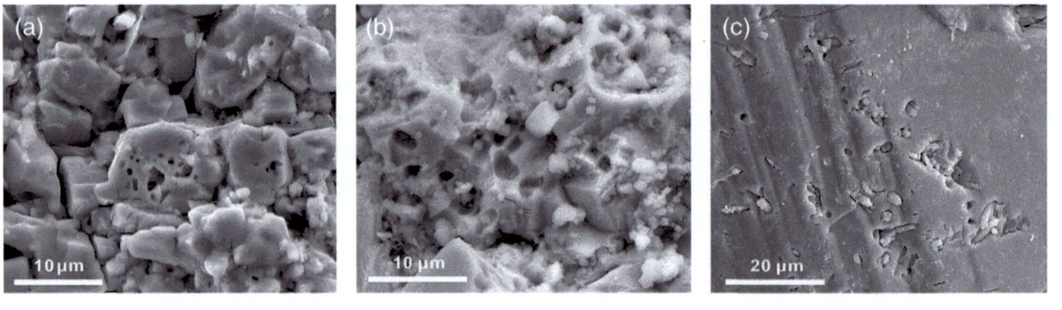

Figure 9.18 Pitted etching marks on pyrite surfaces produced by iron-oxidizing bacteria. (a) From the Ediacaran Doushantuo Formation of South China; (b) modern marine deposits in East Pacific Rise; and (c) laboratory incubation. *Source:* Reprinted with permission from Zhang et al. 2021, © 2021 Zhang et al.

Isomers that differ in three-dimensional configurations of their atoms are stereoisomers, including enantiomers and diastereomers; although many configurations are possible chemically, biogeochemical reactions only favor one specific configuration (e.g. cholesterol). For lipids, biological sourced molecules usually have an even-over-odd preference, while abiotic systems generate those with random carbon distribution. Such a structural preference can be an important biosignature.

Isotopic fractionation is another important biosignature. In metabolic reactions, enzymes prefer to use lighter isotopes, making the biological product depleted of heavier isotopes and the surrounding environment enriched in heavier isotopes. Carbon, hydrogen, oxygen, nitrogen, and sulfur isotopic systems can be impacted by such biological processes and can be detected using a mass spectrometer. **Biological pigments** are light-reactive compounds that participate in photosynthesis and photoprotection. For example, photosynthetic pigments (e.g. chlorophyll) transfer carbon dioxide to carbohydrate using light, and photoprotective pigments (e.g. scytonemin) could protect DNA by absorbing UV radiation. Detection of biological pigments can be achieved by spectroscopy from remote observation on telescopes. **Thermodynamic disequilibrium** exists in Earth's atmosphere, as suggested by the coexistence of oxygen and methane. Life on Earth is in close contact with its environment and thus drives the system out of equilibrium. Similarly, the thermodynamic disequilibrium in the atmosphere and oceans of extraterrestrial bodies, such as co-located oxidants and reductants, is a useful biosignature for ongoing and recent biological activities.

9.4.3 Case Studies

9.4.3.1 Viking Biological Investigation

In 1976, the NASA Viking project was carried out by two identical spacecraft, which both consist of an orbiter and a lander. Viking 1 landed at the western slope of Chryse Planitia near the equator, and Viking 2 landed further north at Utopia Planitia. Gas chromatography–mass spectrometry (GC-MS), which is able to identify different molecules in test samples, was loaded on each spacecraft. Three biological experiments, including gas exchange (GEx), labeled release (LR), and pyrolytic release (PR), were conducted to detect signs of biological activity on Mars. In LR experiment, soil was collected and inoculated with an aqueous nutrient solution that is labeled with radioactive ^{14}C. The atmosphere above the soil is monitored. If there is any organism that could utilize the nutrient, CO_2 enriched in ^{14}C is expected to be detected. In PR experiment, soil was collected and CO_2 and CO that are labeled with ^{14}C were provided as a carbon source. After a period of incubation, the soil was heated and the product was tested for ^{14}C. If there is any biological process that transforms inorganic carbon into biomass, ^{14}C-enriched product is supposed to be found. The experiments results are ambiguous. LR detected radioactive ^{14}C in CO_2, which was initially considered as a sign of life, but GC-MS was not able to detect organic molecules at the surface of sampling sites, suggesting the absence of biomass. In 2008, the Phoenix lander found Martian soil contains perchlorate that can oxidize organic molecules. Under such an environment, most results from the Viking experiments are compatible with abiotic processes, and thus not considered as signs of life.

9.4.3.2 ALH84001 Martian Meteorite

ALH84001 is a Martian meteorite found in Allan Hills in Antarctica in 1984, an igneous orthopyroxenite (Figure 9.19a). In 1996, McKay and his colleagues announced that possible past Martian life was found in a fragment of ALH84001. It stimulated wide public discussion and even President Bill Clinton of the United States addressed the result on Television. The most important evidence is the

Figure 9.19 ALH 84001 Martian meteorite (a) and microstructures resembling nanobacteria on Earth are shown in the middle of panel (b). *Sources:* (a) NASA / JSC / Stanford University / NASA / Public Domain; (b) NASA / NASA / Public Domain.

microstructures on the fresh fractured surface of ALH84001 that resemble nanobacteria on Earth (Figure 9.19b), and thus they were considered as microfossils of Martian bacteria (McKay et al. 1996). Additionally, abundant polycyclic aromatic hydrocarbons (PAHs) were considered as degradation products of past microorganisms; highly ordered magnetite crystals were related to magnetotactic bacteria; carbonate globules were thought to be products of some bacteria (Martel et al. 2012; McKay et al. 1996).

However, following studies found that potential microstructure of nanobacteria can be reproduced via abiotic processes, such as amorphous-to-crystalline mineral transformation (Martel et al. 2012). The presence of PAHs may be attributed to contamination or abiotic origin and magnetite crystals can be formed inorganically at relatively high temperature. Therefore, at this stage, the majority of the astrobiological community believes the ALH84001 meteorite does not contain any fossils, and life has not yet been found beyond Earth.

9.5 Lessons from Explorations

Through the effort of exploring extraterrestrial life, human beings will be able to answer a series of fundamental scientific questions, improve human's ability of adapting to different environments, discover more resources and living spaces, stimulate the advance of technology, and promote international corporations and communications. Exploring life beyond Earth, which could be at various stages of evolution and fundamentally different from what we are familiar with, helps us to understand the essence and origin of life. As environments beyond Earth are often extreme, compared to those on Earth, understanding extraterrestrial life would help humans to be more adaptive to changes in environmental conditions related to climate change, geological processes, and extraterrestrial factors. Even if not able to discover signs of life beyond Earth, potential habitable extraterrestrial bodies would provide abundant resources for excavation and living spaces for immigration if they are exhausted on Earth. Astrobiological exploration requires technologies that are more advanced than the average level, and therefore, it would stimulate the development of cutting-edge theories and technologies, and in turn give them back to our society. Last, since astrobiology research always involves different nations regardless of their boundaries, it would help the international community stay closer.

References

Cable, M.L., Neveu, M., Hsu, H.-W., and Hoehler, T.M. 2020. Enceladus. In: *Planetary astrobiology* (eds. V.S. Meadows et al.), pp. 217–246. Tucson, University of Arizona. 10.2458/azu_uapress_9780816540068-ch009.

Canfield, D.E., Erik, K., and Bo, T. 2005. The phosphorus cycle. In: *Advances in marine biology* (eds. D.E. Canfield, E. Kristensen, and B. Thamdrup), pp. 419–440. London, Academic Press.

Carrier, B.L., Beaty, D.W., Meyer, M.A. et al. 2020. Mars extant life: what's next? conference report. *Astrobiology*, 20, pp. 785–814.

Clifford, S.M., Lasue, J., Heggy, E., Boisson, J., McGovern, P., and Max, M.D. 2010. Depth of the Martian cryosphere: revised estimates and implications for the existence and detection of subpermafrost groundwater. *Journal of Geophysical Research: Planets*, 115, e07001.

Cockell, C.S. 2016. *Astrobiology: understanding life in the universe*. West Sussex, UK, John Wiley & Sons.

Cockell, C.S., Bush, T., Bryce, C., Direito, S., Fox-Powell, M., Harrison, J.P., Lammer, H., Landenmark, H., Martin-Torres, J., and Nicholson, N. 2016. Habitability: a review. *Astrobiology*, 16, pp. 89–117.

Colón-Santos, S., Vázquez-Salazar, A., Adams, A., Campillo-Balderas, J.A., Hernández-Morales, R., Jácome, R., Muñoz-Velasco, I., Rodriguez, L.E., Schaible, M.J., Schaible, G.A., Szeinbaum, N., Thweatt, J.L., and Trubl, G. 2024. Chapter 2: what is life? *Astrobiology*, 24, pp. S40–S56.

Davila, A., Kahre, M.A., Quinn, R., and Des Marais, D.J. 2020. The biological potential of present-day Mars. In: *Planetary Astrobiology* (eds. V. Meadows et al.), pp. 169–184. Tucson, Univ. of Arizona. 10.2458/azu_uapress_9780816540068-ch007.

Grefenstette, N., Chou, L., Colón-Santos, S., Fisher, T.M., Mierzejewski, V., Nural, C., Sinhadc, P., Vidaurri, M., Vincent, L., and Weng, M.M. 2024. Chapter 9: life as we don't know it. *Astrobiology*, 24, pp. S186–S201.

Haberle, R.M. 2015. Solar System/Sun, atmospheres, evolution of atmospheres | Planetary atmospheres: Mars. In: *Encyclopedia of atmospheric sciences (Second Edition)* (eds. G.R. North, J. Pyle, and F. Zhang), pp. 168–177. Oxford, Academic Press.

Hoehler, T.M. 2007. An energy balance concept for habitability. *Astrobiology*, 7, pp. 824–838.

Hörst, S.M. 2017. Titan's atmosphere and climate. *Journal of Geophysical Research, Planets*, 122, pp. 432–482.

Kasting, J.F. 1988. Runaway and moist greenhouse atmospheres and the evolution of Earth and Venus. *Icarus*, 74, pp. 472–494.

Kasting, J.F., and Pollack, J.B. 1983. Loss of water from Venus. I. Hydrodynamic escape of hydrogen. *Icarus*, 53, pp. 479–508.

Kasting, J.F., Whitmire, D.P., and Reynolds, R.T. 1993. Habitable zones around main sequence stars. *Icarus*, 101, pp. 108–128.

Kopparapu, R.K., Ramirez, R., Kasting, J.F., Eymet, V., Robinson, T.D., Mahadevan, S., Terrien, R.C., Domagal-Goldman, S., Meadows, V., and Deshpande, R. 2013. Habitable zones around main-sequence stars: new estimates. *The Astrophysical Journal*, 765.

Liu, W., and Zhang, X.L. 2011. Evidence for microbial dissolution of pyrite from the Lower Cambrian oolitic limestone, South China. *Biogeosciences Discussions*, 2011, pp. 2035–2056.

Liu, W., and Zhang, X. 2015. Experimental study of microbial pyrite oxidation: a suggestion for geologically useful biosignatures. *Geomicrobiology Journal*, 32, pp. 466–471.

Liu, W., Zhang, X., and Wang, W. 2020. Microbial imprints on sulfide minerals in submarine hydrothermal deposits of the east pacific rise. *Journal of Geophysical Research: Biogeosciences*, 125, pp. e2020JG005736.

Lunine, J.I., Cable, M.L., Hörst, S.M., and Rahm, M. 2020. The astrobiolcogy of Titan. In: *Planetary Astrobiology* (eds. V. Meadows et al.), pp. 247–266. Tucson, University of Arizona. 10.2458/azu_uapress_9780816540068-ch010.

Martel, J., Young, D., Peng, H.-H., Wu, C.-Y., and Young, J.D. 2012. Biomimetic properties of minerals and the search for life in the Martian meteorite ALH84001. *Annual Review of Earth and Planetary Sciences*, 40, pp. 167–193.

McKay, D.S., Gibson, E.K., Jr., Thomas-Keprta, K.L., Vali, H., Romanek, C.S., Clemett, S.J., Chillier, X.D., Maechling, C.R., and Zare, R.N. 1996. Search for past life on Mars: possible relic biogenic activity in Martian meteorite ALH84001. *Science*, 273, pp. 924–930.

Onstott, T.C., Ehlmann, B.L., Sapers, H., Coleman, M., Ivarsson, M., Marlow, J.J., Neubeck, A., and Niles, P. 2019. Paleo-rock-hosted life on Earth and the search on Mars: a review and strategy for exploration. *Astrobiology*, 19, pp. 1230–1262.

Orosei, R., Lauro, S.E., Pettinelli, E. et al. 2018. Radar evidence of subglacial liquid water on Mars. *Science*, 361, pp. 490–493.

Styczinski, M.J., Cooper, Z.S., Glaser, D.M., Lehmer, O., Mierzejewski, V., and Tarnas, J. 2024. Chapter 7: assessing habitability beyond earth. *Astrobiology*, 24, pp. S143–S163.

Thweatt, J.L., Harman, C.E., Araújo, M.N., Marlow, J.J., Oliver, G.C., Sabuda, M.C., Sevgen, S., and Wilpiszeki, R.L. 2024. Chapter 6: the breadth and limits of life on earth. *Astrobiology*, 24, pp. S124–S142.

Tobie, G., Choblet, G., and Sotin, C. 2003. Tidally heated convection: constraints on Europa's ice shell thickness. *Journal of Geophysical Research, Planets*, 108. 10.1029/2003JE002099.

Zhang, X., Chang, C., Cui, L., and Qiao, Y. 2021. Ecosystem reconstruction during the Cambrian explosion. *Paleontological Research*, 25, 305-314, 310.

Glossary

Acid mine drainage (AMD) The outflow of acidic water from metal or coal mines, where water becomes acidic and polluted as a result of microbial oxidation of sulfide minerals like pyrite; also known as acid and metalliferous drainage (AMD) or acid rock drainage (ARD).

ALH84001 A Martian meteorite found in Allan Hills in Antarctica in 1984, from which evidence for life was claimed by a group of scientists in 1996 but rejected by the scientific community later.

Amorphous calcium carbonate (ACC) The most soluble and least stable form among the solid $CaCO_3$ phases, serving as a precursor to crystalline $CaCO_3$ polymorphs in various natural and synthetic calcification systems.

Anhydrobiosis A form of cryptobiosis, a survival strategy in which organisms enter a state of suspended metabolism due to extreme desiccation.

Assimilation The process of absorbing and integrating information, substances, or cultural elements into a larger system.

Authigenesis The process by which minerals form in place within sedimentary environments rather than being transported from elsewhere.

Autophagy A natural cellular process where cells degrade and recycle unnecessary or damaged components, which plays a crucial role in maintaining cellular health by removing dysfunctional parts and repurposing them for repair.

Banded iron formation (BIF) A type of rock that is characterized by alternating layers of iron-rich and silica-rich minerals, and significant for serving as a major source of iron ore for industrial use.

Basalt An extrusive igneous rock enriched in Mg and Fe silicate minerals, formed under a rapid cooling process near the surface.

Barite A barium sulfate ($BaSO_4$) mineral known for its high specific gravity.

Biocrust Also known as biological soil crust, complex community of living microorganisms inhabiting the surface of soils in arid and semi-arid ecosystems, which form a millimeter-to-centimeter thick layer of stable aggregates of soil particles.

Biogeochemical cycles Natural pathways through which essential elements circulate between living organisms and the Earth's environment. These cycles ensure that biological elements carbon, nitrogen, oxygen, phosphorus, and hydrogen are continuously recycled, supporting life on our planet.

Biological pump The process by which carbon and other nutrients are transported from the surface ocean to the deep ocean through biological activity, playing a key role in biochemical cycles.

Introduction to Geobiology, First Edition. Xingliang Zhang, Zisheng Guo, Wei Liu, Ruliang He, Luoyang Li, and Weiduo Hao.

Companion website: www.wiley.com/go/Zhang_1e

Biomarker A short for biological marker compound, also known as the molecular fossil, refers to an ancient biomolecule preserved in rocks that is able to indicate the existence of a specific life form.

Biomineralization toolkit A set of evolutionarily conserved genes, proteins, and molecules inherited from a common ancestor that were independently co-opted across lineages, resulting in the multiple and parallel origins of biomineralization.

Biosignature A series of physical or chemical signals that are related to essential attributes of life (e.g. metabolism, growth, reproduction) and distinct from the abiotic environmental background.

Biostabilization A biological process by which biological activities enhance the stability of sediments and reduce their susceptibility to erosion by binding, adhering, and sealing sediments.

Bioturbation The process by which living organisms, particularly animals, disturb and mix the unconsolidated sediment or soil in their environment through activities such as burrowing, feeding, and moving.

Black smoker A dark-colored chimney-like structure composed of metal sulfides typically found on the ocean floor along mid-ocean ridges, where superheated (>300 °C), mineral-rich fluids containing dissolved metals like iron, copper, and zinc, precipitate metal sulfides upon contact with cold seawater, forming the black smoker.

Bog iron A naturally occurring form of iron ore that develops in wetlands, bogs, and swamps through the oxidation of ferrous iron-rich groundwater.

Boring Billion A geological time period between 1.8 and 0.8 billion years ago during Earth's Proterozoic Eon, marked by tectonic stability, climatic stasis, and slow biological evolution.

Borings Structures of tunnels or holes excavated by organisms in indurated rock or shell material.

Burrows Biogenic structures of tunnels or holes dug by organisms and emplaced in unconsolidated sediments.

C_3 plants Plants that use the C_3 carbon fixation pathway, where the enzyme RuBisCO fixes CO_2 into a 3-carbon molecule called 3-phosphoglyceric acid (3-PGA). This process is the most common photosynthetic pathway among plants.

C_4 plants Plants that use a specialized photosynthetic pathway for carbon fixation in two steps: CO_2 is first fixed into a 4-carbon compound (oxaloacetate) in mesophyll cells and the 4-carbon compound is transported to bundle sheath cells, where CO_2 is released for the Calvin cycle. The process enhances their efficiency in hot, dry environments.

Cable bacteria Filamentous microbes that conduct electricity across sediment layers, enabling long-distance electron transport.

Calcite sea vs. aragonite sea Geological periods when different polymorphs of calcium carbonate—calcite and aragonite—dominated marine carbonate deposition, which were generally driven by variations in seawater Mg:Ca ratio, aragonite favored when the ratio >2 and calcite favored when the ratio <2.

Canfield Ocean A hypothetical ocean model proposed by geochemist Donald Canfield, who argued that the deep ocean was anoxic and sulfidic (also known as euxinic) during the time of the Boring Billion (1.8–0.8 billion years ago).

Carbonate platform A sedimentary body with a flat top and steep flank, which formed in shallow, warm marine environments where carbonate sediments accumulate over time.

Chasmolith An endolithic organism that colonizes in pre-existing cracks and fissures in rocks.

Chert A fine-grained, dense sedimentary rock made primarily of microcrystalline silicon dioxide (SiO_2).

Cleavage The tendency of a crystal to split along a specific plane that is weakly bonded in crystallographic perspective.

Cold seeps Areas on the ocean floor where hydrogen sulfide, methane, and other hydrocarbon-rich fluids escape from the Earth's crust, typically along submarine continental margins.

Collophane A fine-grained, amorphous (lacking a crystalline structure) form of the mineral apatite typically found in phosphatic sediments.

Congruent dissolution Dissolution of a mineral completely into its constituent ions.

Craton A large, stable block of the Earth's lithosphere that forms the foundation of a continent and typically composed of Precambrian rocks, often more than a billion years old.

Crossed-bladed microstructure A type of shell microstructure produced in molluscs, consisting of intersecting, blade-like calcium carbonate crystals, typically aragonite.

Cryosphere All the frozen water on Earth, including glaciers, ice sheets, sea ice, snow cover, and permafrost.

Cryptobiosis A survival mechanism in which an organism enters a state of suspended metabolism, allowing it to endure extreme environmental conditions.

Cyanobacteria Also known as blue-green algae, a group of photosynthetic bacteria that play a crucial role in Earth's ecosystems.

Daisyworld A concept developed by James Lovelock and Andrew Watson as part of the Gaia theory to explore how life can self-regulate environmental conditions to maintain a stable and habitable planet. The hypothetical Daisyworld consists of black and white daisies whose growth is influenced by temperature; black daisies absorb heat and warm the planet, while white daisies reflect heat and cool the planet.

Dendrolite A type of microbialite characterized by dendritic, shrub-like structures created by microbial calcification.

Deep homology A concept in phylogenetic systematics referring to that certain genetic, molecular, or phenotypic traits are conserved across extensive evolutionary distances, originating from a common ancestor.

Diamictite A type of non-sorted sedimentary rock composed of a wide size and composition range of particles in a muddy matrix, which may form in various settings like glacial environments, submarine landslides, or even from debris flows.

Diapause A phenomenon where an organism temporarily halts its development to survive unfavorable environmental conditions, which allows organisms to endure extreme temperatures, drought, or food scarcity by reducing metabolic activity.

Dissolved inorganic carbon (DIC) The total concentration of inorganic carbon species present in water, including carbon dioxide (CO_2), carbonic acid (H_2CO_3), bicarbonate (HCO_3^-), and carbonate (CO_3^{2-}) ions.

Dissolved organic carbon (DOC) The fraction of organic carbon that is found in water in a dissolved form, typically as particles smaller than 0.45 microns.

Earth's energy imbalance The difference between the amount of energy the planet receives from the Sun and the amount it radiates back into space. When this balance is disrupted, Earth retains more energy than it emits, leading to global warming.

Endolith An organism living in the internal habitats of rocks and shells.

Endosymbiosis A symbiotic phenomenon between two cells or organisms, one engulfing another and both benefiting from the relationship.

Epilith An organism that grows on the surface of rocks.

Euendoliths Microorganisms such as cyanobacteria, algae, fungi that bore into and inhabit within the mineralized or rocky substrates, resulting in dense boreholes even collapse of skeletons, minerals, or rocks.

Eukaryogenesis The evolutionary process that led to the origin of eukaryotic cells from ancestral prokaryotic cells. The most popular theory is endosymbiosis which states that key organelles such as mitochondria and chloroplasts originated from free-living prokaryotes forming endosymbiotic relationships with ancestral host cells and eventually integrated as permanent components of the eukaryotic cell.

Extracellular Polymeric Substances (EPS) A complex mixture of polysaccharides, proteins, lipids, nucleic acids, and other biomolecules that are secreted by microbial cells into their external environment.

Extremozyme A specialized enzyme that functions in extreme environmental conditions and is often produced by extremophiles.

Eutrophication The process where excess nutrients, primarily nitrogen and phosphorus, accumulate in a water body, leading to excessive algal growth and ecosystem disruption.

Feedback loop A chain of processes where outputs would be routed back as inputs, allowing a self-regulated system.

Feldspar A group of aluminum tectosilicate minerals containing varying amounts of K, Na, and Ca. The common minerals are plagioclase (with Na and Ca) and K-feldspar.

Fischer–Tropsch-type (FTT) synthesis A chemical process used to convert carbon monoxide (CO) and hydrogen (H_2) gases into liquid hydrocarbons, which was invented by German chemists Franz Fischer and Hans Tropsch.

Gaia hypothesis Proposes that Earth's living organisms interact with their inorganic surroundings to form a self-regulating system that maintains conditions favorable for life.

Gigantism A phenomenon where certain species evolve to become significantly larger than their relatives due to specific environmental factors like the high oxygen level, as seen in the late Paleozoic era.

Glaciation A geological process by which glaciers form and spread, shaping the Earth's surface in dramatic ways.

Global Boundary Stratotype Section and Point (GSSP) An internationally agreed-upon reference point on a stratigraphic section, defining the lower boundary of a stratigraphic unit on the International Chronostratigraphic Chart by a spike in a rock succession coincident with available biological and other markers.

Global Standard Stratigraphic Age (GSSA) A chronological reference point and criterion used to define a lower boundary of a geological Eon, Era, and Period on the International Chronostratigraphic Chart, particularly when biological and geological criteria are insufficient for precise boundary definitions.

Global phosphogenesis The episodic formation of phosphorite deposits (phosphate-rich sediments and rocks) on a worldwide scale.

Goldich Weathering Sequence A weathering sequence of different minerals and rocks, during which igneous silicate minerals crystallize at higher temperature (e.g. olivine) in magma weather first, while silicates crystallized at lower temperature (e.g. K-feldspar and quartz) are more resistant.

Graphite A naturally occurring mineral of crystalline carbon with a layered, planar structure, each layer composed of hexagonal carbon atoms.

Great Ordovician Biodiversification Event (GOBE) A major evolutionary radiation of marine fauna during the latest Cambrian to early Ordovician, marked by a dramatic surge in biodiversity, whereby the Cambrian fauna fizzled out to be replaced by the Paleozoic fauna.

Greenhouse gases Atmospheric gases that trap heat, contributing to the greenhouse effect and global warming, which include CO_2, CH_4, N_2O, O_3, and water vapor, along with synthetic gases like chlorofluorocarbons (CFCs) and hydrofluorocarbons (HFCs).

Greenstone belt A geological formation consisting of ancient volcanic and sedimentary rocks that have undergone low-grade metamorphism, which is typically found within a craton.

Habitability The ability of an environment to support the metabolic activity that allows for survival, maintenance, growth, or reproduction for at least one known organism.

Habitable zone A range of orbits surrounding a star where liquid water can be stably sustained on the planetary surfaces, also called the Goldilocks zone.

Homeostasis The process by which living organisms maintain a stable internal environment despite external changes.

Hot origin hypothesis A concept in the field of abiogenesis (the origin of life) proposing that life on Earth originated in extreme, high-temperature environments such as hydrothermal vents on the ocean floor, or hot springs.

Hydrothermal alteration Chemical and mineralogical changes in rocks caused by mineral-rich fluids from magmatic, metamorphic, or seawater sources, altering composition and structure through dissolution, leaching, and mineral precipitation.

Hydrothermal vent A geothermal feature on the ocean floor where superheated water, rich in minerals, is expelled from the Earth's crust, typically found in areas where tectonic plates are diverging or converging, such as at mid-ocean ridges or subduction zones.

Hypolithic communities Microbial communities that live beneath rocks in arid or desert environments.

Icy moons Natural satellites of planets with surfaces mostly composed of water ice, many of which have subsurface oceans and silicate cores.

Incongruent dissolution Partial dissolution of a mineral with formation of secondary minerals.

Isotope fractionation The process where isotopes of an element are distributed unevenly due to physical, chemical, or biological processes.

Late Heavy Bombardment (LHB) A hypothesized astronomical event occurred during $\sim$4 to 3.8 billion years before the present when a significant number of asteroids or comets bombarded the inner planets of the solar system, including Earth and the Moon. This event left behind abundant craters, some of which are still visible on the Moon and Mercury.

Lichen A symbiont, consisting of at least two organisms, including a fungus (mycobiont) and photosynthetic partner (photobiont).

Magnetofossils Nanometer scale, fossilized magnetic mineral particles that were produced by magnetotactic bacteria to create magnetosomes—structures that help the bacteria navigate using Earth's magnetic field.

Magnetotactic bacteria A fascinating group of bacteria containing well-organized, intracellular, nano-scale magnetic particles (magnetosomes) that have the ability to orient themselves and navigate using Earth's magnetic field.

Mantle drip A geological phenomenon where a dense, cold portion of the Earth's lithosphere sinks into the more fluid upper mantle.

Mantle secretome Macromolecules secreted by the molluscan mantle that exert control over shell formation and other biological processes.

Marine redox A dynamic oxidation-reduction (redox) process that occurs in marine environments, particularly in seawater, sediments, and marine ecosystems.

Mass extinction A biological crisis driven by dramatic environmental changes (such as volcanic eruptions, asteroid impacts, and climate shifts), which causes extinction of a large proportion of taxa across a range of lineages in various ecosystems during a relatively short period of time.

Mass-independent fractionation (MIF) A phenomenon where isotopic ratios are altered in a way that deviates from the normal mass-dependent fractionation based on isotopic masses. MIF is typically caused by non–mass-dependent processes, such as photochemical reactions, nuclear volume effects, or magnetic effects.

Mesosphere The third layer of Earth's atmosphere between stratosphere and thermosphere, extending from about 50 to 85 km above Earth's surface, within which temperature decreases as altitude increases.

Methane hydrates Also known as methane clathrates, ice-like structures where methane gas is trapped within a lattice of water molecules. They form under high-pressure, low-temperature conditions, typically in deep-sea sediments and beneath permafrost.

Methanophenazine A redox-active cofactor found in certain methanogenic archaea, which plays a crucial role in electron transport within these microorganisms.

Microbialite A carbonate buildup formed by the activity of microorganisms, particularly photosynthesis of cyanobacterial and algal mats that trap, bind, and induce the precipitation of sediments, leading to the formation of carbonate-rich deposits.

Microbioerosion The biological erosion of hard substrates, primarily carbonates, caused by microorganisms such as bacteria, fungi, and algae, which play a crucial role in sediment production, nutrient cycling, and the destruction of skeletal and non-skeletal carbonates in environments.

Mineral polymorphs Minerals that have the same chemical composition but differ in crystal structure and physical properties like hardness, density, or crystal habit. For example, calcite and aragonite are polymorphs of calcium carbonate ($CaCO_3$).

Molecular clock An approach to estimate the timing of lineage divergence or genetic mutation rate by assuming that DNA mutations accumulate at a relatively constant rate over time.

Monounsaturated fatty acids A type of unsaturated fat that contains one double bond in their carbon chain.

Mycobiont The fungus component of lichen, which obtains energy from organic matter synthesized by the photobiont.

Nacre Also known as the "mother-of-pearl," a highly structured organic-mineral biocomposite that forms the inner layer of some molluscan shells and pearls.

Nematodes Also known as roundworms, are animals of the phylum Nematoda that are among the most abundant animals on Earth and inhabit a wide range of environments, from soil and freshwater to marine ecosystems and even extreme habitats like deep underground water-filled cracks.

Nickel-Strunz classification A system used to categorize minerals based on their chemical composition and crystal structure.

Non-skeletal carbonates Carbonate minerals, sediments, and cements that form through abiotic or microbially mediated processes, rather than being directly produced by aggregations of mineralized bioclasts.

Ocean acidification The process by which the pH of the ocean decreases over time, making the seawater more acidic, caused primarily by the absorption of excess carbon dioxide in seawater.

Oxygen minimum zone (OMZ) A region in the ocean where oxygen levels are at their lowest due to a combination of biological, chemical, and physical processes, typically occurring at depths of 200–1500 m.

Paleosol An ancient soil that has been preserved in the geological record.

Periostracum An organic, non-mineralized layer that covers the outer surface of the mineral shell of many molluscs and brachiopods, typically made of proteins and polysaccharides.

Permeability The ability of sediments or other material to allow fluids or gases to pass through it.

Photobiont A photosynthetic symbiont usually found in lichens, which can be green algae, diatoms, or cyanobacteria.

Photorespiration A process in plants where the enzyme RuBisCO mistakenly binds oxygen instead of carbon dioxide during photosynthesis, leading to a wasteful cycle that reduces the efficiency of carbon fixation and energy production.

Phyllosilicates A group of silicate minerals that have layered structure and hydroxyl (−OH) which is usually considered as a result of water-rock reactions.

Polyunsaturated fatty acids A type of unsaturated fat that contains multiple double bonds in their carbon chain.

Primary fluid inclusion A microscopic bubble of liquid and/or gas trapped within minerals during their formation, recording primary signatures of environments.

Protodolomite Also known as the "disordered dolomite," an imperfectly ordered, metastable form of dolomite ($CaMg(CO_3)_2$) that often serves as a precursor to certain sedimentary dolomite.

Pseudofeces Particles that are rejected during filter-feeding by molluscs, which are usually not digested.

Pyrite framboids Microscopic, spheroidal to sub-spheroidal aggregates of equidimensional, equimorphic pyrite microcrystals, often exhibiting remarkable degrees of self-ordering.

Reactive oxygen species (ROS) Highly reactive molecules derived from oxygen (O_2), water, and hydrogen peroxide. They play a dual role in biological systems—acting as signaling molecules in normal cellular processes while also contributing to oxidative stress when their levels become excessive.

Red bed A sedimentary rock layer, typically sandstone or mudstone, that is reddish in color due to the presence of iron oxides, primarily hematite, which is significant in geology for providing insights into Earth's past climate conditions and environments.

Rhizine A root-like structure that is commonly found in lichens, which can penetrate into the substrate and loosen fragments.

Rhizosphere An area in soil influenced by plant root secretions and inhabited by microorganisms.

Rock cycle A series of geological processes, such as sedimentation, volcanism, and metamorphism that transform the three types of rock: igneous, sedimentary, and metamorphic rocks.

Runaway greenhouse effect A positive feedback loop caused by excess accumulation of greenhouse gases in the atmosphere, which blocks shortwave radiation and thus increases the temperature of the planet steadily.

Saltern An area or installation designed for salt production, typically through evaporation of seawater or brine.

Saturated fatty acids A type of fat in which all the carbon atoms in the fatty acid chain are connected by single bonds, meaning they are "saturated" with hydrogen atoms. This structure makes them solid at room temperature and more resistant to oxidation compared to unsaturated fats.

Self-assembly A process by which molecules or components spontaneously organize themselves into ordered structures without the need for external guidance or template.

Serpentinization A geological process of hydrothermal alteration in which ultramafic rocks (rich in Mg and Fe) chemically react with water at high temperatures. The reaction transforms the original minerals (such as olivine and pyroxene) into serpentine minerals (lizardite or chrysotile) forming serpentinite and producing alkaline, less hot fluids (40–90 °C) filled with hydrogen and methane gas.

Snowball Earth A hypothesis suggesting that Earth experienced several extreme ice ages on a global scale in the Proterozoic Eon, during which the planet may have been almost entirely covered with ice and shows a "snowball-like" appearance.

Soda lake A type of alkaline lake with a high pH level, typically between 9 and 12.

Solid Earth The planet's solid surface and interior, excluding fluid envelopes like the atmosphere and hydrosphere, and encompassing Earth's crust, mantle, and core.

Source and sink Places where elements are released or stored in biogeochemical cycles.

Stagnant lid A tectonic state of early Earth, where the outermost layer is a single, solid, unbroken crust and remains immobile and does not undergo plate tectonics as present Earth.

Sterols Organic compounds characterized by a hydroxyl group attached to a steroid backbone, making them structurally similar to cholesterol and a subgroup of steroids.

Stratosphere The second layer of Earth's atmosphere, located above the troposphere and below the mesosphere, extending from about 10–50 km above Earth's surface. It is composed of stratified temperature zones and contains the ozone layer, which absorbs harmful UV radiation from the Sun.

Stromatolite A type of microbialite characterized by distinct laminated structures.

Subduction A geological process that occurs at convergent plate boundaries, where one tectonic plate wedges under another and is forced into the mantle beneath the Earth's crust, which is a key driver of Earth's tectonic activity, rock cycle, volcanic activity, and formation of mountain ranges.

Sulfate-reducing bacteria (SRB) A diverse group of anaerobic bacteria that use sulfate (SO_4^{2-}) as a terminal electron acceptor during respiration, producing hydrogen sulfide (H_2S) as a byproduct.

Solfataric gases Volcanic emissions consisting of steam, sulfur compounds, and other volcanic gases released from solfataras. The solfataric gases are composed of hydrogen sulfide (H_2S), sulfur dioxide (SO_2), carbon dioxide (CO_2), and sometimes hydrochloric acid (HCl) and hydrofluoric acid (HF).

Supernova A powerful and luminous explosion of a massive star that marks its end.

System A set of interconnected components or elements that work together to form a whole.

Syntrophy A cooperative interaction between at least two microbial species that work together to degrade a single substrate.

Tardigrades A phylum of microscopic, segmented, eight-legged animals within the Ecdysozoa, famous for their extreme resilience.

Tetraether lipids A unique class of membrane lipids that are able to create unconventional, monolayer membranes, providing exceptional stability and resistance to extreme conditions.

Tetrahydropyrimidine A partially reduced form of pyrimidine, composed of a six-membered ring with two nitrogen atoms.

Theia The goddess of the Moon in Greek mythology. In planetary science, Theia is a hypothesized planet about the size of Mars or possibly larger in the early Solar System, which collided with the early Earth ~4.5 billion years ago. The debris from this collision eventually coalesced to form the Moon.

Thallus The entire body of plant, algae, fungi, and moss that has not been differentiated into organs.

Thrombolite A type of microbialite exhibiting a clotted, irregular texture due to the way microbial communities trap and bind sediment or bioturbation.

Tidal heating Also known as tidal flexing, internal frictional heat generated by deformations of an astronomic body with an elliptical orbit caused by the gravitational force of a massive object.

Tidal locking An astronomic body spins on its axis at a same rate as it rotates on its orbit, which means the same side of the body always faces its partner; for example, all the large moons in the solar system are tidally locked to their planets.

Trace fossil A trace preserved in rocks that is created by the activities of organisms rather than their physical remains; also known as ichnofossil and serving as indirect evidence of prehistoric life.

Troposphere The lowest layer of Earth's atmosphere, extending from the surface up to about 8–15 km, depending on latitude and season, which contains 80% of the atmosphere's mass and 99% of its water vapor, making it the primary layer for weather phenomena.

Tun state A survival mechanism used by tardigrades; when faced with harsh conditions such as desiccation, freezing, or oxygen deprivation, tardigrades enter a cryptobiotic state called the tun state, and their metabolism drops to 0.01% of its normal rate.

Vesicles A small, membrane-bound sac or compartment within or outside cells, composed of a lipid bilayer and performing crucial functions in the transport, storage, and digestion of substances.

Viking biological investigation A series of biological experiments carried out by Viking landers on the surface of Mars in 1976 to detect signs of life.

Volcanic island arc A geological feature formed along a convergent tectonic plate boundary, where one oceanic plate subducts beneath another and creates a chain of volcanic islands with a curved shape.

Water activity A measure of the availability of water in a substance, which is defined as the ratio of the vapor pressure of water in a sample to the vapor pressure of pure water at the same temperature. The scale ranges from 0 (completely dry) to 1 (pure water).

White smoker A light-colored chimney-like structure composed of barium, calcium, and silicon deposits typically found on the ocean floor, particularly in deep-sea environments, where hot, alkaline (pH >7), light mineral-rich fluid is expelled into the surrounding cool ocean water, resulting in precipitation of light-colored minerals and forming white smoke.

Zircon A zirconium silicate ($ZrSiO_4$) mineral that is renowned for its durability and its role as a "geological clock."

Index

Page locators in **bold** indicate tables. Page locators in *italics* indicate figures. This index uses letter-by-letter alphabetization.

a

Introduction to Geobiology, First Edition. Xingliang Zhang, Zisheng Guo, Wei Liu, Ruliang He, Luoyang Li, and Weiduo Hao.

Companion website: www.wiley.com/go/Zhang_1e

C

d

h

i

p

t